OPTIMIZATION CONCEPTS AND APPLICATIONS IN ENGINEERING

Second Edition

It is vitally important to meet or exceed previous quality and reliability standards while at the same time reducing resource consumption. This textbook addresses this critical imperative integrating theory, modeling, the development of numerical methods, and problem solving, thus preparing the student to apply optimization to real-world problems. This text covers a broad variety of optimization problems using the following: unconstrained, constrained, gradient, and nongradient techniques; duality concepts; multiobjective optimization; linear, integer, geometric, and dynamic programming with applications; and finite element-based optimization. In this revised and enhanced second edition of *Optimization Concepts and Applications in Engineering*, the already robust pedagogy has been enhanced with more detailed explanations and an increased number of solved examples and end-of-chapter problems. The source codes are now available free on multiple platforms. It is ideal for advanced undergraduate or graduate courses and for practicing engineers in all engineering disciplines, as well as in applied mathematics.

Ashok D. Belegundu has been a Professor of Mechanical Engineering at The Pennsylvania State University, University Park, since 1986. Prior to this, he taught at GMI, now Kettering University, in Michigan. He received his B. Tech. degree from IIT Madras and his Ph.D. from the University of Iowa. He has been a principal investigator on research projects involving optimization for several agencies including the National Science Foundation, Army Research Office, NASA, SERC (UK), MacNeal-Schwendler Corporation, Gentex Corporation, and Ingersoll-Rand. He has organized two international conferences on optimization in industry and has authored or edited four books and written a chapter in a book. A detailed list of his publications and projects can be found at http://www.mne.psu.edu/Directories/Faculty/Belegundu-A.html. He has advised more than 50 graduate students. He has given short courses on finite elements and optimization to the Forging Industry Association, Hazleton Pumps, and Infosys (India). He has served as an associate editor for *AIAA Journal* and for *Mechanics Based Design of Structures and Machines*. He teaches a distance education course on optimal design through Penn State.

Tirupathi R. Chandrupatla has been a Professor and Founding Chair of Mechanical Engineering at Rowan University since 1995. He started his career as a design engineer with Hindustan Machine Tools (HMT), Bangalore. He then taught at IIT Bombay. Professor Chandrupatla also taught at the University of Kentucky, Lexington, and GMI Engineering and Management Institute (now Kettering University), before joining Rowan. He received the Lindback Distinguished Teaching Award at Rowan University in 2005. He is also the author of *Quality and Reliability in Engineering* and two books on finite element analysis. Professor Chandrupatla has broad research interests, which include design, optimization, manufacturing engineering, finite element analysis, and quality and reliability. He has published widely in these areas and serves as an industry consultant. Professor Chandrupatla is a registered Professional Engineer and also a Certified Manufacturing Engineer. He is a member of ASEE, ASME, SAE, and SME. For further information, visit http://users.rowan.edu/~chandrupatla.

Optimization Concepts and Applications in Engineering

Second Edition

Ashok D. Belegundu

The Pennsylvania State University

Tirupathi R. Chandrupatla

Rowan University

CAMBRIDGE UNIVERSITY PRESS
Cambridge, New York, Melbourne, Madrid, Cape Town,
Singapore, São Paulo, Delhi, Tokyo, Mexico City

Cambridge University Press
32 Avenue of the Americas, New York, NY 10013-2473, USA

www.cambridge.org
Information on this title: www.cambridge.org/9780521878463

First published by Prentice Hall 1999
Second edition published 2011

Printed in the United States of America

A catalog record for this publication is available from the British Library.

Library of Congress Cataloging in Publication data

Belegundu, Ashok D., 1956–
Optimization concepts and applications in engineering / Ashok D. Belegundu,
Tirupathi R. Chandrupatla. – 2nd ed.
p. cm.
Includes index.
ISBN 978-0-521-87846-3 (hardback)
1. Engineering – Mathematical models. 2. Engineering design – Mathematics.
3. Mathematical optimization. 4. Engineering models. I. Chandrupatla,
Tirupathi R., 1944– II. Title.
TA342.B45 2011
519.602′462–dc22 2010049376

ISBN 978-0-521-87846-3 Hardback

To our families

Contents

Preface

This book is a revised and enhanced edition of the first edition. The authors have identified a clear need for teaching engineering optimization in a manner that integrates theory, algorithms, modeling, and hands-on experience based on their extensive experience in teaching, research, and interactions with students. They have strived to adhere to this pedagogy and reinforced it further in the second edition, with more detailed explanations, an increased number of solved examples and end-of-chapter problems, and source codes on multiple platforms.

The development of the software, which parallels the theory, has helped to explain the implementation aspects in the text with greater insight and accuracy. Students have integrated the optimization programs with simulation codes in their theses. The programs can be tried out by researchers and practicing engineers as well. Programs on the CD-ROM have been developed in Matlab, Excel VBA, VBScript, and Fortran. A battery of methods is available for the user. This leads to effective solution of problems since no single method can be successful on all problems.

The book deals with a variety of optimization problems: unconstrained, constrained, gradient, and nongradient techniques; duality concepts; multiobjective optimization; linear, integer, geometric, and dynamic programming with applications; and finite element–based optimization. Matlab graphics and optimization toolbox routines and the Excel Solver optimizer are presented in detail. Through solved examples, problem-solving strategies are presented for handling problems where the number of variables depends on the number of discretization points in a mesh and for handling time-dependent constraints. Chapter 8 deals exclusively with treatment of the objective function itself as opposed to methods for minimizing it.

This book can be used in courses at the graduate or senior-undergraduate level and as a learning resource for practicing engineers. Specifically, the text can be used in courses on engineering optimization, design optimization, structural optimization, and nonlinear programming. The book may be used in mechanical, aerospace, civil, industrial, architectural, chemical, and electrical engineering, as well as in applied mathematics. In deciding which chapters are to be covered in a course, the instructor may note the following. Chapters 1, 2, 3.1–3.5, and 8 are fundamental. Chapters 4, 9, and 11 focus on linear problems, whereas Chapters 5–7 focus on nonlinear problems. Even if the focus is on nonlinear problems, Sections 4.1–4.6 present important concepts related to constraints. Chapters 10 and 12 are specialized topics. Thus, for instance, a course on structural optimization (i.e., finite element-based optimization) may cover Chapters 1–2, 3.1–3.5, 4.1–4.6, 5, 6, 7.7–7.10, 8, and 12.

We are grateful to the students at our respective institutions for motivating us to develop this book. It has been a pleasure working with our editor, Peter Gordon.

OPTIMIZATION CONCEPTS AND APPLICATIONS IN ENGINEERING

Second Edition

1

Preliminary Concepts

1.1 Introduction

Optimization is the process of maximizing or minimizing a desired objective function while satisfying the prevailing constraints. Nature has an abundance of examples where an optimum system status is sought. In metals and alloys, the atoms take positions of least energy to form unit cells. These unit cells define the crystalline structure of materials. A liquid droplet in zero gravity is a perfect sphere, which is the geometric form of least surface area for a given volume. Tall trees form ribs near the base to strengthen them in bending. The honeycomb structure is one of the most compact packaging arrangements. Genetic mutation for survival is another example of nature's optimization process. Like nature, organizations and businesses have also strived toward excellence. Solutions to their problems have been based mostly on judgment and experience. However, increased competition and consumer demands often require that the solutions be optimum and not just feasible solutions. A small savings in a mass-produced part will result in substantial savings for the corporation. In vehicles, weight minimization can impact fuel efficiency, increased payloads, or performance. Limited material or labor resources must be utilized to maximize profit. Often, optimization of a design process saves money for a company by simply reducing the developmental time.

In order for engineers to apply optimization at their work place, they must have an understanding of both the theory and the algorithms and techniques. This is because there is considerable effort needed to apply optimization techniques on practical problems to achieve an improvement. This effort invariably requires tuning algorithmic parameters, scaling, and even modifying the techniques for the specific application. Moreover, the user may have to try several optimization methods to find one that can be successfully applied. To date, optimization has been used more as a design or decision aid, rather than for concept generation or detailed design. In this sense, optimization is an engineering tool similar to, say, finite element analysis (FEA).

This book aims at providing the reader with basic theory combined with development and use of numerical techniques. A CD-ROM containing computer programs that parallel the discussion in the text is provided. The computer programs give the reader the opportunity to gain hands-on experience. These programs should be valuable to both students and professionals. Importantly, the optimization programs with source code can be integrated with the user's simulation software. The development of the software has also helped to explain the optimization procedures in the written text with greater insight. Several examples are worked out in the text; many of these involve the programs provided. User subroutines to solve some of these examples are also provided on the CD-ROM.

1.2 Historical Sketch

The use of a gradient method (requiring derivatives of the functions) for minimization was first presented by Cauchy in 1847. Modern optimization methods were pioneered by Courant's [1943] paper on penalty functions, Dantzig's paper on the simplex method for linear programming [1951]; and Karush, Kuhn, and Tucker who derived the "KKT" optimality conditions for constrained problems [1939, 1951]. Thereafter, particularly in the 1960s, several numerical methods to solve nonlinear optimization problems were developed. Mixed integer programming received impetus by the branch and bound technique originally developed by Land and Doig [1960] and the cutting plane method by Gomory [1960]. Methods for unconstrained minimization include conjugate gradient methods of Fletcher and Reeves [1964] and the variable metric methods of Davidon–Fletcher–Powell (DFP) in [1959]. Constrained optimization methods were pioneered by Rosen's gradient projection method [1960], Zoutendijk's method of feasible directions [1960], the generalized reduced gradient method by Abadie and Carpenter [1969] and Fiacco and McCormick's SUMT techniques [1968]. Multivariable optimization needed efficient methods for single variable search. The traditional interval search methods using Fibonacci numbers and Golden Section ratio were followed by efficient hybrid polynomial-interval methods of Brent [1971] and others. Sequential quadratic programming (SQP) methods for constrained minimization were developed in the 1970s. Development of interior methods for linear programming started with the work of Karmarkar in 1984. His paper and the related US patent (4744028) renewed interest in interior methods (see the IBM Web site for patent search: http://patent.womplex.ibm.com/).

Also in the 1960s, alongside developments in gradient-based methods, there were developments in nongradient or "direct" methods, principally Rosenbrock's method of orthogonal directions [1960], the pattern search method of Hooke and

Jeeves [1961], Powell's method of conjugate directions [1964], the simplex method of Nelder and Meade [1965], and the method of Box [1965]. Special methods that exploit some particular structure of a problem were also developed. Dynamic programming originated from the work of Bellman who stated the principle of optimal policy for system optimization [1952]. Geometric programming originated from the work of Duffin, Peterson, Zener [1967]. Lasdon [1970] drew attention to large-scale systems. Pareto optimality was developed in the context of multiobjective optimization. More recently, there has been focus on stochastic methods, which are better able to determine global minima. Most notable among these are genetic algorithms [Holland 1975, Goldberg 1989], simulated annealing algorithms that originated from Metropolis [1953], and differential evolution methods [Price and Storn, http://www.icsi.berkeley.edu/~storn/code.html].

In operations research and industrial engineering, use of optimization techniques in manufacturing, production, inventory control, transportation, scheduling, networks, and finance has resulted in considerable savings for a wide range of businesses and industries. Several operations research textbooks are available to the reader. For instance, optimization of airline schedules is an integer program that can be solved using the branch and bound technique [Nemhauser 1997]. Shortest path routines have been used to reroute traffic due to road blocks. The routines may also be applied to route messages on the Internet.

The use of nonlinear optimization techniques in structural design was pioneered by Schmit [1960]. Early literature on engineering optimization are Johnson [1961], Wilde [1967], Fox [1971], Siddall [1972], Haug and Arora [1979], Morris [1982], Reklaitis, Ravindran and Ragsdell [1983], Vanderplaats [1984], Papalambros and Wilde [1988], Banichuk [1990], Haftka and Gurdal [1991]. Several authors have added to this collection including books on specialized topics such as structural topology optimization [Bendsoe and Sigmund 2004], design sensitivity analysis [Haug, Choi and Komkov 1986], optimization using evolutionary algorithms [Deb 2001] and books specifically targeting chemical, electrical, industrial, computer science, and other engineering systems. We refer the reader to the bibliography at end of this book. These, along with several others that have appeared in the last decade, have made an impact in educating engineers to apply optimization techniques. Today, applications are everywhere, from identifying structures of protein molecules to tracing of electromagnetic rays. Optimization has been used for decades in sizing airplane wings. The challenge is to increase its utilization in bringing out the final product.

Widely available and relatively easy to use optimization software packages, popular in universities, include the MATLAB optimization toolbox and the EXCEL SOLVER. Also available are GAMS modeling packages (http://gams.nist.gov/) and CPLEX software (http://www.ilog.com/). Other resources include Web sites maintained by Argonne national labs (http://www-fp.mcs.anl.gov/OTC/Guide/

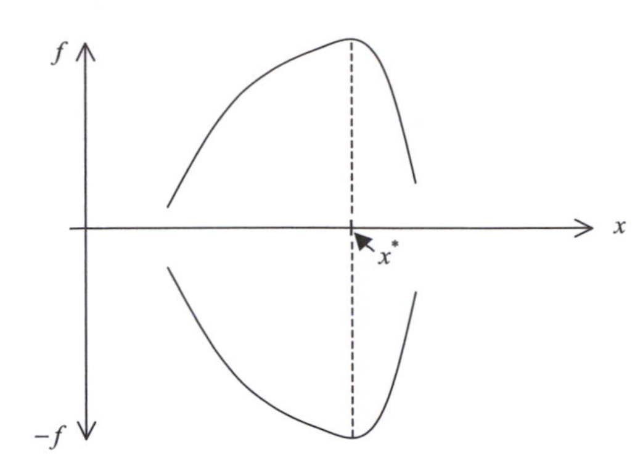

Figure 1.1. Maximization of f is equivalent to minimization of $-f$.

SoftwareGuide/) and by SIAM (http://www.siam.org/). GAMS is tied to a host of optimizers.

Structural and simulation-based optimization software packages that can be procured from companies include ALTAIR (http://www.altair.com/), GENESIS (http://www.vrand.com/), iSIGHT (http://www.engineous.com/), modeFRONTIER (http://www.esteco.com/), and FE-Design (http://www.fe-design.de/en/home.html). Optimization capability is offered in analysis commercial packages such as ANSYS and NASTRAN.

1.3 The Nonlinear Programming Problem

Most engineering optimization problems may be expressed as minimizing (or maximizing) a function subject to inequality and equality constraints, which is referred to as a nonlinear programming (NLP) problem. The word "programming" means "planning." The general form is

$$\begin{array}{lll} \text{minimize} & f(\mathbf{x}) & \\ \text{subject to} & g_i(\mathbf{x}) \leq 0 & i = 1, \ldots, m \\ \text{and} & h_j(\mathbf{x}) = 0 & j = 1, \ldots, \ell \\ \text{and} & \mathbf{x}^{\mathrm{L}} \leq \mathbf{x} \leq \mathbf{x}^{\mathrm{U}} & \end{array} \tag{1.1}$$

where $\mathbf{x} = (x_1, x_2, \ldots, x_n)^{\mathrm{T}}$ is a column vector of n real-valued *design variables*. In Eq. (1.1), f is the *objective* or *cost* function, g's are *inequality constraints*, and h's are *equality constraints*. The notation $\mathbf{x}^0$ for the starting point, $\mathbf{x}^*$ for optimum, and $\mathbf{x}^k$ for the (current) point at the kth iteration will be generally used.

Maximization versus Minimization

Note that maximization of f is equivalent to minimization of $-f$ (Fig. 1.1).

Problems may be manipulated so as to be in the form (1.1). Vectors $\mathbf{x}^{\mathrm{L}}$, $\mathbf{x}^{\mathrm{U}}$ represent explicit lower and upper bounds on the design variables, respectively, and

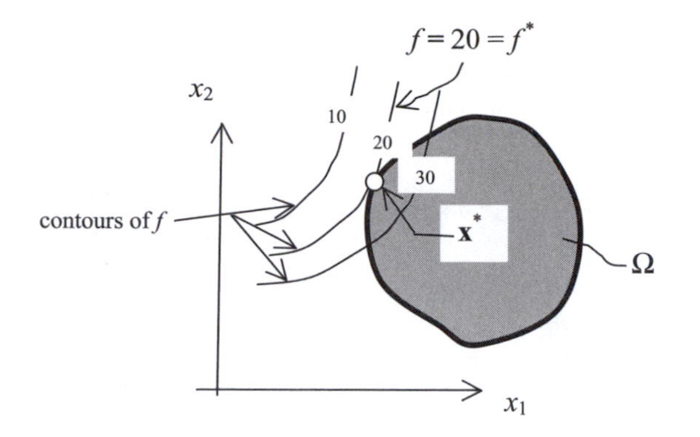

Figure 1.2. Graphical representation of NLP in **x**-space.

are also inequality constraints like the g's. Importantly, we can express (1.1) in the form

$$\begin{array}{ll} \text{minimize} & f(\mathbf{x}) \\ \text{subject to} & \mathbf{x} \in \Omega \end{array} \tag{1.2}$$

where $\Omega = \{\mathbf{x} : \mathbf{g}(\mathbf{x}) \leq \mathbf{0}, \mathbf{h}(\mathbf{x}) = \mathbf{0}, \mathbf{x}^L \leq \mathbf{x} \leq \mathbf{x}^U\}$. Ω, a subset of R^n, is called the *feasible region*.

In unconstrained problems, the constraints are not present – thus, the feasible region is the entire space R^n. Graphical representation in design-space (or **x**-space) for $n = 2$ variables is given in Fig. 1.2. Curves of constant f value or *objective function contours* are drawn, and the optimum is defined by the highest contour curve passing through Ω, which usually, but not always, is a point on the boundary Ω.

Example 1.1

Consider the constraints $\{g_1 \equiv x_1 \geq 0, g_2 \equiv x_2 \geq 0, g_3 \equiv x_1 + x_2 \leq 1\}$. The associated feasible set Ω is shown in Fig. E1.1.

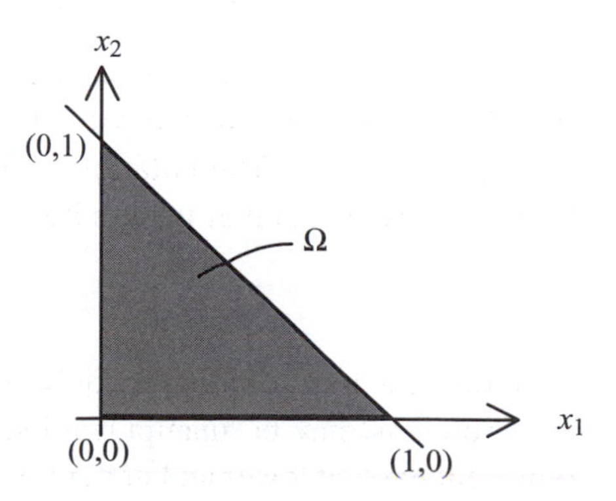

Figure E1.1. Illustration of feasible set Ω.

Upper Bound

It is important to understand the following inequality, which states that any feasible design provides an upper bound to the optimum objective function value:

$$f(\mathbf{x}^*) \le f(\hat{\mathbf{x}}) \qquad \text{for } any\ \hat{\mathbf{x}} \in \Omega$$

Minimization over a Superset

Given sets (i.e., feasible regions) S_1 and S_2 with $S_1 \subseteq S_2$; that is, S_1 is a subset of S_2 (or contained within S_2). If f_1^*, f_2^* represent the minimum values of a function f over S_1 and S_2, respectively, then

$$f_2^* \le f_1^*$$

To illustrate this, consider the following example. Let us consider monthly wages earned among a group of 100 workers. Among these workers, assume that Mr. Smith has the minimum earnings of $800. Now, assume a new worker joins the group. Thus, there are now 101 workers. Evidently, the minimum wages among the 101 workers will be less than or equal to $800 depending on the wages of the newcomer.

Types of Variables and Problems

Additions restrictions may be imposed on a variables x_j as follows:

x_j is continuous (default).
x_j is binary (equals 0 or 1).
x_j is integer (equals 1 or 2 or 3, . . . , or N)
x_j is discrete (takes values 10 mm, 20 mm, or 30 mm, etc.)

Specialized names are given to the NLP problem in (1.1) as follows:

Linear Programming (LP): when all functions (objective and constraints) are linear (in $\mathbf{x}$).

Integer Programming (IP): an LP when all variables are required to be integers.

0–1 Programming: special case of an IP where variables are required to be 0 or 1.

Mixed Integer Programming (MIP): an IP where some variables are required to be integers, others are continuous.

MINLP: an MIP with nonlinear functions.

Quadratic Programming (QP): when an objective function is a quadratic function in $\mathbf{x}$ and all constraints are linear.

Convex Programming: when the objective function is convex (for minimization) or concave (for maximization) and the feasible region Ω is a convex set. Here, any local minimum is also a global minimum. Powerful solution techniques that can handle a large number of variables exist for this category. Convexity of Ω is guaranteed when all inequality constraints g_i are convex functions and all equality constraints h_j are linear.

Combinatorial Problems: These generally involve determining an optimum permutation of a set of integers, or equivalently, an optimum choice among a set of discrete choices. Some combinatorial problems can be posed as LP problems (which are much easier to solve). Heuristic algorithms (containing *thumb rules*) play a crucial role in solving large-scale combinatorial problems where the aim is to obtain near-optimal solutions rather than the exact optimum.

1.4 Optimization Problem Modeling

Modeling refers to the translation of a physical problem into mathematical form. While modeling is discussed throughout the text, a few examples are presented below, with the aim of giving an immediate idea to the student as to how variables, objectives, and constraints are defined in different situations. Detailed problem descriptions and exercises and solution techniques are given throughout the text.

Example 1.2 (Shortest Distance from a Point to a Line)

Determine the shortest distance d between a given point $\mathbf{x}^0 = (x_1^0, x_2^0)$ and a given line $a_0 + a_1x_1 + a_2x_2 = 0$ (Fig. E1.2). If $\mathbf{x}$ is a point on the line, we may pose the optimization problem:

$$\begin{aligned} &\text{minimize} \quad f = (x_1 - x_1^0)^2 + (x_2 - x_2^0)^2 \\ &\text{subject to} \quad h(\mathbf{x}) \equiv a_0 + a_1x_1 + a_2x_2 = 0 \end{aligned}$$

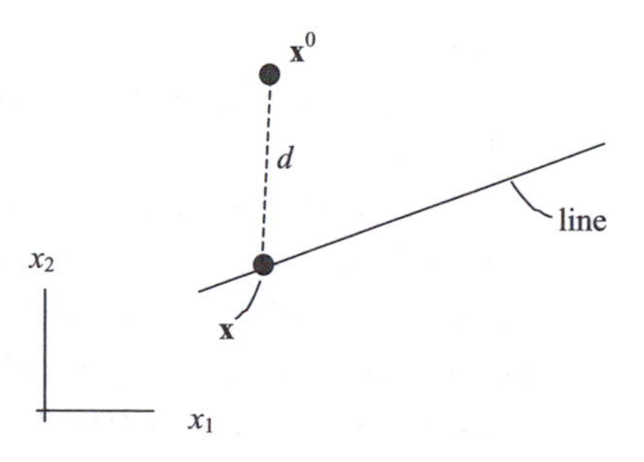

Figure E1.2. Shortest distance problem posed as an optimization problem.

where f is the objective function denoting the square of the distance (d^2), $\mathbf{x} = (x_1, x_2)^T$ are two variables in the problem, and h represents a linear *equality* constraint. The reader is encouraged to understand the objective function. At optimum, we will find that the solution $\mathbf{x}^*$ will lie at the foot of the perpendicular drawn from $\mathbf{x}^0$ to the line.

In fact, the preceding problem can be written in matrix form as

$$\begin{aligned} &\text{minimize} \quad f = (\mathbf{x} - \mathbf{x}^0)^T(\mathbf{x} - \mathbf{x}^0) \\ &\text{subject to} \quad h(\mathbf{x}) \equiv \mathbf{a}^T\mathbf{x} - b = 0 \end{aligned}$$

where $\mathbf{a} = [a_1 \ a_2]$, $b = -a_0$. Using the method of *Lagrange multipliers* (Chapter 5), we obtain a closed-form solution for the point $\mathbf{x}^*$:

$$\mathbf{x}^* = \mathbf{x}^0 - \frac{(\mathbf{a}^T\mathbf{x}^0 - b)}{(\mathbf{a}^T\mathbf{a})}\mathbf{a}$$

Note that $\mathbf{a}^T\mathbf{a}$ is a scalar. The shortest distance d is

$$d = \frac{|\mathbf{a}^T\mathbf{x}^0 - b|}{\sqrt{\mathbf{a}^T\mathbf{a}}}$$

Extensions

The problem can be readily generalized to finding the shortest distance from a point to a plane, in two or three dimensions.

Example 1.3 (Beam on Two Supports)

First, consider a uniformly loaded beam on two supports as shown in Fig. E1.3a. The beam length is $2L$ units, and the spacing between supports is $2a$ units. We wish to determine the half-spacing a/L so as to minimize the maximum deflection that occurs in the beam. One can assume $L = 1$.

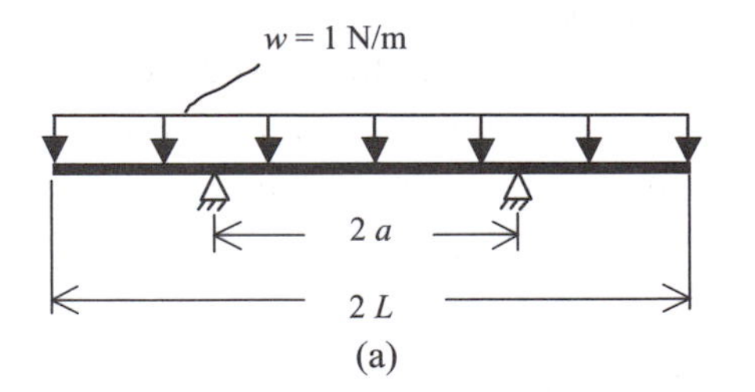

Figure E1.3a. Uniformly loaded beam on two supports.

This simple problem takes a little thought to formulate and solve using an available optimization routine. To provide insight, consider the deflected shapes when the support spacing is too large (Fig. E1.3a), wherein the maximum deflection occurs at the center, and when the spacing is too small (Fig. E1.3b), wherein the maximum deflection occurs at the ends. Thus, the graph of maximum deflection versus spacing is convex or cup-shaped with a well-defined minimum.

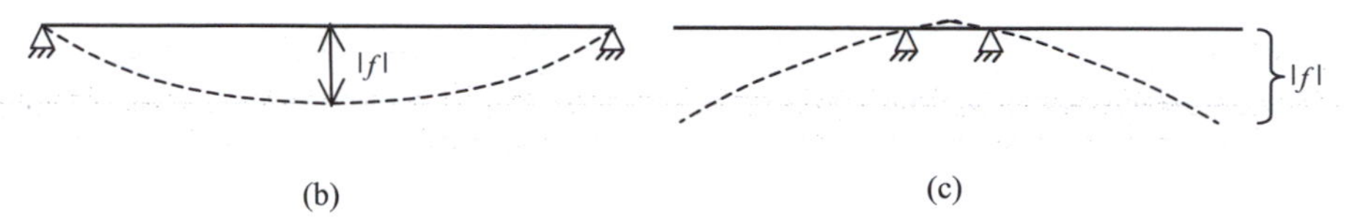

Figure E1.3b–c. Effect of support spacing on maximum deflection in beam.

Thus, we may state that the maximum deflection δ at any location in the beam, can be reduced to checking the maximum at just two locations. With this insight, the objective function f, which is to be minimized, is given by

$$f(a) \equiv \max_{0 \le x \le 1} \delta(x, a) = \max\{\delta(0, a), \delta(1, a)\} = \max(\delta_{\text{center}}, \delta_{\text{end}})$$

Absolute values for δ are to be used. Beam theory provides the relationship between a and $\delta\,(x, a)$. We now have to solve the optimization problem

$$\begin{aligned} &\text{minimize} \quad f(a) \\ &\text{subject to} \quad 0 \le a \le 1 \end{aligned}$$

We may use the Golden Section Search or other techniques discussed in Chapter 2 to solve this unconstrained one-dimensional problem.

Extension – I (Minimize Peak Stress in Beam)

The objective function in the beam support problem may be changed as follows: determine a to minimize the maximum bending stress. Also, the problem may be modified by considering multiple equally spaced supports.

Further, the problem of supporting above-ground, long, and continuous pipelines such as portions of the famous Alaskan oil pipeline is considerably more complicated owing to supports with foundations, code specifications, wind and seismic loads, etc.

Extension – II (Plate on Supports)

The following (more difficult) problem involves supporting a plate rather than a beam as in the preceding example. Given a fixed number of supports N_s, where $N_s \ge 3$ and an integer, determine the optimum support locations to minimize the maximum displacement due to the plate's self-weight. This problem occurs, for example, when a roof panel or tile has to be attached to a ceiling with a discrete number of pins. Care must be taken that all supports do not line up in a straight line to avoid instability. Generalizing this problem still further will lead to optimum *fixturing* of three-dimensional objects to withstand loads in service, handling, or during manufacturing.

Example 1.4 (Designing with Customer Feedback)

In this example, we show one way in which customer feedback is used to develop an objective function f for subsequent optimization. We present a rather simple example to focus on concepts. A fancy outdoor cafe is interested in designing a unique beer mug. Two criteria or *attributes* are to be chosen. The first attribute is the volume, V in oz, and the second attribute is the aspect ratio, H/D, where $H =$ height and $D =$ diameter. To manufacture a mug, we need to know the *design variables* H and D. Lower and upper limits have been identified on each of the attributes, and within these limits, the attributes can take on continuous values. Let us choose

$$8 \leq V \leq 16, \qquad 0.6 \leq H/D \leq 3.0$$

To obtain customer feedback in an economical fashion, three discrete levels, LMH or low/medium/high, are set for each attribute, leading to only 9 different types of mugs. For further economy, prototypes of only a subset of these 9 mugs may be made for customer feedback. However, in this example, all 9 mugs are made, and ratings of these from a customer (or a group) is obtained as shown in Table E1.4. For example, an ML mug corresponds to a mug of volume 12 oz and aspect ratio 0.6, and is rated at 35 units. The linearly scaled ratings in the last column correspond to a range of 0–100.

Table E1.4. *Sample customer ratings of a set of mugs.*

Sample mugs (V, H/D)	Customer rating	Scaled rating
L L	50.00	27.27
L M	60.00	45.45
L H	75.00	72.73
M L	35.00	0.00
M M	90.00	100.00
M H	70.00	63.64
H L	35.00	0.00
H M	85.00	90.91
H H	50.00	27.27

Under reasonable assumptions, it is possible to use the feedback in Table E1.4 to develop a *value* or *preference* function $P(V, H/D)$ whose maximization leads to an optimum. While details are given in Chapter 8, we outline the main aspects in the following. An additive value function $P = P_1(V) + P_2(H/D)$ can be developed using least-squares, where P_i are the individual part-worth (or part-value) functions. On the basis of data in Table E1.4, we obtain the functions in Fig. E1.4. Evidently, the customer likes 12-oz, medium aspect ratio mugs. We now pose the problem of maximizing the total value as

$$\begin{aligned}
\text{maximize} \quad & P = P_1(V) + P_2(H/D) \\
\text{subject to} \quad & 8 \le V \le 16 \\
& 0.6 \le H/D \le 3 \\
& H \ge 0, \quad D \ge 0, \quad \text{and} \quad V = \left(\frac{\pi D^2 H}{4}\right)\left(\frac{8}{250}\right) \text{oz}
\end{aligned}$$

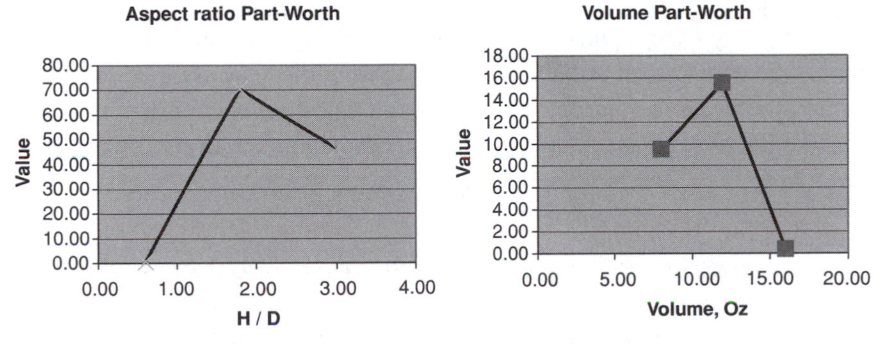

Figure E1.4. Part-worth (or part-value) functions P_i.

whose solution yields an optimum $H^* = 11.6$ cm, $D^* = 6.4$ cm, $V^* = 12$ oz, $H^*/D^* = 1.8$, as was expected from the part-worths. Care must be taken to obtain a global maximum while solving the preceding NLP by repeated optimizations of the numerical algorithm from different starting values of H and D.

Example 1.5 (Problem Involving Binary Variables)

A simple example involving binary variables is given below. Suppose there are three constraints $g_1(\mathbf{x}) \le 0, g_2(\mathbf{x}) \le 0, g_3(\mathbf{x}) \le 0$, and at least one of these must be satisfied. We may formulate this logic by introducing binary variables $\{y_i\}$ that take on values 0 or 1 and require

$$g_1(\mathbf{x}) \le My_1, \quad g_2(\mathbf{x}) \le My_2, \quad g_3(\mathbf{x}) \le My_3 \quad \text{and} \quad y_1 + y_2 + y_3 \le 2$$

where M is a large positive number compared to magnitude of any g_i. The last inequality forces at least one of the binary y's to be zero. For example, if $y_2 = 0$, $y_1 = y_3 = 1$, $M = 500$, then we have $g_2 \le 0$ (must be satisfied), $g_1 \le 500$ (need not be satisfied), $g_3 \le 500$ (need not be satisfied).

Example 1.6 (VLSI Floor Planning or Kitchen Layout Problem)
VLSI (very large-scale integration) is a process used to build electronic components such as microprocessors and memory chips comprising millions of transistors. The first stage of the VLSI design process typically produces a set of indivisible rectangular blocks called *cells*. In a second stage, interconnection information is used to determine the relative placements of these cells. In a third stage, implementations are selected for the various cells with the goal of optimizing the total area, which is related to cost of the chips. It is the third stage, *floor plan optimization*, for which we give a simple example below. This problem is analogous to the problem of designing a kitchen. Assume that we have decided on the components the kitchen is to contain (fridge, dishwasher, etc.) and how these components are to be arranged. Assume also that we can choose among several possible models for each of these components, with different models having different rectangular shapes but occupying the same floor area. In the floor plan optimization phase of our kitchen design, we select models so as to make the best use of available floor space.

Data
We are given three rectangular cells. Dimensions of C1 is 5×10, C2 can be chosen as 3×8, 2×12, or 6×4, and C3 can be chosen 5×8 or 8×5. Relative ordering of the cells must satisfy the following vertical and horizontal ordering:

Vertical: C2 must lie above C3.
Horizontal: C1 must lie to the left of C2 and C3.

This problem may be solved by different approaches. We present one approach, namely, a MINLP or mixed-integer nonlinear program, and encourage the student to follow the *modeling* of this problem. Let (w_i, h_i), $i = 1, 2, 3$, denote the width and height of cell i, and (x_i, y_i) denote the coordinates of the left bottom corner of cell i. Also, let W, H represent the width and height, respectively, of the bounding rectangle. We have the constraints

$$y_1 \ge 0, \quad y_1 + h_1 \le H, \quad y_3 \ge 0, \quad y_3 + h_3 \le y_2, \quad y_2 + h_2 \le H \qquad \text{vertical ordering}$$

$$x_1 \ge 0, \quad x_1 + w_1 \le x_2, \quad x_1 + w_1 \le x_3, \quad x_2 + w_2 \le W, \quad x_3 + w_3 \le W \qquad \text{horizontal ordering}$$

Introducing the binary (or 0/1) variables δ_{ij} to implement discrete selection,

$$w_2 = 8\delta_{21} + 12\delta_{22} + 4\delta_{23}, \quad h_2 = 3\delta_{21} + 2\delta_{22} + 6\delta_{23}$$
$$w_3 = 5\delta_{31} + 8\delta_{32}, \quad h_3 = 8\delta_{31} + 5\delta_{32}$$
$$\delta_{21} + \delta_{22} + \delta_{23} = 1, \quad \delta_{31} + \delta_{32} = 1$$
$$\delta_{ij} = 0 \quad \text{or} \quad 1$$

Thus, there are 13 variables in the problem are (x_i, y_i), δ_{ij}, W, H, and the objective function is

$$\text{minimize} \quad f = WH$$

Solution (a branch-and-bound procedure explained in Chapter 9 can be used) is

$$[x_1, y_1, w_1, h_1, x_2, y_2, w_2, h_2, x_3, y_3, w_3, h_3] = [0, 0, 5, 10, 5, 7, 8, 3, 5, 0, 8, 5].$$

The associated optimum objective is f = Area = 130 units^2, with a floor plan shown in Fig. E1.6.

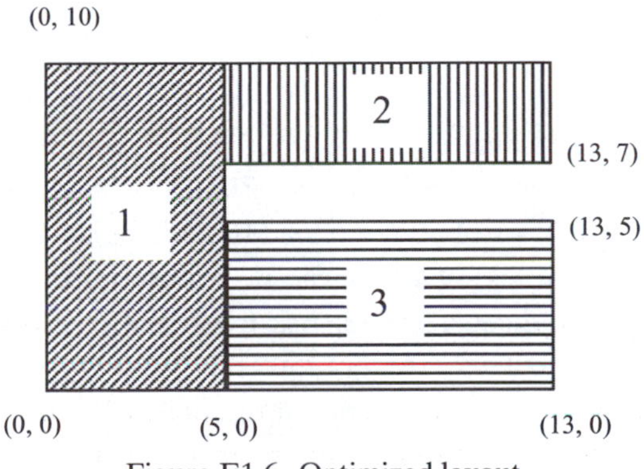

Figure E1.6. Optimized layout.

The solution is not unique in that cell 2 can be moved a little vertically. However, the objective function value is unchanged and is a global minimum.

Example 1.7 (Portfolio Selection)
We discuss this problem because it has many important features of considerable importance in engineering, specifically, *multiple attributes* and *statistical considerations*, owing to having to balance profit with risk. The problem is as follows: Assume you have \$100 and you want to invest it in 4 stocks – how much will you

invest in each? Obviously, this will depend on your overall balance between profit and risk, as some stocks may yield high return with high risk, others may yield lower return with lower risk. Mathematical modeling of this problem is as follows [Markowitz].

Let c_i, $i = 1, \ldots, n$, represent the average return (over, say, a 2 month period) of stock i, with n being the total number of stocks. Let σ_i^2 represent the variance of each stock where σ_i is the standard deviation, and x_i is the money invested in stock i, expressed as a percentage of total investment. It follows from statistics that

$$\text{Expected (average) return on total investment} = \sum_{i=1}^{n} c_i x_i$$

$$\text{Variance of total investment} = \sum_{i,j=1}^{n} \sigma_i \sigma_j \rho_{ij} x_i x_j$$

where ρ_{ij} is the correlation coefficient between a pair of stocks i and j. The investor's objective is to maximize return with limited risk (as indicated by variance). This balance is achieved by a *penalty function* as

$$\text{Objective function } f = \sum_{i=1}^{n} c_i x_i - \frac{1}{\lambda} \sum_{i,j=1}^{n} \sigma_{ij} x_i x_j \rightarrow \text{maximize}$$

where λ is a penalty parameter, say, equal to 10. The implication of this parameter in the preceding equation is that a variance of, say, 50 units is equivalent to a "certain or sure" loss of $50/\lambda =$ \$5 on the normalized return; a variance of 200 units will be like a loss of \$20. λ can be chosen on the basis of experience. A higher value of λ indicates greater risk tolerance. An investor for a pension fund must choose a smaller λ. The optimization problem may be summarized in matrix form as

$$\begin{aligned} &\text{maximize} \quad f = \mathbf{c}^{\mathrm{T}}\mathbf{x} - \tfrac{1}{\lambda}\mathbf{x}^{\mathrm{T}}\mathbf{R}\mathbf{x} \\ &\text{subject to} \quad \sum_{i=1}^{n} x_i = 1, \quad \mathbf{x} \geq \mathbf{0} \end{aligned}$$

which is a quadratic programming (QP) problem, owing to a quadratic objective function and linear constraints. **R** is a square, symmetric covariance matrix. Solution **x** gives the percent portfolio investment. We will revisit this problem in Chapter 8, specifically the construction of the objective function f, and introduce important concepts relating to multiattribute optimization, preferences, and tradeoffs.

An Extension

The investor may require that any investment x_i must either equal zero or be at least 2% (i.e., 0.02) in magnitude. That is, very small nonzero investments are to be avoided. (How will the reader formulate this?)

Example 1.8 (Combinatorial Problems with Integer Ordering)

Some "combinatorial" example problems are presented in the following wherein the objective and constraint functions depend on the *ordering* of a set of integers. For example, {1, 2, 3} can be reshuffled into {1, 3, 2}, {2, 1, 3}, {2, 3, 1}, {3, 1, 2}, {3, 2, 1}. In general, N integers can be ordered in $N!$ ways. Even with a small $N = 10$, we have 3,628,800 possibilities. For $N = 15$, we have billions of orderings. Thus, optimization algorithms need to be used rather than just trying out every possibility.

Traveling Salesman Problem

Given a collection of N cities and the cost of travel between each pair of them, the traveling salesman problem (TSP) is to find the cheapest way of visiting all of the cities and returning to your starting point. As discussed in the preceding text, this is a deceptively simple problem. In practice, there are further twists or constraints such as the salesman has to stay overnight and not all cities have motels. Here, the ordering $\{1, 4, 13, 7, \ldots, 1\}$ means that the salesman will start from city 1 and visit cities in the order 4, 13, 7, and so on.

Refueling Optimization (Courtesy: Professor S. Levine)

Here, the problem has nonlinear objective and constraint functions. In contrast, the aforementioned TSP had a linear objective. A reactor core consists of number of fuel assemblies (FAs). Power produced by each FA depends upon the enrichment of the fissile material it contains and its location in the core. Enrichment of FA degrades as it continues to burn up to deliver power. Thus, a fresh FA always gives more power than used FA located at the same position in a reactor core. At a particular instant, the reactor core contains FAs of different enrichment. When an FA's enrichment falls below a particular limit, it is replaced with a fresh FA, which is known as refueling. To minimize the cost of operation of the reactor, i.e., minimize refueling in the present context, we try to find an optimum configuration of the FAs in the core so as to maximize the cycle time (time between refueling), and at the same time restricting the power produced by each FA below a certain limit to avoid its mechanical and thermal damage. A reactor is stable when it operates under a critical condition, which means the neutron population remains relatively constant. For continued operation of the reactor, we always produce more neutrons than required and put the right amount of boron to absorb the excess neutrons such that the reactor is stable. Thus, the amount of boron used in a reactor is directly proportional to the amount of excess neutrons. As the reactor operates and the fuel burns up, the quantity of excess neutron decreases and the operator cuts down the boron amount such that reactor is stable. When the quantity of excess neutron approaches zero, the reactor needs to be refueled (end of cycle). Thus, it can be stated that the amount of boron present in a stable reactor directly indicates cycle time. In other words, our objective function to *maximize* is the amount of boron left at the end after a specified number of depletion steps. Computer programs are available to evaluate cycle time and normalized power (NP) for any given ordering of the FAs. With this background, we may state the problem as follows.

Problem: A 1/4 symmetry model of the reactor is shown in Fig. E1.8, which contains 37 FAs. There are 11 fresh FAs and others are used fuels with different burn up. Shuffle the locations of FAs in the core to find optimum configuration that maximizes the boron concentration (BC) at the end of a specified number of depletion steps, with the constraint that the NP of *each* FA is below 1.5.

```
                    x   x   x   x
            x   x   x   x   x   x   x   x
        x   x   x   x   x   x   x   x   x   x
    x   x   x   x   x   x   x   x   x   x   x   x
    x   x   x   x   x   x   x   x   x   x   x   x
x   x   x   x   x   x   x   x   x   x   x   x   x   x
x   x   x   x   x   x   x   x   x   x   x   x   x   x
x   x   x   x   x   x   x   1   2   3   4   5   6   7
x   x   x   x   x   x   x   8   9   10  11  12  13  14
    x   x   x   x   x   x   15  16  17  18  19  20
    x   x   x   x   x   x   21  22  23  24  25  26
        x   x   x   x   x   27  28  29  30  31
            x   x   x   x   32  33  34  35
                    x   x   36  37
```

Figure E1.8. 1/4 symmetry core showing 37 fuel assemblies (FAs) – (FA 37 is fixed in location).

Suggestion: Use the simulated annealing optimizer.

Renumbering for Efficient Equation Solving in Finite Elements
A finite element model consists of elements, with each element connected to a set of nodes. The equilibrium equations, which are a set of simultaneous system of equations, take the form $\mathbf{K}\,\mathbf{U} = \mathbf{F}$. In theory, $\mathbf{K}$ is $(n \times n)$ in size, $\mathbf{U}$ is $(n \times 1)$, and $\mathbf{F}$ is $(n \times 1)$. In computation, $\mathbf{K}$ is known to have a large number of zeros and, hence, $\mathbf{K}$ is stored in various other forms such as banded, skyline, sparse, etc. To give a clearer idea to the reader, let us focus on a banded $\mathbf{K}$ of dimension $(n \times nbw)$, where bandwidth $nbw \ll n$. Now, the smaller the value of nbw, the smaller the size of $\mathbf{K}$ and more efficient is the solution of the system of equations. The value of nbw can be minimized by renumbering the nodes in the finite element model [Chandrupatla and Belegundu, 1999]. In more popular "sparse solvers" today, we again have a combinatorial problem where the "fill-in" as opposed to the bandwidth is to be minimized. More often, specialized heuristic algorithms are used as opposed to general-purpose optimizers to handle these problems owing to $n \geq 10^6$ typically.

Example 1.9 (Structural Optimization Examples)
An important topic today in engineering optimization is to combine simulation codes, such as finite element (FEA) or boundary element (BEA) or computational fluid dynamics, with optimizers. This allows preliminary optimized designs of complex engineering components and subsystems. We give a few simple examples in what follows.

Shape Optimization
A two-dimensional finite element model of a bicycle chain link under tension is shown in Fig. E1.9a. Determine the shape of the outer surface and/or the diameter of the holes to minimize the weight subject to a specified von Mises stress

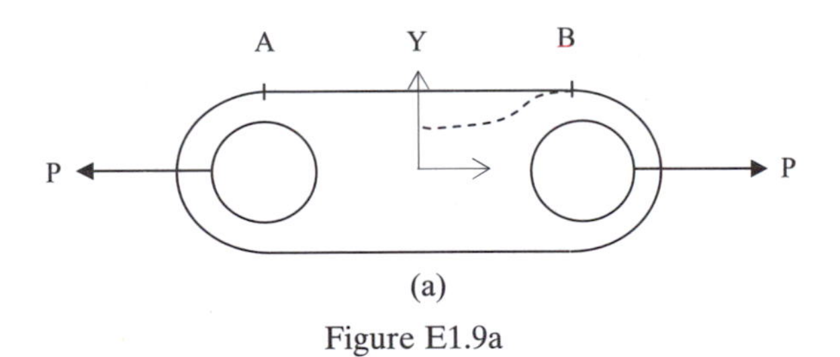

(a)

Figure E1.9a

limit. The thickness of the part may be added as an additional design variable. Note: the initial thickness at the start of the optimization iterations is chosen so as to make the stress constraint *active* – that is, so as to make the stress equal to its limiting value. Optimization involves defining permissible shape variations, linking these to finite element mesh coordinates, and safeguarding against mesh distortion during optimization [Zhang 1992 and Salagame 1997].

As another example, what is the optimum shape of an axisymmetric offshore structure founded on the sea floor, in a seismically active region, to minimize the maximum hydrodynamic forces generated by horizontal ground motion, subject to a volume limit? Here too, closed-form solutions are not obtainable. As another example, what is the optimum shape of a plate that can deflect the waves from a blast load and mitigate damage [Argod et al. 2010]?

Sizing Optimization

A laminated composite is a layered material, where each layer can have a different thickness, volume fraction of the high-cost fibers, and ply angle. The associated parameters need to be optimized to achieve desired characteristics. As an example, consider a 5-ply symmetrical laminate that is cooled from an elevated processing temperature (Fig. E1.9b). Residual stresses are produced.

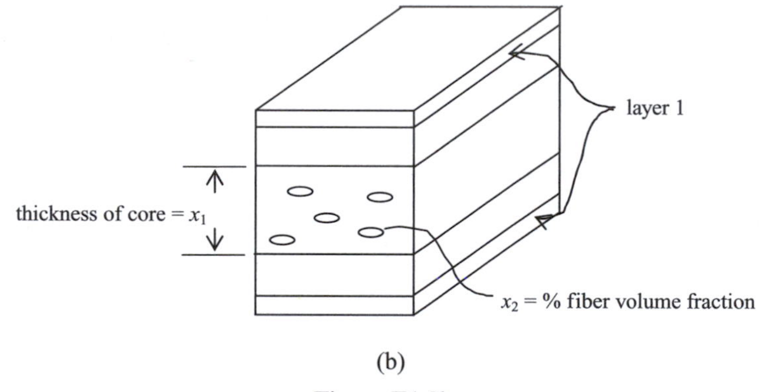

(b)

Figure E1.9b

Now, in the context of designing a cutting tool insert [Belegundu and Salagame 1995], the following optimization problem can be formulated. Thickness of the core layer and percent of silicon carbide (SiC) fiber in the SiC-aluminum oxide matrix are chosen as the two design variables.

minimize {tensile stress in the core AND cost of fiber}
subject to difference between stresses in adjacent layers $\leq \sigma_0$
stress in layer 1 ≤ 0
$\mathbf{x}^L \leq \mathbf{x} \leq \mathbf{x}^U$

There are two objectives, making this a multiobjective problem. The first constraint is aimed at preventing delamination, while the second constraint ensures negative stresses in the surface layers of the cutting tool insert to resist the generation and propagation of flaws. The functional relationship between variables x_i and the stresses is obtained from mechanics.

Noise Reduction Problems

(a) Given a vibrating plate or shell, driven harmonically at some location, determine the location of a given mass m_0 on a selected region on the structure to minimize radiated sound power. Interestingly, in such problems, the loading and structure may be symmetrical while the optimum location of the mass need not be [Constans et al. 1998].
(b) Problem is same as (a), but instead of a mass, the attachment device is a spring-mass-damper absorber [Grissom et al. 2006]. Applications include side panels of a refrigerator or a washing machine, various flat and curved panels in an automobile. Solution of this category of problems requires a finite element code to obtain vibration response, an acoustic code that takes vibration response and outputs acoustic response, and a suitable optimizer (usually nongradient type).

1.5 Graphical Solution of One- and Two-Variable Problems

For $n = 1$ or $n = 2$, it is possible to graph the NLP problem and obtain the solution. This is an important exercise for a beginner. Guidelines and examples are given in the following.

Example 1.10

Plot $f = (x - 1)^2$, over the domain $x \in [0, 2]$. We may code this in MATLAB as

```
x = [0: .01 :2];{% this defines an array x = 0, 0.01, 0.02,
                   ..., 1.99, 2}
f = (x-1).^2;{%  note the period (.) before the exponent
    (^) - this evaluates %  an array f  at each
    value of x as f = (0-1)^2, (0.01-1)^2, ...}
plot(x, f ) {%  provides the plot, whose minimum occurs
    at x = 1.}
```

where "%" refers to a comment within the code. In EXCEL (spreadsheet), we can define two columns as shown in Table E1.10, and use the "SCATTER PLOT" in the chart wizard to plot x versus f.

Table E1.10. *Use of EXCEL to plot a function of one variable.*

x	f
0	= (0–1)^2
0.01	= (0.01–1)^2
0.02	= (0.02–1)^2
...	...
1.99	= (1.99–1)^2
2	= (2–1)^2

Example 1.11

Consider the "corrugated spring" function $f = -\cos(kR) + 0.1\ R^2$ where R is the radius, $R = \sqrt{(x_1 - c_1)^2 + (x_2 - c_2)^2}$, $c_1 = 5$, $c_2 = 5$, $k = 5$. Draw

(i) a 3-D plot of this function and
(ii) a contour plot of the function in variable space or x_1–x_2 space.

The following code in MATLAB will generate both the plots:

```
[x1, x2] = meshgrid(2:.1:8,1:.1:8);
c1=5; c2=5; k=5;
R = sqrt( (x1-c1).^2 + (x2-c2).^2);
f = -cos(k*R) + 0.1*(R.^2);
surf (x1, x2, f)
figure(2)
contour(x1,x2,f)
```

In EXCEL, we must define a matrix in the spreadsheet, and then, after highlighting the entire box (note: the top left corner must be left blank), select "surface" chart with suboption 3-D or Contour. The number of contour levels may be increased by double-clicking the legend and, under *scale*, reducing the value of the major unit (Table E1.11).

Table E1.11. *Use of EXCEL to obtain a contour plot of a function in two variables.*

(BLANK)	2	**2.1**	...	8
2		$\mathbf{f(x_1, x_2) = f(2, 2.1)}$	...	
2.1			...	
...	...	...	...	...
8			...	$f(8, 8)$

Example 1.12

This example illustrates plotting of feasible region Ω and objective function contours for the NLP in (1.2) with $n = 2$, in particular:

$$\begin{array}{lll} \text{Minimize} & (x_1 + 2)^2 - x_2 & \equiv f \\ \text{subject to} & x_1^2/4 + x_2 - 1 \leq 0 & \equiv g_1 \\ & 2 + x_1 - 2x_2 \ \leq 0 & \equiv g_2 \end{array}$$

Method I

Here, for each g_i, we plot the curve obtained by setting $g_i(x_1, x_2) = 0$. This can be done by choosing various values for x_1 and obtaining corresponding values of x_2 by solving the equation, followed by an x_1–x_2 plot. Then, the feasible side of the curve can be readily identified by checking where $g_i\ (\mathbf{x}) < 0$. Pursuing this technique, for g_1, we have $x_2 = 1 - x_1^2/4$, a parabola, and for g_2, we have $x_2 = (2 + x_1)/2$, which is a straight line (Fig. E1.12). The intersection of the two individual feasible regions defines the feasible region Ω. Contours of the objective function correspond to curves $f(\mathbf{x}) = c$. Note that values for c and also the range for x_1 have to be chosen by some trial and error. Contours of f represent a family of parallel or nonintersecting curves. Identify the lowest contour within the feasible region and this will define the optimum point. Visually, we see from Fig. E1.12 that the optimum is $(x_1^*, x_2^*) = (-1.6, 0.25)$. The actual solution, obtained by an optimizer, is $\mathbf{x}^* = (-1.6, 0.36)$, with $f^* = -0.2$. Importantly, at the optimum point, the constraint g_1 is *active* (or "tight" or "binding") while g_2 is *inactive* (or "slack"). The MATLAB code to generate the plot is as follows:

```
x1=[-3:.1:1];
x2g1=1-x1.^2/4;
x2g2=1+.5*x1;
plot(x1,x2g1)
hold
plot(x1,x2g2,'-.')
axis([-3,1,-3,3])
for i=1:3
c=-1+0.25*i;
x2f = (x1+2).^2-c;
plot(x1,x2f,':')
end
```

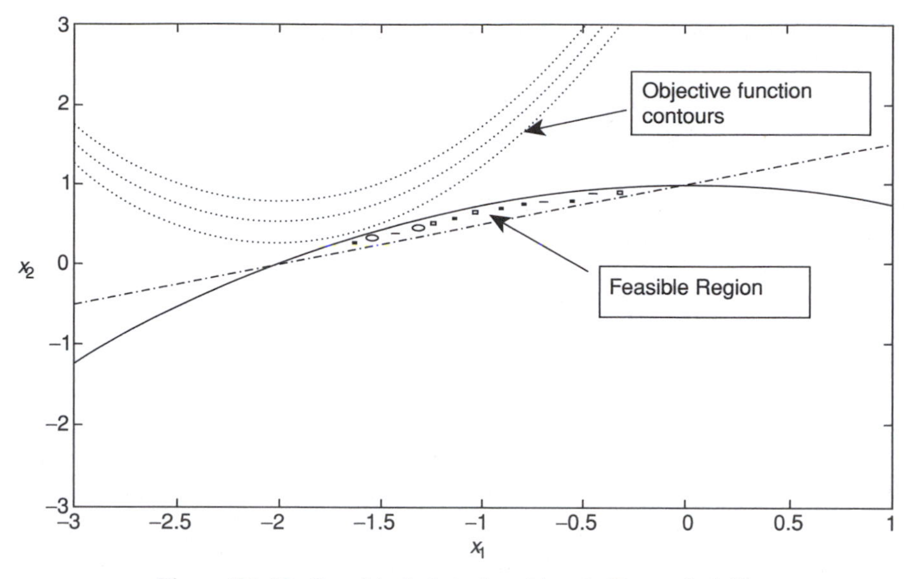

Figure E1.12. Graphical plot of problem in Example 1.11.

Method II

Recognizing that a plot of $g_i = 0$ is essentially a *contour* with level or value = 0, the following MATLAB code provides a similar plot:

```
[x1,x2]= meshgrid (-3:.01:1, -3:.01:3);
g1 = x1.^2/4 + x2 -1;
g2 = 2 + x1 - 2*x2;
f = (x1+2).^2 - x2;
contour (x1,x2,g1,[0 0]) % g1 = 0 contour
hold
contour (x1,x2,g2,[0 0]) % g2 = 0 contour
%now obtain f= -0.5, f= -0.25,... contours
contour (x1,x2,f,[-.5 -.25 -.1 0],'b');
% f= -0.5,-0.25,-.1,0.
```

The reader is urged to understand both approaches presented here.

1.6 Existence of a Minimum and a Maximum: Weierstrass Theorem

A central theorem concerns the existence of a solution to our problem in (1.1) or (1.2). Let $\mathbf{x}^*$ denote the solution – if one exists – to the minimization problem

$$\begin{aligned} &\text{minimize} \quad f(\mathbf{x}) \\ &\text{subject to} \quad \mathbf{x} \in \Omega \end{aligned}$$

and let $\mathbf{x}^{**}$ denote the solution to the maximization problem

$$\begin{array}{ll} \text{maximize} & f(\mathbf{x}) \\ \text{subject to} & \mathbf{x} \in \Omega \end{array}$$

Wierstraas theorem states conditions under which $\mathbf{x}^*$ and $\mathbf{x}^{**}$ exist. If the conditions are satisfied, then a solution exists. While it is true that a solution may exist even though the conditions are not satisfied, this is highly unlikely in practical situations owing to the simplicity of the conditions.

Wierstraas Theorem

Let f be a *continuous function* defined over a *closed* and *bounded* set $\Omega \subseteq D(f)$. Then, there exist points $\mathbf{x}^*$ and $\mathbf{x}^{**}$ in Ω where f attains its minimum and maximum, respectively. That is, $f(\mathbf{x}^*)$ and $f(\mathbf{x}^{**})$ are the minimum and maximum values of f in the set. This theorem essentially states the conditions under which an extremum *exists*.

We will briefly discuss the terms "closed set," "bounded set," and "continuous function," prior to illustrating the theorem with examples. Firstly, $D(f)$ means domain of the function. The set $\{\mathbf{x} : |\mathbf{x}| \leq 1\}$ is an example of a *closed* set, while $\{\mathbf{x} : |\mathbf{x}| < 1\}$ is an *open* set. A set is *bounded* if it is contained within some sphere of finite radius, i.e., for any point $\boldsymbol{a}$ in the set, $\boldsymbol{a}^{\mathrm{T}}\boldsymbol{a} < c$, where c is a finite number. For example, the set of all positive integers, $\{1, 2, \ldots\}$, is not bounded. A set that is both closed and bounded is called a *compact* set.

A *function* can be viewed as a mapping, whereby an element $\boldsymbol{a}$ from its domain is mapped into a scalar $b = f(\boldsymbol{a})$ in its range. Importantly, for a given $\boldsymbol{a}$, there is a unique b. However, different $\boldsymbol{a}$'s may have the same b. We may also think of x as a "knob setting" and $f(x)$ as a "measurement." Evaluating $f(x)$ for a given x will usually involve a computer in engineering problems. We say that the function f is *continuous* at a point $\boldsymbol{a}$ if : given *any* sequence $\{\boldsymbol{x}_k\}$ in $D(f)$ which converges to $\boldsymbol{a}$, then $f(\boldsymbol{x}_k)$ must converge to $f(\boldsymbol{a})$. In one variable, this reduces to a statement commonly found in elementary calculus books as "left limit = right limit = $f(a)$." Further, f is continuous over a set S implies that it is continuous at each point in S.

We now illustrate Wierstraas theorem through examples.

Example 1.13

Consider the problem

$$\begin{array}{ll} \text{minimize} & f = x \\ \text{subject to} & 0 < x \leq 1 \end{array}$$

This simple problem evidently does not have a solution. We observe that the constraint set Ω is not closed. Rewriting the constraint as $0 \leq x \leq 1$ results in $x = 0$ being the solution. This transforms the constraint set from an open set to a closed set and satisfies conditions in the Wierstraas theorem.

Example 1.14

Consider the cantilever beam in Fig. E1.14a with a tip load P and tip deflection δ. The cross-section is rectangular with width and height equal to x_1, x_2, respectively. It is desired to minimize the tip deflection with given amount of material, or

$$\begin{aligned} &\text{minimize} \quad \delta \\ &\text{subject to} \quad A \leq A_0 \end{aligned}$$

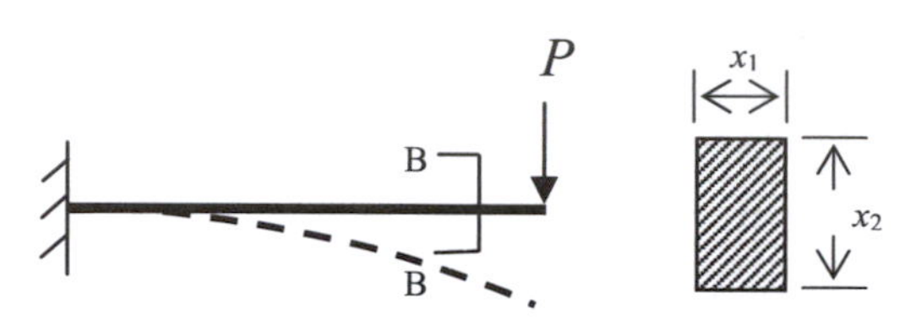

Section B-B

Figure E1.14a. Cantilever beam.

This problem does not have a solution – that is, it is *ill-posed*. This can be seen by substituting the beam equation $\delta = \frac{PL^3}{3EI} = \frac{c}{x_1 x_2^3}$, where c is a known scalar. Owing to the fact that x_2 is cubed in the denominator of the δ expression, with a given $A = x_1\, x_2$, the solution will tend to increase x_2 and reduce x_1. That is, the beam will tend to be infinitely slender with $x_1 \to 0$, $x_2 \to \infty$. A look at the feasible region Ω shows that it is unbounded, violating the simple conditions in the Wierstraas theorem (Fig. E1.14b). Again, a simple modification can be made that renders Ω bounded.

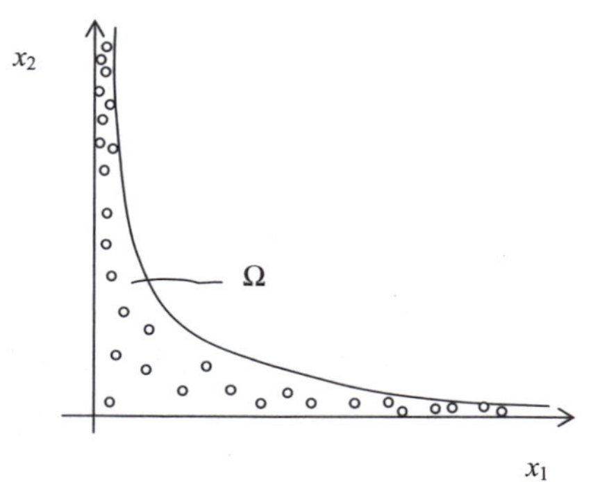

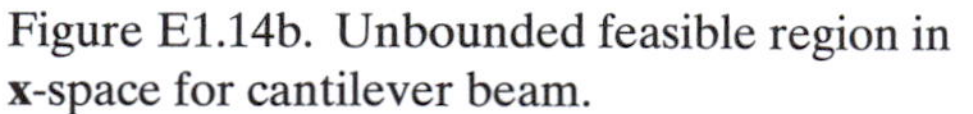

Figure E1.14b. Unbounded feasible region in **x**-space for cantilever beam.

1.7 Quadratic Forms and Positive Definite Matrices

Consider a function f given by $f = x_1^2 - 6x_1 x_2 + 9x_2^2$. We call f a quadratic form in two variables. We can write f in matrix notation as

$$f = (x_1 \quad x_2) \begin{bmatrix} 1 & -3 \\ -3 & 9 \end{bmatrix} \begin{pmatrix} x_1 \\ x_2 \end{pmatrix}$$

If $\mathbf{x}$ is an n-vector, then the general expression for a quadratic form is $f = \mathbf{x}^{\mathrm{T}}\mathbf{A}\,\mathbf{x}$, where $\mathbf{A}$ is a symmetric matrix. In quadratic forms, we can assume that the coefficient matrix $\mathbf{A}$ is symmetric without loss of generality since $\mathbf{x}^{\mathrm{T}}\mathbf{A}\,\mathbf{x} = \mathbf{x}^{\mathrm{T}}\,(\frac{\mathbf{A}+\mathbf{A}^{\mathrm{T}}}{2})\mathbf{x}$ with $(\frac{\mathbf{A}+\mathbf{A}^{\mathrm{T}}}{2})$ a symmetric matrix. A quadratic form can also be expressed in index notation as $\sum_i \sum_j x_i\, a_{ij}\, x_j$.

Definition

A quadratic form is *positive definite* if

$\mathbf{y}^{\mathrm{T}}\,\mathbf{A}\,\mathbf{y} > 0$ for *any* nonzero vector $\mathbf{y}$, and
$\mathbf{y}^{\mathrm{T}}\,\mathbf{A}\,\mathbf{y} = 0$ if and only if $\mathbf{y} = \mathbf{0}$.

In this case, we also say that $\mathbf{A}$ is a positive definite matrix. All eigenvalues of a positive definite matrix are strictly positive. $\mathbf{A}$ is negative definite if $(-\mathbf{A})$ is positive definite. For a small-sized matrices, Sylvester's test, given below is useful for checking whether a matrix is positive definite.

Positive Semidefinite

A quadratic form is positive semidefinite if $\mathbf{y}^{\mathrm{T}}\mathbf{A}\,\mathbf{y} \geq 0$ for *any* nonzero $\mathbf{y}$. All eigenvalues of a positive semidefinite matrix are nonnegative.

Sylvester's Test for a Positive Definite Matrix

Let $\mathbf{A}^i$ denote the submatrix formed by deleting the last $n - i$ rows and columns of $\mathbf{A}$, and let det $(\mathbf{A}^i)$ = determinant of $\mathbf{A}^i$. Then, $\mathbf{A}$ is positive definite if and only if det $(\mathbf{A}^i) > 0$ for $i = 1, 2, \ldots, n$. That is, the determinants of all principal minors are positive.

Example 1.15

Use Sylvestor's test to check the positive definiteness of **A** in the following:

$$\mathbf{A} = \begin{bmatrix} 1 & 2 & 3 \\ 2 & 5 & -1 \\ 3 & -1 & 2 \end{bmatrix}$$

We have

$$\mathbf{A}^1 = 1 > 0,\ \mathbf{A}^2 = \det\begin{bmatrix} 1 & 2 \\ 2 & 5 \end{bmatrix} = 5 - 4 = 1 > 0,\ \mathrm{A}_3 = \det\begin{bmatrix} 1 & 2 & 3 \\ 2 & 5 & -1 \\ 3 & -1 & 2 \end{bmatrix} = -56$$

Thus, **A** is not positive definite. Further, the command eig(A) in MATLAB gives the eigenvalues as [−2.2006, 4.3483, 5.8523] from which we can deduce the same conclusion.

1.8 C^n Continuity of a Function

We say that the function f is C^0 *continuous* at a point $\boldsymbol{a}$ if : given *any* sequence $\{\boldsymbol{x}_k\}$ in $D(f)$ which converges to $\boldsymbol{a}$, then $f(\boldsymbol{x}_k)$ must converge to $f(\boldsymbol{a})$. Further, f is continuous over a set S implies that it is continuous at each point in S.

Example 1.16

Consider the function f with $D(f) = R^1$:

$$\begin{aligned} f(x) &= 0, \quad x \leq 0 \\ &= 1, \quad x > 0 \end{aligned}$$

The sequence $x_k = 1/k$, $k = 1, 2, 3, \ldots$, converges (from the right) to $a = 0$, but the sequence $f(x_k) = 1$ for all k and does not converge to $f(a) \equiv f(0) = 0$. Thus, f is discontinuous at $a = 0$.

Example 1.17

Consider the function g along with its domain $D(g)$ defined by

$$\begin{aligned} g(x) &= 1, \quad 0 \leq x \leq 1 \\ &= 2, \quad 2 \leq x \leq 3 \end{aligned}$$

While a plot of the function g versus x is "disconnected," the function g is continuous at each point in its domain by virtue of the way in which it is defined. Note that any sequence we construct must lie in the domain.

C^1 and C^2 Continuity

Let A be an open set of R_n and $f: R_n \rightarrow R^1$. If each of the functions $\frac{\partial f}{\partial x_i}$, $i = 1, \ldots, n$, is continuous on this set, then we write $f \in C^1$ or that f is C^1 continuous, or state that f is "*smooth.*" If each of the functions $\frac{\partial^2 f}{\partial x_i \partial x_j}$, $1 \leq i, j \leq n$, is continuous on the set, then we write $f \in C^2$.

Example 1.18
Consider

$$f = \tfrac{1}{2}\max(0, x-5)^2, x \in R^1.$$

We have (see Fig. E1.18)

$$\frac{\partial f}{\partial x} = \max(0,\ x\ -\ 5)$$

$$\frac{\partial^2 f}{\partial x^2} = 0,\ \text{for}\ x < 5$$

$$= 1, \text{for}\ x > 5$$

The first derivative is a continuous function, while the second derivative is not continuous at $x = 0$. Thus f is only C^1 (and not C^2) continuous on R^1.

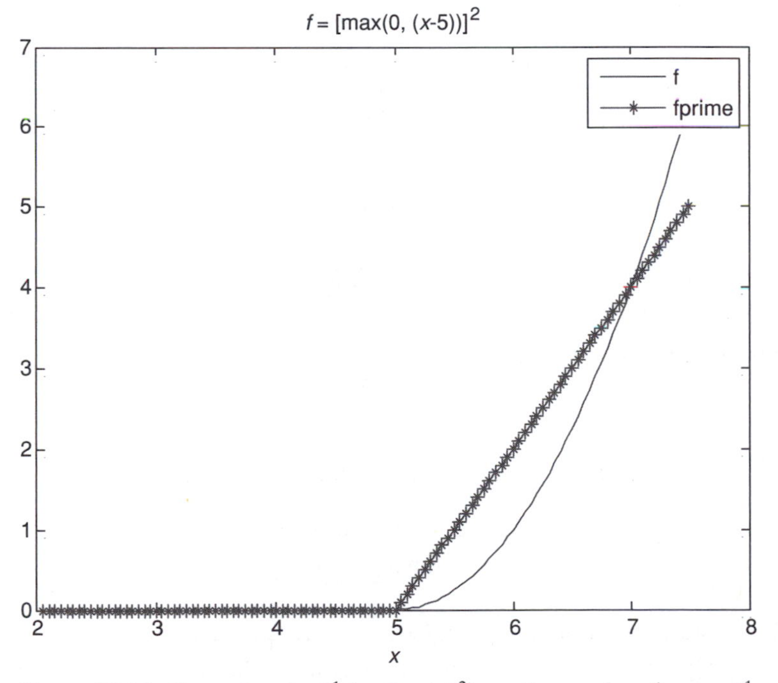

Figure E1.18. Example of a C^1 (and not C^2) continuous function on R^1.

1.9 Gradient Vector, Hessian Matrix, and Their Numerical Evaluation Using Divided Differences

Gradient Vector

Given a function $f(\mathbf{x}) \in C^1$ (i.e., C^1 continuous), we introduce the *gradient vector* ∇f, a *column* vector, as

$$\nabla f = \left[\frac{\partial f}{\partial x_1}, \frac{\partial f}{\partial x_2}, \ldots, \frac{\partial f}{\partial x_n}\right]^{\mathrm{T}} \tag{1.3}$$

The gradient evaluated at a point $\mathbf{c}$ is denoted by $\nabla f(\mathbf{c})$. The geometric significance of the gradient vector is discussed in Chapter 3 – that the vector is normal to the contour surface $f(\mathbf{x}) = f(\mathbf{c})$. The gradient of a vector of m-functions, $\nabla \mathbf{g}$, is defined to be a matrix of dimension $(n \times m)$ as

$$\nabla \mathbf{g} = \begin{bmatrix} \frac{\partial g_1}{\partial x_1} & \frac{\partial g_2}{\partial x_1} & \cdots & \frac{\partial g_m}{\partial x_1} \\ \frac{\partial g_1}{\partial x_2} & \frac{\partial g_2}{\partial x_2} & & \frac{\partial g_m}{\partial x_2} \\ \vdots & \vdots & \ddots & \vdots \\ \frac{\partial g_1}{\partial x_n} & \frac{\partial g_2}{\partial x_n} & & \frac{\partial g_m}{\partial x_n} \end{bmatrix} \tag{1.4}$$

Directional Derivative

This is a very important concept. The derivative of f in any direction $\mathbf{s}$ at a point $\mathbf{c}$, termed the "directional derivative," can be obtained as

$$D_{\mathbf{s}} f(\mathbf{c}) = \lim_{t \to 0} \frac{1}{t} \{f(\mathbf{c} + t\mathbf{s}) - f(\mathbf{c})\} = \nabla f(\mathbf{c})^{\mathrm{T}} \mathbf{s} \tag{1.5a}$$

To be precise, $\mathbf{s}$ must be a unit vector in order to use the term "directional derivative." We can express the directional derivative in an alternate manner. We may introduce a scalar variable α and denote points along the vector $\mathbf{s}$ emanating from $\mathbf{c}$ as $\mathbf{x}(\alpha) = \mathbf{c} + \alpha\mathbf{s}$. Then, denoting the function $f(\alpha) \equiv f(\mathbf{x}(\alpha))$, we have

$$D_{\mathbf{s}} f(\mathbf{c}) = \left.\frac{df}{d\alpha}\right|_{\alpha=0} = \nabla f(\mathbf{c})^{\mathrm{T}} \mathbf{s} \tag{1.5b}$$

Example 1.19

Given $f = x_1 x_2$, $\mathbf{c} = [1, 0]^T$, $\mathbf{s} = [-1, 1]^T$, find: (i) the gradient vector $\nabla f(\mathbf{c})$, (ii) the directional derivative using $D_\mathbf{s} f(\mathbf{c}) = \nabla f(\mathbf{c})^T \mathbf{s}$, (iii) plot of $f(\alpha)$, and (iv) the directional derivative using

$$D_\mathbf{s} f(\mathbf{c}) = \left.\frac{df}{d\alpha}\right|_{\alpha=0}$$

(i) $\nabla f(\mathbf{x}) = [x_2, x_1]$. Thus, $\nabla f(\mathbf{c}) = [0, 1]^T$.
(ii) $D_\mathbf{s} f(\mathbf{c}) = [0\ 1]\left[\begin{smallmatrix}-1\\1\end{smallmatrix}\right] = 1$.
(iii) $\mathbf{x}(\alpha) = [1 - \alpha, 0 + \alpha]^T$. Thus, $f(\alpha) = \alpha(1 - \alpha)$. This function is plotted in Fig. E1.19.
(iv) $D_\mathbf{s} f(\mathbf{c}) = \left.\frac{df}{d\alpha}\right|_{\alpha=0} = 1 - 2(0) = 1$.

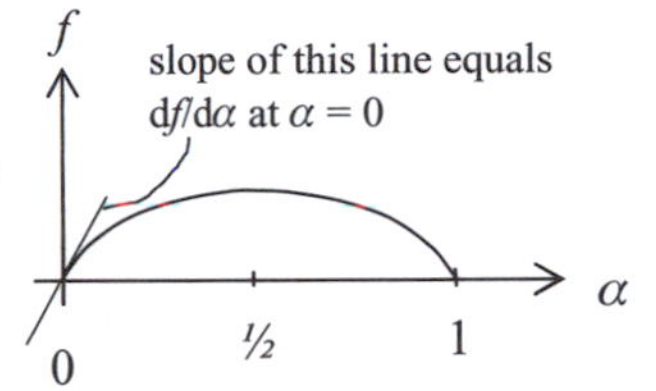

Figure E1.19. Illustration of directional derivative in the example.

Hessian Matrix

Given a function $f \in C^2$, we define the matrix of second partial derivatives

$$\nabla^2 f \equiv \begin{bmatrix} \frac{\partial^2 f}{\partial x_1 \partial x_1} & \frac{\partial^2 f}{\partial x_1 \partial x_2} & \cdots & \frac{\partial^2 f}{\partial x_1 \partial x_n} \\ & \frac{\partial^2 f}{\partial x_2 \partial x_2} & \cdots & \frac{\partial^2 f}{\partial x_2 \partial x_n} \\ & & \ddots & \\ \textit{symmetric} & & & \frac{\partial^2 f}{\partial x_n \partial x_n} \end{bmatrix} \tag{1.6}$$

as the *Hessian* of the function. Note that $\frac{\partial^2 f}{\partial x_i \partial x_i} = \frac{\partial^2 f}{\partial x_i^2}$. The Hessian is related to the convexity of the function as discussed in Chapter 3. The Hessian evaluated at a point $\mathbf{c}$ is denoted as $\nabla^2 f(\mathbf{c})$.

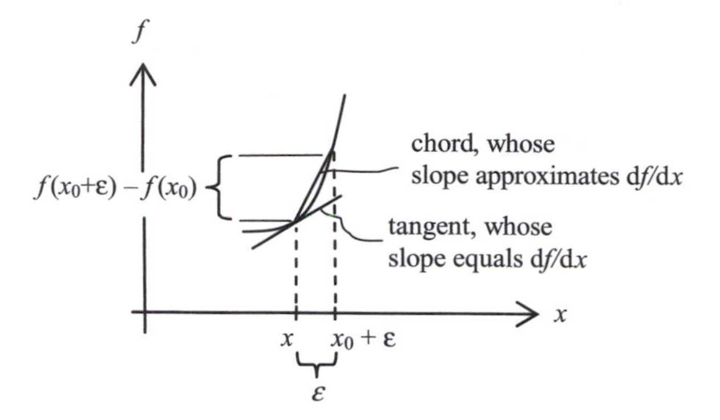

Figure 1.3. Illustration of forward difference formula.

Numerical Evaluation of Gradient and Hessian Matrices

Newton's forward difference formula is often utilized to obtain an approximate expression for ∇f (and also for $\nabla^2 f$) at a given point. The forward difference formula is

$$\frac{\partial f}{\partial x_i} \approx \frac{f(x_1^0, x_2^0, \ldots, x_i^0 + \varepsilon, \ldots, x_n^0) - f(\mathbf{x}^0)}{\varepsilon}, \quad i = 1, \ldots, n \tag{1.7}$$

where ε is the divided difference parameter. In (1.7), each variable is perturbed one at a time and f re-evaluated. Too small a choice of ε leads to a loss of significance error since two very nearly equal quantities will be subtracted from one another. Small ε is also not recommended in problems where there is "noise" in the function evaluation, as occurs when using a nonlinear iterative simulation code to evaluate f. Too large an ε results in a large truncation error, which is the error due to truncating the Taylor series to just the first two terms; graphically, this is the error between the slope of the tangent and the chord at the point (Fig. 1.3). A thumb rule such as ε = maximum ($\varepsilon_{\text{minimum}}$, 0.01 x_i^0) works well for most functions. Note that n function evaluations are needed to compute $\nabla f(\mathbf{x}^0)$. Thus, this scheme is not recommended in problems where each function evaluation is computationally time consuming, except to debug code based on analytical derivatives.

For greater accuracy, the central difference formula is given by

$$\frac{\partial f}{\partial x_i} \approx \frac{f(x_1^0, x_2^0, \ldots, x_i^0 + 0.5\,\varepsilon, \ldots, x_n^0) - f(x_1^0, x_2^0, \ldots, x_i^0 - 0.5\,\varepsilon, \ldots, x_n^0)}{\varepsilon},$$
$$i = 1, \ldots, n \tag{1.8}$$

Computer implementation of forward differences to obtain ∇f from Eq (1.8) is relative straightforward.

Forward Difference Computer Logic

```
  % first evaluate f0 = f (x0) at the given point x0
[NFE,F0] = GETFUN(X,NFE,N);
for i=1:N
  epsil=.01*abs(X(i))+0.0001;
    xold=X(i);
  X(i)=X(i)+epsil;
  [NFE,F] = GETFUN(X,NFE,N);
  DF(i)= (F-F0)/epsil;
  X(i)=xold;
 end
  % NFE = number of function calls
```

Example 1.20

Consider the generalized eigenvalue problem

$$\begin{bmatrix} 1 & -1 \\ -1 & 1 \end{bmatrix} y^i = \lambda_i \begin{bmatrix} (c+x) & 0 \\ 0 & c \end{bmatrix} y^i, \quad i = 1, 2$$

where $c = 10^{-8}$. Evaluate the derivative of the second or highest eigenvalue with respect to x, $\frac{d\lambda_2}{dx}$, at the point $x_0 = 0$. Note that the first or lowest eigenvalue equals zero for all x. Forward difference is based on

$$\frac{d\lambda}{dx} \approx \frac{\lambda(x_0 + \varepsilon) - \lambda(x_0)}{\varepsilon}$$

while the analytical expression (given in Chapter 12) is not discussed here. Results using the MATLAB code in the following are shown in Fig. E1.20. Note that 100% in Fig. E1.20 means perfect agreement between forward difference and analytical ("exact") formula. As per the theory, results show that very low values of $\varepsilon\,(\leq 10^{-21})$ lead to cancellation error, while higher values ($\geq 10^{-10}$) lead to truncation error. It is generally difficult to choose a wide range of values for ε, in eigenvalue or other iterative analyses, particularly when there are several variables, as a choice for ε that gives good agreement for x_j may give poor agreement for x_k.

```
clear all; close all;
A= [1 -1;-1 1];
c= 1e-8;
B0= c*[1 0;0 1];
[V0,D0] = eig(A,B0);
epsil = 1e-10;  % forward diff. parameter
deltaB= [epsil 0;0 0];
B= B0 + deltaB;
[V,D] = eig(A,B);
snumer= (D(2,2)-D0(2,2))/epsil;
% analytical derivative below
denom= V0(:,2)'*B0*V0(:,2);
dBdx= [1 0;0 0];
sanaly= -D0(2,2)/denom*V0(:,2)'*dBdx*V0(:,2);
% print
epsil, snumer, sanaly
```

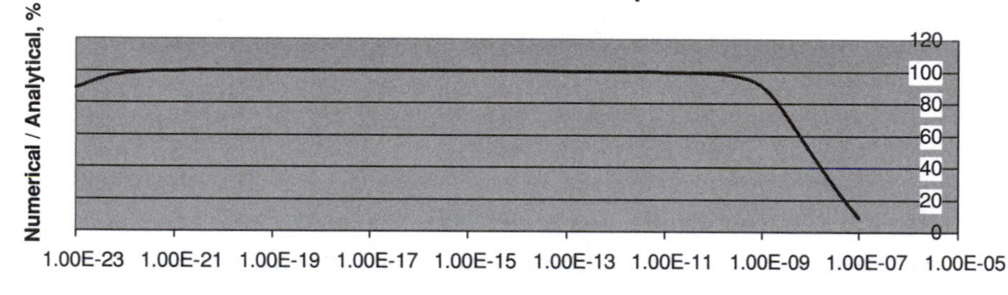

Figure E1.20. Dependence of numerical derivative on choice of ε for example.

Hessian

Numerical evaluation of the Hessian matrix follows along the same lines. A forward difference scheme for the *ij*th component of the Hessian yields

$$\frac{d}{dx_j}\left(\frac{df}{dx_i}\right) = \frac{1}{\varepsilon_j}\left[\frac{f(\ldots, x_i^0+\varepsilon_i, \ldots, x_j^0+\varepsilon_j, \ldots) - f(\ldots, x_j^0+\varepsilon_j, \ldots)}{\varepsilon_i} - \frac{f(\ldots, x_i^0+\varepsilon_i, \ldots) - f(\mathbf{x}^0)}{\varepsilon_i}\right] \tag{1.9}$$

Programs GRADIENT and HESSIAN: These are provided for the reader to experiment with and to compare numerical and analytical derivatives with different choices of ε. These programs (especially GRADIENT) can also be used in gradient-based techniques discussed in later chapters by either insertion of the relevant code or to debug an analytical expression for ∇f. Note that the programs require the following user-supplied subroutines: "Getfun" and "Gradient." "Getfun" provides $f(\mathbf{x})$ for an input value of $\mathbf{x}$ while "Gradient" provides an *analytical* value of $\nabla f(\mathbf{x})$ for an input value of $\mathbf{x}$. This analytical value is printed out along with the forward, backward, and central difference results. Program HESSIAN generates numerical Hessian and compares to user-supplied analytical expressions.

1.10 Taylor's Theorem, Linear, and Quadratic Approximations

This theorem is used everywhere in developing optimization theory and algorithms.

One Dimension: We first state the theorem in one dimension. Suppose that $f \in C^p$ on an interval $J = [a, b]$. If x_0, x belong to J, then there exists a number γ between x_0 and x such that

$$f(x) = f(x^0) + \frac{f'(x_0)}{1!}(x - x_0) + \frac{f''(x_0)}{2!}(x - x_0)^2 + \cdots + \frac{f^{(p-1)}(x_0)}{(p-1)!}(x - x_0)^{(p-1)} + \frac{f^{(p)}(\gamma)}{p!}(x - x_0)^p \tag{1.10}$$

The last term on the right-hand side is called the "remainder term." If we take just the first two terms, we get a linear expansion or approximation to f at the point x_0:

$$f(x) \approx f(x_0) + f'(x_0)(x - x_0) \tag{1.11}$$

Note that the "=" sign has been replaced by a "≈" sign as only two terms are considered. We can rewrite Eq. (1.11) as $f(x) \approx \ell(x)$ by defining a linear function $\ell(x)$ as

$$\ell(x) = f(x) + f'(x_0)(x - x_0) \tag{1.12}$$

Similarly, a quadratic approximating function can be defined as

$$q(x) = f(x_0) + f'(x_0)(x - x_0) + \tfrac{1}{2} f''(x_0)(x - x_0)^2 \tag{1.13}$$

and, again, recognize that generally $f \approx q$ only in a neighborhood of the expansion point x_0.

n-Dimensions

We now state the theorem in n dimensions but with only two terms, as it fulfills our requirement. Note that $\mathbf{x} = [x_1, x_2, \ldots, x_n]^{\mathrm{T}}$ is a column vector, and $\mathbf{x}^0 = [x_1^0, x_2^0, \ldots, x_n^0]^{\mathrm{T}}$. Suppose that f is a function that is C^2 continuous with open domain Ω in R^n. If $\mathbf{x}^0$ and $\mathbf{x}$ belong to Ω then there exists a point $\mathbf{z}$ on the line segment connecting $\mathbf{x}^0$ and $\mathbf{x}$ (i.e., there is a point $\mathbf{z} = \alpha\, \mathbf{x}^0 + (1 - \alpha)\, \mathbf{x}$ with $0 < \alpha < 1$) such that

$$f(\mathbf{x}) = f(\mathbf{x}^0) + \nabla f(\mathbf{x}^0)^{\mathrm{T}}(\mathbf{x} - \mathbf{x}^0) + \tfrac{1}{2}(\mathbf{x} - \mathbf{x}^0)^{\mathrm{T}}[\nabla^2 f(\mathbf{z})](\mathbf{x} - \mathbf{x}^0) \qquad (1.14)$$

Taylor's theorem is used to derive optimality conditions and develop numerical procedures based on constructing linear and quadratic approximations to nonlinear functions. Example 1.21 illustrates this. Equations (1.12) and (1.13) for the one variable case generalize to

$$\ell(\mathbf{x}) = f(\mathbf{x}^0) + \nabla f(\mathbf{x}^0)^{\mathrm{T}}(\mathbf{x} - \mathbf{x}^0) \qquad (1.15)$$

$$q(\mathbf{x}) = f(\mathbf{x}^0) + \nabla f(\mathbf{x}^0)^{\mathrm{T}}(\mathbf{x} - \mathbf{x}^0) + \tfrac{1}{2}(\mathbf{x} - \mathbf{x}^0)^{\mathrm{T}}\,[\nabla^2 f(\mathbf{x}^0)](\mathbf{x} - \mathbf{x}^0) \qquad (1.16)$$

Example 1.21

Given $f = 2\,x_1 + \frac{x_2}{x_1}$, and a point in $\mathbf{x}^0 = (1, 0.5)^{\mathrm{T}}$ in R_2. Construct linear and quadratic approximations to the original function f at $\mathbf{x}^0$ and plot contours of these functions passing through $\mathbf{x}^0$. We have the linear approximation

$$\begin{aligned}\ell(\mathbf{x}) &= f(\mathbf{x}^0) + \nabla f(\mathbf{x}^0)^{\mathrm{T}}\,(\mathbf{x} - \mathbf{x}^0)\\ &= 2.5 + (1.5, 1) \cdot (x_1 - 1, x_2 - 0.5)^{\mathrm{T}}\\ &= 0.5 + 1.5x_1 + x_2\end{aligned}$$

and the quadratic approximation

$$\begin{aligned}q(\mathbf{x}) &= \ell(\mathbf{x}) + \tfrac{1}{2}(\mathbf{x} - \mathbf{x}^0)^{\mathrm{T}}[\nabla^2 f(\mathbf{x}^0)](\mathbf{x} - \mathbf{x}^0)\\ &= 0.5 + x_1 + 2x_2 - x_1x_2 + \tfrac{1}{2}x_1^2\end{aligned}$$

Plotting

We need to plot, in variable-space or $\mathbf{x}$-space, the contours

$$f(\mathbf{x}) = c,\ \ell(\mathbf{x}) = c,\ q(\mathbf{x}) = c, \text{ where } c = f(\mathbf{x}^0) = 2.5$$

Thus, from the preceding equations, we have to plot the contours

$$2x_1 + \frac{x_2}{x_1} = 2.5$$
$$0.5 + 1.5x_1 + x_2 = 2.5$$
$$0.5 + x_1 + 1.5x_2 - x_1x_2 + \tfrac{1}{2}x_1^2 = 2.5$$

These contour plots are shown in Fig. E1.21 using the following MATLAB code:

```
clear all; close all;
[x1,x2] = meshgrid(0.5:.1:1.5,-0.25:.1:1.5);
f = 2*x1 + x2./x1;
L = 0.5 + 1.5*x1 + x2;
q = .5 + x1 + 2*x2 -x1.*x2 + 0.5*x1.^2;
contour(x1,x2,f,[2.5 2.5],'Linewidth',1.5)
hold
contour(x1,x2,L,[2.5 2.5],'--')
contour(x1,x2,q,[2.5 2.5],':')
legend('f','L','q')
```

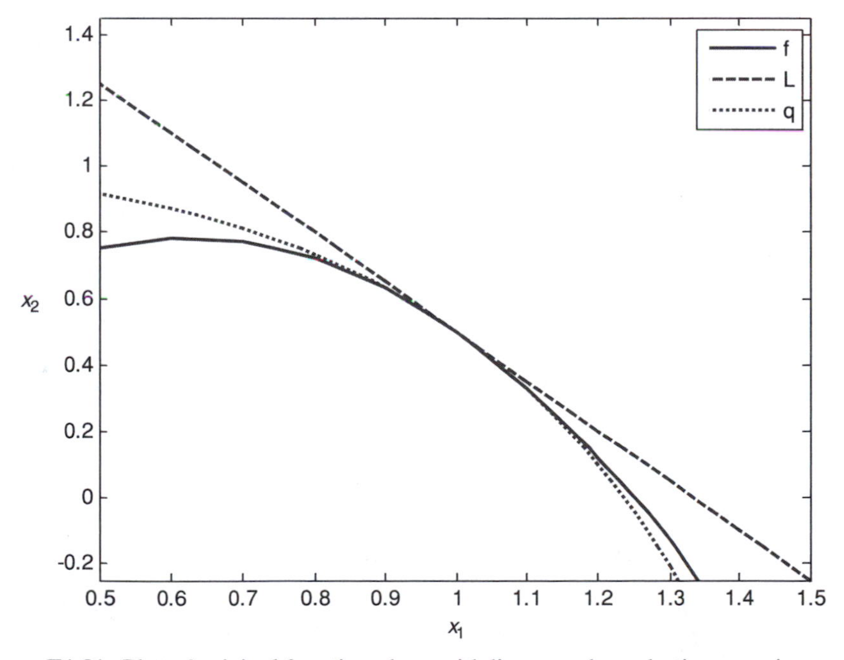

Figure E1.21. Plot of original function along with linear and quadratic expansions at $\mathbf{x}^0$.

1.11 Miscellaneous Topics

Linear Independence of Vectors

Let $\{\mathbf{x}^k\}$, $k = 1, \ldots, m$ be a set of m vectors. Then, we say that the set of vectors are linearly independent if the linear combination $\sum_{k=1}^{m} \alpha_k \mathbf{x}^k = 0$ *implies that* all α_k are zero. For, if the above holds with, say, $\alpha_p \neq 0$, then we can divide through by α_p and obtain $\mathbf{x}^p = \sum_{k \neq p} \frac{\alpha_k}{\alpha_p} \mathbf{x}^k$. This means that the pth vector has been expressed in terms of the rest whence the vectors are linearly dependent.

Norms

Given a vector $\mathbf{y} = [y_1, y_2, \ldots, y_n]^{\mathrm{T}}$, then the pth norm given by $\|\mathbf{y}\|_p = (\sum_i |y_i|^p)^{1/p}$. With $p = 2$, we get the commonly used L_2-norm $\|\mathbf{y}\|_2 = \sqrt{y_1^2 + y_2^2 + \cdots + y_n^2} = \sqrt{\mathbf{y}^{\mathrm{T}}\mathbf{y}}$. Further, $p = 1$ gives the L_1-norm $\|\mathbf{y}\|_1 = \sum_i |y_i|$. The infinity norm is $\|\mathbf{y}\|_\infty = \max_i[|y_1|, |y_2|, \ldots, |y_n|]$. As an example, $\mathbf{y} = [1, 2, -3]$ gives $\|\mathbf{y}\|_1 = 6$, $\|\mathbf{y}\|_2 = \sqrt{14}$, $\|\mathbf{y}\|_\infty = 3$.

Elementary Row Operations (ERO) and Pivoting

A procedure used widely in linear programming (Chapter 4) will be given in the following. The reader may skip this material until having begun reading Chapter 4. Consider a system of linear simultaneous equations

$$\mathbf{A}\mathbf{x} = \mathbf{b}$$

where $\mathbf{A}$ is $(m \times n)$, $\mathbf{b}$ is $(m \times 1)$, and $\mathbf{x}$ is $(n \times 1)$. That is, we have m equations in n unknowns, with $m < n$. Let $\mathbf{A}$ be partitioned as $\mathbf{A} = [\mathbf{I}, \mathbf{B}]$, where I is the identity matrix and $\mathbf{B}$ is an $(m \times m)$ nonsingular matrix. Then, a solution is readily available by setting the $(n - m)$ "nonbasic" variables to zero and solving for the m "basic" variables from $\mathbf{B}^{-1}\,\mathbf{b}$. Our interest is to switch a nonbasic variable with a basic variable using ERO on the matrix $\mathbf{A}$. For illustration, consider the system

$$\begin{array}{cccrrrl} x_1 & & & +x_4 & +x_5 & -x_6 = & 5 \\ & x_2 & & +2x_4 & -3x_5 & +x_6 = & 3 \\ & & x_3 & -x_4 & +2x_5 & -x_6 = & -1 \end{array}$$

We may set the nonbasic variables $x_4 = x_5 = x_6 = 0$ and immediately obtain $x_1 = 5$, $x_2 = 3$, $x_3 = -1$. Now suppose that we wanted to switch the nonbasic variable x_5 with the basic variable x_2. This can be achieved using ERO. The system can be expressed in the form of a tableau, with the last column representing the **b** vector as

$$\begin{array}{c} \\ x_1 \\ x_2 \\ x_3 \end{array} \begin{array}{c} \begin{array}{ccccccc} x_1 & x_2 & x_3 & x_4 & x_5 & x_6 & \mathbf{b} \end{array} \\ \begin{bmatrix} 1 & 0 & 0 & 1 & 1 & -1 & 5 \\ 0 & 1 & 0 & 2 & -3 & 1 & 3 \\ 0 & 0 & 1 & -1 & 2 & -1 & -1 \end{bmatrix} \end{array}$$

The $x_2 - x_5$ switch results in the (2, 5) element being the *pivot* (= −3 here). If we now premultiply the tableau by a matrix **T** given by

$$\begin{bmatrix} 1 & \frac{1}{3} & 0 \\ 0 & -\frac{1}{3} & 0 \\ 0 & \frac{2}{3} & 1 \end{bmatrix}$$

then we arrive at

$$\begin{array}{c} \\ x_1 \\ x_5 \\ x_3 \end{array} \begin{array}{c} \begin{array}{ccccccc} x_1 & x_2 & x_3 & x_4 & x_5 & x_6 & \mathbf{b} \end{array} \\ \begin{bmatrix} 1 & \frac{1}{3} & 0 & \frac{5}{3} & 0 & -\frac{2}{3} & 6 \\ 0 & -\frac{1}{3} & 0 & -\frac{2}{3} & 1 & -\frac{1}{3} & -1 \\ 0 & \frac{2}{3} & 1 & \frac{1}{3} & 0 & -\frac{1}{3} & 1 \end{bmatrix} \end{array}$$

from which $x_2 = x_4 = x_6 = 0, x_1 = 6, x_5 = -1, x_3 = 1$. The premultiplication by **T** is equivalent to ERO on the tableau. The structure of the **T** matrix can be understood by considering a general situation, wherein the elements of the tableau are denoted by a_{ij}, and we wish to interchange a basic variable x_p with a nonbasic variable x_q. Then, a_{pq} is the pivot and **T** is formed with 1's along the diagonal, 0's elsewhere, and with the qth column defined as

$$\left[\frac{-a_{1q}}{a_{pq}} \quad \bullet \quad \bullet \quad \frac{-a_{(p-1)q}}{a_{pq}} \quad \frac{1}{a_{pq}} \quad \frac{-a_{(p+1)q}}{a_{pq}} \quad \bullet \quad \bullet \quad \frac{-a_{mq}}{a_{pq}} \right]^{\mathrm{T}}$$

In view of the above, we may express the coefficients of the new tableau $\boldsymbol{a}'$ in terms of the old tableau $\boldsymbol{a}$ as

$$\begin{aligned} & a'_{ij} = a_{ij} - \frac{a_{iq}}{a_{pq}} a_{pj} \quad \text{and} \quad b'_i = b_i - \frac{a_{iq}}{a_{pq}} b_p, \qquad i \neq p, i = 1 \text{ to } m, \ j = 1 \text{ to } n \\ & a'_{pj} = \frac{a_{pj}}{a_{pq}} \quad \text{and} \quad b'_p = \frac{b_p}{a_{pq}}, \quad j = 1 \text{ to } n \end{aligned} \tag{1.17}$$

A FEW LEADS TO OPTIMIZATION SOFTWARE

Book by J.J. More and S.J. Wright, *Optimization Software Guide*, 1993; see: http://www.siam.org/catalog/mcc12/more.htm (Series: Frontiers in Applied Mathematics 14, ISBN 0-89871-322-6)

www.mcs.anl.gov/	(Argonne National Labs Web site)
www.gams.com	(GAMS code)
www.uni-bayreuth.de/departments/math/org	(by K. Schittkowski)
www.altair.com	(Altair corp. – structural optimization)
www.mscsoftware.com	(MSC/Nastran code for structural optimization)
www.altair.com	(includes topology optimization)
www.vrand.com	(by G.N. Vanderplaats – DOC and GENESIS, the latter code for structural optimization)

SOME OPTIMIZATION CONFERENCES

INFORMS (ORSA-TIMS)	Institute for operations research and the management sciences
SIAM Conferences	Society for industrial and applied math
NSF Grantees Conferences	
WCSMO	World Congress on Structural and Multidisciplinary Optimization
ASMO UK/ISSMO	Engineering Design Optimization
AIAA/USAF/NASA Symposium on Multidisciplinary Analysis and Optimization	
ASME Design Automation Conference	(American Society of Mechanical Engineers)
EngOpt (2008) International Conference on Engineering Optimization. Rio de Janeiro, Brazil Optimization In Industry (I and II), 1997, 1999	(Industry applications)

SOME OPTIMIZATION SPECIFIC JOURNALS

Structural and Multidisciplinary Optimization (official Journal for ISSMO)
AIAA Journal
Journal of Global Optimization
Optimization and Engineering
Evolutionary Optimization (Online)
Engineering Optimization
Interfaces
Journal of Optimization Theory and Applications (SIAM)
Journal of Mathematical Programming (SIAM)

COMPUTER PROGRAMS

1) Program GRADIENT – computes the gradient of a function supplied by user in "Subroutine Getfun" using forward, backward, and central differences, and compares this with analytical gradient provided by the user in "Subroutine Gradient"
2) Program HESSIAN – similar to GRADIENT, except a Hessian matrix is output instead of a gradient vector

PROBLEMS

P1.1. Consider the problem

$$\begin{aligned} &\text{minimize} \quad f = (x_1 - 3)^2 + (x_2 - 3)^2 \\ &\text{subject to} \quad x_1 \geq 0, \quad x_2 \geq 0, \quad x_1 + 2x_2 \leq 1 \end{aligned}$$

Let the optimum value of the objective to the problem be f^*. Now if the third constraint is changed to $x_1 + 2x_2 \leq 1.3$, will the optimum objective decrease or increase in value?

P1.2. Consider the problem

$$\begin{aligned} &\text{minimize} \quad && f = x_1 + x_2 + x_3 \\ &\text{subject to} \quad && x_1^{x_2} - x_3 \geq 1 \\ & && x_i \geq 0.1, \quad i = 1, 2, 3 \end{aligned}$$

Quickly, determine an upper bound M to the optimum solution f^*.

P1.3. Are the conditions in the Weirstrass theorem necessary or sufficient for a minimum to exist?

P1.4. For the cantilever problem in Example 1.14 in the text, redraw the feasible region (Figure E1.14b) with the additional constraint: $x_2 \leq x_1/3$. Does the problem now have a solution? Are Wierstraas conditions satisfied?

P1.5. Using formulae in Example 1.2, obtain $\mathbf{x}^*$ and the shortest distance d from the point $(1, 2)^T$ to the line $y = x - 1$. Provide a plot.

P1.6. In Example 3, beam on two supports, develop a test based on physical insight to verify an optimum solution.

P1.7. Formulate two optimization problems from your everyday activities, and state how the solution to the problems will be beneficial to you.

P1.8. Using the definition, determine if the following vectors are linearly independent:

$$\begin{pmatrix} 1 \\ -5 \\ 2 \end{pmatrix}, \begin{pmatrix} 0 \\ 6 \\ -1 \end{pmatrix}, \begin{pmatrix} 3 \\ -3 \\ 4 \end{pmatrix}$$

P1.9. Determine whether the following quadratic form is positive definite:

$$f = 2x_1^2 + 5x_2^2 + 3x_3^2 - 2x_1x_2 - 4x_2x_3, \quad \mathbf{x} \in R^3$$

P1.10. Given $f = [\max(0, x - 3)]^4$, determine the highest value of n wherein $f \in C^n$

P1.11. If $f = \mathbf{g}^T \mathbf{h}$, where $f: R^n \to R^1$, $\mathbf{g}: R^n \to R^m$, $\mathbf{h}: R^n \to R^m$, show that

$$\nabla f = [\nabla \mathbf{h}]\mathbf{g} + [\nabla \mathbf{g}]\mathbf{h}$$

P1.12. Referring to the following graph, is f a function of x in domain [1, 3]?

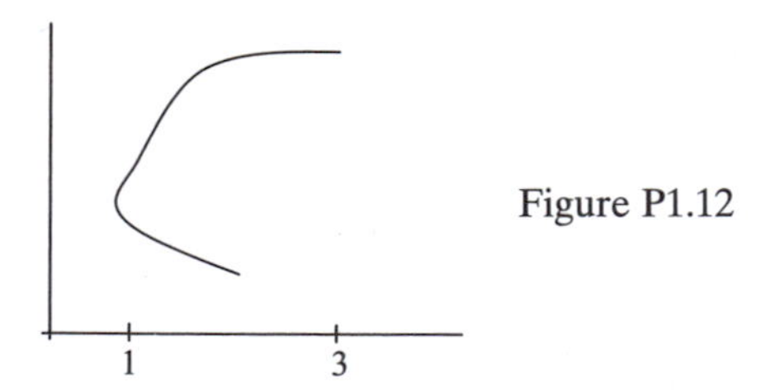

Figure P1.12

P1.13. Show why the diagonal elements of a positive definite matrix must be strictly positive. *Hint:* use the definition given in Section 1.7 and choose appropriate $\mathbf{y}$'s.

P1.14. If n function evaluations are involved in obtaining the gradient ∇f via forward differences, where n equals the number of variables, how many function evaluations are involved in (i) obtaining ∇f via central differences as in Eq. (1.8) and (ii) obtaining the Hessian as per Eq. (1.9)?

P1.15. Compare analytical, forward difference, backward difference and central difference derivatives of f with respect to x_1 and x_2, using program GRADIENT for the function $f = \frac{4\,x_2^2 - x_1\,x_2}{10000\,(x_2\,x_1^3 - x_1^4)}$ at the point $\mathbf{x}^0 = (0.5,\ 1.5)^{\mathrm{T}}$. *Note:* provide the function in Subroutine Getfun and the analytical gradient in Subroutine Gradient.

P1.16. Given $f = 100(x_2 - x_1^2)^2 + (1 - x_1)^2$, $\mathbf{x}_0 = (2, 1)^{\mathrm{T}}$, $\mathbf{s} = (-1, 0)^{\mathrm{T}}$.

(i) Write the expression for $f(\alpha) \equiv f(\mathbf{x}(\alpha))$ along the direction **s**, from the point $\mathbf{x}^0$.

(ii) Use forward differences to estimate the directional derivative of f along **s** at $\mathbf{x}^0$; do not evaluate ∇f.

P1.17. Plot the derivative $\frac{d\lambda_2}{dx}$ versus ε using a forward difference scheme. Do not consider analytical expression. Take $x_0 = 0$. For what range of ε do you see stable results?

$$\begin{bmatrix} 1 & -1 \\ -1 & 1 \end{bmatrix} y^i = \lambda_i \begin{bmatrix} c & x \\ x & c \end{bmatrix} y^i, \quad i = 1, 2$$

P1.18. Repeat P1.17 with central differences, and plot forward and central differences on same figure. Compare your results.

P1.19. Given the inequality constraint $g \equiv 100 - x_1\,x_2^2 \le 0$, and a point on the boundary $\mathbf{x}^0 = (4, 5)^{\mathrm{T}}$, develop an expression for:

(i) a linear approximation to g at $\mathbf{x}^0$ and

(ii) a quadratic approximation to g at $\mathbf{x}^0$.

Plot the original g and the approximations and indicate the feasible side of the curves.

P1.20. Given the inequality constraint $g \equiv 100 - x_1\,x_2 \le 0$ develop an expression for a linear approximation to g at $\mathbf{x}^0$. Plot the original g and the approximation and indicate the feasible side of the curves. Do the problem including plots for two different choices of $\mathbf{x}^0$:

(i) a point on the boundary $\mathbf{x}^0 = (4, 25)^{\mathrm{T}}$ and

(ii) a point not on the boundary $\mathbf{x}^0 = (5, 25)^{\mathrm{T}}$.

P1.21. Consider the forward difference approximation in Eq. (1.7) and Taylor's theorem in Eq. (1.10). Assume that the second derivative of *f* is bounded by *M* on R^1, or $\left|\frac{\partial^2 f}{\partial x^2}\right| \leq M$.

(i) Provide an expression for the "truncation" error in the forward difference formula.

(ii) What is the dependence of the truncation error on the divided difference parameter *h*.

P1.22. Using Taylor's theorem, show why the central difference formula, Eq. (1.8), gives smaller error compared with the forward difference formula in Eq. (1.7).

P1.23. Consider the following structure. Determine the maximum load *P* subject to cable forces being within limits, when:

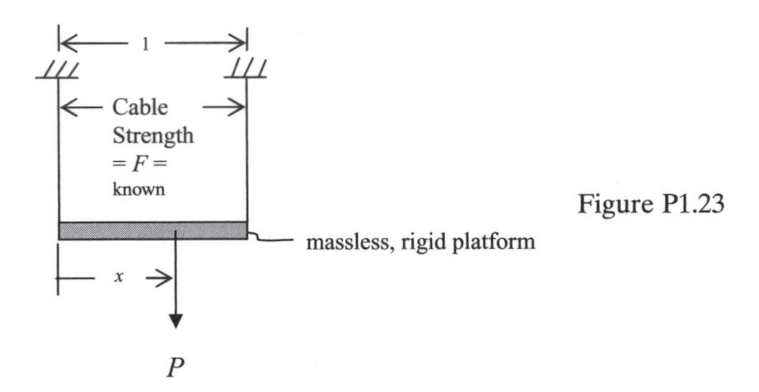

Figure P1.23

(i) the position of the load *P* can be anywhere between $0 \leq x \leq 1$ and

(ii) the load *P* is fixed at the location $x = 0.5$ m.

(Solve this by any approach.)

REFERENCES

Argod, V., Nayak, S.K., Singh, A.K. and Belegundu, A.D., Shape optimization of solid isotropic plates to mitigate the effects of air blast loading, *Mechanics Based Design of Structures and Machines*, 38:362–371, 2010.

Atkinson, K.E., *An Introduction to Numerical Analysis*, Wiley, New York, 1978.

Arora, J.S., *Guide to Structural Optimization*, ASCE Manuals and Reports an Engineering Practice No. 90, 1997.

Arora, J.S., *Introduction to Optimum Design*, Elsevier, New York, 2004.

Arora, J.S. (Ed.), *Optimization of Structural and Mechanical Systems*, World Scientific Publishing, Hackensack, NJ, 2007.

Banichuk, N.V., *Introduction to Optimization of Structures*, Springer, New York, 1990.

Bazaraa, M.S., Sherali, H.D., and Shetty, C.M., *Nonlinear Programming: Theory and Algorithms*, John Wiley & Sons, Hoboken, NJ, 2006.

Belegundu, A.D. and Salagame, R., Optimization of laminated ceramic composites for minimum residual stress and cost, *Microcomputers in Civil Engineering*, vol. 10, Issue 4, pp. 301–306, 1995.

Bendsoe, M. and Sigmund, O., *Topology Optimization*, Springer, New York, 2002.

Bhatti, M.A., *Practical Optimization Methods: With Mathematica Applications*, Springer-Verlag, New York, 2000.

Boyd, S. and Vandenberghe, L., *Convex Optimization*, Cambridge University Press, Cambridge, 2004.

Brent, R.P., *Algorithms of Minimization without Derivatives*, Chapter 4, Prentice-Hall, Englewood Cliffs, NJ, 1973, pp. 47–60.

Cauchy, A., Methode generale pour la resolution des systemes d'equations simultanes, C.R. Acad. Sci. Par., **25**, 536–538, 1847.

Chandrupatla, T.R. and Belegundu, A.D., *Introduction to Finite Elements in Engineering*, 3rd Edition Prentice-Hall, Englewood Cliffs, NJ, 2002.

Constans, E., Koopmann, G.H., and Belegundu, A.D., The use of modal tailoring to minimize the radiated sound power from vibrating shells: theory and experiment, *Journal of Sound and Vibration*, **217**(2), 335–350, 1998.

Courant, R., Variational methods for the solution of problems of equilibrium and vibrations, *Bulletin of the American Mathematical Society*, **49**, 1–23, 1943.

Courant, R. and Robbins, H., *What Is Mathematics? An Elementary Approach to Ideas and Methods*, Oxford University Press, Oxford, 1996.

Dantzig, G.B., "Maximization of a Linear Function of Variables Subject to Linear Inequalities", in T.C. Koopmans (Ed.), *Activity Analysis of Production and Allocation*, Wiley, New York, 1951, Chapter 21.

Dantzig, G.B., Linear programming – the story about how it began: some legends, a little about its historical significance, and comments about where its many mathematical programming extensions may be headed, *Operations Research*, **50**(1), 42–47, 2002.

Deb, K., *Multi-Objective Optimization Using Evolutionary Algorithms*, Wiley, New York, 2001.

Fiacco, A.V. and McCormick, G.P., *Nonlinear Programming: Sequential Unconstrained Minimization Techniques*, SIAM, Philadelphia, PA, 1990.

Fletcher, R. and Powell, M.J.D, A rapidly convergent descent method for minimization, *The Computer Journal*, **6**, 163–168, 1963.

Fletcher, R. and Reeves, C.M., Function minimization by conjugate gradients, *The Computer Journal*, **7**, 149–154, 1964.

Fox, R.L., *Optimization Methods for Engineering Design*, Addison-Wesley, Boston, MA, 1971.

Goldberg, D.E., *Genetic Algorithms in Search, Optimisation and Machine Learning*, Addison-Wesley, Boston, MA, 1989.

Golden, B.L. and Wasil, E.A., Nonlinear programming on a microcomputer, *Computers and Operations Research*, **13**(2/3), 149–166, 1986.

Gomory, R.E., An algorithm for the mixed integer problem, Rand Report, R.M. 25797, July 1960.

Grissom, M.D., Belegundu, A.D., Rangaswamy, A., Koopmann, G.H., Conjoint analysis based multiattribute optimization: application in acoustical design, *Structural and Multidisciplinary Optimization*, (**31**), 8–16, 2006.

Haftka, R.T., *Elements of Structural Optimization*, 3rd Edition, Kluwer, Dordrecht, 1992.

Haug, E.J. and Arora, J.S., *Applied Optimal Design*, Wiley, New York, 1979.

Haug, E.J., Choi, K.K., and Komkov, V., *Design Sensitivity Analysis of Structural Systems*, Academic Press, New York, 1985.

Hooke, R. and Jeeves, T.A., Direct search solution of numerical and statistical problems, *Journal of the ACM*, **8**, 212–229, 1961.

Johnson, R.C., *Optimum Design of Mechanical Elements*, Wiley-Interscience, New York, 1961.

Karmarkar, N., A new polynomial time algorithm for linear programming, *Combinatorica*, **4**, 373–395, 1984.

Karush, W., Minima of functions of several variables with inequalities as side conditions, MS Thesis, Department of Mathematics, University of Chicago, Chicago, IL, 1939.

Koopmann, G.H. and Fahnline, J.B., *Designing Quiet Structures: A Sound Power Minimization Approach*, Academic Press, New York, 1997.

Kuhn, H.W. and Tucker, A.W., "Non-linear programming", in J. Neyman (Ed.), *Proceedings of the Second Berkeley Symposium on Mathematical Statistics and Probability*, University of California Press, Berkeley, 1951, pp. 481–493.

Land, A.H. and Doig, A., An automatic method of solving discrete programming problems, *Econometrica*, **28**, 497–520, 1960.

Lasdon, L., *Optimization Theory for Large Systems*, Macmillan, New York, 1970. (Reprinted by Dover, 2002.)

Luenberger, D.G., *Introduction to Linear and Nonlinear Programming*, Addison-Wesley, Reading, MA, 1973.

Morris, A.J. (ed.), *Foundations of Structural Optimization*, John Wiley & Sons, New York, 1982.

Murty, K.G., *Linear Programming*, Wiley, New York, 1983.

Nash, S.G. and Sofer, A., *Linear and Nonlinear Programming*, McGraw-Hill, New York, 1995.

Nahmias, S., *Production and Operations Analysis*, McGraw-Hill, New York, 2004.

Nemhauser, G., Optimization in Airline Scheduling, Keynote Speaker, Optimization in Industry, Palm Coast, FL, March 1997.

Papalambros, P.Y. and Wilde, D.J., *Principles of Optimal Design*, 2nd Edition, Cambridge University Press, Cambridge, 2000.

Pareto, V., *Manual of Political Economy*, Augustus M. Kelley Publishers, New York, 1971. (Translated from French edition of 1927.)

Ravindran, A., Ragsdell, K.M., and Reklaitis, G.V., *Engineering Optimization – Methods and Applications*, 2nd Edition, Wiley, New York, 2006.

Rao, S.S., *Engineering Optimization*, 3rd Edition, Wiley, New York, 1996.

Roos, C., Terlaky, T., and Vial, J.P., *Theory and Algorithms for Linear Optimization: An Interior Point Approach*, Wiley, New York, 1997.

Rosen, J., The gradient projection method for nonlinear programming, II. Non-linear constraints, *Journal of the Society for Industrial and Applied Mathematics*, **9**, 514–532, 1961.

Salagame, R.R. and Belegundu, A.D., A simple *p*-adaptive refinement procedure for structural shape optimization, *Finite Elements in Analysis and Desing*, **24**, 133–155, 1997.

Schmit, L.A., Structural design by systematic synthesis, *Proceedings of the 2nd Conference on Electronic Computation, ASCE*, New York, pp. 105–122, 1960.

Siddall, J.N., *Analytical Decision-Making in Engineering Design*, Prentice-Hall, Englewood Cliffs, NJ, 1972.

Vanderplaats, G.N., *Numerical Optimization Techniques for Engineering Design: With Applications*, McGraw-Hill, New York, 1984.

Waren, A.D. and Lasdon, L.S., The status of nonlinear programming software, *Operations Research*, **27**(3), 431–456, 1979.

Wilde, D.J. and Beightler, C.S., *Foundations of Optimization*, Prentice-Hall, Englewood Cliffs, NJ, 1967.

Zhang, S. and Belegundu, A.D., Mesh distortion control in shape optimization, *AIAA Journal*, **31**(7), 1360–1362, 1992.

Zoutendijk, G., *Methods of Feasible Directions*, Elsevier, New York, 1960.

2

One-Dimensional Unconstrained Minimization

2.1 Introduction

Determination of the minimum of a real valued function of one variable, and the location of that minimum, plays an important role in nonlinear optimization. A one-dimensional minimization routine may be called several times in a multivariable problem. We show later that at the minimum point of a sufficiently smooth function, the slope is zero. If the slope and curvature information is available, minimum may be obtained by finding the location where the slope is zero and the curvature is positive. The need to determine the zero of a function occurs frequently in nonlinear optimization. Reliable and efficient ways of finding the minimum or a zero of a function are necessary for developing robust techniques for solving multivariable problems. We present the basic concepts involved in single variable minimization and zero finding.

2.2 Theory Related to Single Variable (Univariate) Minimization

We present the minimization ideas by considering a simple example. The first step is to determine the *objective function* that is to be optimized.

Example 2.1

Determine the objective function for building a minimum cost cylindrical refrigeration tank of volume 50 m^3, if the circular ends cost \$10 per m^2, the cylindrical wall costs \$6 per mm^2, and it costs \$80 per m^2 to refrigerate over the useful life.

Solution

Cylindrical tank of diameter x and volume V is shown in Fig. E2.1.

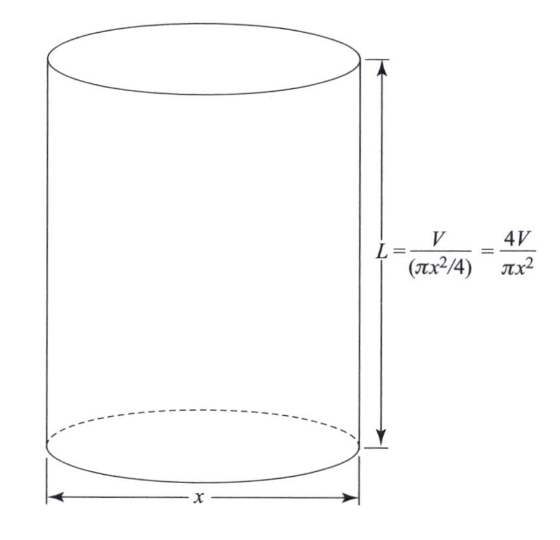

Figure E2.1. Refrigerated tank.

We denote the total cost as f. We have

$$f = (10)(2)\left(\frac{\pi x^2}{4}\right) + (6)(\pi x L) + 80\left(2.\frac{\pi x^2}{4} + \pi x L\right)$$
$$= 45\pi x^2 + 86\pi x L,$$

Substituting

$$L = \frac{(50)(4)}{\pi x^2} = \frac{200}{\pi x^2},$$

we get

$$f = 45\pi x^2 + \frac{17200}{x}$$

The function $f(x)$ is the objective function to be minimized. One variable plot of the function over a sufficiently large range of x shows the distinct characteristics.

The one variable problem may be stated as

$$\text{minimize } f(x) \qquad \text{for all real } x \tag{2.1}$$

The point x^* is a *weak* local minimum if there exists a $\delta > 0$, such that $f(x^*) \leq f(x)$ for all x such that $|x - x^*| < \delta$, that is, $f(x^*) \leq f(x)$ for all x in a δ-neighborhood of x^*. The point x^* is a *strong* (or *strict*) local minimum if $f(x^*) \leq f(x)$ is replaced by $f(x^*) < f(x)$ in the preceding statement. Further, x^* is a *global minimum* if $f(x^*) < f(x)$ for all x. These cases are illustrated in Fig. 2.1. If a minimum does not exist, the function is not bounded below.

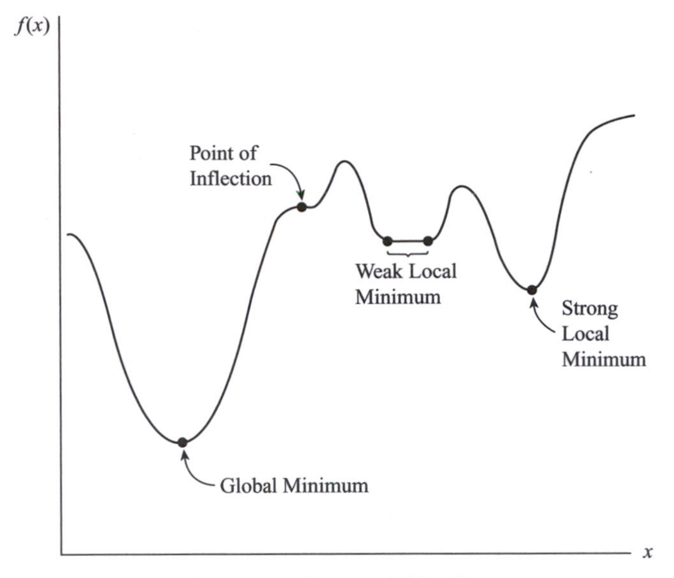

Figure 2.1. One-variable plot.

Optimality Conditions

If the function f in Eq. (2.1) has continuous second-order derivatives everywhere (i.e., C^2 continuous), the necessary conditions for x^* to be a local minimum are given by

$$f'(x^*) = 0 \tag{2.2}$$

$$f''(x^*) \geq 0 \tag{2.3}$$

where $f'(x^*)$ represents the derivative of f evaluated at $x = x^*$. In deriving these conditions, we make use of Taylor's expansion in the neighborhood of x^*. For a small real number $h > 0$, we use the first-order expansion at $x^* + h$ and $x^* - h$,

$$f(x^* + h) = f(x^*) + hf'(x^*) + O(h^2) \tag{2.4a}$$

$$f(x^* - h) = f(x^*) - hf'(x^*) + O(h^2) \tag{2.4b}$$

where the term $O(h^2)$ can be understood from the definition: $O(h^n)/h^r \to 0$ as $h \to 0$ for $0 \leq r < n$. For sufficiently small h, the remainder term can be dropped in comparison to the linear term. Thus, for $f(x^*)$ to be minimum, we need

$$f(x^* + h) - f(x^*) \approx \quad hf'(x^*) \geq 0$$

$$f(x^* - h) - f(x^*) \approx -hf'(x^*) \geq 0$$

Since h is not zero, these inequalities are satisfied when $f'(x^*) = 0$, which is Eq. (2.2). Condition given in Eq. (2.3) can be obtained using the second-order expansion

$$f(x^* + h) = f(x^*) + hf'(x^*) + \frac{h^2}{2} f''(x^*) + O(h^3)$$

At the minimum, $f'(x^*)$ is zero. Also the term $\frac{h^2}{2} f''(x^*)$dominates the remainder term $O(h^3)$. Thus,

$$f(x^* + h) - f(x^*) \approx \frac{h^2}{2} f''(x^*) \geq 0$$

Since h^2 is always positive, we need $f''(x^*) \geq 0$.

Points of inflection or flat regions shown in Fig. 2.1 satisfy the necessary conditions.

The *sufficient conditions* for x^* to be a strong local minimum are

$$f'(x^*) = 0 \tag{2.5}$$

$$f''(x^*) > 0 \tag{2.6}$$

These follow from the definition of a strong local minimum.

Example 2.2

Find the dimensions of the minimum cost refrigeration tank of Example 2.1 using optimality conditions.

Solution

We derived the objective function

$$\min f(x) = 45\pi x^2 + \frac{17200}{x}$$

$$f'(x) = 90\pi x - \frac{17200}{x^2} = 0$$

$$x^3 = \frac{17200}{90\pi} = 60.833$$

Diameter $\quad x = 3.93\,\text{m}$

Length $\quad L = \frac{200}{\pi x^2} = 4.12\,\text{m}$

Cost $\quad f = 45\pi x^2 + \frac{17200}{x} = \$6560$

Also $\quad f''(x) = 90\pi + \frac{(3)(17200)}{x^3}$

which is strictly positive at $x = 3.93$ m. Thus, the solution is a strict or strong minimum.

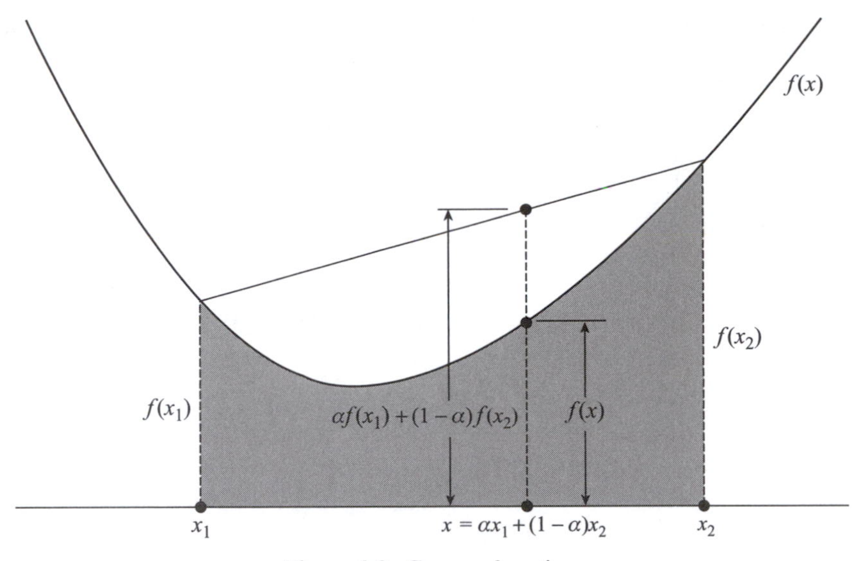

Figure 2.2. Convex function.

Convexity ideas can be used in defining optimality. A set S is called a *convex set* if for any two points in the set, every point on the line joining the two points is in the set. Alternatively, the set S is *convex* if for every pair of points x_1 and x_2 in S, and every α such that $0 \le \alpha \le 1$, the point $\alpha\, x_1 + (1 - \alpha)\, x_2$ is in S. As an example, the set of all real numbers R^1 is a convex set. Any closed interval of R^1 is also a convex set.

A function $f(x)$ is called a *convex function* defined on the convex set S if for every pair of points x_1 and x_2 in S, and $0 \le \alpha \le 1$, the following condition is satisfied

$$f(\alpha x_1 + (1-\alpha)x_2) \le \alpha f(x_1) + (1-\alpha) f(x_2) \tag{2.7}$$

Geometrically, this means that the graph of the function between the two points lies below the line segment joining the two points on the graph as shown in Fig. 2.2. We observe that the function defined in Example 2.1 is convex for $x > 0$. The function f is *strictly convex* if $f(\alpha x_1 + (1-\alpha)x_2) < \alpha f(x_1) + (1-\alpha) f(x_2)$ for $0 < \alpha < 1$. A function f is said to be *concave* if $-f$ is *convex*.

Example 2.3

Prove that $f = |x|,\ x \in R^1$, is a convex function.

Using the triangular inequality $|x + y| \le |x| + |y|$, we have, for any two real numbers x_1 and x_2 and $0 < \alpha < 1$,

$$\begin{aligned} f(\alpha x_1 + (1-\alpha)x_2) = \left|\alpha x_1 + (1-\alpha)\, x_2\right| &\le \alpha\, |x_1| + (1-\alpha)\, |x_2| \\ &= \alpha f(x_1) + (1-\alpha) f(x_2) \end{aligned}$$

which is the inequality defining a convex function in (2.7).

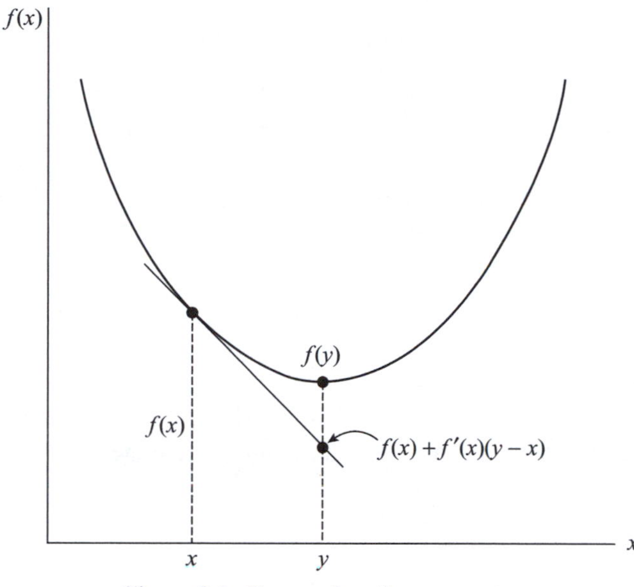

Figure 2.3. Convex function property.

Properties of Convex Functions

1. If f has continuous first derivatives then f is convex over a convex set S if and only if for every x and y in S,

$$f(y) \geq f(x) + f'(x)(y - x) \tag{2.8}$$

   This means that the graph of the function lies above the tangent line drawn at point as shown in Fig. 2.3.
2. If f has continuous second derivatives, then f is convex over a convex set S if and only if for every x in S,

$$f''(x) \geq 0 \tag{2.9}$$

3. If $f(x^*)$ is a local minimum for a convex function f on a convex set S, then it is also a global minimum.
4. If f has continuous first derivatives on a convex set S and for a point x^* in S, $f'(x^*)(y - x^*) \geq 0$ for every y in S, then x^* is a global minimum point of f over S. This follows from property (2.8) and the definition of global minimum.

Thus, we have used convexity ideas in characterizing the global minimum of a function. Further, the sufficiency condition (2.6) is equivalent to f being locally strictly convex.

Maximum of a Function

The problem of finding the maximum of a function ϕ is the same as determining the minimum of $f = -\phi$ as seen in Fig. 2.4. While the minimum location x^* is the same, the function value $\phi(x^*)$ is obtained as $-f(x^*)$. When f is positive, another transformation may be used to maximize f, namely, minimize $1/(1 + f)$. In this case, care must be taken as convexity aspects of the original function are entirely changed.

Example 2.4

Determine the dimensions of an open box of maximum volume that can be constructed from an A4 sheet (210 mm × 297 mm) by cutting four squares of side x from the corners and folding and gluing the edges as shown in Fig. E2.4.

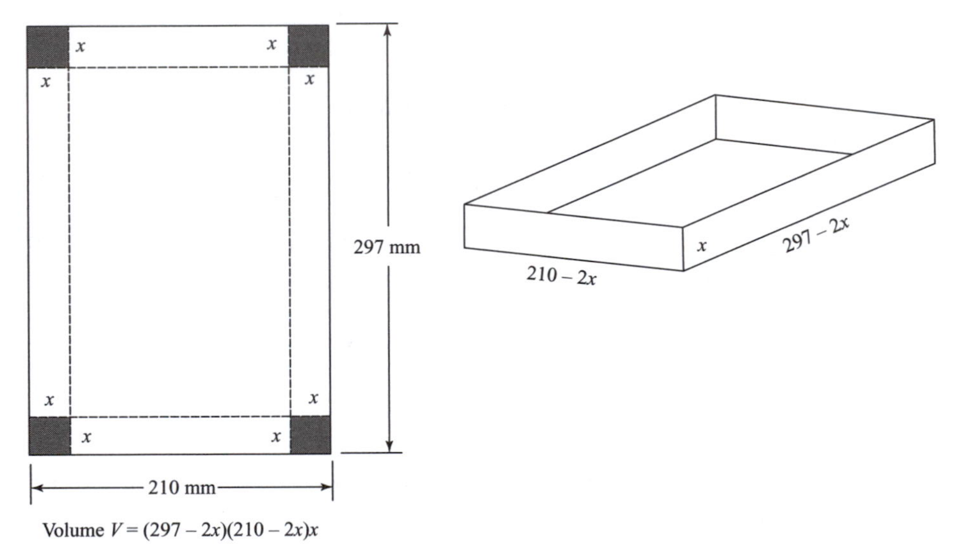

Figure E2.4. Open box problem.

Solution

The problem is to

$$\text{maximize} \quad V = (297 - 2x)(210 - 2x)\,x = 62370x - 1014x^2 + 4x^3$$

$$\text{We set} \quad f = -V = -62370x + 1014x^2 - 4x^3$$

Setting $f'(x) = 0$, we get

$$f'(x) = -62370 + 2028x^2 - 12x^2 = 0$$

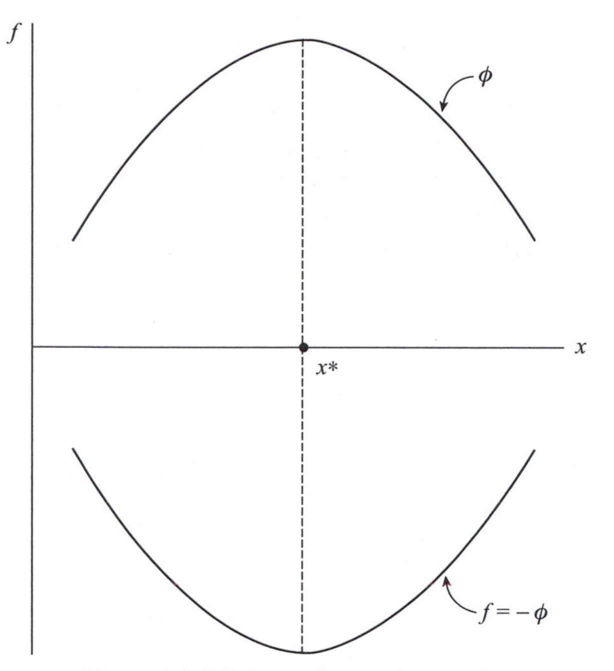

Figure 2.4. Minimum f at maximum of ϕ.

Using the solution for the quadratic equation,

$$x = \frac{-2028 \pm \sqrt{2028^2 - 4\,(12)\,(62370)}}{(2)\,(-12)}$$

The two roots are

$$x_1 = 40.423\,\text{mm} \qquad x_1 = 128.577\,\text{mm}$$

A physically reasonable cut is the first root $x^* = 40.423\,\text{mm}$. The second derivative $f''(x^*) = 2028 - 24x^* = 1057.848$. $f''(x^*) > 0$ implies that x^* is a strict minimum of f, or maximum of V.

The dimensions of the box are

Height $x^* = 40.423\,\text{mm}$
Length $= 297 - 2x^* = 216.154\,\text{mm}$
Width $= 210 - 2x^* = 129.154\,\text{mm}$

Maximum volume $= 40.423 \times 216.154 \times 129.154 = 1128495.1\ \text{mm}^3 = 1128.5\ \text{cm}^3$.

We now proceed to discuss various methods of finding the interval that brackets the minimum, called the *interval of uncertainty*, and the techniques of locating the minimum.

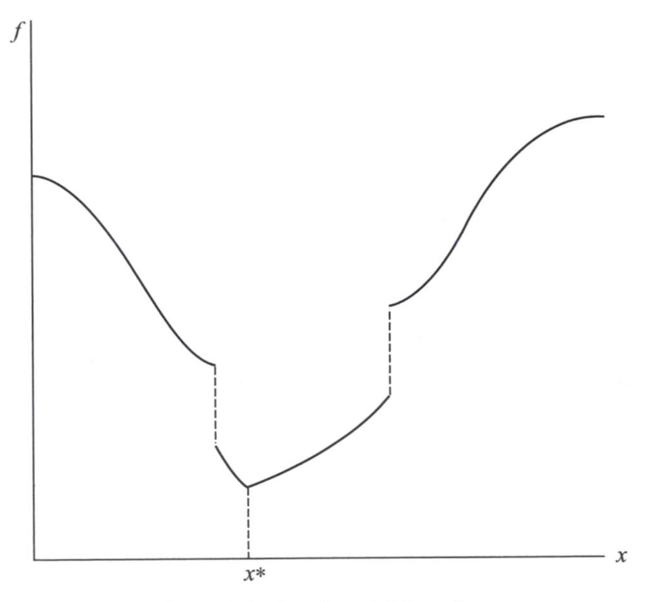

Figure 2.5. A unimodal function.

2.3 Unimodality and Bracketing the Minimum

Several of the methods discussed here require that the function f be unimodal, that is, it has a single local minimum. A function f is said to be *unimodal* in an interval S if there is a unique x^* in S such that for every pair of points x_1 and x_2 in S and $x_1 < x_2$: if $x_2 < x^*$, then $f(x^1) > f(x^2)$, or if $x_1 > x^*$ then $f(x^1) < f(x^2)$. In other words, the function monotonically increases on either side of the minimum point x^*. An example of a unimodal function is shown in Fig. 2.5.

Bracketing the Minimum

The first step in the process of the determination of the minimum is to bracket it in an interval. We choose a starting point 1 with coordinate x_1 and a step size Δ as shown in Fig. 2.6. We choose an expansion parameter $\gamma \geq 1$. $\gamma = 1$ corresponds to uniform spacing of points where functions are evaluated. A common practice is to choose $\gamma = 2$, i.e., doubling the step size at each successive iteration, or choose $\gamma = 1.618$, the Golden Section ratio. The bracketing algorithm is outlined in the following.

Bracketing Algorithm/Three-Point Pattern

1. Set $x_2 = x_1 + \Delta$
2. Evaluate f_1 and f_2

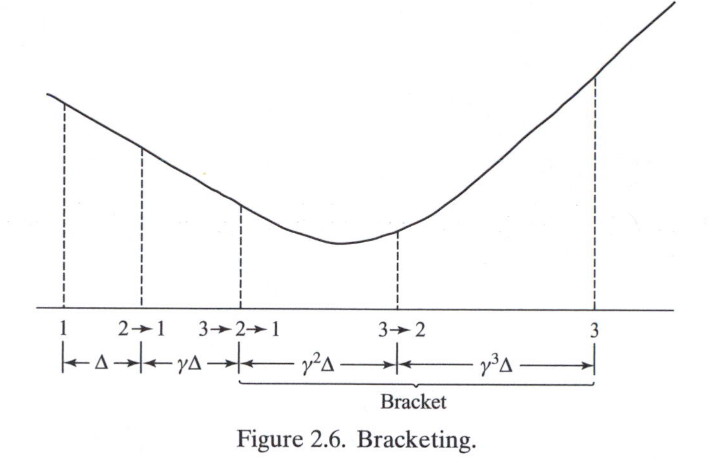

Figure 2.6. Bracketing.

3. If $f_2 \leq f_1$ go to step 5
4. Interchange f_1 and f_2, and x_1 and x_2, and set $\Delta = -\Delta$
5. Set $\Delta = \gamma\Delta$, $x_3 = x_2 + \Delta$, and evaluate f_3 at x_3
6. If $f_3 > f_2$, go to step 8
7. Rename f_2 as f_1, f_3 as f_2, x_2 as x_1, x_3 as x_2, go to step 5
8. Points 1, 2, and 3 satisfy $f_1 \geq f_2 < f_3$ (three-point pattern)

Note that if the function increases at the start, we interchange points 1 and 2 at step 4, and change the sign of Δ. We then keep marching in increasing increments till the function rises.

Interval 1–3 brackets the minimum. The three points 1, 2, and 3 with point 2 located between 1 and 3 and satisfying $f_1 \geq f_2 < f_3$ (or $f_1 > f_2 \leq f_3$) form the *three-point pattern*. Establishing the interval 1–3 is necessary for the Fibonacci method. The three-point pattern is needed for the Golden Section method, and for the version of the quadratic search presented here.

We emphasize here that each function evaluation has a certain time or cost associated with it. Take, for example, the setting of a valve in a process control operation. Each setting requires time and operator interaction. In a large stress or deformation analysis problem, each evaluation for a trial geometry may be a full-scale finite element analysis involving considerable time and cost. In some situations, the number of trials may be limited for various reasons. We present here minimum and zero finding techniques which are robust and efficient.

2.4 Fibonacci Method

Fibonacci method is the best when the interval containing the minimum or the interval of uncertainty is to be reduced to a given value in the least number of trials

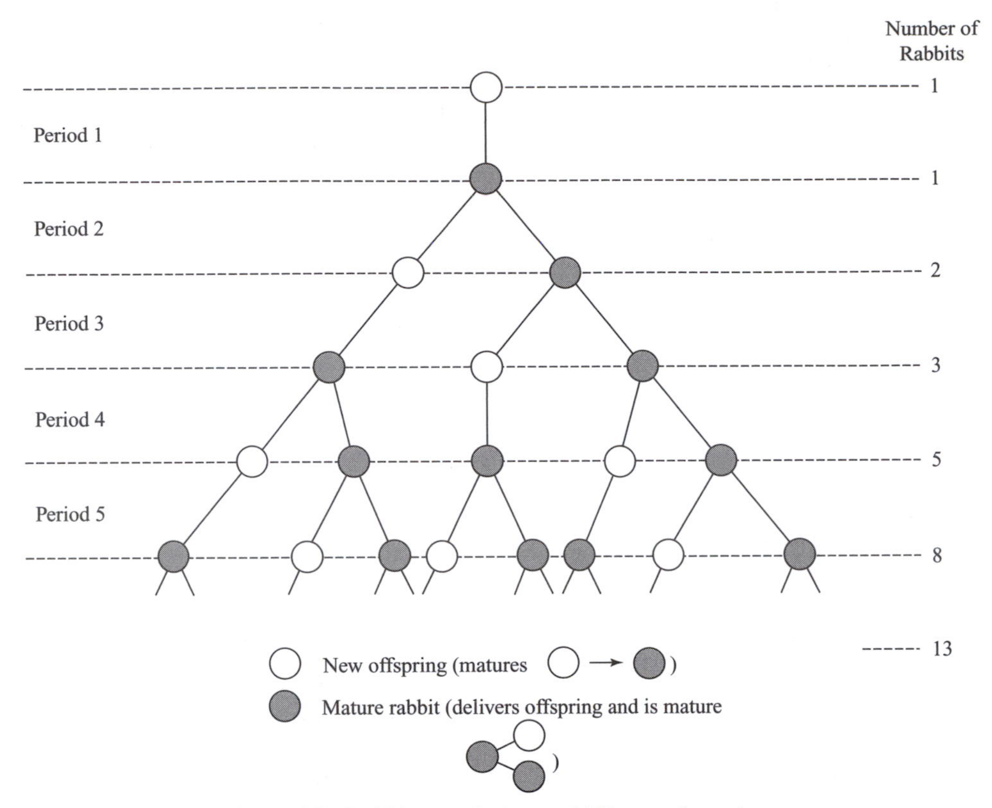

Figure 2.7. Rabbit population and Fibonacci numbers.

or function evaluations. This can also be interpreted as maximum reduction of the interval of uncertainty in the given number of trials. Use of Fibonnaci numbers for interval reduction is one of the earliest applications of genetic ideas in optimization. Leonardo of Pisa (nicknamed Fibonacci) developed the Fibonacci numbers in the study of reproduction of rabbits. His observations showed that a new rabbit offspring matures in a certain period and that each mature rabbit delivers an offspring in the same period and remains mature at the end of the period. This process is shown in Fig. 2.7. If the total population is counted at the end of each period, starting from the first rabbit, we get the sequence of numbers 1, 1, 2, 3, 5, 8, 13, 21, ... We observe that each number starting from the third number is the sum of the preceding two numbers.

If Fibonacci numbers are denoted $F_0, F_1, F_2, \ldots, F_n, \ldots$, the sequence can be generated using

$$F_0 = 1 \quad F_1 = 1$$

$$F_i = F_{i-1} + F_{i-2} \quad i = 2 \text{ to } n \tag{2.10}$$

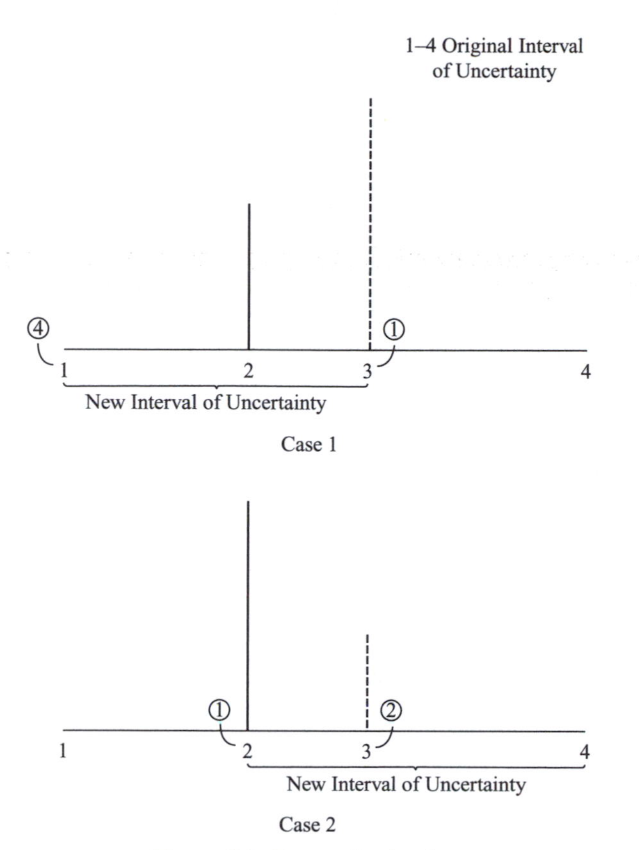

Figure 2.8. Interval reduction.

We turn our attention to the interval reduction strategy and the use of Fibonacci numbers. Consider the starting interval of uncertainty 1–4 in Fig. 2.8. We introduce point 2, and then introduce point 3. Two cases arise. If $f_2 < f_3$ (case 1) then 1–3 is the new interval of uncertainty and if $f_2 > f_3$ (case 2) then 2–4 is the new interval of uncertainty. $f_2 = f_3$ can be clubbed with either of the two cases. Since the two cases are equally likely to occur, it is desirable that the points 2 and 3 be placed symmetrically with respect to the center of the interval. If case 1 occurs, let us rename points 1 and 3 as 4 and 1 respectively, or if case 2 occurs, let us rename points 2 and 3 as 1 and 2 respectively. In case 2, we need to set $f_2 = f_3$. We now seek to introduce a new point 3 and be able to reduce the interval of certainty further. This requires that the new point 3 be located such that length 1–3 is the same as length 4–2. Thus, the new point 3 is related to the initial division. Without loss in generality, we assume that the interval of uncertainty keeps occurring to the left as illustrated in Fig. 2.9. We start with an interval I_1 and the final interval of uncertainty after $n - 1$ interval reductions is, say, I_n. At the final stage, the intervals may be overlapped by a small

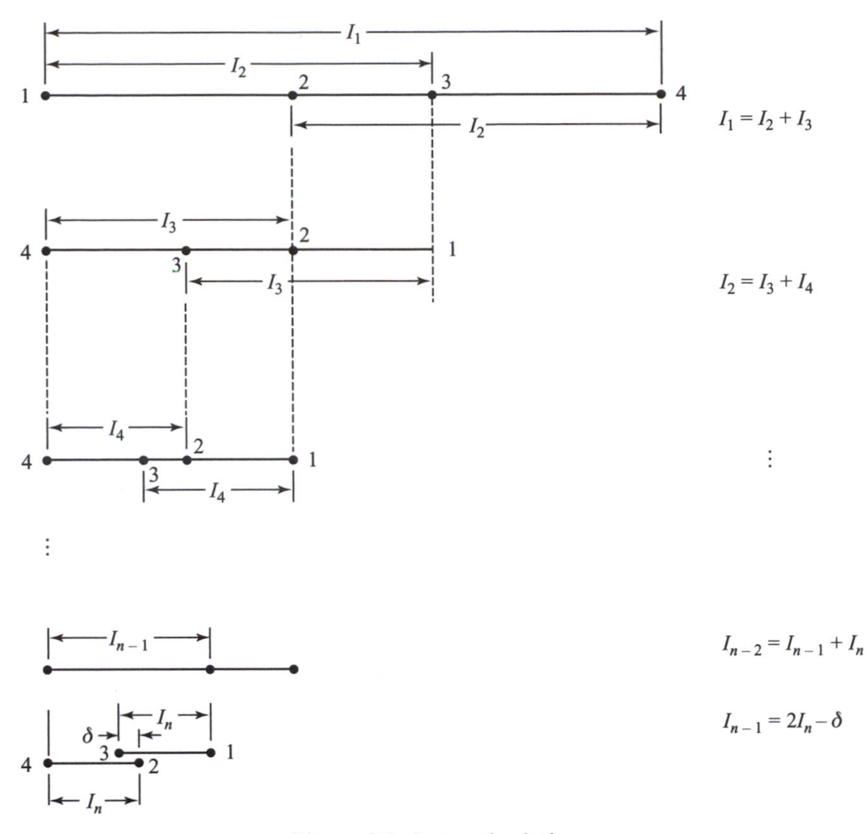

Figure 2.9. Interval relations.

quantity, say δ. δ may be chosen as the machine precision in a computer calculation, or as precision used in a measuring instrument. All interval relations are developed by setting $\delta = 0$. A small δ ($\delta \ll I_n$) can be introduced at the final stage to decide the interval of uncertainty. The interval relations follow from Fig. 2.9.

$$
\begin{aligned}
I_1 &= I_2 + I_3 \\
I_2 &= I_3 + I_4 \\
&\vdots \\
I_j &= I_{j+1} + I_{j+2} \\
&\vdots \\
I_{n-2} &= I_{n-1} + I_n \\
I_{n-1} &= 2I_n
\end{aligned}
$$

By proceeding in the reverse order and expressing each interval in terms of I_n, we get

$$I_{n-1} = 2I_n$$
$$I_{n-2} = I_{n-1} + I_n = 3I_n$$
$$I_{n-3} = I_{n-2} + I_{n-1} = 5I_n$$
$$I_{n-3} = I_{n-3} + I_{n-2} = 8I_n$$
$$\vdots$$

The coefficients 2, 3, 5, 8, ... appearing above are Fibonnaci numbers. Making use of the Fibonacci number definition of Eq. (2.10), we have

$$I_{n-j} = F_{j+1}\, I_n \quad j = 1, 2, \ldots, n-1$$

For $j = n - 1$ and $n - 2$,

$$I_1 = F_n I_n \tag{2.11}$$

$$I_2 = F_{n-1} I_n \tag{2.12}$$

From Eq. (2.11), we get the size of the final interval

$$I_n = \frac{I_1}{F_n}$$

Furthermore, it may be observed that exactly n trials or function calls are involved in reducing I_1 to I_n, starting from two trials at the first iteration and adding one trial at each successive iteration. Thus, the above equation states that the ratio of final to initial interval lengths after n trials is

$$\frac{I_n}{I_1} = \frac{1}{F_n} \tag{2.13}$$

Eqs. (2.11) and (2.12) yield

$$I_2 = \frac{F_{n-1}}{F_n} I_1 \tag{2.14}$$

For a given interval I_1, I_2 is calculated from above. Then the relations $I_3 = I_1 - I_2$, $I_4 = I_2 - I_3$, etc. follow.

Example 2.5

Consider the interval [0, 1], and number of trials $n = 5$. Eq. (2.14) yields $I_2 = {}^5/_8$; thus, the set of four points defining the new intervals is given by [0, $^3/_8$, $^5/_8$, 1]. Upon making two function evaluations at $^3/_8$ and $^5/_8$, the new interval will be either [0, $^5/_8$] or [$^3/_8$, 1]. Without loss in generality, assume that the interval of uncertainty is the left interval, namely, [0, $^5/_8$]. Proceeding in this manner, the successive intervals defined by sets of four points are $[0, \mathbf{2/8}, 3/8, 5/8]$, $[0, \mathbf{1/8}, 2/8, 3/8]$, $[0, 1/8, 1/8, 2/8]$. At this stage, the two central points coincide; thus, we will not be able to make a decision in the selection. The strategy is to choose a parameter δ based on the computational precision and to make the final four points as $[0, 1/8, \mathbf{1/8} + \delta, 2/8]$. The final interval will be $[0, 1/8 + \delta]$ or $[1/8, 2/8]$. The boldfaces above refer to new trials – we count a total of $n = 5$, and, as given in Eq. (2.13), $I_5/I_1 = 1/8 = 1/F_5$.

We note here that Fibonacci numbers grow rapidly. It is advantageous to use Binet's formula for Fibonacci numbers, especially when the ratio in Eq. (2.14) is to be evaluated. Binet's formula for the ith Fibonacci number is as follows:

$$F_i = \frac{1}{\sqrt{5}}\left[\left(\frac{1+\sqrt{5}}{2}\right)^{i+1} - \left(\frac{1-\sqrt{5}}{2}\right)^{i+1}\right] \tag{2.15}$$

The reader is encouraged to calculate the Fibonacci numbers using the above relation and check with the earlier series. The ratio of two consecutive Fibonacci numbers can be written as as follows:

$$\frac{F_{i-1}}{F_i} = \left(\frac{\sqrt{5}-1}{2}\right)\left(\frac{1-s^i}{1-s^{i+1}}\right) \quad i = 2, 3, \ldots, n \tag{2.16}$$

$$s = \frac{1-\sqrt{5}}{1+\sqrt{5}}$$

Calculation of this ratio is stable even for large values of i since $|s| = 0.38196$.

We now present the minimization algorithm based on Fibonacci numbers.

Fibonacci Algorithm for Function Minimization

1. Specify the interval x_1, x_4 $(I_1 = |x_4 - x_1|)$
2. Specify the number of interval reductions n_i; $n = n_i + 1$ or desired accuracy ε.
3. If ε is given, find the smallest n such that $\frac{1}{F_n} < \varepsilon$
4. Evaluate $c = \frac{\sqrt{5}-1}{2}$, $s = \frac{1-\sqrt{5}}{1+\sqrt{5}}$, $\alpha = \frac{c(1-s^n)}{(1-s^{n+1})}$
5. Introduce point x_3, $x_3 = \alpha\, x_4 + (1-\alpha)\, x_1$, and evaluate f_3
6. **DO** $i = 1, n-1$

Introduce point x_2 as

if $i = n - 1$ then

$$x_2 = 0.01\, x_1 + 0.99\, x_3$$

else

$$x_2 = \alpha\, x_1 + (1-\alpha)\, x_4$$

endif

Evaluate f_2

if $f_2 < f_3$ then

$$x_4 = x_3,\ x_3 = x_2,\ f_3 = f_2$$

else

$$x_1 = x_4,\ x_4 = x_2$$

endif

$$\alpha = \frac{c\left(1 - s^{n-i}\right)}{\left(1 - s^{n-i+1}\right)}$$

ENDDO

The algorithm has been implemented in the program FIBONACI.

Example 2.6

In an interval reduction problem, the initial interval is given to be 4.68 units. The final interval desired is 0.01 units. Find the number of interval reductions using Fibonacci method.

Solution

Making use of Eq. (2.13), we need to choose the smallest n such that

$$\frac{1}{F_n} < \frac{0.01}{4.68}$$

$$\text{or } F_n > 468$$

From the Fibonacci sequence $F_0 = 1, F_1 = 1, F_2 = 2, \ldots, F_{13} = 377, F_{14} = 610$, we get $n = 14$. The number of interval reductions is $n - 1 = 13$.

Example 2.7

A projectile released from a height h at an angle θ with respect to the horizontal in a gravitational field g, shown in Fig. E2.7, travels a distance D when it hits the ground. D is given by

$$D = \left(\frac{V \sin\theta}{g} + \sqrt{\frac{2h}{g} + \left(\frac{V \sin\theta}{g}\right)^2} \right) V \cos\theta$$

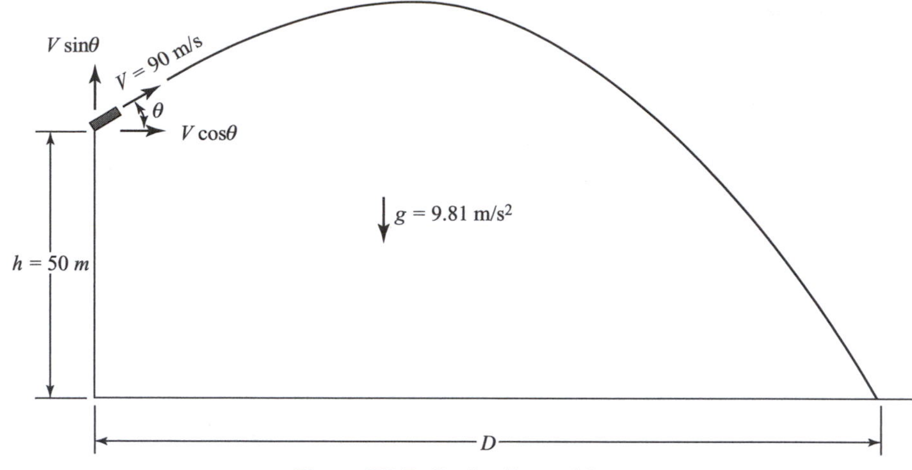

Figure E2.7. Projectile problem.

If $h = 50\,\text{m}$, $V = 90\,\text{m/s}$, $g = 9.81\,\text{m/s}^2$, determine the angle θ, in degrees, for which the distance D is a maximum. Also, calculate the maximum distance D in meters. Use the range for θ of 0° to 80°, and compare your results for 7 and 19 Fibonacci interval reductions.

Solution

We set up the problem as minimizing F = −*D*.

In the subroutine defining the function, we define

```
PI = 3.14159      G = 9.81      V = 90      H = 50
Y = PI * X / 180          (Degrees to radians)
F = −(V*SIN(Y)/G+SQR(2*H/G + (V*SIN(Y)/G)^2))*V*COS(Y))
```

Range 0, 80

7 interval reductions $\rightarrow$ $n = 8$

Distance D (−F) = 873.80 m

X = 42.376°

Final interval 42.353° – 44.706°

19 interval reductions $\rightarrow$ $n = 20$

Distance D (−F) = 874.26 m

X = 43.362°

Final interval 43.362° – 43.369°

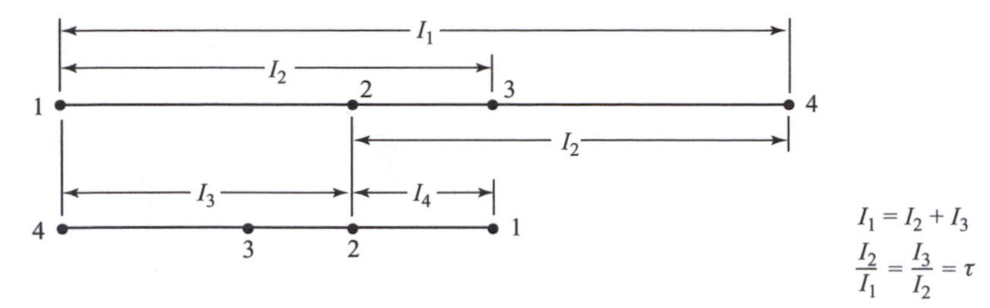

Figure 2.10. Golden section.

2.5 Golden Section Method

While the Fibonacci sectioning technique is good for a given number of evaluations, as the number of iterations is increased, the ratio $\frac{F_{n-1}}{F_n}$ reaches the limit $(\frac{\sqrt{5}-1}{2}) = 0.61803$. For number of iterations greater than 12, the ratio is within $\pm 0.0014\%$ of the limit $(\frac{\sqrt{5}-1}{2})$, which is called the golden ratio. The golden ratio forms the central idea in the golden section search, where the interval reduction strategy is independent of the number of trials. A uniform reduction strategy maintaining the relations given in Fig. 2.9 (excluding the last interval relation) is used. Referring to Fig. 2.10, we wish to maintain

$$
\begin{aligned}
&I_1 = I_2 + I_3 \\
&I_2 = I_3 + I_4 \\
&\vdots \\
&\text{and } \frac{I_2}{I_1} = \frac{I_3}{I_2} = \frac{I_4}{I_3} = \cdots = \tau
\end{aligned}
$$

Substituting $I_2 = \tau I_1$, and $I_3 = \tau I_2 = \tau^2 I_1$ into the first relation, we get

$$\tau^2 + \tau - 1 = 0 \tag{2.17}$$

The positive solution of this equation is given by

$$\tau = \left(\frac{\sqrt{5}-1}{2}\right) = 0.61803 \tag{2.18}$$

This value is referred to as the golden ratio. We also observe that the interval ratio in Fibonacci search as discussed in the previous section also reaches this value in the limit. The golden ratio was used extensively in Greek architecture. If a square formed by the smaller side is cut from a golden rectangle shown in Fig. 2.11, the remaining rectangle is a golden rectangle with sides of golden proportion. The aesthetic appearance of this rectangle was exploited by the medieval painters. Leonardo da Vinci used the golden ratio in defining the geometric forms

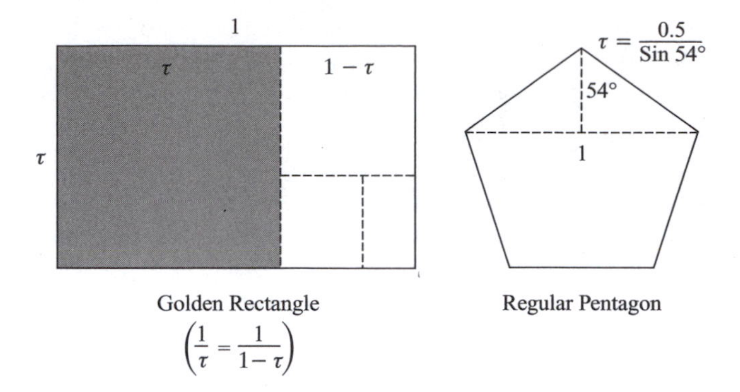

Figure 2.11. Golden section ratio.

and proportions of the human body. In a regular pentagon, shown in Fig. 2.11, the ratio of the side to the distance between two alternate corners is the golden ratio τ, which can be expressed as $0.5/\sin 54^\circ$. We also note that $\tau^2 = 1 - \tau, 1 + \tau = 1/\tau$.

The final interval I_n can be expressed in terms of I_1 as

$$I_n = \tau^{n-1} I_1$$

Exactly n trials or function calls are involved in reducing I_1 to I_n, starting from two trials at the first iteration and adding one trial at each successive iteration. Thus, the above equation states that the ratio of final to initial interval lengths after n trials is

$$\frac{I_n}{I_1} = \tau^{n-1} \tag{2.19}$$

For a given initial interval, if the final interval desired i.e. I_n is known, the number of intervals n can be expressed as

$$n = \text{int}\left(1.5 + \frac{\ln I_n - \ln I_1}{\ln \tau}\right)$$

where int (.) defines the integer part.

The interval reduction strategy in golden section algorithm follows the steps used in Fibonacci algorithm. The point introduction is similar to the strategy given in Fig. 2.8. The basic golden section algorithm is similar to the Fibonacci algorithm except that α at step 4 is replaced by the golden ratio τ $(= \frac{\sqrt{5}-1}{2})$. This algorithm starting from an initial interval of uncertainty is in the program GOLDINTV.

The golden section method is a robust method that can be efficiently integrated with the three-point bracketing algorithm presented earlier if the expansion parameter is chosen as $\frac{1}{\tau}$ $(= 1 + \tau)$ or 1.618034. This integration of the bracketing with the golden section enables one to start at some starting point with a given initial step size to march forward. Point 3 in the final three points of Fig. 2.6 can be called 4 and then we follow the strategy of introducing a new point as presented in the Fibonacci algorithm. The integrated bracketing and golden section algorithm is given below.

Integrated Bracketing and Golden Section Algorithm

1. Give start point x_1, step size s
2. $x_2 = x_1 + s$, Evaluate f_2
3. If $f_2 > f_1$ then interchange 1 and 2; set $s = -s$
4. $s = \frac{s}{\tau}$; $x_4 = x_2 + s$; Evaluate f_4
5. If $f_4 > f_2$ go to 7
6. Rename 2, 4 as 1, 2, respectively: go to 4
7. $x_3 = \tau\, x_4 + (1 - \tau)\, x_1$; Evaluate f_3
8. if $f_2 < f_3$ then

   $$x_4 = x_1, \quad x_1 = x_3$$

   else

   $$x_1 = x_2, \quad x_2 = x_3, \quad f_2 = f_3$$

   endif

9. < *check stopping criterion* > ok then go to 10; not ok go to 7
10. End: print optimum x_2, f_2

This algorithm has been implemented in the program GOLDLINE. We now discuss some details of the stopping criterion needed at step 9.

Stopping Criteria

In the process of reducing the interval of uncertainty, we keep updating the interval forming the three-point pattern. The basic criteria involve checking the interval size and the function reduction. We first specify the tolerance values ε_x for the interval and ε_F for the function. We may choose ε_x = 1e−4 and ε_F = 1e−6. Other values may be tried depending on the precision and accuracy needed.

The tolerance ε_x is an absolute limit. We also need a relative part, which is based on the machine precision, and the x value of the current best point, which is the middle point in the three-point pattern. The stopping criterion based on the interval size may be stated as

$$|x_1 - x_3| \leq \varepsilon_R\,|x_2| + \varepsilon_{\text{abs}} \tag{2.20}$$

where | . | is the absolute value and ε_R can be taken equal to $\sqrt{\varepsilon_m}$, where ε_m is the machine precision defined as the smallest number for which the computer recognizes $1 + \varepsilon_m > 1$. For the function check, we use the descent function $\bar{f}$ defined below, which is meaningful for a three-point pattern.

$$\bar{f} = \frac{f_1 + f_2 + f_3}{3}$$

We keep the value of $\bar{f}$ from the previous step and call it $\bar{f}_{\text{old}}$. The current value $\bar{f}$ is then compared with $\bar{f}_{\text{old}}$. We use the current best value f_2 for determining the relative part, of the tolerance. Thus, the criterion based on the function value is

$$\left|\bar{f} - \bar{f}_{\text{old}}\right| \leq \varepsilon_R \,|f_2| + \varepsilon_{\text{abs}} \tag{2.21}$$

It is a good idea to check this over two successive iterations. We stop when either of the two criteria Eq. (2.20) or Eq. (2.21) is met. These criteria are used in GOLDLINE.

Example 2.8
Compare the ratio of final to initial interval lengths of Fibonacci and Golden Section search for $n = 5$ and $n = 10$ function evaluations, respectively.

From Eqs. (2.13) and (2.19), we have

	$I_{n=5}/I_{\text{initial}}$	$I_{n=10}/I_{\text{initial}}$
Fibonacci search:		
$I_n/I_{\text{initial}} = 1/F_n$	0.125	0.011236
Golden Section search:		
$I_n/I_{\text{initial}} = \tau^{n-1}$	0.1459	0.013156

As per theory, Fibonacci search gives the smallest interval of uncertainty for fixed n, among sectioning methods.

Example 2.9
A company plans to borrow x thousand dollars to set up a new plant on equal yearly installments over the next $n = 8$ years. The interest charged is $r_c = c_1 + c_2 x$ per year, where $c_1 = 0.05$ and $c_2 = 0.0003$ per thousand dollars. The money earned can be reinvested at a rate of $r_e = 0.06$ per year. The expected future value of the return is

$$f_r = c_3\left(1 - e^{-c_4 x}\right)$$
$$c_3 = \$300000, \quad c_4 = 0.01 \text{ per } \$1000$$

The future value of the payment is

$$f_p = \left[\frac{(1+r_e)^n - 1}{r_e}\right]\left[\frac{r_c(1+r_c)^n}{(1+r_c)^n - 1}\right]x$$

Neglecting the taxes and other charges, determine the amount of money x thousands of dollars to be borrowed for maximum future value of profit

$$f = f_r - f_p$$

Solution

The nature of the problem is clear. When $x = 0$, there is no return or payment. The given data and the function are input into the program in the GETFUN subroutine as

```
C1 = 0.05 C2 = 0.0003
 N = 8 RE = 0.06 C3 = 300 C4 = 0.01
RC = C1 + C2 * X
FR = C3* (1 - EXP( - C4*X)
FP = (((1 + RE)^N - 1)* (RC* (1 + RC)^N/((1 + RC)^N - 1))*X
 F = -(FR - FP)
```

Note that since profit is to be maximized, we put F = −(FR − FP). The sign of F from the result must be changed to get the profit. The starting point is $x_1 = 0$ thousand dollars. Initial step size STP = 0.2 thousand dollars.

The solution is

Number of function evaluations = 27
Amount borrowed in thousands of dollars = 54.556
Profit (future value in thousands of dollars) = 36.987

(check the nature of the profit function by plotting f versus x).

Sectioning methods are robust and reliable. In applications where the number of function evaluations is critical, we look for other methods. Polynomial fit methods fall in this category.

2.6 Polynomial-Based Methods

As the interval becomes small, a smooth function can be approximated by a polynomial. The minimum point of the polynomial fit, which falls in the interval of uncertainty, is a good candidate for the minimum value of the function. When only function values at discrete points are known, the popular method is using three points and fitting by a quadratic polynomial. When both function value and its

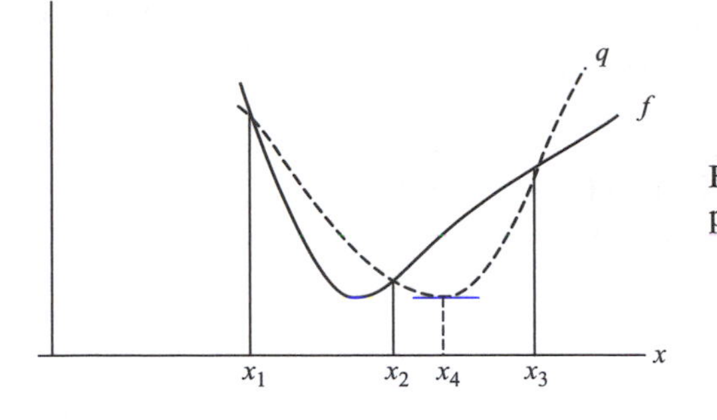

Figure 2.12. Quadratic fit with three-point pattern.

derivative are known at each point, the complete information of two points can be used in fitting a cubic order polynomial. The basic idea is to start from an initial interval bracketing the minimum and reducing the interval of uncertainty using polynomial information. Robust algorithms are obtained when conditioning and sectioning ideas are integrated.

Quadratic Fit Algorithm

Assume that the function is unimodal, and start with an initial three-point pattern, obtainable using the algorithm given in Section 2.3. Let the points be numbered as $[x_1, x_2, x_3]$ with $[f_1, f_2, f_3]$ being the corresponding function values, where $x_1 < x_2 < x_3$ and $f_2 \leq \min(f_1, f_3)$. Figure 2.12 depicts the situation, where we see that the minimum can be either in $[x_1, x_2]$ or in $[x_2, x_3]$. Thus, $[x_1, x_3]$ is the interval of uncertainty. A quadratic can be fitted through these points as shown in Fig. 2.12. An expression for the quadratic function can be obtained by letting $q(x) = a + bx + cx^2$ and determining coefficients a, b and c using $[x_1, f_1]$, $[x_2, f_2]$, $[x_3, f_3]$. Alternatively, the Lagrange polynomial may be fitted as

$$q(x) = f_1 \frac{(x - x_2)(x - x_3)}{(x_1 - x_2)(x_1 - x_3)} + f_2 \frac{(x - x_1)(x - x_3)}{(x_2 - x_1)(x_2 - x_3)} + f_3 \frac{(x - x_1)(x - x_2)}{(x_3 - x_1)(x_3 - x_2)} \tag{2.22}$$

The minimum point, call it x_4, can be obtained by setting $\mathrm{d}q/\mathrm{d}x = 0$, which yields

$$x_4 = \frac{\frac{f_1}{A}(x_2 + x_3) + \frac{f_2}{B}(x_1 + x_3) + \frac{f_3}{C}(x_1 + x_2)}{2\left(\frac{f_1}{A} + \frac{f_2}{B} + \frac{f_3}{C}\right)},$$

$$\text{where } A = (x_1 - x_2)(x_1 - x_3),\ B = (x_2 - x_1)(x_2 - x_3),$$
$$C = (x_3 - x_1)(x_3 - x_2) \tag{2.23}$$

The value $f_4 = f(x_4)$ is evaluated and, as per Fig. 2.8, a new interval of uncertainty is identified after comparing f_2 with f_4, as

$$[x_1, x_2, x_3]_{new} = \begin{cases} [x_1, x_2, x_4] & if \quad x_4 > x_2 \quad and \quad f(x_4) \geq f_2 \\ [x_2, x_4, x_3] & if \quad x_4 > x_2 \quad and \quad f(x_4) \geq f_2 \\ [x_4, x_2, x_3] & if \quad x_4 < x_2 \quad and \quad f(x_4) \geq f_2 \\ [x_1, x_4, x_2] & if \quad x_4 < x_2 \quad and \quad f(x_4) < f_2 \end{cases}$$

The process above is repeated until convergence. See stopping criteria in Eqs. (2.20) and (2.21). However, pure quadratic fit procedure just described is prone to fail often [Robinson 1979; Gill et al. 1981]. This is because *safeguards* are necessary, as detailed in the following.

Safeguards in Quadratic (Polynomial) Fit

Three main safeguards are necessary to ensure robustness of polynomial fit algorithms:

(1) The point x_4 must lie within the interval $[x_1, x_3]$ – this will be the case for the quadratic fit above with the stated assumptions, but needs to be monitored in general.
(2) The point x_4 must not be too close to any of the three existing points else the subsequent fit will be ill-conditioned. This is especially needed when the polynomial fit algorithm is embedded in a n-variable routine (Chapter 3). This is taken care of by defining a measure δ and moving or bumping the point away from the existing point by this amount. The following scheme is implemented in Program Quadfit.

$$\text{if} \begin{cases} |x_4 - x_1| < \delta, & then\ set \quad x_4 = x_1 + \delta \\ |x_4 - x_3| < \delta, & then\ set \quad x_4 = x_3 - \delta \\ |x_4 - x_2| < \delta, & and \quad x_2 > .5^*(x_1 + x_3) \quad then\ set \quad x_4 = x_2 - \delta \\ |x_4 - x_2| < \delta, & and \quad x_2 \leq .5^*(x_1 + x_3) \quad then\ set \quad x_4 = x_2 + \delta \end{cases}$$

where $\delta > 0$ is small enough. For instance, $\delta = s \min\{|x_3 - x_2|, |x_2 - x_1|\}$ where $0 < s < ½$ (see *sfrac* parameter in program, default = 1/8).
(3) The safeguard has to do with very slow convergence as measured by reductions in the length of the intervals. Figure 2.13 illustrates the situation. The old three-point pattern is $[x_1, x_2, x_3]$. Upon fitting a quadratic and minimizing, a new point x_4 is obtained. If $f(x_4) < f(x_2)$, then the new three-point pattern is $[x_2, x_4, x_3]$. However, since x_4 is close to x_2, $L_{\text{new}}/L_{\text{old}}$ is close to unity. For instance, if this ratio equals 0.95, then only a 5% reduction in interval length has been achieved.

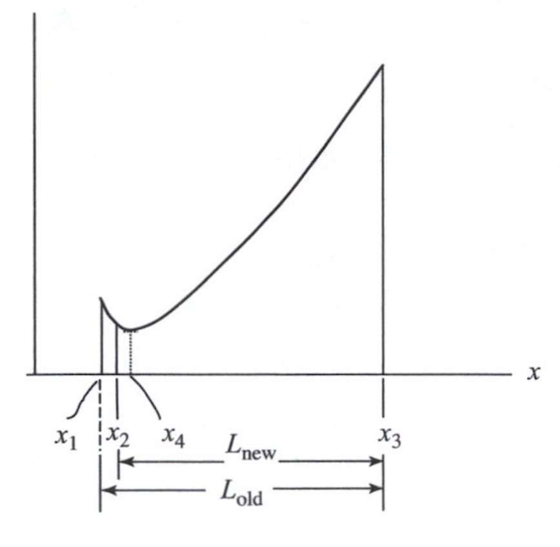

Figure 2.13. Slow convergence in a polynomial fit algorithm.

In fact, a golden section step would have reduced the interval by $\tau = 0.618$. This motivates the logic that if $L_{new}/L_{old} > \tau$ at any step, or for two consecutive steps, then a golden section step is implemented as

if $x_2 \le (x_1 + x_3)/2$ then
$x_4 = x_2 + (1 - \tau)(x_3 - x_2);$
else
$x_4 = x_3 - (1 - \tau)(x_2 - x_1);$
endif

The reader may verify that turning off this safeguard in program quadfit leads to failure in minimizing $f = \exp(x) - 5\,x$, starting from an initial interval [−10, 0, 10].

Stopping Criteria

(i) Based on total number of function evaluations
(ii) Based on x-value: stop if the interval length is sufficiently small as

$$xtol = \sqrt{\varepsilon_m}\,(1. + \text{abs}(x_2))$$

stop if $\text{abs}(x_3 - x_1) \le 2.0^*\, xtol$, for two consecutive iterations
(iii) Based on f-value: stop if the change in average function value within the interval is sufficiently small as

$$ftol = \sqrt{\varepsilon_m}\,(1. + abs(f_2))$$

stop if $\text{abs}(\bar{f}_{new} - \bar{f}_{old}) \le ftol$, for two consecutive iterations

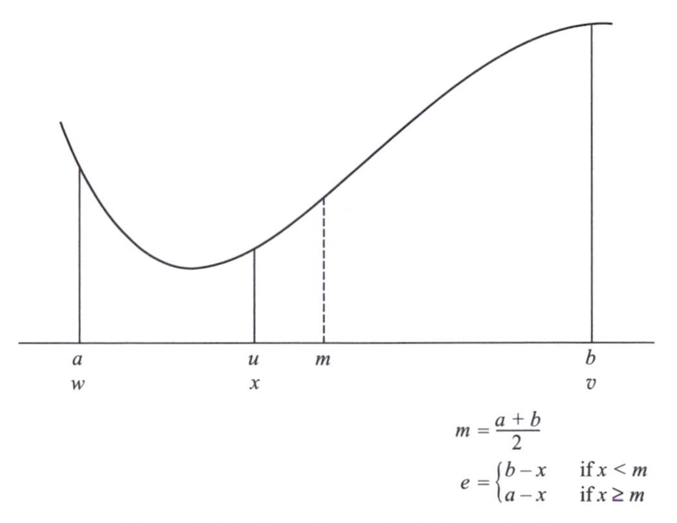

Figure 2.14. Brent's quadratic fit method.

Brent's Quadratic Fit – Sectioning Algorithm

The basic idea of the algorithm is to fit a quadratic polynomial when applicable and to accept the quadratic minimum if certain criteria are met. Golden sectioning is carried out otherwise. Brent's method starts with bracketing of an interval that has the minimum [Brent 1973]. At any stage, five points a, b, x, v, w are considered. These points may not all be distinct. Points a and b bracket the minimum. x is the point with the least function value. w is the point with the second least function value. v is the previous value of w, and u is the point where the function has been most recently evaluated. See Fig. 2.14. Quadratic fit is tried for x, v, and w whenever they are distinct. The quadratic minimum point q is at

$$q = x - 0.5\frac{(x-w)^2\left[f(x) - f(v)\right] - (x-v)^2\left[f(x) - f(w)\right]}{(x-w)\left[f(x) - f(v)\right] - (x-v)\left[f(x) - f(w)\right]} \tag{2.24}$$

The quadratic fit is accepted when the minimum point q is likely to fall inside the interval and is well conditioned otherwise the new point is introduced using golden sectioning. The new point introduced is called u. From the old set a, b, x, v, w and the new point u introduced, the new set of a, b, x, v, w are established and the algorithm proceeds.

Brent's Algorithm for Minimum

1. Start with three-point pattern a, b, and x with a, b forming the interval and the least value of function at x.
2. w and v are initialized at x.

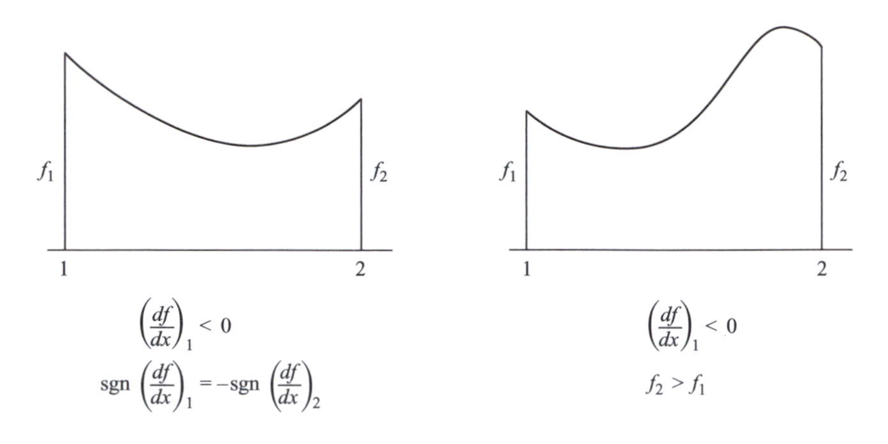

Figure 2.15. Two-Point pattern.

3. If the points x, w, and v are all distinct then goto 5.
4. Calculate u using golden sectioning of the larger of the two intervals x–a or x–b. Go to 7.
5. Try quadratic fit for x, w, and v. If the quadratic minimum is likely to fall inside a–b, then determine the minimum point u.
6. If the point u is close to a, b, or x, then adjust u into the larger of x–a or x–b such that u is away from x by a minimum distance *tol* chosen based on machine tolerance.
7. Evaluate the function value at u.
8. From among a, b, x, w, v, and u, determine the new a, b, x, w, v.
9. If the larger of the intervals x–a or x–b is smaller than 2**tol* then convergence has been achieved then *exit* else go to 3.

Brent's original program uses convergence based on the interval size only. In the program BRENTGLD included here, a criterion based on the function values is also introduced.

Other variations on when to switch from a quadratic fit to a sectioning step have been published [Chandrupatla 1988].

Cubic Polynomial Fit for Finding the Minimum

If the derivative information is available, cubic polynomial fit can be used. The first step is to establish a two point pattern shown in Fig. 2.15, such that

$$\left(\frac{df}{dx}\right)_1 < 0 \quad \text{and} \quad \text{sgn}\left(\frac{df}{dx}\right)_1 = -\text{sgn}\left(\frac{df}{dx}\right)_2$$

$$\text{or} \tag{2.25}$$

$$\left(\frac{df}{dx}\right)_1 < 0 \quad \text{and} \quad f_2 > f_1$$

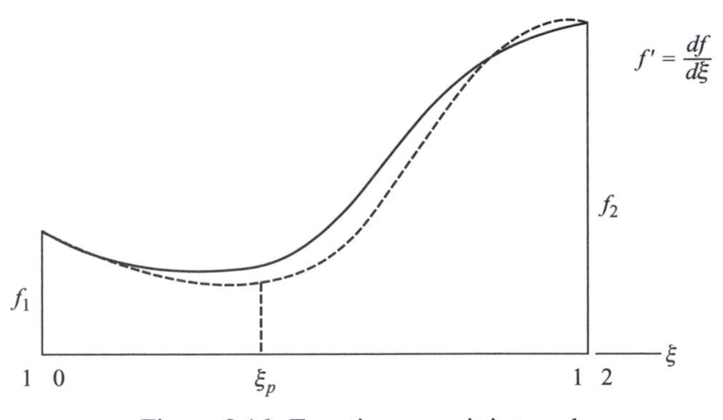

Figure 2.16. Function on unit interval.

We make use of the unit interval shown in Fig. 2.16, where the following transformation is used

$$\xi = \frac{x - x_1}{x_2 - x_1}$$

Denoting $f' = \frac{df}{d\xi}$, we have

$$f' = \frac{df}{d\xi} = (x_2 - x_1)\frac{df}{dx}$$

The cubic polynomial fit is then represented by

$$f = a\xi^3 + b\xi^2 + f_1'\xi + f_1$$

where

$$a = f_2' + f_1' - 2(f_2 - f_1)$$
$$b = 3(f_2 - f_1) - f_2' - 2f_1'$$

By equating $\frac{df}{d\xi} = 0$, and checking the condition $\frac{d^2 f}{d\xi^2} > 0$ for the minimum, we obtain the location of the minimum at

$$\xi_p = \frac{-b + \sqrt{b^2 - 3af_1'}}{3a} \tag{2.26}$$

Above expression is used when $b < 0$. If $b > 0$, the alternate expression for accurate calculation is $\xi_p = f_1'/(-b - \sqrt{b^2 - 3af_1'})$. The new point is introduced at $x_p = x_1 + (x_2 - x_1)\,\xi_p$, by making sure that it is at a minimum distance away from the end points of the interval. The new point is named as 1 if $\mathrm{sgn}(f') = \mathrm{sgn}(f_1')$ and named as 2 otherwise. Convergence is established by checking the interval $|x_2 - x_1|$, or two consecutive values of $0.5(f_1 + f_2)$.

Example 2.10

Obtain the solution for the following problems using the computer programs Quadfit_S and Cubic2P:

$$f_1 = x + \frac{1}{x}$$

$$f_2 = e^x - 5x$$

$$f_3 = x^5 - 5x^3 - 20x + 5$$

$$f_4 = 8x^3 - 2x^2 - 7x + 3$$

$$f_5 = 5 + (x - 2)^6$$

$$f_6 = 100(1 - x^3)^2 + (1 - x^2) + 2(1 - x)^2$$

$$f_7 = e^x - 2x + \frac{0.01}{x} - \frac{0.000001}{x^2}$$

Solution

Computer programs BRENTGLD and CUBIC2P are used in arriving at the results that are tabulated as follows.

Function	x/f
1	1.0000 2.0000
2	1.6094 −3.0472
3	2.0000 −43.000
4	0.6298 −0.2034
5	2.0046 5.0000
6	1.0011 −1.1e-3
7	0.7032 0.6280

The reader is encouraged to experiment with other starting points, step sizes, convergence parameters, and functions.

2.7 Shubert–Piyavskii Method for Optimization of Non-unimodal Functions

The techniques presented in the preceding text, namely, Fibonacci, Golden Section, Polynomial fit methods, require the function to be unimodal. However, engineering functions are often multimodal, and further, their modality cannot be ascertained *a priori*. Techniques for finding the global minimum are few, and can be broadly classified on the basis of deterministic or random search. A deterministic technique developed by Shubert and Piyavskii [Shubert 1972] is presented here. The problem is posed in the form

$$\begin{aligned} &\text{maximize } f(x) \\ &a \leq x \leq b \end{aligned} \tag{2.27}$$

It is assumed that the function f is Lipschitz continuous. That is, there is a constant $L > 0$ whose value is assumed to be known where

$$\left|f(x) - f(y)\right| \leq L|x - y| \quad \text{for any } x, y \in [a, b] \tag{2.28}$$

Thus, L is some upper bound on the derivative of the function in the interval. A sequence of points, $x_0, x_1, x_2, \ldots$, converging to the global maximum is generated by sequentially constructing a *saw-tooth* cover over the function. The algorithm is presented below, along with a computer program *Shubert*.

Starting Step: In the following, We denote $y_n \equiv f(x_n)$. To start with, a first sampling point x_0 is chosen at the mid-point of the interval as $x_0 = 0.5(a + b)$. At x_0, a pair of lines are drawn with slope L given by the equation $y(x) = y_0 + L|x - x_0|$, as shown in Fig. 2.17. Note that this pair of lines will not intersect the graph of $f(x)$ owing to L representing an upper bound on the derivative. The intersection of this pair of lines with the end points of the interval gives us an initial saw-tooth consisting of two points, namely, $\{(b, y_0 + \frac{L}{2}(b - a)), (a, y_0 + \frac{L}{2}(b - a))\} \equiv \{(t_1, z_1), (t_2, z_2)\}$.

Typical Step: Let the saw-tooth vertices be $\{(t_1, z_1), (t_2, z_2), \ldots, (t_n, z_n)\}$ with $z_1 \leq z_2 \leq \cdots \leq z_n$. The function f is now sampled at the location of the largest peak in the saw-tooth cover, namely, at t_n, and a pair of lines with slope L are drawn from $y_n = f(t_n)$ whose intersection with the existing saw-tooth cover yields two new peaks which replace the last point (t_n, z_n). Thus, the new saw-tooth cover is given by the vector $\{(t_1, z_1), (t_2, z_2), \ldots, (t_{n-1}, z_{n-1}), (t_l, z_l), (t_r, z_r)\}$. These coordinates are sorted in ascending values of the second component. Expressions for the new two points are

$$z_l = z_r = \frac{1}{2}(z_n + y_n), \quad t_l = t_n - \frac{1}{2L}(z_n - y_n), \quad t_r = t_n + \frac{1}{2L}(z_n - y_n)$$

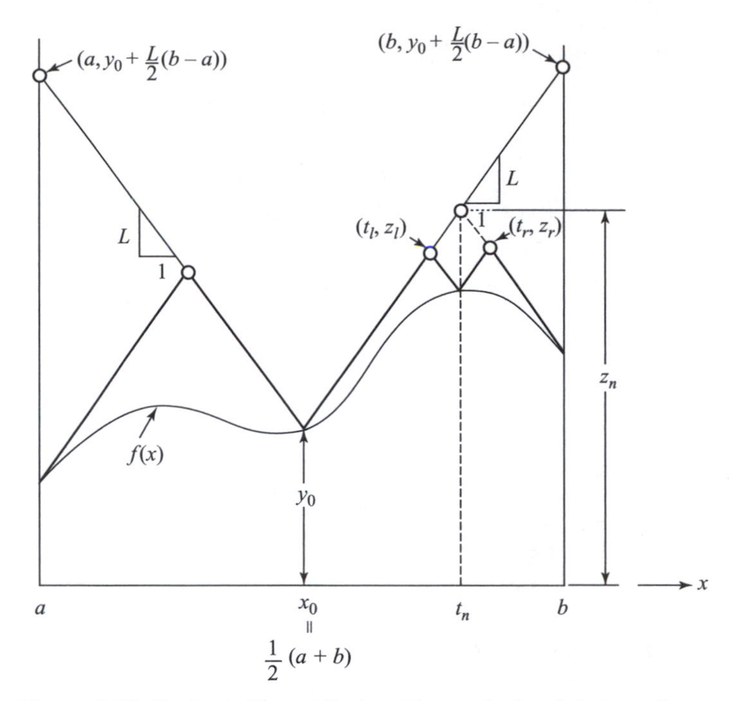

Figure 2.17. Shubert–Piyavskii algorithm to find a global maximum.

After the initial step, (t_r, z_r) is added, then (t_l, z_l), and thereafter two points at each step (see Fig. 2.17).

Stopping Criterion and Intervals of Uncertainty. The algorithm is stopped when the difference in height between the maximum saw-tooth peak and the corresponding function value is less than a specified tolerance, or when

$$z_n - y_n \leq \varepsilon \tag{2.29}$$

Uncertainty intervals may be determined from Eq. (2.29) as shown in Fig. 2.18. Thus, the uncertainty set is the union of the intervals

$$\left[t_i - \frac{1}{L}(z_i - y_n), \quad t_i + \frac{1}{L}(z_i - y_n)\right] \quad \text{for which}$$

$$z_i \geq y_n, \quad i = 1, \ldots, n \tag{2.30}$$

Coding the union of the intervals involves a tolerance.

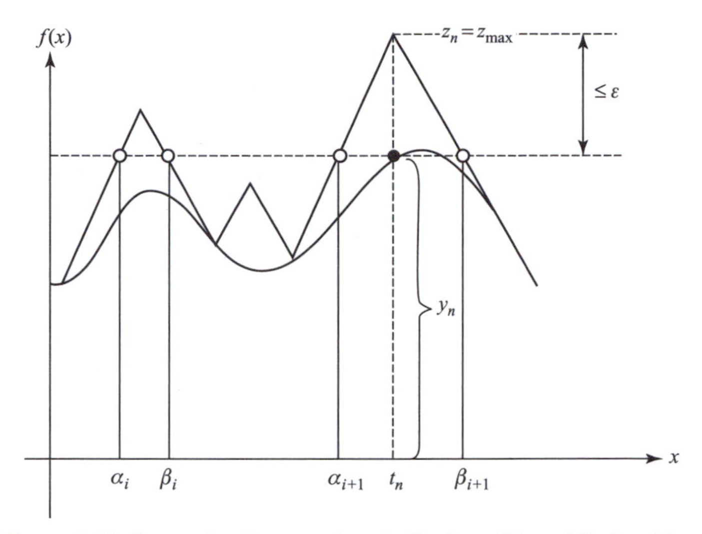

Figure 2.18. Intervals of uncertainty in Shubert–Piyavskii algorithm.

Example 2.11

(i) Consider the function $f = -\sin x - \sin \frac{10}{3}x$, to be maximized in the interval
$[-2.7, 7.5]$, $L = 6.5$, $\varepsilon = 0.01$. We obtain, using Program Shubert, the optimum $x^* = 5.1457, f^* = 1.8996$.

(ii) Consider the function $f = \sum_{k=1}^{5} k \sin((k+1)x + k)$, to be maximized in the interval
$[-10, 10]$, $L = 70$, $\varepsilon = 0.01$. We obtain the optimum $f^* = 12.031$ with the intervals of uncertainty being $[-6.78, -6.7635]$, $[-0.4997, -0.4856]$, $[5.7869, 5.8027]$.

2.8 Using MATLAB

The one-dimensional program in Matlab is *fminbnd*. It is recommended to use Matlab optimization routines using a "user subroutine" to evaluate f for a given x. That is, a user-subroutine "GETFUN." For example, consider the function $f = 2 - 2x + e^x$, with initial interval [0, 2]. We first create a file, "getfun.m" consisting of the code

```
function [f] = getfun(x)
f = 2 - 2*x + exp(x);
```

Then, Matlab is executed with the command:

```
[xopt, fopt, ifl, out] = fminbnd('getfun',0,2)
```

which produces the desired solution.

To see the default parameters associated with "fminbnd," type

```
optimset('fminbnd')
```

For instance, we see that tolX has a default value of 1.0e−4. This may be changed by

```
options=optimset('tolX',1.e-5)
```

and then executing fminbnd with options in the calling arguments.

2.9 Zero of a Function

In some functions, when the derivative information is available, an unconstrained minimum may be more efficiently found by finding the zero of the derivative. Also, a robust method of finding the zero of a function is a necessary tool in constrained nonlinear optimization algorithms discussed later in Chapter 7.

The problem of finding the root or zero of a function $f(x)$ is same as finding x such that

$$f(x) = 0$$

The problem of finding $g(x) = a$ can be put in the above form by setting $f(x) = g(x) - a$. Such problems frequently occur in design and analysis. The problem of finding the minimum of a smooth function of one variable leads to finding the zero of its derivative. The zero finding algorithms may be classified into two basic categories – one where we start from an arbitrary point, and the second where an initial interval of uncertainty has been established. Newton's method and the secant method fall in the first category. Bisection and other hybrid methods fall in the second category.

Newton's Method

In the Newton's method, we start at a point x_k where f_k and f'_k are evaluated. The next approximation x_{k+1} to the zero of the function $f(x)$, as seen from Fig. 2.19, is given by

$$x_{k+1} = x_k - \frac{f_k}{f'_k} \tag{2.31}$$

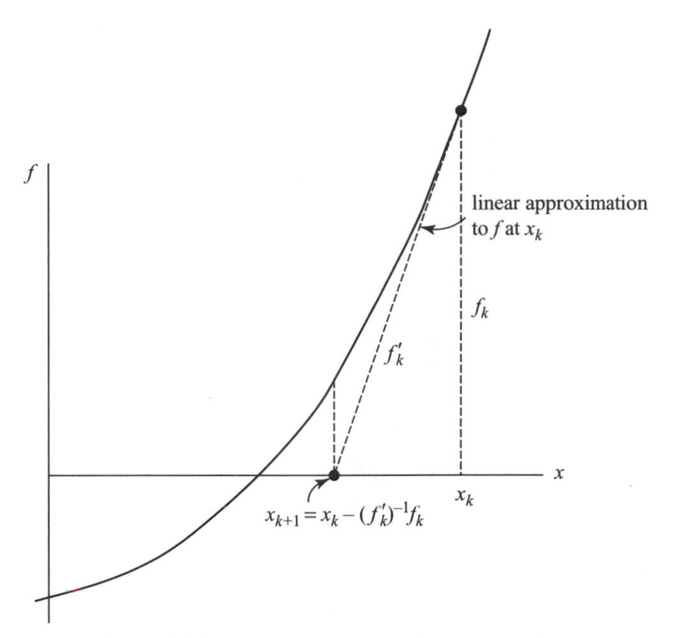

Figure 2.19. Newton's method for zero-finding.

Example 2.12

An endless V-belt with a pitch diameter of 500 mm is tightly wrapped around pulleys of pitch radii 60 mm and 40 mm as shown in Fig. E2.12. Determine the center distance of the pulleys.

Solution

Designating R and r as the larger and smaller pitch radii of pulleys respectively, and θ as half contact angle on the smaller pulley, L as belt length and C as the center distance, following relationships can be derived easily:

$$\tan\theta - \theta = \frac{0.5L - \pi R}{R - r}$$

$$C = \frac{R - r}{\cos\theta}$$

Denoting $A = \frac{0.5L - \pi R}{R - r}$, since $L = 500$ mm, $R = 60$ mm, and $r = 40$ mm, we determine $A = 3.075$. Now designating $f = \tan\theta - \theta - A$, we need to determine θ such that $f = 0$. Noting that $f' = \tan^2\theta$, we apply Eq. (2.31) starting with a value of $\theta_1 = 0.45\pi$ and the corresponding $f_1 = 1.825$. Newton's algorithm yields the result $\theta = 1.3485$ radians (77.26°) in four iterations. The corresponding center distance C is then calculated as 90.71 mm.

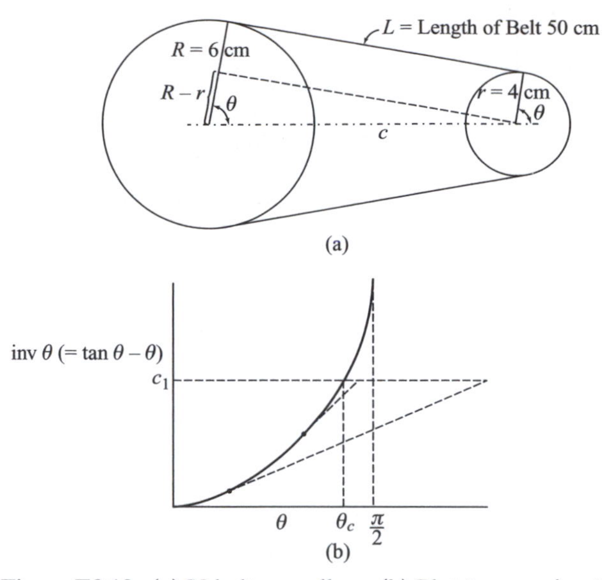

Figure E2.12. (a) V-belt on pulleys. (b) Plot θ versus inv θ.

We note here that if we start with a small value of θ_1, then the next estimation θ_2 may be larger than 0.5π, where the function f changes its sign. Some strategy can be introduced to safeguard this.

The term $\tan\theta - \theta$ is special in involute gear geometry. This entity is designated as *involute function of* θ ($\text{inv}\,\theta = \tan\theta - \theta$) and plays an important role in gear design.

Newton's method for finding a zero of a function is not guaranteed to work for arbitrary starts. If we start sufficiently close to a zero, the method converges to it at a quadratic rate. If x_k approaches x^* as k goes to infinity, and there exist positive numbers p and β such that

$$\lim_{k\to\infty} \frac{|x_{k+1} - x^*|}{|x_k - x^*|^p} = \beta \tag{2.32}$$

we say that *the rate of convergence* is of order p. The larger the value of p, the higher is the order of convergence. For Newton's method, $p = 2$ (quadratic convergence). If $p = 1$ and $\beta < 1$, the convergence is said to be *linear* with a *convergence ratio* of β. The limiting case when $\beta = 0$ is referred to as *superlinear convergence*.

In the *secant method*, the derivative f'_k is approximated using the last two function values. Thus,

$$x_{k+1} = x_k - \frac{x_k - x_{k-1}}{f_k - f_{k-1}} f_k \tag{2.33}$$

Newton's method can be used for minimization. In this case, we find the zero of the derivative of the function.

Method of Bisection

In this method, we first establish an initial bracket x_1, x_2 where the function values f_1, f_2 are opposite in sign. The function value f is then evaluated at the mid point $x = 0.5\ (x_1 + x_2)$. If f and f_1 have opposite signs, then x and f are renamed as x_2 and f_2 respectively, and if f and f_1 have same signs, then x and f are renamed as x_1 and f_1 respectively. This is carried out till $f = 0$ or abs $(x_2 - x_1)$ is smaller than a predetermined value. Solution of Example 2.8 using the bisection approach is left as an exercise. We note that f has a negative value at $\theta = 0$ and a positive value at $\theta = 0.5\pi$. Bisection is applied to this interval.

Bisection can be combined with polynomial fit approach. Brent's algorithm for zero finding uses bisection and quadratic fit ideas. See [Forsythe et al. 1977; Press et al. 1992] for an alternate algorithms and discussions.

COMPUTER PROGRAMS

FIBONACI, GOLDINTV, GOLDLINE, QUADFIT, BRENTGLD, CUBIC2P, SHUBERT

PROBLEMS

P2.1. Prove that $f = x^2$ is a convex function using the following:

(i) The definition in Eq. (2.7)
(ii) Property 1 of a convex function in Eq. (2.8)
(iii) Property 2 of a convex function in Eq. (2.9)

P2.2. Write down the necessary and sufficient conditions for a local *maximum* to f.

P2.3. A function of one variable is graphed in the following:

(i) Is f convex?
(ii) Is f unimodal?

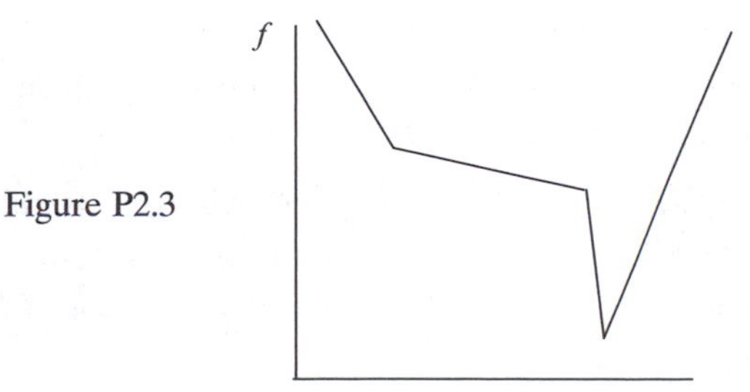

Figure P2.3

P2.4. A part is produced on a lathe in a machine shop. The cost of the part includes machining cost, tool related cost, and cost of idle time. The cost for the machining time is inversely proportional to the cutting speed V m/min. The tool related costs are proportional to $V^{1.5}$. The cost c in dollars is given by

$$c = \frac{240}{V} + 10^{-4} V^{1.5} + 0.45$$

Determine the cutting speed for minimum cost and the associated minimum cost using optimality conditions in both (2.5) and (2.6).

P2.5. In a solution of potassium (K) and chlorine (Cl), the equilibrium distance r in nanometers (10^{-9} m) between the two atoms is obtained as the minimum of the total energy E (sum of the energy of attraction and the energy of repulsion) in electron volts (eV) given by

$$E = -\frac{1.44}{r} + \frac{5.9 \times 10^{-6}}{r^9}$$

Determine the equilibrium spacing and the corresponding energy using optimality conditions in both (2.5) and (2.6).

P2.6. State whether Fibonacci and Golden Section search methods will work on the following problem: minimize f, where

$$\begin{aligned} f &= (x-1)^2 + 2 \quad \text{if } x \leq 1 \\ &= (x-1)^2, \qquad \text{if } x > 1 \end{aligned}$$

Justify your answer.

P2.7. Consider $f = x_1^4 + x_2^4$, with x_1 and x_2 real. To minimize f, I set $\nabla f = \mathbf{0}$ which yields $\mathbf{x}^* = (0, 0)^T$. I claim that this is the global minimum. Explain my reasoning.

P2.8. Given $f(x) = 2 - 2x + e^x$ on R^1, initial interval of uncertainty = [0, 2], determine the following:

(i) The first two points placed within the interval as per the Fibonacci method.

(ii) The first two points placed within the interval as per the Golden Section method.

(iii) The ratio of I_1/I_{10} as per Fibonacci and Golden Section search methods, respectively.

P2.9. A) Consider the function $f = -V$, where $V = x(210 - 2x)(297 - 2x)$ = volume of open box. See Example 2.4. Determine whether x is convex on: (a) R^1, and (b) on the subset $\{0 \leq x \leq 105\}$. Also, provide a plot of f vs x over, say, the interval [0, 250] to support your comments.

B) With exactly four (4) function evaluations, determine the smallest interval of uncertainty on the problem by *hand calculations*: minimize $f = e^{3x} + 5e^{-2x}$, on the initial interval [0, 1] as per:

(i) Fibonacci search
(ii) Golden Section search

P2.10. Solve the unconstrained problem: minimize $f = e^{3x} + 5e^{-2x}$, using MATLAB fminbnd. Check the optimality conditions at the optimum obtained from the program. Is the solution a local minimum or a global minimum?

P2.11. Solve P2.4 using Program FIBONACI.

P2.12. Solve P2.5 using Program GOLDLINE and GOLDINTV.

P2.13. Solve P2.4 using Matlab "fminbnd" routine.

P2.14. Two disks of diameters 10 cm and 20 cm are to be placed inside a rectangular region. Determine the region (a) of least perimeter, (b) of least area.

P2.15. Three disks each of diameter 10 cm are to be placed inside a rectangular region. Determine the region (a) of least perimeter, (b) of least area.

P2.16. A trapezoidal region *ABCD* is to be built inside a semicircle of radius 10m with points *A* and *B* diametrically opposite and points *C* and *D* on the semicircle with *CD* parallel to *AB* as shown in Fig. P2.16. Determine the height *h* which maximizes the area of *ABCD*, and evaluate the area.

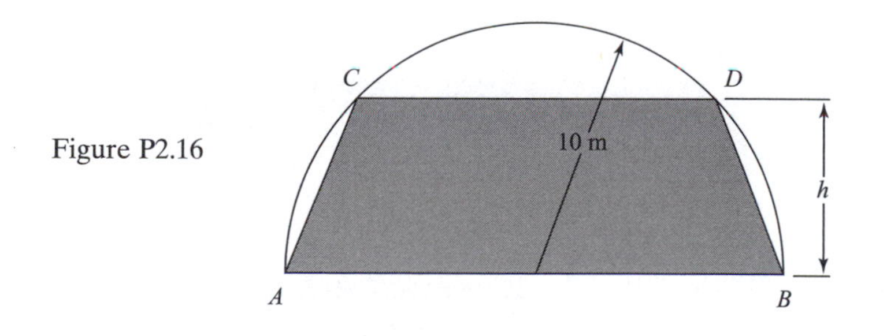

Figure P2.16

P2.17. A frustum of a pyramid is to be inscribed in a hemisphere of radius 10 m. The base of the pyramid is a square with its corners touching the base of the hemisphere. The top surface is a square parallel to the base as shown in Fig. P2.17. Determine the dimensions that maximize the volume. What is the volume in m^3?

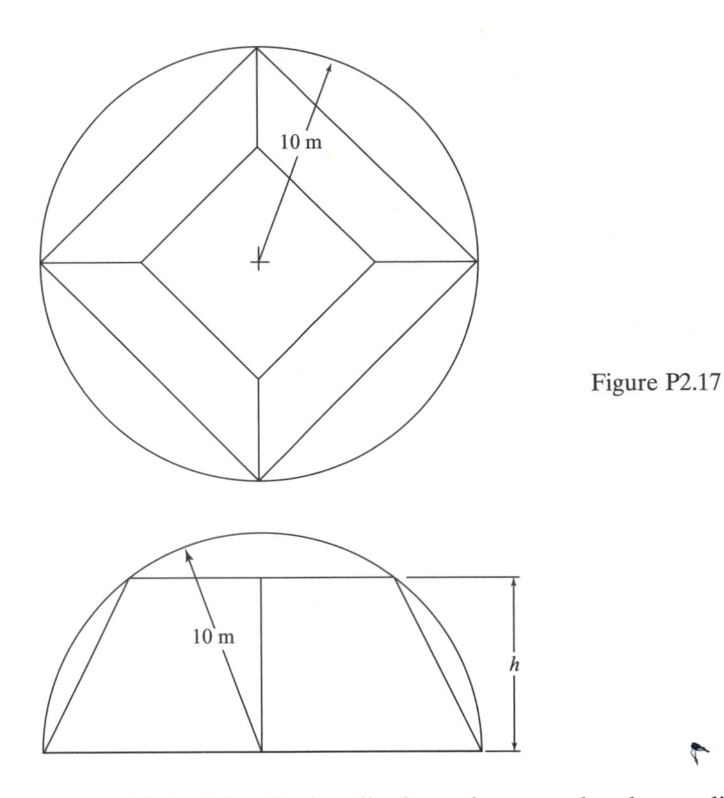

Figure P2.17

P2.18. A water canal is built by digging dirt from the ground and spreading it on the banks as shown in Fig. P2.18. If 50 m^2 of earth is the cross sectional area dug, determine the dimensions of the canal which minimize the wetted perimeter. Assume that the canal will flow full. (*Hint*: The wetted perimeter is the length of the bottom side and the sum of the lengths of the two slant sides.)

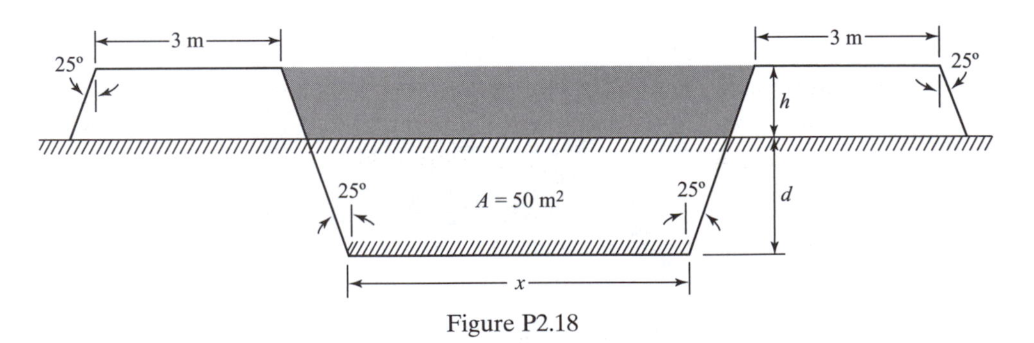

Figure P2.18

P2.19. An open box is made by cutting out equal squares from the corners, bending the flaps, and welding the edges (Fig. P2.19). Every cubic inch of the volume of the open box brings in a profit of \$0.10, every square inch of corner waste results in a cost of \$0.04, every inch of welding length costs \$0.02. Determine the box dimensions for maximum profit.

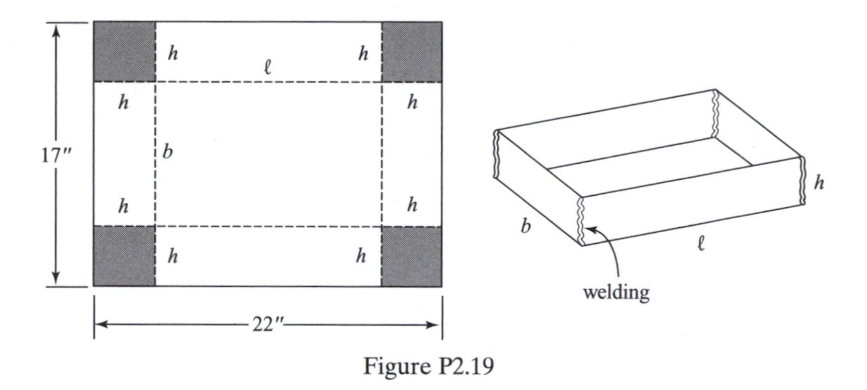

Figure P2.19

P2.20. In the toggle type mechanism shown in Fig. P2.20, the spring is at its free length. The potential energy π is given by

$$\pi = \frac{1}{2}kx^2 - Py$$

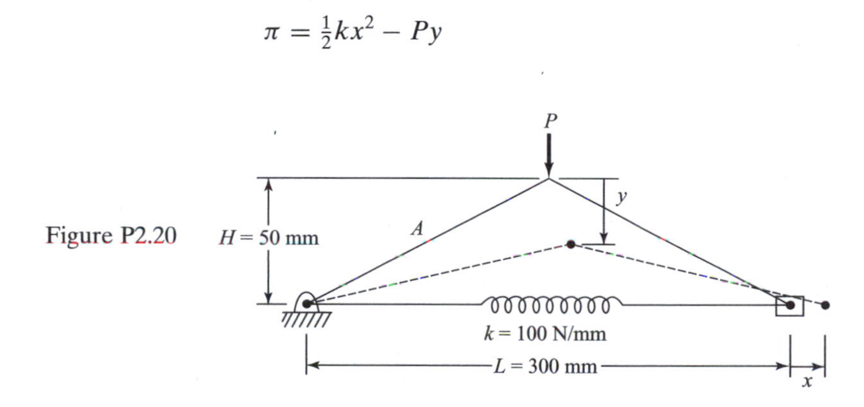

Figure P2.20

If the equilibrium configuration is the one corresponding to minimum potential energy, determine x (and y). Solve for load P ranging from [100 N to 500 N] in steps of 100 N. That is, $P(1) = 100$, $P(2) = 200, \ldots, P(5) = 500$.

(I) Plot P versus x and P vs y, with P the ordinate.
(II) Until what value of P is the response linear?

Hints:

i. First compute the constant A from $A = \sqrt{\left(\frac{L}{2}\right)^2 + H^2}$.
ii. Within the loop $i = 1{:}5$, call *fminbnd* or other code to minimize π using the relation $A^2 = \left(\frac{L+x}{2}\right)^2 + (H - y)^2$ to express y in terms of x. Thus, x is the independent variable. Importantly, the initial interval of uncertainty for x that must be supplied to the code must be $0 \le x \le (2^*A - L)$. The latter

comes from the fact that A is the hypotenuse and, hence, the longest side of the triangle. Obtain and store $x(i)$ and corresponding $y(i)$.

iii. Outside the loop, plot P versus x, P versus y.

P2.21. A machine tool spindle is supported by two rolling contact bearings as shown in Fig. P2.21. The front bearing has a stiffness S_1 of 1000 N/μm, and the rear bearing has a stiffness S_2 of 600 N/μm, the overhang a is 100 mm and the spindle diameter d is 75 mm. The modulus of elasticity E of the spindle material (steel) is 210×10^3 N/mm^2. The distance between the bearings is b and we denote $\alpha = a/b$. The deflection per unit load f is given by

$$f = \frac{a^3}{3EI}\left(1 + \frac{1}{\alpha}\right) + \frac{1}{S_1}(1+\alpha)^2 + \frac{1}{S_2}\alpha^2$$

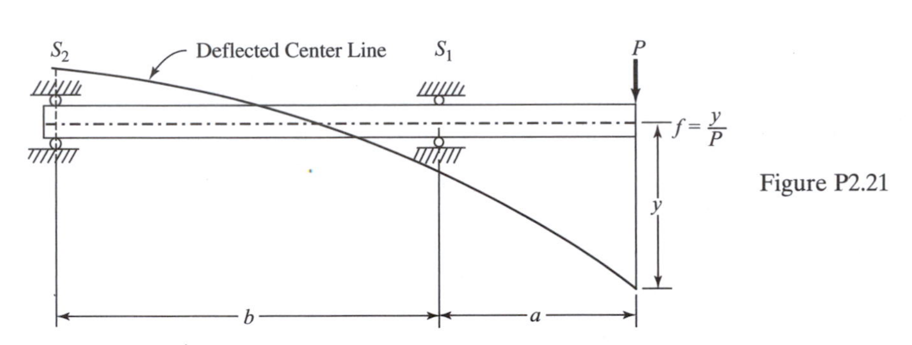

Figure P2.21

where $I = \frac{\pi d^4}{64}$ is the moment of inertia of the spindle section. Determine the spindle spacing b for minimum deflection per unit load (i.e., maximum stiffness).

P2.22. A beam of length $2L$ is to be supported by two symmetrically placed point supports. Determine the spacing for minimizing the maximum deflection of the beam. See Fig. P2.22.

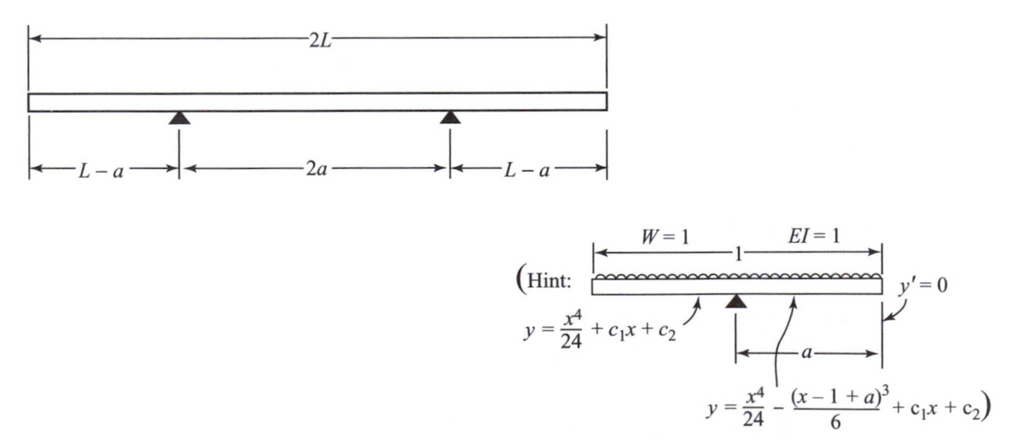

Figure P2.22

P2.23. Repeat P2.22, but with the objective to *minimize the maximum bending stress* in the beam. (Assume $L = 1$). Use MATLAB fminbnd or a code provided on the CD-ROM. *Hint*: bending stress is proportional to bending moment. Considering ½-symmetry, with origin $\xi = 0$ at the left end, $0 \le \xi \le 1$, expression for moment is $M(\xi) = -0.5\,\xi^{\,2} + \max(0, \xi - 1 + a)$. Thus, determine optimum spacing a to minimize the maximum moment abs(M) in the beam.

P2.24. Use Program SHUBERT to find the global maximum of $f = (-3\,x + 1.4)\sin(18x)$, $x \in [0, 1.2]$ Also, determine all the local maxima and intervals of uncertainty. Validate your solution by plotting the function.

P2.25. Conduct a study on the functions in Example 2.10 comparing quadfit, cubic fit, and Matlab fminbnd programs.

P2.26. In a four bar linkage arrangement shown in Fig. P2.26, angles θ and ϕ are related by Freudenstein's equation given by

$$R_1 \cos\phi - R_2 \cos\theta + R_3 - \cos(\phi - \theta) = 0$$

$$\text{where } R_1 = \frac{d}{a}, \quad R_2 = \frac{d}{b}, \quad R_3 = \frac{d^2 + a^2 + b^2 - c^2}{2ab}$$

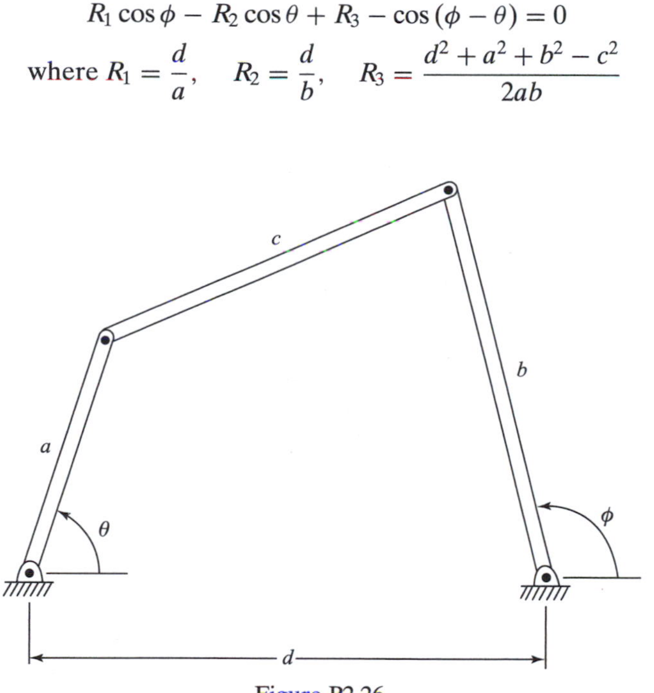

Figure P2.26

If $d = 16$ cm, $a = 8$ cm, $b = 12$ cm, and $c = 12$ cm, use a zero-finding algorithm to find ϕ for $\theta = 10°, 20°, 30°, \ldots, 90°$.

REFERENCES

Brent, R.P., *Algorithms of Minimization without Derivatives*, Prentice-Hall, Englewood Cliffs, NJ, 1973, pp. 47–60.

Chandrupatla, T.R., An efficient quadratic fit-Sectioning algorithm for minimization without derivatives, *Computer Methods in Applied Mechanics and Engineering*, **152**, 211–217, 1988.

Forsythe, G.E., Malcolm, M.A., and Moler, C.B., *Computer Methods for Mathematical Computations*, Prentice-Hall, Englewood Cliffs, NJ, 1977.

Gill, P.E., Murray, W., and Wright, M.H., *Practical Optimization*, Academic Press, New York, 1981.

Press, W.H., Teukolsky, S.A., Vetterling, W.T., and Flannery, B.P., *Numerical Recipes*, Cambridge University Press, New York, 1992.

Robinson, S.M., Quadratic interpolation is risky, *SIAM Journal on Numerical Analysis*, **16**, 377–379, 1979.

Shubert, B.O., A sequential method seeking the global maximum of a function, *SIAM Journal Numerical on Analysis*, **9**(3), 1972.

3

Unconstrained Optimization

3.1 Introduction

Many engineering problems involve the unconstrained minimization of a function of several variables. Such problems arise in finding the equilibrium of a system by minimizing its energy, fitting an equation to a set of data points using least squares, or in determining the parameters of a probability distribution to fit data. Unconstrained problems also arise when the constraints are eliminated or accounted for by suitable penalty functions. All these problems are of the form

$$\text{minimize} \quad f(x_1, x_2, \ldots, x_n)$$

where $\mathbf{x} = [\, x_1, x_2, \ldots, x_n]^{\mathrm{T}}$ is a column vector of n real-valued design variables. In this chapter, we discuss the necessary and sufficient conditions for optimality, convex functions, and gradient-based methods for minimization. Note that nongradient or search techniques for minimization are given in Chapter 7.

An example problem that occurs in mechanics is presented in the following.

Example 3.1

Consider a nonlinear spring system as shown in Fig. E3.1. The displacements Q_1 and Q_2 under the applied load can be obtained by minimizing the potential energy Π given by

$$\Pi = \tfrac{1}{2}k_1(\Delta L_1)^2 + \tfrac{1}{2}\,k_2\,(\Delta L_2)^2 - F_1\,Q_1 - F_2\,Q_2$$

where the spring extensions ΔL_1 and ΔL_2 are related to the displacements Q_1 and Q_2, respectively, as

$$\Delta L_1 = \sqrt{(Q_1 + 10)^2 + (Q_2 - 10)^2} - 10\sqrt{2}$$
$$\Delta L_2 = \sqrt{(Q_1 - 10)^2 + (Q_2 - 10)^2} - 10\sqrt{2}$$

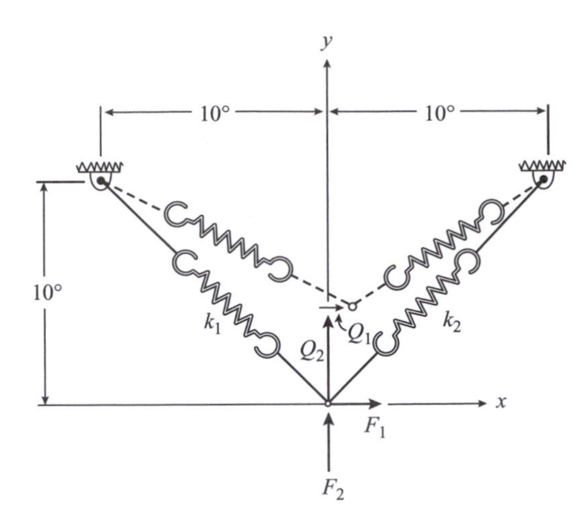

Figure E3.1. A nonlinear spring system.

The problem reduces to the form

$$\text{minimize} \quad \Pi(Q_1, Q_2)$$

Using numerical techniques that will be discussed in this chapter, we obtain, for $k_1 = k_2 = 1$ lb/in, $F_1 = 0$, $F_2 = 2$, the solution $Q_1 = 0$ and $Q_2 = 2.55$ in. The solution for increasing values of F_2 brings out an interesting phenomenon (see exercise problem).

3.2 Necessary and Sufficient Conditions for Optimality

A point is a local minimum if all other points in its neighborhood have a higher function value. This statement can be expressed more accurately as follows. The point $\mathbf{x}^* = [x_1{}^*, x_2{}^*, \ldots, x_n{}^*]^{\mathrm{T}}$ is a *weak* local minimum if there exists a $\delta > 0$ such that $f(\mathbf{x}^*) \leq f(\mathbf{x})$ for all $\mathbf{x}$ such that $||\mathbf{x} - \mathbf{x}^*|| < \delta$. We may also state that $\mathbf{x}^*$ is a *strong* local minimum if $f(\mathbf{x}^*) < f(\mathbf{x})$ for all $\mathbf{x}$ such that $||\mathbf{x} - \mathbf{x}^*|| < \delta$. Further, $\mathbf{x}^*$ is a *global* minimum if $f(\mathbf{x}^*) \leq f(\mathbf{x})$ for all $\mathbf{x} \in R^n$.

Necessary Conditions for Optimality

If $f \in C^1$, the necessary condition for $\mathbf{x}^*$ to be a local minimum is:

$$\nabla f(\mathbf{x}^*) = \mathbf{0} \tag{3.1}$$

Equation (3.1) states that the gradient of the function equals zero or that the derivatives $\frac{\partial f(\mathbf{x}^*)}{\partial x_i} = 0$, $i = 1, 2, \ldots, n$. A point $\mathbf{x}^*$ that satisfies Eq. (3.1) is called

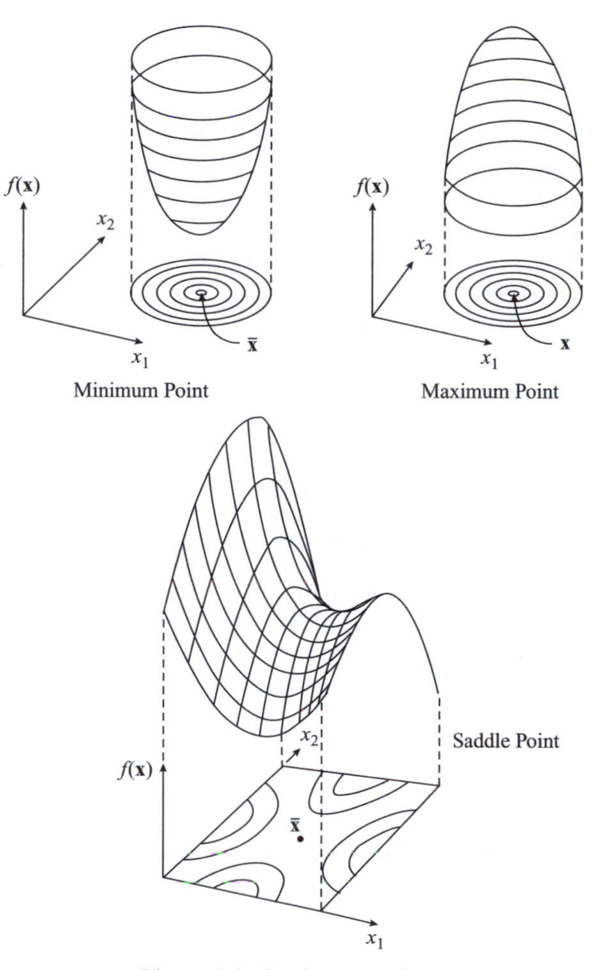

Figure 3.1. Stationary points.

a *stationary* point – it can be a minimum or maximum, or a saddle point (see Fig. 3.1).

The first-order necessary condition in (3.1) may be derived as follows. Let $\mathbf{u}^i = [0, \ldots, 1, \ldots, 0]^{\mathrm{T}}$ be a unit vector with 1 in the ith location. Then, $(\mathbf{x}^* + h\mathbf{u}^i)$ with $h > 0$ will represent a perturbation of magnitude h in $\mathbf{x}^*$. For $\mathbf{x}^*$ to be a minimum, we require both

$$f(\mathbf{x}^* + h\mathbf{u}^i) - f(\mathbf{x}^*) \geq 0 \tag{3.2a}$$

$$f(\mathbf{x}^* - h\mathbf{u}^i) - f(\mathbf{x}^*) \geq 0 \tag{3.2b}$$

for h sufficiently small. Now, $f(\mathbf{x}^* + h\mathbf{u}^i)$ and $f(\mathbf{x}^* - h\mathbf{u}^i)$ can be expanded about $\mathbf{x}^*$ using Taylor series

$$f(\mathbf{x}^* + h\mathbf{u}^i) = f(\mathbf{x}^*) + h\,\partial f(\mathbf{x}^*)/\partial x_i + O(h^2) \tag{3.3a}$$

$$f(\mathbf{x}^* - h\mathbf{u}^i) = f(\mathbf{x}^*) - h\,\partial f(\mathbf{x}^*)/\partial x_i + O(h^2) \tag{3.3b}$$

For small enough h, the term $h\,\partial f(\mathbf{x}^*)/\partial x_i$ will dominate the remainder term $O(h^2)$. Choosing h sufficiently small, we deduce that $\partial f(\mathbf{x}^*)/\partial x_i \geq 0$ as well as $-\partial f(\mathbf{x}^*)/\partial x_i \geq 0$, which gives $\partial f(\mathbf{x}^*)/\partial x_i = 0$. This is true for $i = 1, \ldots, n$ or equivalently

$$\nabla f(\mathbf{x}^*) = \mathbf{0}.$$

Sufficiency Conditions

The sufficient conditions for $\mathbf{x}^*$ to be a *strict* local minimum are

$$\nabla f(\mathbf{x}^*) = \mathbf{0}. \tag{3.4}$$

$$\nabla^2 f(\mathbf{x}^*) \text{ is positive definite} \tag{3.5}$$

The sufficiency condition may be derived as follows. Let $\mathbf{y}$ be any vector in R^n. Since $\mathbf{x}^*$ is a strict minimum we require

$$f(\mathbf{x}^* + h\mathbf{y}) - f(\mathbf{x}^*) > \mathbf{0} \tag{3.6}$$

for h sufficiently small. Expanding $f(\mathbf{x}^* + h\mathbf{y})$ about $\mathbf{x}^*$, we get

$$f(\mathbf{x}^* + h\mathbf{y}) = f(\mathbf{x}^*) + h\nabla f(\mathbf{x}^*)^{\mathrm{T}}\mathbf{y} + \tfrac{1}{2}h^2\mathbf{y}^{\mathrm{T}}\,\nabla^2 f(\mathbf{x}^*)\mathbf{y} + \mathbf{O}(h^3) \tag{3.7}$$

Noting that $\nabla f(\mathbf{x}^*) = \mathbf{0}$, and choosing h small enough, we can see that the term $\mathbf{y}^{\mathrm{T}}$ $\nabla^2 f(\mathbf{x}^*)\,\mathbf{y}$ dominates the remainder term $O(h^3)$. Thus, we deduce that

$$\mathbf{y}^T\,\nabla^2 f(\mathbf{x}^*)\mathbf{y} > 0$$

Since the aforementioned has to hold for *any* $\mathbf{y}$, by definition, $\nabla^2 f(\mathbf{x}^*)$ must be positive definite. The geometric significance of (3.5) is tied to the concept of convexity, which is discussed subsequently.

Example 3.2

An equation of the form $y = a + \frac{b}{x}$ is used to provide a best fit in the sense of least squares for the following (x, y) points: $(1, 6)$, $(3, 10)$, and $(6, 2)$. Determine a, b using necessary and sufficient conditions.

The problem of least squares is

$$\text{minimize } f = \sum_{1}^{3}\left(a + \frac{b}{x_i} - y_i\right)^2$$

There are two variables, a and b. Upon substituting for (x_i, y_i), we get

$$\text{minimize } f = 3a^2 + 41/36b^2 + 3ab - 58/3b - 36a + 140$$

The necessary conditions require

$$\partial f/\partial a = 6a + 3b - 36 = 0$$
$$\partial f/\partial b = 41/18b + 3a - 58/3 = 0$$

which yield

$$a^* = 5.1428, \quad b^* = 1.7143, \quad \text{and} \quad f^* = 30.86$$

Thus, the best fit equation is $y(x) = 5.1428 + 1.7143/x$. The sufficient condition to check whether (a^*, b^*) is a minimum is:

$$\nabla^2 f = \begin{bmatrix} \dfrac{\partial^2 f}{\partial a^2} & \dfrac{\partial^2 f}{\partial a \partial b} \\ \dfrac{\partial^2 f}{\partial b \partial a} & \dfrac{\partial^2 f}{\partial b^2} \end{bmatrix}_{(a^*,b^*)}$$
$$= \begin{bmatrix} 6 & 3 \\ 3 & \frac{41}{18} \end{bmatrix}$$

Using Sylvester's check given in Chapter 1, we have $6 > 0$ and $(6)\,(41/18) - 3^2 > 0$; hence, $\nabla^2 f$ is positive definite. Thus, the solution (a^*, b^*) is a strong local minimum. Further, from Section 3.3, f is strictly convex and, hence, the solution is a strong *global* minimum.

Example 3.3

Consider the function $f = 2x_1^2 + 8x_2^2 + 8x_1x_2 - 5x_1$. Discuss the necessary conditions for optimality for this function in R^2. We have

$$\partial f/\partial x_1 = 4x_1 + 8x_2 - 5 = 0 \quad \text{or} \quad 4x_1 + 8x_2 = 5$$
$$\partial f/\partial x_2 = 8x_1 + 16x_2 = 0 \quad \text{or} \quad 4x_1 + 8x_2 = 0$$

which are obviously incompatible. No solution exists (in an unbounded domain), which implies that there is no local minimum to this function. A graph of this function is shown in Fig. E3.3.

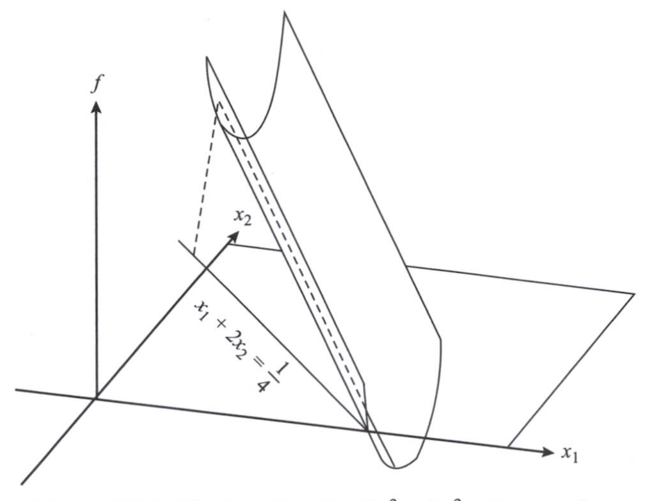

Figure E3.3. The function $f = 2x_1^2 + 8x_2^2 + 8x_1x_2 - 5x_1$.

3.3 Convexity

We have seen that conditions on the Hessian matrix enters into the optimality conditions. In fact, the Hessian matrix $\nabla^2 f$ is related to the convexity of the function as we will discuss in this section. First, we will define convex sets, followed by convex functions.

A set S in R^n is convex if for every $\mathbf{x}^1, \mathbf{x}^2 \in S$ and every real number α, $0 < \alpha < 1$, the point $\alpha\mathbf{x}^1 + (1 - \alpha)\mathbf{x}^2 \in S$. That is, a set S is convex if, given any two points in the set, every point on the line segment joining these two points is also in the set (see Fig. 3.2). We note that R^n is a convex set.

The definition and properties of a convex function defined over R^1 were discussed in Chapter 2. These ideas carry over to functions of n variables. A function

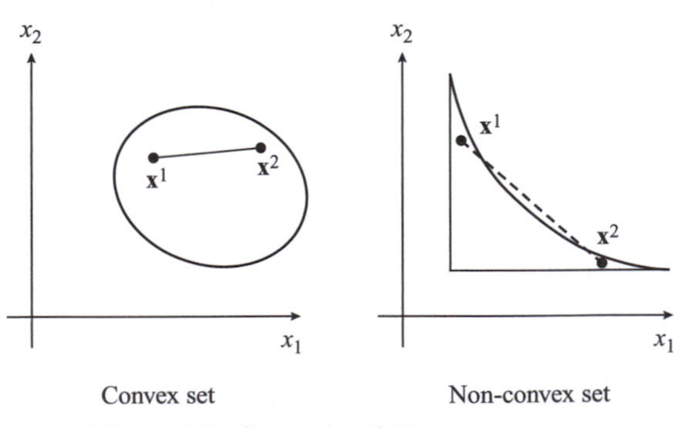

Figure 3.2. Convex and Nonconvex sets.

$f(\mathbf{x})$ defined over a convex set S is called a convex function if for every pair of points $\mathbf{x}^1, \mathbf{x}^2 \in S$, and every α, $0 \le \alpha \le 1$, there holds

$$f(\alpha\mathbf{x}^1 + (1-\alpha)\mathbf{x}^2) \le \alpha f(\mathbf{x}^1) + (1-\alpha) f(\mathbf{x}^2) \tag{3.8}$$

The geometrical significance of (3.8) is shown in Fig. 2.2 for a function of one variable.

Properties of Convex Functions

(1) Let $f \in C^1$ (i.e., the first derivative of f is a continuous function). Then, f is convex over a convex set S if and only if

$$f(\mathbf{y}) \ge f(\mathbf{x}) + \nabla f(\mathbf{x})^{\mathrm{T}}(\mathbf{y} - \mathbf{x}) \tag{3.9}$$

for all $\mathbf{x}, \mathbf{y} \in S$.

(2) Let $f \in C^2$ (see Chapter 1 for the meaning of this assumption). Then, f is convex over a convex set S, containing an interior point if and only if $\nabla^2 f$ is positive semidefinite throughout S. The sufficiency condition in (3.5) states, in effect, that $\mathbf{x}^*$ is a strong local minimum if the function f is locally strictly convex.

(3) If f is a convex function, then any relative minimum of f is a global minimum. This makes intuitive sense.

(4) If a function f is convex and it satisfies the necessary condition $\nabla f(\mathbf{x}^*) = \mathbf{0}$, then $\mathbf{x}^*$ is a global minimum of f. This follows from (3.9) upon choosing $\mathbf{x}^* = \mathbf{x}$. In other words, for a convex function, the necessary condition is also sufficient for global optimality.

Example 3.4

Consider the function

$$f = x_1 x_2$$

over the convex set $S = \{(x_1, x_2) \in R^2 : x_1 > 0, x_2 > 0\}$. Is f convex over the set S?

Since the function f is twice continuously differentiable, we can use the Hessian test. We have

$$\nabla^2 f = \begin{bmatrix} 0 & 1 \\ 1 & 0 \end{bmatrix}$$

The Hessian matrix has eigenvalues $-1, 1$ and is therefore not positive semidefinite and, thus, f is not convex. A plot of the function in Fig. E3.4 also shows why f is not convex.

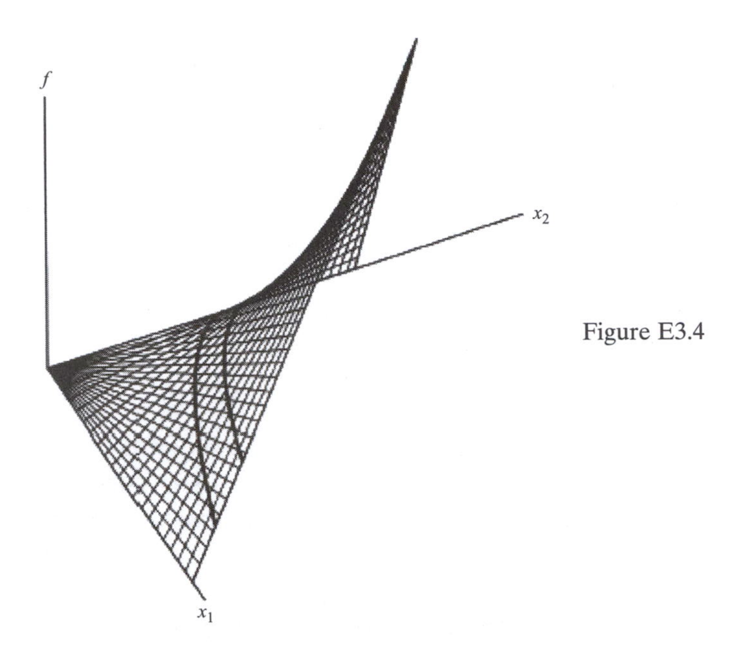

Figure E3.4

3.4 Basic Concepts: Starting Design, Direction Vector, and Step Size

So far, we have discussed the optimality conditions and convexity. Prior to presenting numerical methods to minimize $f(\mathbf{x})$, we need to understand certain basic concepts.

Most numerical methods require a starting design or point, which we will call $\mathbf{x}_0$. From this point, a "direction of travel" $\mathbf{d}_0$ is then determined. A "step size" α_0 is then determined based on minimizing f as much as possible, and the design point is updated as $\mathbf{x}_1 = \mathbf{x}_0 + \alpha_0 \, \mathbf{d}_0$. Then, the process of where to go and how far to go are repeated from $\mathbf{x}_1$. The actual choice of the direction vector and step size differ among the various numerical methods, which will be discussed subsequently.

In nonconvex problems with multiple local minima, the solution obtained by gradient methods in this chapter will only find the local minimum nearest to the starting point. For problems of small dimension, a *grid search* may be employed to determine a good starting point. Here, we discretize each design variable between its lower and upper limits into $(N_d - 1)$ parts and evaluate the objective f at each grid point. With $n = 2$, we have a total of p^2 grid points. In general, we have p^n

points. Design of experiments may also be used to select the grid points. The point with the least value of f is chosen as the starting point for the gradient methods discussed in the following. Further, several nongradient techniques are presented in Chapter 7.

Example 3.5

Given $f = x_1^2 + 5x_2^2$, a point $\mathbf{x}^0 = (3, 1)^{\mathrm{T}}, f_0 \equiv f(\mathbf{x}^0) = 14$.

(i) Construct $f(\alpha)$ along the direction $\mathbf{d} = (-3, -5)^{\mathrm{T}}$ and provide a plot of f versus $\alpha, \alpha \geq 0$.
(ii) Find the slope $\frac{\mathbf{d}f(\alpha)}{\mathbf{d}\alpha}\big|_{\alpha=0}$. Verify that this equals $\nabla f(\mathbf{x}^0)^{\mathrm{T}}\,\mathbf{d}$
(iii) Minimize $f(\alpha)$ with respect to α, to obtain step size α_0. Give the corresponding new point $\mathbf{x}^1$and value of $f_1 \equiv f(\mathbf{x}^1)$.
(iv) Provide a plot showing contour(s) of the function, direction $\mathbf{d}$, $\mathbf{x}^0$ and $\mathbf{x}^1$.

We have $\mathbf{x}(\alpha) = (3 - 3\alpha,\ 1 - 5\alpha)^{\mathrm{T}}$, and $f(\alpha) = (3 - 3\alpha)^2 + 5\ (1 - 5\alpha)^2$. To obtain a minimum, set the derivative $\frac{\mathbf{d}f(\alpha)}{\mathbf{d}\alpha} = 0$, to obtain $\alpha_0 = 0.253731$. The second derivative test may be used to ensure that this is a minimum. We get $\mathbf{x}^1 = \mathbf{x}^\circ + \alpha_0\ \mathbf{d} = (2.238806, -0.268657)^{\mathrm{T}}$ and $f_1 = 5.37314$ as compared with the starting value of $f_0 = 14$. The initial slope (i.e., slope at $\alpha = 0$) of $f(\alpha)$ versus α graph is readily calculated as -68. The dot product $\nabla\ f(\mathbf{x}^0)^{\mathrm{T}}\,\mathbf{d}$ gives $(6, 10).\ (-3, -5)^{\mathrm{T}} = -68$, also. The negative value indicates a downhill or descent direction. Plots using MATLAB are given in Figs. E3.5a, b. Note that the contour at $\mathbf{x}^1$ is tangent to $\mathbf{d}$. This follows from the fact that a smaller or larger step will end at a point on a higher contour value.

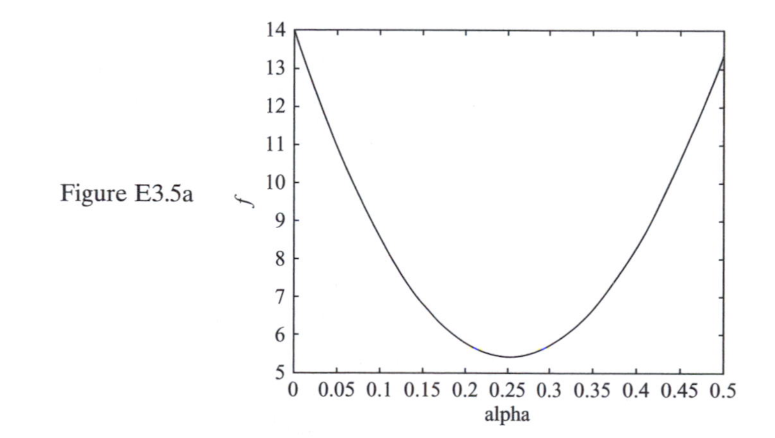

Figure E3.5a

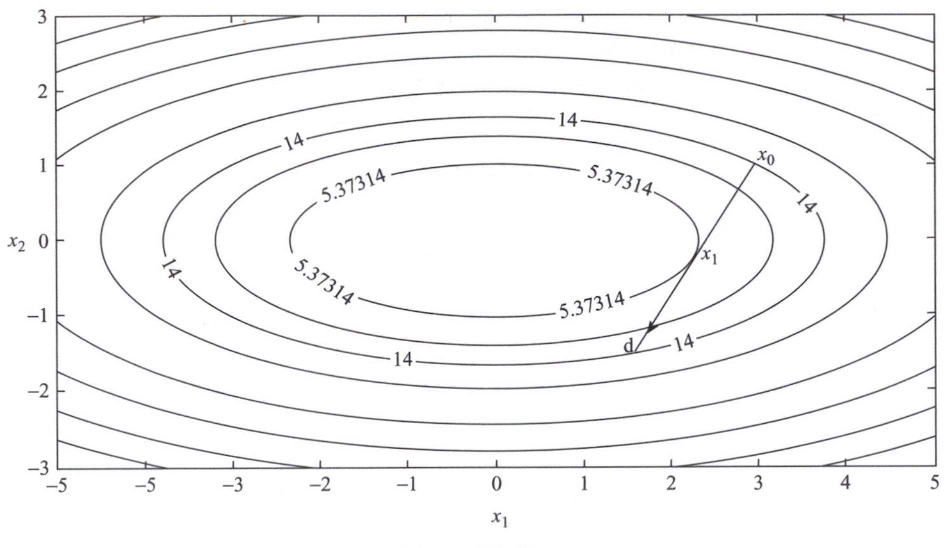

Figure E3.5b

Example 3.6

Consider the cantilever beam shown in Fig. E3.6. The design variables are taken to be the width and height of the rectangular cross-section. Thus, $\mathbf{x} = (w, h)^{\mathrm{T}}$. The current design is chosen as $\mathbf{x}_0 = (1, 3)^{\mathrm{T}}$ in. Find the bending stress at the initial design. Further, given that $\mathbf{d}_0 = (-1/\sqrt{5}, -2/\sqrt{5})^{\mathrm{T}}$ and $\alpha_0 = 0.2$, find the updated design and the updated maximum bending stress.

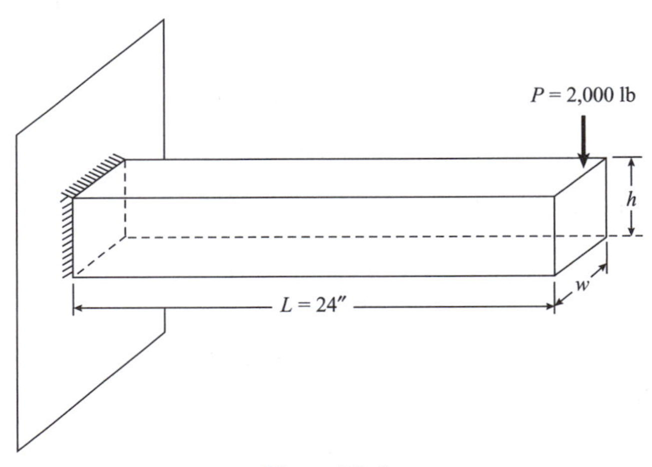

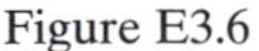

Figure E3.6

The current value of the bending stress is

$$\sigma_0 = \frac{6M}{wh^2} = \frac{6.(2000 \times 24)}{1.3^2} = 32{,}000 \text{ psi}$$

The updated design is $\mathbf{x}_1 = \mathbf{x}_0 + \alpha_0\, \mathbf{d}_0$ or

$$\mathbf{x}^1 = \begin{pmatrix} 1 \\ 3 \end{pmatrix} + 0.2 \begin{pmatrix} -1/\sqrt{5} \\ -2/\sqrt{5} \end{pmatrix} = \begin{pmatrix} 0.91056 \\ 2.82111 \end{pmatrix}$$

The updated value of the bending stress is given by $\sigma_1 = 71342$ psi.

3.5 The Steepest Descent Method

We will begin our discussion of numerical methods for unconstrained minimization of a function of n variables by considering the steepest descent method. Many of the concepts discussed in connection with this method are of relevance when discussing other numerical methods in the text.

Direction Vector

Let $\mathbf{x}_k$ be the current point at the kth iteration. $k = 0$ corresponds to the starting point. We need to choose a downhill direction $\mathbf{d}$ and then a step size $\alpha > 0$ such that the new point $\mathbf{x}_k + \alpha\mathbf{d}$ is better. That is, we desire $f(\mathbf{x}_k + \alpha\mathbf{d}) < f(\mathbf{x}_k)$. To see how $\mathbf{d}$ should be chosen, we will expand f about $\mathbf{x}_k$ using Taylor's expansion

$$f(\mathbf{x}_k + \alpha\mathbf{d}) = f(\mathbf{x_k}) + \alpha\, \nabla f(\mathbf{x_k})^{\mathrm{T}}\mathbf{d} + \mathbf{O}(\alpha^2)$$

Thus, the change δf in f given as $\delta f = f(\mathbf{x}_k + \alpha\mathbf{d}) - f(\mathbf{x}_k)$ is

$$\delta f = \alpha \nabla f(\mathbf{x}_k)^{\mathrm{T}}\mathbf{d} + O(\alpha^2)$$

For small enough α the term $O(\alpha^2)$ is dominated. Consequently, we have

$$\delta f \cong \alpha \nabla f(\mathbf{x}_k)^{\mathrm{T}}\mathbf{d}$$

For a reduction in f or $\delta f < 0$, we require $\mathbf{d}$ to be a *descent direction* or a direction that satisfies

$$\nabla f(\mathbf{x}_k)^{\mathrm{T}}\mathbf{d} < 0 \tag{3.10}$$

The steepest descent method is based on choosing $\mathbf{d}$ at the kth iteration, which we will denote as $\mathbf{d}_k$, as

$$\mathbf{d}_k = -\nabla f(\mathbf{x}_k) \tag{3.11}$$

This choice of the direction vector evidently satisfies (3.10), since $\nabla f(\mathbf{x}_k)^{\mathrm{T}}\, \mathbf{d}_k = -\,||\nabla f(\mathbf{x}_k)||^2 < 0$. The direction in (3.11) will be hereafter referred to as the *steepest descent direction*. Further insight can be gained by recognizing the geometrical significance of the gradient vector ∇f as given in the following.

Geometric Significance of the Gradient Vector

In $x_1 - x_2$ space (when $n = 2$) the gradient vector $\nabla f(\mathbf{x}_k)$ is normal or perpendicular to the constant objective curves passing through $\mathbf{x}_k$. More generally, if we construct a tangent hyperplane to the constant objective hypersurface, then the gradient vector will be normal to this plane. To prove this, consider any curve on the constant objective surface emanating from $\mathbf{x}_k$. Let s be the arc length along the curve. Along the curve we can write $\mathbf{x} = \mathbf{x}(s)$ with $\mathbf{x}(0) \equiv \mathbf{x}_k$. Since $f(\mathbf{x}(s)) = \text{constant}$,

$$\frac{d}{ds} \; f(x(s))]_{s=0} = \nabla f(\mathbf{x}^k)^{\mathrm{T}}\dot{\mathbf{x}}(0) = 0$$

where $\dot{\mathbf{x}}(0) = \lim\limits_{s \to 0} \frac{\mathbf{x}(s)-\mathbf{x}(0)}{s}$. Since $\mathbf{x}(s) - \mathbf{x}(0)$ is the chord joining $\mathbf{x}_k$ to $\mathbf{x}(s)$, the limit of this chord as $s \to 0$ is a vector that is tangent to the constant objective surface. Thus, $\dot{\mathbf{x}}\,(0)$ is a vector in the tangent plane. The preceding equation states that $\nabla f(\mathbf{x}_k)$ is normal to this vector. We arrived at this result considering a single curve through $\mathbf{x}_k$. Since this is true for all curves through $\mathbf{x}_k$, we conclude that $\nabla f(\mathbf{x}_k)$ is normal to the tangent plane. See Fig. 3.3.

Further, $\nabla f(\mathbf{x}_k)$ points in the direction of increasing function value.

Example 3.7
Given $f = x_1\, x_2^2$, $\mathbf{x}_0 = (1, 2)^{\mathrm{T}}$.

(i) Find the steepest descent direction at $\mathbf{x}_0$.
(ii) Is $\mathbf{d} = (-1, 2)^{\mathrm{T}}$ a direction of descent?

We have the gradient vector $\nabla f = (x_2^2, 2x_1\, x_2)^{\mathrm{T}}$. The steepest descent direction at $\mathbf{x}_0$ is the negative of the gradient vector that gives $(-4, -4)^{\mathrm{T}}$. Now, in (ii), to check if $\mathbf{d}$ is a descent direction, we need to check the dot product of $\mathbf{d}$ with the gradient vector as given in (3.10). We have $\nabla f(\mathbf{x}_0)^{\mathrm{T}}\, \mathbf{d} = (4, 4)\,(-1, 2)^{\mathrm{T}}$ $= +4 > 0$. Thus, $\mathbf{d}$ is not a descent direction.

Line Search

We now return to the steepest descent method. After having determined a direction vector $\mathbf{d}_k$ at the point $\mathbf{x}_k$, the question is "how far" to go along this direction. Thus, we need to develop a numerical procedure to determine the step size α_k along $\mathbf{d}_k$. This problem can be resolved once we recognize that it is one-dimensional – it involves only the variable α. Examples 3.5 and 3.6 illustrate this point. Specifically, as we move along $\mathbf{d}_k$, the design variables and the objective function depend only on α as

$$\mathbf{x}_k + \alpha \mathbf{d}_k \equiv \mathbf{x}(\alpha) \tag{3.12}$$

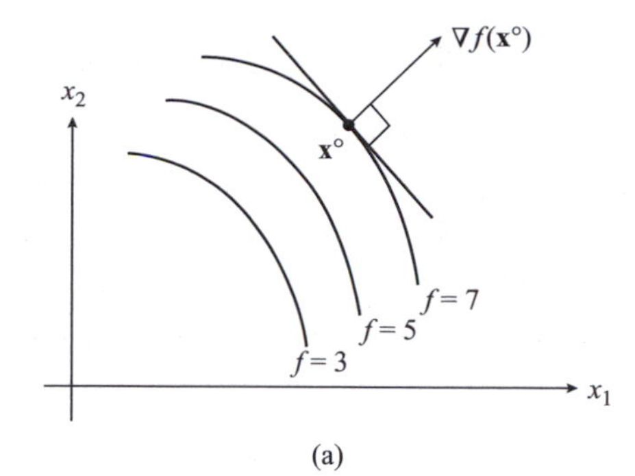

(a)

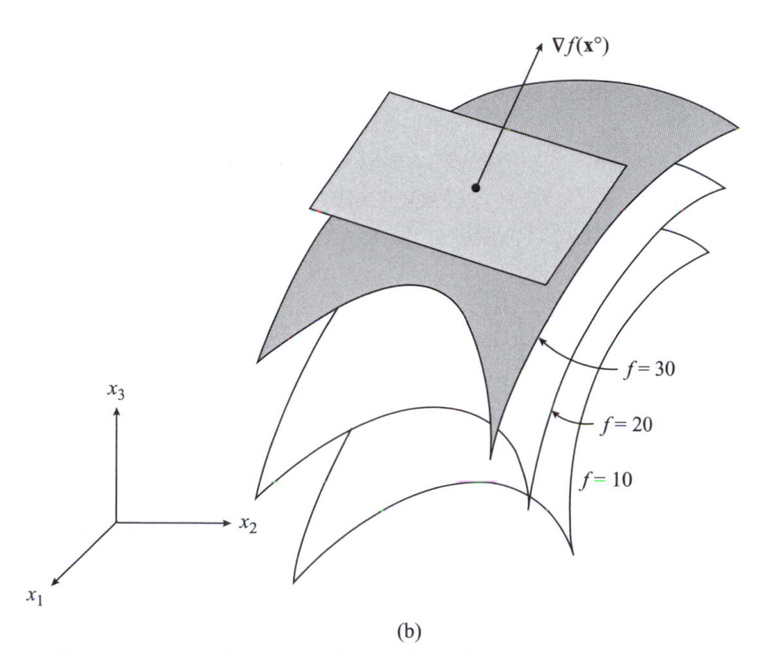

(b)

Figure 3.3. Geometric significance of the gradient vector in two and three dimensions.

and

$$f(\mathbf{x}_k + \alpha \mathbf{d}_k) \equiv f(\alpha) \tag{3.13}$$

The notation $f(\alpha) \equiv f(\mathbf{x}_k + \alpha \, \mathbf{d}_k)$ will be used in this text. The slope or derivative $f'(\alpha) = \mathrm{d}f/\mathrm{d}\alpha$ is called the *directional derivative* of f along the direction $\mathbf{d}$ and is given by the expression

$$\frac{df(\hat{\alpha})}{d\alpha} = \nabla f(\mathbf{x}_k + \hat{\alpha} \, \mathbf{d}_k)^{\mathrm{T}} \ \ \mathbf{d}^k \tag{3.14}$$

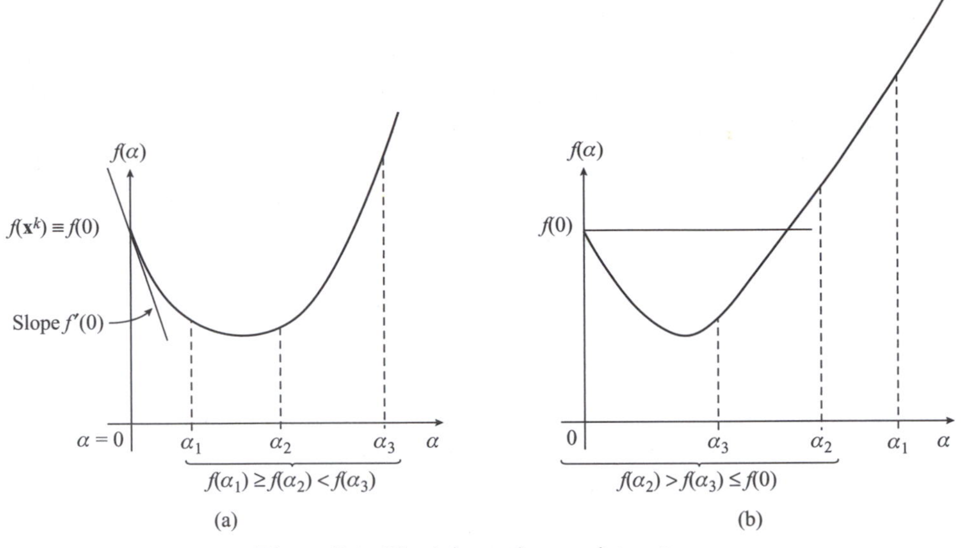

Figure 3.4. Obtaining a three-point pattern.

In the steepest descent method, the direction vector is $-\nabla f(\mathbf{x}_k)$ resulting in the slope at the current point $\alpha = 0$ being

$$\frac{df(0)}{d\alpha} = [\nabla f(\mathbf{x}_k)]^{\mathrm{T}}\ [-\nabla f(\mathbf{x}_k)]\ = -\ \|\ \nabla f(\mathbf{x}_k)\ \|^2 < 0 \tag{3.15}$$

implying a move in a downhill direction. See also Eq. (1.5b) in Chapter 1.

Exact Line Search: Here, we may choose α so as to minimize $f(\alpha)$:

$$\underset{\alpha \geq 0}{\text{minimize}}\ \ f(\alpha) \equiv f(\mathbf{x}_k + \alpha\,\mathbf{d}_k) \tag{3.16}$$

We will denote α_k to be the solution to the preceding problem. The problem in (3.16) is called "exact line search" since we are minimizing f along the line $\mathbf{d}_k$. It is also possible to obtain a step size α_k from approximate line search procedures as discussed in Section 3.9. We give details in performing exact line search with "Quadratic Fit."

The main step is to obtain a "three-point pattern." Once the minimum point is bracketed, the quadratic fit or other techniques discussed in Chapter 2 may be used to obtain the minimum. Consider a plot of $f(\alpha)$ versus α as shown in Fig. 3.4a. Bear in mind that $f(0) \equiv f(\mathbf{x}_k)$. Since the slope of the curve at $\alpha = 0$ is negative, we know that the function immediately decreases with $\alpha > 0$. Now, choose an initial step $\alpha_1 \equiv s$. If $f(\alpha_1) < f(0)$, then we march forward by taking steps of $\alpha_{i+1} = \alpha_i + \frac{s}{\tau^i}$, $i = 1, 2, \ldots$, where $\tau = 0.618034$ is the golden section ratio. We continue this march until the function starts to increase and obtain $f(\alpha_{i+1}) \geq f(\alpha_i) < f(\alpha_{i-1})$ whereupon $[\alpha_{i-1}, \alpha_i, \alpha_{i+1}]$ is the three-point pattern.

However, if the initial step α_1 results in $f(\alpha_1) > f(0)$ as depicted in Fig. 3.4b, then we define points within $[0, \alpha_1]$ by $\alpha_{i+1} = \tau\, \alpha_i, i = 1, 2, \ldots$ until $f(\alpha_i) > f(\alpha_{i+1}) \leq f(0)$. We will then have $[0, \alpha_{i+1}, \alpha_i]$ as our three-point pattern.

Again, once we have identified a three-point pattern, the quadratic fit or other techniques (see Chapter 2) may be used to determine the minimum α_k. The reader can see how the aforementioned logic is implemented in program STEEPEST.

Exact line search as described previously, based on solving Problem in (3.16), may entail a large number of function calls. A *sufficient decrease* or Armijo condition is sometimes more efficient and considerably simplifies the computer code. Further details on this are given later in Section 3.9, so as not to disturb the presentation of the methods. However, all computer codes accompanying this text use exact line search strategy, which is the most robust approach for general problems.

Stopping Criteria

Starting from an initial point, we determine a direction vector and a step size, and obtain a new point as $\mathbf{x}_{k+1} = \mathbf{x}_k + \alpha_k\, \mathbf{d}_k$. The question now is to know when to stop the iterative process. Two stopping criteria are discussed in the following and are implemented in the computer programs.

(I) Before performing line search, the necessary condition for optimality is checked:

$$\| \nabla f(\mathbf{x}_k) \| \leq \varepsilon_G \tag{3.17}$$

where ε_G is a tolerance on the gradient and is supplied by the user. If (3.17) is satisfied, then the process is terminated. We note that the gradient can also vanish for a local maximum point. However, we are using a descent strategy where the value of f reduces from iteration to iteration – as opposed to a root-finding strategy. Thus, the chances of converging to a local maximum point are remote.

(II) We should also check successive reductions in f as a criterion for stopping. Here, we check if

$$\left| f(\mathbf{x}_k + 1) - f(\mathbf{x}_k) \right| \leq \varepsilon_A + \varepsilon_R \left| f(\mathbf{x}_k) \right| \tag{3.18}$$

where ε_A = absolute tolerance on the change in function value and ε_R = relative tolerance. Only if (3.18) is satisfied for two consecutive iterations, is the descent process stopped.

The reader should note that both a gradient check and a function check are necessary for robust stopping criteria. We also impose a limit on the number of iterations. The algorithm is presented as follows.

Steepest Descent Algorithm

1. Select a starting design point $\mathbf{x}_0$ and parameters $\varepsilon_G, \varepsilon_A, \varepsilon_R$. Set iteration index $k = 0$.
2. Compute $\nabla f(\mathbf{x}_k)$. Stop if $||\nabla f(\mathbf{x}_k)|| \le \varepsilon_G$. Otherwise, define a normalized direction vector $\mathbf{d}_k = -\nabla f(\mathbf{x}_k)/||\nabla f(\mathbf{x}_k)||$.
3. Obtain α_k from exact or approximate line search techniques. Update $\mathbf{x}_{k+1} = \mathbf{x}_k + \alpha_k\, \mathbf{d}_k$.
4. Evaluate $f(\mathbf{x}_{k+1})$. Stop if (3.18) is satisfied for two successive iterations. Otherwise set $k = k + 1$, $\mathbf{x}_k = \mathbf{x}_{k+1}$, and go to step 2.

Convergence Characteristics of the Steepest Descent Method

The steepest descent method zigzags its way towards the optimum point. This is because each step is orthogonal to the previous step. This follows from the fact that α_k is obtained by minimizing $f(\mathbf{x}_k + \alpha\, \mathbf{d}_k)$. Thus, $\mathrm{d}/\mathrm{d}\alpha\ \{f(\mathbf{x}_k + \alpha\, \mathbf{d}_k)\} = 0$ at $\alpha = \alpha_k$. Upon differentiation, we get $\nabla f(\mathbf{x}_k + \alpha_k\, \mathbf{d}_k)^{\mathrm{T}}\, \mathbf{d}_k = 0$ or $\nabla f(\mathbf{x}_{k+1})^{\mathrm{T}}\, \mathbf{d}_k = 0$. Since $\mathbf{d}_{k+1} = -\nabla f(\mathbf{x}_{k+1})$, we arrive at the result $\mathbf{d}_{k+1}^{\mathrm{T}} \mathbf{d}_k = 0$. Note that this result is true only if the line search is performed exactly.

Application of the steepest descent method to quadratic functions leads to further understanding of the method. Consider the quadratic function

$$f = x_1^2 + a\, x_2^2$$

The steps taken by the steepest descent method to the optimum are shown in Fig. 3.5. Larger is the value of a, more elongated are the contours of f, and slower is the convergence rate. It has been proven that the speed of convergence of the method is related to the spectral condition number of the Hessian matrix. The spectral condition number κ of a symmetric positive definite matrix $\mathbf{A}$ is defined as the ratio of the largest to the smallest eigenvalue, or as

$$\kappa = \frac{\lambda_{\max}}{\lambda_{\min}} \tag{3.19}$$

For well-conditioned Hessian matrices, the condition number is close to unity, contours are more circular, and the method is at its best. Higher the condition number, more ill-conditioned is the Hessian, contours are more elliptical, more is the amount of zigzagging as the optimum is approached, smaller is the step sizes and, thus, poorer is the rate of convergence. For the function $f = x_1^2 + a\, x_2^2$ considered

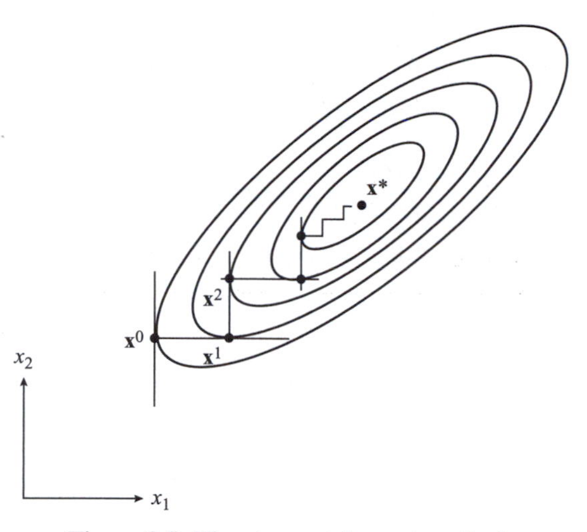

Figure 3.5. The steepest descent method.

previously, we have

$$\nabla^2 f = \begin{bmatrix} 2 & 0 \\ 0 & 2a \end{bmatrix}$$

Thus, $\kappa = a$. Convergence is fastest when $a = 1$. For a nonquadratic function, the conditioning of the Hessian matrix at the optimum $\mathbf{x}^*$ affects the convergence rate. While details may be found in references given at the end of the chapter, and specific proofs must incorporate the line search strategy as well, the main aspects regarding convergence of the steepest descent method are:

(i) The method is convergent from *any* starting point under mild conditions. The term "globally convergent" is used, but this is not to be confused with finding the global minimum – the method only finds a local minimum. Luenberger [1965] contains an excellent discussion of global convergence requirements for unconstrained minimization algorithms. In contrast, pure Newton's method discussed next does not have this theoretical property.

(ii) The method has a *linear rate* of convergence. Specifically, if the Hessian at the optimum is positive definite with spectral condition number κ, then the sequence of function values at each iteration satisfy

$$f_{k+1} - f^* \approx \frac{(\kappa - 1)^2}{(\kappa + 1)^2} [f_k - f^*] \tag{3.20}$$

where f^* is the optimum value. The preceding inequality can be written as $e_{k+1} \leq Ce_k^r$ where e represents error and $r = 1$ indicates linear rate of convergence.

Typically, the steepest descent method takes only a few iterations to bring a far-off starting point into the "optimum region" but then takes hundreds of iterations to make very little progress toward the solution.

Scaling

The aforementioned discussion leads to the concept of scaling the design variables so that the Hessian matrix is well conditioned. For the function $f = x_1^2 + a\,x_2^2$, we can define new variables y_1 and y_2 as

$$y_1 = x_1,\ y_2 = \sqrt{a}x_2$$

which leads to the function in y variables given by $g = y_1^2 + y_2^2$. The Hessian of the new function has the best conditioning possible with contours corresponding to circles.

In general, consider a function $f = f(\mathbf{x})$. If we scale the variables as

$$\mathbf{x} = \mathbf{T}\ \mathbf{y} \tag{3.21}$$

then the new function is $g(\mathbf{y}) \equiv f(\mathbf{T}\ \mathbf{y})$, and its gradient and Hessian are given by

$$\nabla g = \mathbf{T}^{\mathrm{T}}\,\nabla f \tag{3.22}$$

$$\nabla^2 g = \mathbf{T}^{\mathrm{T}}\,\nabla^2 f\,\mathbf{T} \tag{3.23}$$

Usually $\mathbf{T}$ is chosen as a diagonal matrix. The idea is to choose $\mathbf{T}$ such that the Hessian of g has a condition number close to unity.

Example 3.8

Consider $f = (x_1 - 2)^4 + (x_1 - 2x_2)^2$, $\mathbf{x}_0 = (0, 3)^{\mathrm{T}}$. Perform one iteration of the steepest descent method.

We have $f(\mathbf{x}_0) = 52$, $\nabla f(\mathbf{x}) = [4(x_1 - 2)^3 + 2(x_1 - 2x_2), -4(x_1 - 2x_2)]^{\mathrm{T}}$. Thus, $\mathbf{d}_0 = -\ \nabla f(\mathbf{x}_0) = [44, -24]^{\mathrm{T}}$. Normalizing the direction vector to make it a unit vector, we have $\mathbf{d}_0 = [0.8779, -0.4789]^{\mathrm{T}}$. The solution to the line search problem minimize $f(\alpha) = f(\mathbf{x}_0 + \alpha\ \mathbf{d}_0)$ with $\alpha > 0$ yields $\alpha_0 = 3.0841$. Thus, the new point is $\mathbf{x}_1 = \mathbf{x}_0 + \alpha_0\ \mathbf{d}_0 = [2.707, 1.523]^{\mathrm{T}}$, with $f(\mathbf{x}_1) = 0.365$. The second iteration will now proceed by evaluating $\nabla(\mathbf{x}_{k+1})$, etc.

3.6 The Conjugate Gradient Method

Conjugate gradient methods are a dramatic improvement over the steepest descent method. Conjugate gradient methods can find the minimum of a quadratic function of n variables in n iterations. In contrast, no such property exists with the steepest descent method. It turns out that conjugate gradient methods are also powerful on general functions. Conjugate gradient methods were first presented in [Fletcher and Powell, 1963].

Consider the problem of minimizing a quadratic function

$$\text{minimize } q(\mathbf{x}) = \tfrac{1}{2}\,\mathbf{x}^{\mathrm{T}}\,\mathbf{A}\,\mathbf{x} + \mathbf{c}^{\mathrm{T}}\,\mathbf{x} \tag{3.24}$$

where we assume $\mathbf{A}$ is symmetric and positive definite. We define conjugate directions, or directions that are mutually conjugate with respect to $\mathbf{A}$, as vectors that satisfy

$$\mathbf{d}^{i\mathrm{T}}\mathbf{A}\mathbf{d}^{j} = 0, \quad i \neq j, \quad 0 \leq i,\ j \leq n \tag{3.25}$$

The method of conjugate directions is as follows. We start with an initial point $\mathbf{x}_0$ and a set of conjugate directions $\mathbf{d}^0, \mathbf{d}^1, \ldots, \mathbf{d}^{n-1}$. We minimize $q(\mathbf{x})$ along $\mathbf{d}^0$ to obtain $\mathbf{x}^1$. Then, from $\mathbf{x}^1$, we minimize $q(\mathbf{x})$ along $\mathbf{d}^2$ to obtain $\mathbf{x}^2$. Lastly, we minimize $q(\mathbf{x})$ along $\mathbf{d}^{n-1}$ to obtain $\mathbf{x}^n$. The point $\mathbf{x}^n$ is the minimum solution. That is, the minimum to the quadratic function has been found in n searches. In the algorithm that follows, the *gradients* of q are used to generate the conjugate directions.

We denote $\mathbf{g}$ to be the gradient of q, with $\mathbf{g}_k = \nabla q(\mathbf{x}_k) = \mathbf{A}\,\mathbf{x}_k + \mathbf{c}$. Let $\mathbf{x}_k$ be the current point with $k =$ an iteration index. The first direction $\mathbf{d}_0$ is chosen as the steepest descent direction, $-\mathbf{g}_0$. We proceed to find a new point $\mathbf{x}_{k+1}$ by minimizing $q(\mathbf{x})$ along $\mathbf{d}_k$. Thus

$$\mathbf{x}_{k+1} = \mathbf{x}_k + \alpha_k \mathbf{d}_k \tag{3.26}$$

where α_k is obtained from the line search problem: minimize $f(\alpha) = q(\mathbf{x}_k + \alpha\mathbf{d}_k)$. Setting $\mathrm{d}q(\alpha)/\mathrm{d}\alpha = 0$ yields

$$\alpha_k = -\frac{\mathbf{d}_k^{\mathrm{T}}\mathbf{g}_k}{\mathbf{d}_k^{\mathrm{T}}\mathbf{A}\,\mathbf{d}_k} \tag{3.27}$$

Also, the exact line search condition $\mathrm{d}q(\alpha)/\mathrm{d}\alpha = 0$ yields

$$\mathbf{d}_k^{\mathrm{T}}\mathbf{g}_{k+1} = 0 \tag{3.28}$$

Now the key step: we choose $\mathbf{d}_{k+1}$ to be of the form

$$\mathbf{d}_{k+1} = -\mathbf{g}_{k+1} + \beta_k \mathbf{d}_k \tag{3.29}$$

The aforementioned represents a "deflection" in the steepest descent direction $-\mathbf{g}_{k+1}$. This is illustrated in Fig. 3.6. Requiring $\mathbf{d}_{k+1}$ to be conjugate to $\mathbf{d}_k$ or $\mathbf{d}_{k+1}^{\mathrm{T}}\,\mathbf{A}\,\mathbf{d}_k = 0$, gives

$$\mathbf{g}_{k+1}^{\mathrm{T}}\mathbf{A}\,\mathbf{d}_k + \beta_k \mathbf{d}_k^{\mathrm{T}}\mathbf{A}\,\mathbf{d}_k = 0$$

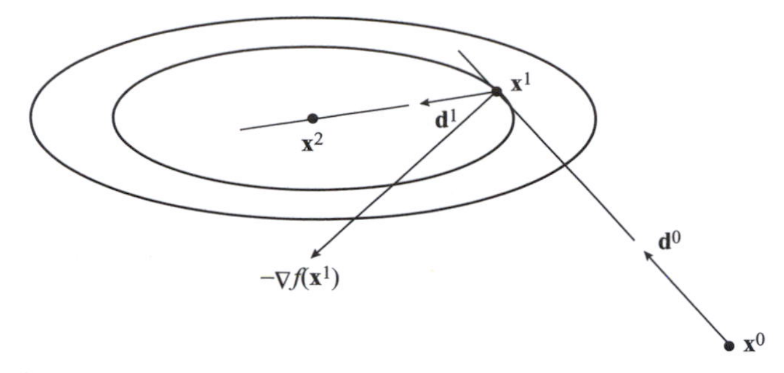

Figure 3.6. Conjugate directions for a quadratic in two variables.

From (3.20), $\mathbf{d}_k = (\mathbf{x}_{k+1} - \mathbf{x}_k)/\alpha_k$. Thus, $\mathbf{A}\,\mathbf{d}_k = (\mathbf{g}_{k+1} - \mathbf{g}_k)/\alpha_k$. The preceding equation now gives

$$\beta_k = \frac{\mathbf{g}_{k+1}^{\mathrm{T}}\,(\mathbf{g}_{k+1} - \mathbf{g}_k)}{\alpha_k\,\mathbf{d}_k^{\mathrm{T}}\,\mathbf{A}\,\mathbf{d}_k} \tag{3.30}$$

Using (3.28) and (3.29) with k replaced by $k-1$, we get

$$\mathbf{d}_k^{\mathrm{T}}\mathbf{g}_k = -\mathbf{g}_k^{\mathrm{T}}\mathbf{g}_k$$

Thus, (3.27) gives

$$\alpha_k = \frac{\mathbf{g}_k^{\mathrm{T}}\mathbf{g}_k}{\mathbf{d}_k^{\mathrm{T}}\,\mathbf{A}\,\mathbf{d}_k} \tag{3.31}$$

Substituting for α_k from the preceding equation into (3.30) yields

$$\beta_k = \frac{\mathbf{g}_{k+1}^{\mathrm{T}}\,(\mathbf{g}_{k+1} - \mathbf{g}_k)}{\mathbf{g}_k^{\mathrm{T}}\,\mathbf{g}_k} \tag{3.32}$$

ALGORITHM: We may now implement the conjugate gradient method as follows. Starting with $k = 0$, an initial point $\mathbf{x}_0$ and $\mathbf{d}_0 = -\nabla q(\mathbf{x}_0)$, we perform line search – that is, determine α_k from (3.31) and then obtain $\mathbf{x}_{k+1}$ from (3.26). Then, β_k is obtained from (3.32) and the next direction $\mathbf{d}_{k+1}$ is given from (3.29).

Since the calculation of β_k in (3.32) is independent of the $\mathbf{A}$ and $\mathbf{c}$ matrices of the quadratic function, it was natural to apply the preceding method to nonquadratic functions. However, numerical line search needs to be performed to find α_k instead of the closed form formula in (3.31). Of course, the finite convergence in n steps is valid only for quadratic functions. Further, in the case of general functions, a *restart* is made every n iterations wherein a steepest descent step is taken. The use of (3.32) is referred to as the *Polak–Rebiere* algorithm. If we consider

$$\begin{aligned}\mathbf{g}_{k+1}^{\mathrm{T}}\mathbf{g}_k &= \mathbf{g}_{k+1}^{\mathrm{T}}(-\mathbf{d}_k + \beta_{k-1}\mathbf{d}_{k-1}) = \beta_{k-1}\ \mathbf{g}_{k+1}^{\mathrm{T}}\mathbf{d}_{k-1} \\ &= \beta_{k-1}(\mathbf{g}_k^{\mathrm{T}} + \alpha_k\, \mathbf{d}_k^{\mathrm{T}}\mathbf{A})\,\mathbf{d}_{k-1} \\ &= 0\end{aligned}$$

we obtain

$$\beta_k = \frac{\mathbf{g}_{k+1}^{\mathrm{T}}\mathbf{g}_{k+1}}{\mathbf{g}_k^{\mathrm{T}}\,\mathbf{g}_k} \tag{3.33}$$

which is the *Fletcher–Reeves* version [Fletcher and Reeves 1964].

Example 3.9

Consider $f = x_1^2 + 4x_2^2$, $\mathbf{x}_0 = (1, 1)^{\mathrm{T}}$. We will perform two iterations of the conjugate gradient algorithm. The first step is the steepest descent iteration. Thus,

$$\mathbf{d}_0 = -\nabla f(x_0) = -(2, 8)^{\mathrm{T}}$$

In the example here, the direction vectors are not normalized to be unit vectors, although this is done in the programs:

$$f(\alpha) = f(\mathbf{x}_0 + \alpha\mathbf{d}_0) = (1 - 2\alpha)^2 + 4(1 - 8\alpha)^2$$

which yields $\alpha_0 = 0.1307692$, $\mathbf{x}_1 = \mathbf{x}_0 + \alpha_0\ \mathbf{d}_0 = (0.7384615, -0.0461538)^{\mathrm{T}}$. For the next iteration, we compute

$$\beta_0 = \frac{\left\|\nabla f(\mathbf{x}^1)\right\|^2}{\left\|\nabla f(\mathbf{x}^0)\right\|^2} = 2.3176/68 = 0.0340828$$

$$\mathbf{d}_1 = -\nabla f^{\mathrm{T}}(\mathbf{x}_1) + \beta_0\mathbf{d}_0 = \begin{pmatrix} -1.476923 \\ 0.369231 \end{pmatrix} + 0.0340828\begin{pmatrix} -2 \\ -8 \end{pmatrix} = \begin{pmatrix} -1.54508 \\ 0.09656 \end{pmatrix}$$

$$f(\alpha) = f(\mathbf{x}_1 + \alpha\mathbf{d}_1) = (0.7384615 - 1.54508\,\alpha)2 + 4(-0.0461538 + 0.09656\,\alpha)^2$$

which yields

$$\alpha_1 = 0.477941$$

$$\mathbf{x}_2 = \mathbf{x}_1 + \alpha_1\mathbf{d}_1 = \begin{pmatrix} 0.7384615 \\ -0.0461538 \end{pmatrix} + 0.477941\begin{pmatrix} -1.54508 \\ 0.09656 \end{pmatrix} = \begin{pmatrix} 0 \\ 0 \end{pmatrix}$$

As expected from theory, convergence is reached after $n = 2$ searches.

Solution of Simultaneous Equations in Finite Element Analysis

In finite element analysis, the equilibrium condition can be obtained by minimizing the potential energy Π

$$\Pi = \tfrac{1}{2}\mathbf{Q}^{\mathrm{T}}\mathbf{K}\,\mathbf{Q} - \mathbf{Q}^{\mathrm{T}}\mathbf{F} \tag{3.34}$$

with respect to the displacement vector $\mathbf{Q} = [Q_1, Q_2, \ldots, Q_N]^{\mathrm{T}}$, where N = number of nodes or degrees of freedom in the model. For a one-dimensional system, we have $n = N$, while for a three-dimensional system, $n = 3\,N$. $\mathbf{K}$ is a positive definite stiffness matrix and $\mathbf{F}$ is a load vector. The popular approach is to apply the necessary conditions and solve the system of simultaneous equations

$$\mathbf{KQ} = \mathbf{F} \tag{3.35}$$

While the basic idea is to use Gaussian elimination, the special structure of $\mathbf{K}$ and the manner in which it is formed is exploited while solving (3.35). Thus, we have banded, skyline and frontal solvers. However, the conjugate gradient method applied to the function Π is also attractive and is used in some codes, especially when $\mathbf{K}$ is dense – that is, when $\mathbf{K}$ does not contain large number of zero elements, causing sparse Gaussian elimination solvers to become unattractive. The main attraction is that the method only requires two vectors $\nabla\Pi(\mathbf{Q}_k)$ and $\nabla\Pi((\mathbf{Q}_{k+1})$ – where β_k in (3.33) is computed. Furthermore, computation of $\nabla\Pi = \mathbf{KQ} - \mathbf{F}$ does not require the entire $\mathbf{K}$ matrix as it is assembled from "element matrices." A brief example is given in what follows to illustrate the main idea.

Consider a one-dimensional problem in elasticity as shown in Fig. 3.7a. The bar is discretized using finite elements as shown in Fig. 3.7b. The model consists of NE elements and $N = NE + 1$ nodes. The displacement at node I is denoted by Q_I. Each finite element is two-noded as shown in Fig. 3.7c, and the "element displacement vector" of the jth element is denoted by $\mathbf{q}^{(j)} = [Q_j, Q_{j+1}]^{\mathrm{T}}$. The key point is that the $(N \times N)$ global stiffness matrix $\mathbf{K}$ consists of (2×2) element stiffness matrices $\mathbf{k}$ as shown in Fig. 3.7d. Thus, evaluation of $\nabla\Pi = \mathbf{KQ} - \mathbf{F}$ does not require forming the entire $\mathbf{K}$ matrix – instead, for each element j, $j = 1, 2, \ldots, NE$, we can compute the (2×1) vector $\mathbf{k}^{(j)}.\mathbf{q}^{(j)}$ and place it in the jth and $j + 1$th locations of $\nabla\Pi$. This is done for each element, with overlapping entries being added. Thus, we *assemble* the gradient vector $\nabla\Pi$ without forming the global stiffness matrix $\mathbf{K}$. Computation of step size α_k requires computation of $\mathbf{d}^{\mathrm{T}}\,\mathbf{K}\,\mathbf{d}$, where $\mathbf{d}$ is a conjugate direction. This computation can be done at the element level as $\sum_{j=1}^{NE} \mathbf{d}^{(j)^{\mathrm{T}}}\,\mathbf{k}^{(j)}\,\mathbf{d}^{(j)}$, where $\mathbf{d}^{(j)} = [\mathbf{d}_j, \mathbf{d}_{j+1}]^{\mathrm{T}}$.

Preconditioning the $\mathbf{K}$ matrix so that it has a good condition number is important in this context since for very large size $\mathbf{K}$, it is necessary to obtain a near-optimal solution rapidly.

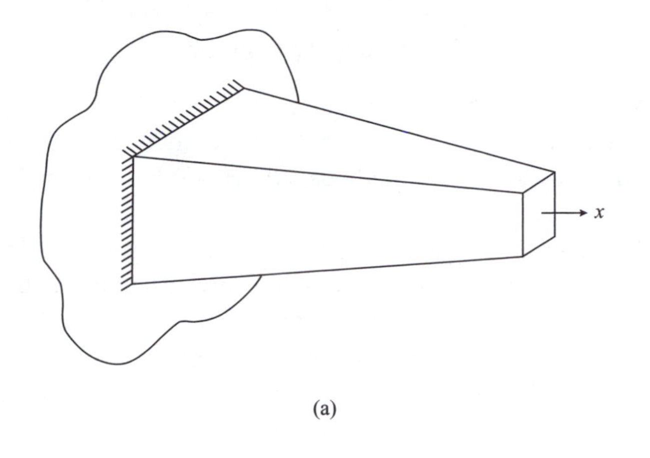

(a)

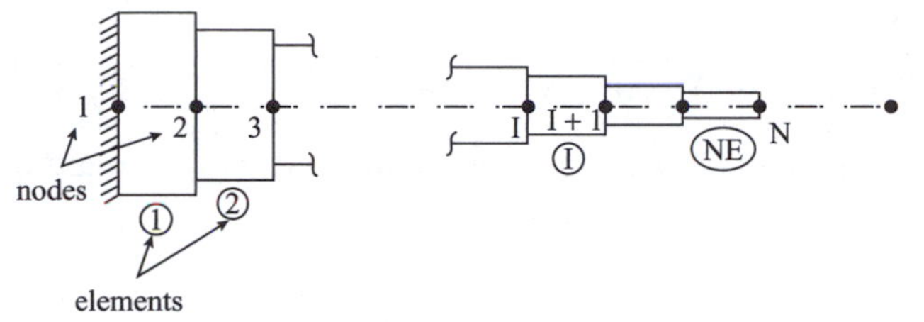

(b)

I ——— I + 1 $\quad \mathbf{k}^{(I)} = \begin{bmatrix} \mathbf{k}_{11} & \mathbf{k}_{12} \\ \mathbf{k}_{21} & \mathbf{k}_{22} \end{bmatrix}$

(c)

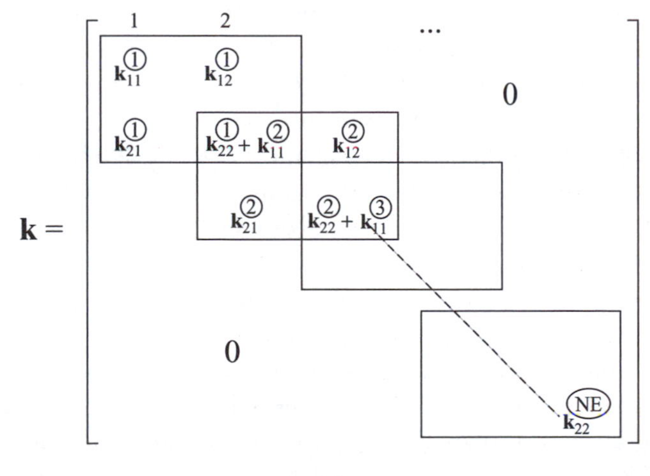

(d)

Figure 3.7. One-dimensional finite elements.

3.7 Newton's Method

Although, as we will show, direct application of Newton's method is not a robust approach to minimization, the concepts behind the method form a stepping stone to other powerful Newton-based methods, which will be discussed in subsequent sections.

First of all, Newton's method uses second derivatives or the Hessian matrix. This is in contrast to the methods of steepest descent and conjugate directions, which are first-order methods as they require only first derivatives or gradient information. Not surprisingly, Newtons's method, *when it converges*, converges at a faster rate than first-order methods. The idea is to construct a quadratic approximation to the function $f(\mathbf{x})$ and minimize the quadratic. Thus, at a current point $\mathbf{x}_k$, we construct the quadratic approximation

$$q(\mathbf{x}) = f(\mathbf{x}_k) + \nabla f(\mathbf{x}_k)^{\mathrm{T}}(\mathbf{x} - \mathbf{x}_k) + \tfrac{1}{2}(\mathbf{x} - \mathbf{x}_k)^{\mathrm{T}}\nabla^2 f(\mathbf{x}_k)(\mathbf{x} - \mathbf{x}_k) \tag{3.36}$$

Assuming $\nabla^2 f(\mathbf{x}_k)$ is positive definite, the minimum of $q(\mathbf{x})$ is found by setting $\nabla q = \mathbf{0}$, which yields

$$[\nabla^2 f(\mathbf{x}_k)]\mathbf{d}_k = -\nabla f(\mathbf{x}_k) \tag{3.37}$$

where $\mathbf{d}_k \equiv \mathbf{x}_{k+1} - \mathbf{x}_k$. Thus, we solve (3.37) for $\mathbf{d}_k$ and then obtain a new point

$$\mathbf{x}_{k+1} = \mathbf{x}_k + \mathbf{d}_k \tag{3.38}$$

We then recompute the gradient and Hessian at the new point, and again use (3.37) and (3.38) to update the point. Thus, given a starting point $\mathbf{x}_0$, we generate a sequence of points $\{\mathbf{x}_k\}$. The limit point of this sequence is the optimum $\mathbf{x}^*$ where $\nabla f(\mathbf{x}^*) = \mathbf{0}$. Of course, if the function f is a quadratic with a positive definite Hessian matrix, then Newton's method will converge in one step. For general functions, however, Newton's method need not converge unless the starting point is "close" to the optimum. There are two main reasons for this.

Firstly, if the function is highly nonlinear, then even a quadratic approximation may be a poor approximation of the function. Such a situation is shown in Fig. 3.8 for a function of one variable. We see from the figure that the minimum of the quadratic leads to a point that has an even higher function value than the current point.

Secondly, the sufficiency conditions that require that $\nabla^2 f(\mathbf{x}^*)$ be positive definite at $\mathbf{x}^*$ place no restrictions on the Hessian at a point $\mathbf{x}_k$ during the iterative process. Thus, if $\nabla^2 f(\mathbf{x}_k)$ is not positive definite or if it is singular, then $q(\mathbf{x})$ need not have a minimum – for example, it can be saddle-shaped or even flat. Equation (3.37) may not be solvable, or if solvable, may yield a worse point than the current point. Of course, if f is strictly convex, then its Hessian is positive definite and Newton's method with line search, based on minimizing $f(\mathbf{x}_k + \alpha\, \mathbf{d}_k)$ can be used effectively.

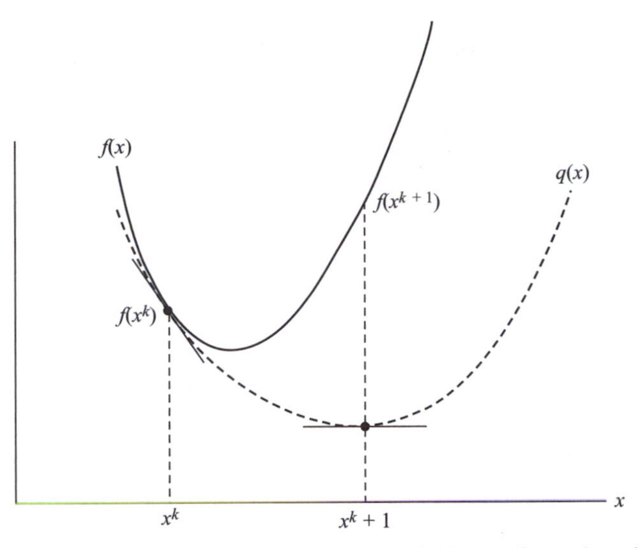

Figure 3.8. Quadratic approximation of a highly nonlinear function.

Example 3.10

Given the function: $f(x) = x_1 - x_2 + 2\ x_1\ x_2 + 2\ x_1{}^2 + x_2^2$, and $\mathbf{x}_0 = (1, 5)^{\mathrm{T}}$, determine the Newton's direction (to minimize f) at $\mathbf{x}_0$.

$$\text{We have: } \nabla\ f = [1 + 2x_2 + 4x_1, -1 + 2x_1 + 2x_2]^{\mathrm{T}}$$

$$\nabla\ f(\mathbf{x}_0) = [15, 11]^{\mathrm{T}}$$

$$\nabla\ ^2 f = \begin{bmatrix} 4 & 2 \\ 2 & 2 \end{bmatrix}$$

Thus,

$$\mathbf{d} = [\nabla^2 f]^{-1}\{-\nabla f\} = \begin{bmatrix} -2 \\ -3.5 \end{bmatrix}$$

Pure Newton's Algorithm

1. Select a starting design point $\mathbf{x}_0$ and parameters $\varepsilon_{\mathrm{G}}, \varepsilon_{\mathrm{A}}, \varepsilon_{\mathrm{R}}$. Set iteration index $k = 0$.
2. Compute $\nabla f(\mathbf{x}_k)$ and $\nabla^2 f(\mathbf{x}_k)$. Stop if $\|\nabla f(\mathbf{x}_k)\| \le \varepsilon_{\mathrm{G}}$. Otherwise, define a direction vector by solving (3.37). Do *not* normalize this direction to be a unit vector.
3. Update $\mathbf{x}_{k+1} = \mathbf{x}_k + \mathbf{d}_k$
4. Evaluate $f(\mathbf{x}_{k+1})$. Stop if (3.18) is satisfied for two successive iterations. Otherwise set $k = k + 1$, $\mathbf{x}_k = \mathbf{x}_{k+1}$, and go to step 2.

Modified Newton Methods

To overcome the two pitfalls in Newton's method as discussed in the preceding text, two modifications can be made. First, to avoid the situation as shown in Fig. 3.8, we can introduce a step-size parameter α_k as

$$\mathbf{x}_{k+1} = \mathbf{x}_k + \alpha_k \mathbf{d}_k$$

where α_k is obtained from line search problem (3.16): minimize $f(\mathbf{x}_k + \alpha\, \mathbf{d}_k)$, where $\mathbf{d}_k$ is obtained as given below, or else based on a sufficient decrease strategy as discussed in Section 3.10.

The second modification is to ensure that the direction vector $\mathbf{d}_k$ is a descent direction to the function f at $\mathbf{x}_k$. If $\mathbf{d}_k$ is a descent direction, then the line search will result in a new point with $f(\mathbf{x}_{k+1}) < f(\mathbf{x}_k)$. Thus, we must ensure that

$$\nabla f(\mathbf{x}_k)^{\mathrm{T}} \mathbf{d}_k < 0$$

or, in view of (3.37), we must ensure that

$$-\nabla f(\mathbf{x}_k)^{\mathrm{T}} [\nabla^2 f(\mathbf{x}_k)]^{-1} \nabla f(\mathbf{x}_k) < 0 \tag{3.39}$$

If $\mathbf{d}_k$ is a descent direction, then the line search will result in a new point with $f(\mathbf{x}_{k+1}) < f(\mathbf{x}_k)$. If $\nabla^2 f(\mathbf{x}_k)$ is positive definite, then so will be its inverse, and (3.39) will be satisfied. One strategy is to replace the Hessian with a symmetric positive definite matrix $\mathbf{F}_k$. A simple scheme is to use a matrix $\mathbf{F}_k$ defined by

$$\mathbf{F}_k = \nabla^2 f(\mathbf{x}_k) + \gamma \mathbf{I} \tag{3.40}$$

where γ is chosen such that all eigenvalues of $\mathbf{F}_k$ are greater than some scalar $\delta > 0$. The direction vector $\mathbf{d}_k$ is now determined from the solution of

$$[\mathbf{F}_k]\mathbf{d}_k = -\nabla f(\mathbf{x}_k) \tag{3.41}$$

Note that if δ is too close to zero, then the $\mathbf{F}_k$ will be ill-conditioned that can lead to round-off errors in the solution of (3.41). The determination of γ can be based on an actual determination of the lowest eigenvalue of $\nabla^2 f(\mathbf{x}_k)$, say by the inverse power method, or by a technique that uses an $\mathbf{L}\,\mathbf{D}\,\mathbf{L}^{\mathrm{T}}$ factorization of the Hessian. In this technique, $\mathbf{D}$ is a diagonal matrix. If all the diagonal elements $D_{ii} > 0$, then the Hessian is positive definite – otherwise, we can modify any negative D_{ii} such that it is positive, which will ensure that $\mathbf{F}_k$ is positive definite. As $\gamma \to \infty$, $\mathbf{F}_k \to \gamma \mathbf{I}$ from Eq. (3.40), and from Eq. (3.41) we have $\mathbf{d}_k \to -\frac{1}{\gamma}\nabla f(\mathbf{x}_k)$ = steepest descent direction. $\gamma = 1$ gives the pure Newton direction. Thus, γ controls both direction and step length. The steepest descent direction, as discussed above, is guaranteed to be a descent or downhill direction: Thus, if the Newton step does not lead to reduction in f, γ is increased in the method. For general functions, there is no special reason to use modified Newton methods – the conjugate gradient method discussed previously and the quasi-Newton methods given in the following are preferable. However, modified Newton methods are advantageous for functions having a special structure for which Hessian matrix can be accurately approximated or easily calculated. An example of this occurs in nonlinear least-squares problems as discussed in the following.

Least Squares and Levenberg–Marquardt Method

Consider the regression problem of fitting a set of m data points ($xdata_i$, $ydata_i$) with a function $F(\mathbf{x}, xdata_i)$, where $\mathbf{x}$ are n variables to be determined, $m \gg n$. For example, if the function is $F = a.\ xdata + b$, then $\mathbf{x} = (a, b)^{\mathrm{T}}$. Ideally, if F at each point $xdata_i$ equals $ydata_i$, then the residual error is zero. The goal of minimizing the residual error can be posed as an unconstrained least-squares problem

$$\underset{\mathbf{x}}{\text{minimize}}\ F = \frac{1}{2}\sum_{i=1}^{m}(f(\mathbf{x}, x\,data_i) - y\,data_i)^2$$

If we assume that a good fit is obtainable, that is, F is small in magnitude at the optimum, and the gradient and Hessian are given by

$$\nabla F(\mathbf{x}) = \mathbf{J}(\mathbf{x})^{\mathrm{T}}\mathbf{f}$$
$$\nabla^2 F(\mathbf{x}) = \mathbf{J}(\mathbf{x})^{\mathrm{T}}\mathbf{J}(\mathbf{x}) + \mathbf{Q}(\mathbf{x}) \approx \mathbf{J}(\mathbf{x})^{\mathrm{T}}\mathbf{J}(\mathbf{x})$$

where $\mathbf{J} = \frac{\partial f}{\partial x_j}$, $j = 1, \ldots, n$ is an $(m \times n)$ matrix, $\mathbf{f}$ is $(m \times 1)$, $\mathbf{Q} = \sum_i f_i\ \nabla^2 f$ where $\mathbf{Q} \to \mathbf{0}$ as the optimum is approached under the small residual assumption in the aforementioned. Thus, only first-order gradient information is needed in the modified Newton algorithm, which is known as the Levenberg–Marquardt algorithm.

Levenberg–Marquardt Algorithm for Least Squares

Select a starting design variable vector $\mathbf{x}_0$, and an initial damping factor $\gamma > 0$. Solve for the direction vector $\mathbf{d}_k$ from

$$\left[\mathbf{J}(\mathbf{x}_k)^{\mathrm{T}}\mathbf{J}(\mathbf{x}_k) + \gamma\ \text{diag}\left(\mathbf{J}(\mathbf{x}_k)^{\mathrm{T}}\mathbf{J}(\mathbf{x}_k)\right)\right]\mathbf{d}_k = -\mathbf{J}(\mathbf{x}_k)^{\mathrm{T}}\mathbf{f} \tag{3.42}$$

Update $\mathbf{x}_{k+1} = \mathbf{x}_k + \mathbf{d}_k$, and evaluate $f(\mathbf{x}_{k+1})$. If the objective function is reduced, that is, if $f(\mathbf{x}_{k+1}) < f(\mathbf{x}_k)$, then reduce the damping factor as, say, $\gamma = \gamma/2$ and begin a new iteration. Otherwise, abandon the direction vector $\mathbf{d}_k$, increase the damping factor as $\gamma = 2\gamma$ and reevaluate $\mathbf{d}_k$ from Eq. (3.42). Stop when reductions in objective are tiny for two consecutive iterations.

3.8 Quasi-Newton Methods

Quasi-Newton methods use a Hessian-like matrix $\mathbf{F}_k$ as discussed previously, but without calculating second-order derivatives – that is, by using gradient information alone. The update formula in modified Newton methods is

$$\mathbf{x}_{k+1} = \mathbf{x}_k - \alpha_k [\mathbf{F}_k]^{-1} \nabla f(\mathbf{x}_k)$$

where $\mathbf{F}_k$ is a positive definite approximation to the Hessian matrix $\nabla^2 f(\mathbf{x}_k)$. In the preceding equation, we see that we can directly work with the inverse of $\mathbf{F}_k$ instead. Denoting the inverse as $\mathbf{H}_k$, we can write the update formula as

$$\mathbf{x}_{k+1} = \mathbf{x}_k - \alpha_k \mathbf{H}_k \nabla f(\mathbf{x}_k) \tag{3.43}$$

with the step size α_{k} determined by minimizing $f(\mathbf{x}_k + \alpha\, \mathbf{d}_k)$ with respect to α, $\alpha > 0$, $\mathbf{d}_k = -\mathbf{H}_k \nabla f(\mathbf{x}_k)$.

The basic idea behind quasi-Newton methods is to start with a symmetric positive definite $\mathbf{H}$, say $\mathbf{H}_0 = \mathbf{I}$, and update $\mathbf{H}$ so that it contains Hessian information. Just as we can use two function values to approximate the first derivative with a forward difference formula, we can also obtain an approximation to second derivatives using the gradient at two points. Specifically, from Taylor series we have

$$\nabla f(\mathbf{x}_{k+1}) = \nabla f(\mathbf{x}_k) + [\nabla^2 f(\mathbf{x}_k)]\boldsymbol{\delta}_k + \cdots \tag{3.44}$$

where $\boldsymbol{\delta}_k = \mathbf{x}_{k+1} - \mathbf{x}_k$. On the grounds that the higher order terms in (3.44) are anyway zero for a quadratic function, we may neglect them to obtain

$$[\nabla^2 f(\mathbf{x}_k)]\boldsymbol{\delta}_k = \boldsymbol{\gamma}_k \tag{3.45}$$

where $\boldsymbol{\gamma}_k = \nabla f(\mathbf{x}_{k+1}) - \nabla f(\mathbf{x}_k)$. The matrix $\mathbf{H}$, which represents an approximation to the inverse of the Hessian, is updated based on satisfying (3.45). Thus, $\mathbf{H}_{k+1}$ must satisfy the so-called "quasi-Newton condition"

$$\mathbf{H}_{k+1}\boldsymbol{\gamma}_k = \boldsymbol{\delta}_k \tag{3.46}$$

The Davidon–Fletcher–Powell (DFP) method, originally presented in [D1] and [F1], is based on updating $\mathbf{H}$ as

$$\mathbf{H}_{k+1} = \mathbf{H}_k + a\mathbf{u}\,\mathbf{u}^{\mathrm{T}} + b\mathbf{v}\,\mathbf{v}^{\mathrm{T}}$$

In view of (3.46),

$$\mathbf{H}_k\boldsymbol{\gamma}_k + a\mathbf{u}\,\mathbf{u}^{\mathrm{T}} + b\mathbf{v}\,\mathbf{v}^{\mathrm{T}} = \boldsymbol{\gamma}_k$$

Choosing $\mathbf{u} = \boldsymbol{\delta}_k$ and $\mathbf{v} = \mathbf{H}_k \boldsymbol{\gamma}_k$ we get $a\mathbf{u}^{\mathrm{T}} \boldsymbol{\gamma}_k = 1$ and $b\mathbf{v}^{\mathrm{T}} \boldsymbol{\gamma}_k = -1$ which determine a and b. The DFP update is now given by

$$\mathbf{H}_{\mathrm{DFP}}^{k+1} = H - \frac{H\gamma\gamma^{\mathrm{T}}H}{\gamma^{\mathrm{T}}H\gamma} + \frac{\delta\delta^{\mathrm{T}}}{\delta^{\mathrm{T}}\gamma} \tag{3.47}$$

The superscript k has been omitted in the matrices on the right-hand side of the preceding equation. We note that $\mathbf{H}$ remains symmetric. Further, it can be shown that $\mathbf{H}$ remains positive definite (assuming $\mathbf{H}_0$ is selected to be positive definite). Thus, the direction vector $\mathbf{d} = -\mathbf{H}_k\ \nabla f(\mathbf{x}_k)$ is a descent direction at every step. Further, when applied to quadratic functions $q(\mathbf{x}) = 1/2\ \mathbf{x}^{\mathrm{T}}\ \mathbf{A}\ \mathbf{x} + \mathbf{c}^{\mathrm{T}}\mathbf{x}$, then $\mathbf{H}^n = \mathbf{A}^{-1}$. This means that, for quadratic functions, the update formula results in an exact inverse to the Hessian matrix after n iterations, which implies convergence at the end of n iterations. We saw that the conjugate gradient method also possessed this property. For large problems, the storage and update of $\mathbf{H}$ may be a disadvantage of quasi-Newton methods as compared to the conjugate gradient method.

This program uses the quadratic fit algorithm for line search.

Another quasi-Newton update formula was suggested by Broyden, Fletcher, Goldfarb, and Shanno, known as the BFGS formula [F3]

$$\mathbf{H}_{\mathrm{BFGS}}^{k+1} = H - \left(\frac{\delta\gamma^{\mathrm{T}}H + H\gamma\delta^{\mathrm{T}}}{\delta^{\mathrm{T}}\gamma}\right) + \left(1 + \frac{\gamma^{\mathrm{T}}H\gamma}{\delta^{\mathrm{T}}\gamma}\right)\frac{\delta\delta^{\mathrm{T}}}{\delta^{\mathrm{T}}\gamma} \tag{3.48}$$

The BFGS update is better suited than the DFP update when using approximate line search procedures. Section 3.9 contains a discussion of approximate line search.

Example 3.11

Consider the quadratic function $f = x_1^2 + 10x_2^2$. We will now perform two iterations using DFP update, starting at (1,1). The first iteration is the steepest descent method since $\mathbf{H}_0 =$ identity matrix. Thus, $\mathbf{d}_0 = (-2, -20)^{\mathrm{T}}$, $\alpha_0 = 0.05045$, and $\mathbf{x}_1 = (0.899, -0.0089)^{\mathrm{T}}$. In program DFP, the direction vector is normalized to be an unit vector; this is not done here. The gradient at the new point is $\nabla f(\mathbf{x}_1) = (1.798, -0.180)$. Thus, $\gamma_0 = \nabla f(\mathbf{x}_1) - \nabla f(\mathbf{x}_0) = (-0.2018, -20.180)^{\mathrm{T}}$, $\delta_0 = \mathbf{x}_1 - \mathbf{x}_0 = (-0.101, -1.1099)^{\mathrm{T}}$, and Eq. (3.47) gives $\mathbf{H}^1 = \left[\begin{smallmatrix} 1.0 & -0.005 \\ -0.005 & 0.05 \end{smallmatrix}\right]$, followed by $\mathbf{d}_1 = (-1.80, 0.018)^{\mathrm{T}}$, $\alpha_1 = 0.5$, $\mathbf{x}^2 = (0.0, 0.0)$. The tolerance on the norm of the gradient vector causes the program to terminate with (0, 0) as the final solution. However, if we were to complete this step, then we find $\mathbf{H}^2 = \left[\begin{smallmatrix} 0.5 & 0 \\ 0 & 0.05 \end{smallmatrix}\right]$, which is the inverse of the Hessian of the quadratic objective function.

Example 3.12

Programs STEEPEST (steepest decent), FLREEV (conjugate gradient method of Fletcher–Reeves) and DFP (quasi–Newton method based on the DFP update), which are provided in the text, are used in this example. Three functions are considered for minimization:

(i) Rosenbrock's function (the minimum is in a parabolic valley):

$$f = 100(x_2 - x_1^2)^2 + (1 - x_1)^2$$
$$\mathbf{x}_0 = (-1.2, 1)^{\mathrm{T}}$$

(ii) Wood's function (this problem has several local minima):

$$\begin{aligned} f = {} & 100(x_2 - x_1^2)^2 + (1 - x_1)^2 + 90(x_4 - x_3^2)^2 + (1 - x_3)^2 \\ & + 10.1[(x_2 - 1)^2 + (x_4 - 1)^2] + 19.8(x_2 - 1)(x_4 - 1) \end{aligned}$$
$$\mathbf{x}_0 = (-3, -1, -3, -1)^{\mathrm{T}}$$

(iii) Powell's function(the Hessian matrix is singular at the optimum):

$$f = (x_1 + 10x_2)^2 + 5(x_3 - x_4)^2 + (x_2 - 2x_3)^4 + 10(x_1 - x_4)^4$$
$$\mathrm{x}_0 = (-3, -1, 0, 1)^{\mathrm{T}}$$

The reader is encouraged to run the computer programs and verify the results in Table 3.1. Note the following changes need to be made when switching from one problem to another in the code:

(i) N (start of the code, = number of variables)
(ii) Starting point (start of the code)
(iii) F = ... (function definition in Subroutine GETFUN)
(iv) DF(1) = __, ..., DF(N) = __ (gradient definition in Subroutine GRADIENT)

The reader may experiment with various parameters at the start of the code too. However, Table 3.1 uses the same parameters for the different methods and problems. NF = # function calls, NG = # gradient calls, in Table 3.1.

Table 3.1. *Results Associated with Example 3.12.*

PROBLEM	STEEPEST (line search based on quadratic fit)	FLREEV (line search based on quadratic fit)	DFP (line search based on quadratic fit)
Rosenbrock's	NF = 8815	NF = 141	NF = 88
	NG = 4359	NG = 28	NG = 18
	$\mathbf{x}^* = (0.996, 0.991)^{\mathrm{T}}$	$\mathbf{x}^* = (1.000, 1.000)^{\mathrm{T}}$	$\mathbf{x}^* = (1.000, 1.000)^{\mathrm{T}}$
	$f^* = 1.91\text{e-}5$	$f^* = 4.53\text{e-}11$	$f^* = 3.70\text{e-}20$
Wood's	NF = 5973	NF = 128	NF = 147
	NG = 2982	NG = 26	NG = 40
	$\mathbf{x}^* = (1.002, 1.003, 0.998, 0.997)^{\mathrm{T}}$	$\mathbf{x}^* = (1.000, 1.000, 1.000, 1.000)^{\mathrm{T}}$	$\mathbf{x}^* = (1.000, 1.000, 1.000, 1.000)^{\mathrm{T}}$
	$f^* = 9.34 \times 10^{-6}$	$f^* = 8.14 \times 10^{-11}$	$f^* = 6.79 \times 10^{-11}$
Powell's	NF = 3103	NF = 158	NF = 52
	NG = 893	NG = 30	NG = 21
	$\mathbf{x}^* = (-.051, 0.005, -0.02515, -.02525)^{\mathrm{T}}$	$\mathbf{x}^* = (-.030, 0.003, -0.011104, -0.01106)^{\mathrm{T}}$	$\mathbf{x}^* = (-.00108, -0.0000105, 0.0034, 0.0034)^{\mathrm{T}}$
	$f^* = 1.35 \times 10^{-5}$	$f^* = 1.66 \times 10^{-6}$	$f^* = 1.28 \times 10^{-9}$

Example 3.13 ("Mesh-Based Optimization")

An important class of optimization problems is introduced through this example. We use the term "mesh-based optimization" problem to refer to problems that require the determination of a function, say $y(x)$, on a mesh. Thus, the number of variables depends on how finely the mesh is discretized. The *Brachistochrone* problem involves finding the path or shape of the curve such that an object sliding from rest and accelerated by gravity will slip from one point to another in the least time. While this problem can be solved analytically, with the minimum time path being a *cycloid*, we will adopt a discretized approach to illustrate mesh-based optimization, which is useful in other problems.

Referring to Fig. E3.13a, there are n discretized points in the mesh, and heights y_i at points $2, 3, \ldots, n-1$ are the variables. Thus,

$$\mathbf{y} = [y_1, y_2, \ldots, y_n]^{\mathrm{T}}$$
$$\mathbf{x} = [y_2, y_3, \ldots, y_{n-1}]^{\mathrm{T}}$$

Importantly, a coarse mesh with, say, $n = 5$ points is first used and corresponding optimum heights are determined. Linear interpolation on the coarse mesh solution to obtain the starting point for a fine mesh is then used. The Matlab function *interp1* makes this task simple indeed. This way, optimum solutions with any size n is obtainable. Directly attempting a large value of n generally fails – that is, the optimizer cannot handle it.

Consider a line segment in the path as shown in Fig. E3.13a. From mechanics, we have

$$-\mu mg(\cos\theta)(\Delta L) = \tfrac{1}{2}m(v_{i+1}^2 - v_i^2) + mg(y_{i+1} - y_i) \tag{a}$$

where m is the mass, v refers to velocity and $g > 0$ is acceleration due to gravity. For frictionless sliding, $\mu = 0$. Further, the time Δt_i taken to traverse this segment is given by

$$\Delta t_i = \frac{(v_{i+1} - v_i)\Delta L}{g(y_i - y_{i+1})}$$

$$\text{where } \Delta L = \sqrt{h^2 + (y_{i+1} - y_i)^2} \tag{b}$$

Equations (a) and (b) provide the needed expressions. The objective function to be minimized is $T = \sum_{i=1}^{n-1} \Delta t_i$. A starting shape (say, linear) is first assumed. Since the object starts from rest, we have $v_1 = 0$. This allows computation of v_2 from Eq. (a), and Δt_1 from Eq. (b). This is repeated for $i = 2, \ldots, n-1$, which then defines the objective function f = total time T.

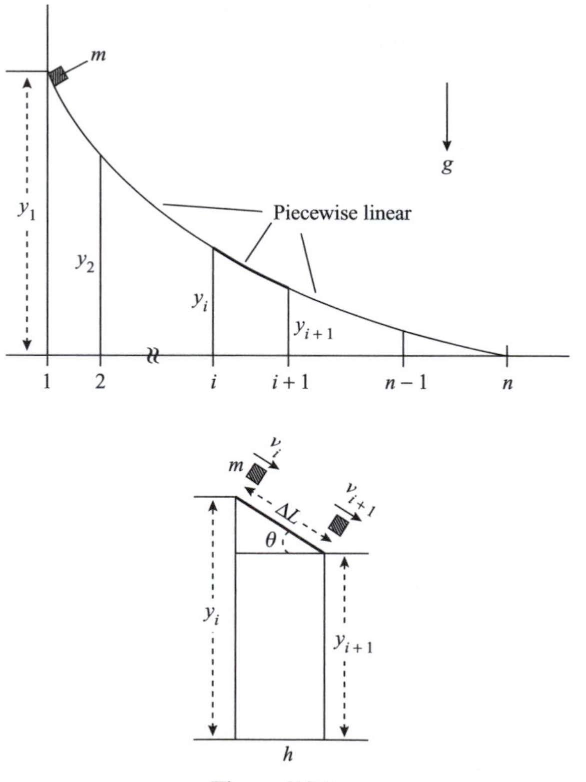

Figure E3.13a

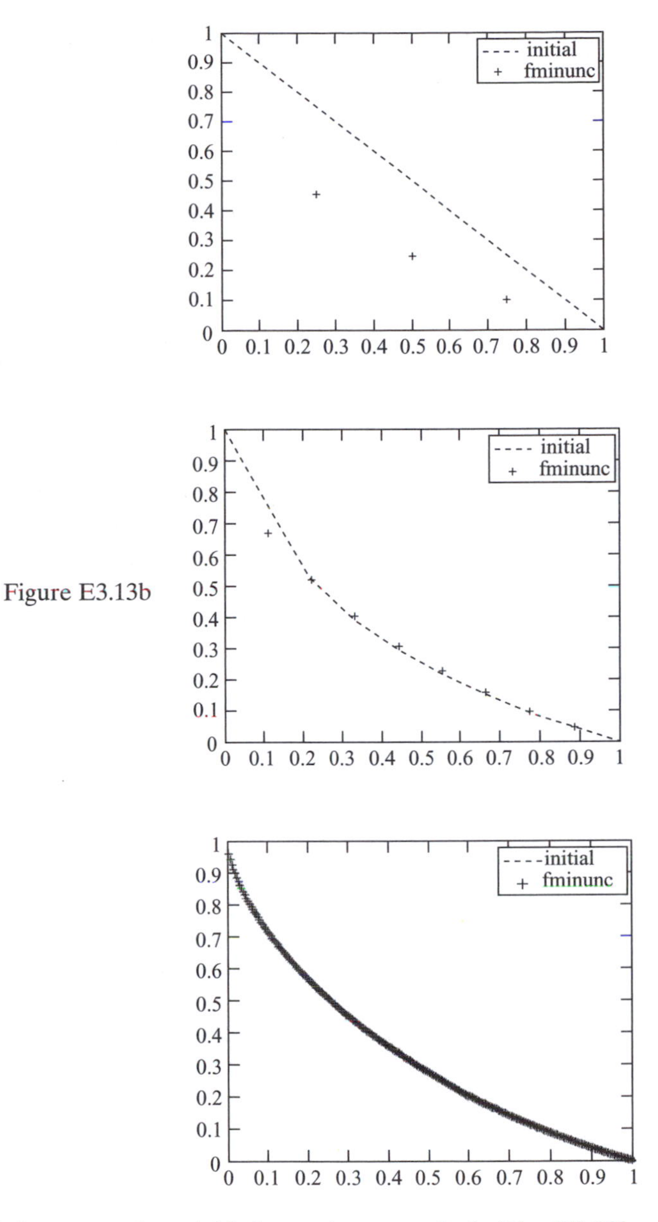

Figure E3.13b

Results using Matlab *fminunc* routine yield the optimum paths in Fig. E3.13b, between two fixed points (0, 1) and (1, 0), with $n = 5, 10, 250$, successively.

3.9 Approximate Line Search

In lieu of solving the "exact" line search problem in (3.16), approximate line search strategies are presented here. The first danger in taking steps that result in very small reductions in f values is safeguarded using a "sufficient decrease" condition

$$f(\mathbf{x}_k + \alpha\, \mathbf{d}_k) \leq f(\mathbf{x}_k) + \beta\alpha \nabla f(\mathbf{x}_k)^{\mathrm{T}} \mathbf{d}_k \tag{3.49}$$

The preceding criterion places an upper limit α_{U}. The preceding inequality may be implemented as given in the algorithm as follows. In (3.49), $0 < \beta < 1$.

Algorithm based on "sufficient decrease" condition ("Armijo" Line Search):

Step 0. Given $0 < \beta < 1$, and current point information $\mathbf{x}_k, f(\mathbf{x}_k), \nabla f(\mathbf{x}_k)$
Step 1. Set $\alpha = 1$
Step 2. Evaluate $f(\mathbf{x}_k + \alpha\, \mathbf{d}_k)$
Step 3. Check condition in (3.49) – if satisfied, then step size for current step is $\alpha_k = \alpha$; else, set $\alpha = \alpha/2$ and go to step 2

Since even very small values of α can also satisfy (3.49), a second requirement for ensuring convergence is to ensure that α is not too close to zero. That is, to ensure that very small steps are not taken. This can be ensured by requiring that the slope $f'(\alpha)$ is smaller in magnitude than $f'(\alpha = 0)$, as

$$f'(a) \geq \sigma\, f'(0) \tag{3.50}$$

The parameter σ is chosen such that $\beta < \sigma < 1$. This choice ensures that there exists a value of α that satisfies both (3.49) and (3.50). Instead of (3.50), a more stringent criterion based on a two-sided limit

$$\left| f'(\alpha) \right| \leq -\sigma f'(0) \tag{3.51}$$

is often used. Conditions in (3.49) and (3.50) are together called Wolfe conditions while (3.49) and (3.51) are called strong Wolfe conditions [Wolfe 1968]. The criterion in (3.50) or (3.51) will specify a lower limit α_{L}.

An example is given in the following to illustrate the two criteria for determining $[\alpha_{\mathrm{L}}, \alpha_{\mathrm{U}}]$.

Example 3.14

Consider $f = (x_1 - 2)^2 + (x_1 - 2x_2)^2$, with $\mathbf{x}_0 = (0, 3)^{\mathrm{T}}$. The steepest descent direction, normalized, is $\mathbf{d}_0 = (0.8779, -0.4789)^{\mathrm{T}}$. We also have

$$f(0) = 52, \quad f'(0) = -50.12 \quad \text{and} \quad f(\alpha) = 4.14\alpha^2 - 25.54\alpha + 40$$

Choosing $\beta = 0.01$ and $\sigma = 0.4$, the upper limit dictated by (3.49) is obtained from

$$4.14\alpha^2 - 25.54\alpha + 40 \leq 52 + 0.01\alpha(-50.12)$$

which yields $\alpha \leq 6.45$. Now, the lower limit dictated by (3.51) is obtained from

$$|8.28\alpha - 25.54| \leq -0.4(-50.12)$$

which yields $\alpha \leq 5.506$ and $\alpha \geq 0.663$. These limits and the limit $\alpha \leq 6.45$ yield the interval

$$0.663 \leq \alpha \leq 5.506$$

Note that an exact line search based on equating $f'(\alpha) = 0$ yields $\alpha = 3.08$.

A final note on line search strategies. The focus here has been on *maintaining descent*. That is, on maintaining that $f(\mathbf{x}_{k+1}) < f(\mathbf{x}_k)$. However, referring to Fig. 3.5, it can be seen that a large step that violates this monotonic decrease followed by a steepest descent step can yield a significant decrease in f. Barzilai has shown such a two-point technique, although theoretically proven only on quadratic functions [Barzilai and Borwein 1988].

3.10 Using MATLAB

The N-D unconstrained gradient-based minimization program in Matlab is *fminunc*. It is recommended to use Matlab optimization routines using a "user subroutine" to evaluate f for a given x. That is, a user subroutine "GETFUN." For example, we first create a file, "testfun.m" consisting of the code

```
function [f] = getfun(X)
f = 100 * (X(1) * X(1) - X(2)) ^ 2 + (1 - X(1)) ^ 2;
```

Then, Matlab is executed with the command:

```
[Xopt,fopt,iflag,output] = fminunc('testfun', X);
```

which produces the desired solution. To see the default parameters type

```
optimset('fminunc')
```

For instance, we can supply the gradient in the user subroutine and avoid the possibly expensive automatic divided difference scheme (the default) by switching on the corresponding feature as

```
options=optimset('GradObj', 'on')
```

fminunc is then executed using the command

```
[Xopt,fopt,iflag,output] = fminunc('testfun', X, options)
```

with a subroutine getfun that provides the analytical gradient in the vector DF as

```
function [f, Df] = getfun(X)
 f = ...
 Df(1) = ...;  Df(2) = ...; Df(N) = ...;
```

COMPUTER PROGRAMS

STEEPEST, FLREEV, DFP

PROBLEMS

P3.1. Plot contours of the function $f = x_1^2 x_2 - x_1 x_2 + 8$, in the range $0 < x_1 < 3$, $0 < x_2 < 10$. You may use Matlab or equivalent program.

P3.2. For the functions given in the following, determine (a) all stationary points and (b) check whether the stationary points that you have obtained are strict local minima, using the sufficiency conditions:

(i) $f = 3x_1 + \dfrac{100}{x_1 x_2} + 5x_2$

(ii) $f = (x_1 - 1)^2 + x_1 x_2 + (x_2 - 1)^2$

(iii) $f = \dfrac{x_1 + x_2}{3 + x_1^2 + x_2^2 + x_1 x_2}$

P3.3. (a) What is meant by a "descent direction"? (Answer this using an inequality.)

(b) If **d** is a solution of **W d** $= -\nabla f$, then state a sufficient condition on **W** that guarantees that **d** is a descent direction. Justify/prove your statement.

P3.4. Given $f = 4x_1^2 + 3x_2^2 - 4x_1x_2 + x_1$, the point $\mathbf{x}_0 = (-1/8, 0)^{\mathrm{T}}$ and the direction vector $\mathbf{d}_0 = -\,(1/5, 2/5)^{\mathrm{T}}$,

(i) Is $\mathbf{d}_0$ a descent direction?
(ii) Denoting $f(\alpha) = f(\boldsymbol{x}_0 + \alpha\boldsymbol{d}_0)$, find $\mathrm{d}f(1)/\mathrm{d}\alpha$, or equivalently, $f'(1)$.

P3.5. State True/False OR circle the correct answer OR fill in the blanks for the following questions:

(a) (T/F?) Conditions in Wierstraas theorem are satisfied for the following problem:

$$\begin{array}{ll} \text{minimize} & f = x_1 + x_2 \\ \text{subject to} & x_1x_2 \geq 1, x_1 \geq 0, x_2 \geq 0 \end{array}$$

(b) (T/F?) The problem in (a) has a solution.
(c) A C^2 continuous function f is convex if and only if its Hessian is ________________.
(d) If f is convex, then a local minimum is also ________________.

P3.6. (Fill in the blanks)
With respect to the problem: minimize $f(\mathbf{x})$, where f is C^2 continuous:

(i) A point $\hat{x}$ is a "candidate point" if ________________
(ii) To determine whether a candidate point corresponds to a local minimum, we need to check the condition ________________

P3.7. Two plots of $f(\alpha) = f(\boldsymbol{x}_0 + \alpha\boldsymbol{d}_0)$ are shown here. Identify which plot corresponds to steepest descent and which plot corresponds to Newton's method.

Figure P3.7

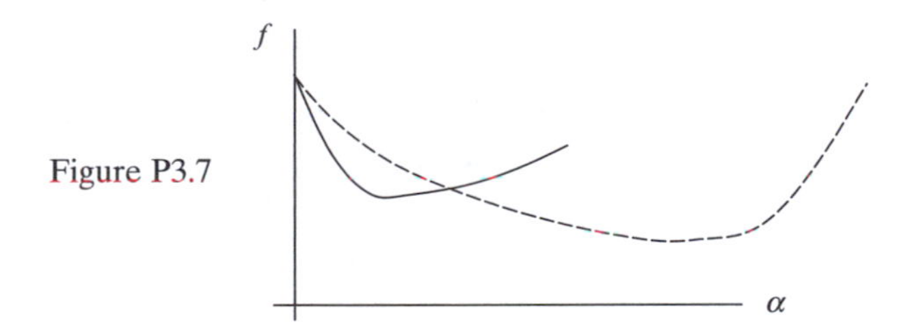

P3.8. (Fill in the blanks)

(a) The problem $f = x_1^2 + 10000x_2^2$ can be more effectively solved using the steepest descent method by ________________.
(b) Preconditioning the Hessian in conjugate gradient methods means to ________________.
(c) Preconditioning the Hessian in conjugate gradient methods helps to ________________.

P3.9. Given $f = x_1^2 + 5x_2^2$, $\mathbf{x}^k = [7, 2]^{\mathrm{T}}$, $k =$ iteration index:

(i) Evaluate current value, $f(\mathbf{x}^k)$. Then, evaluate $f(\mathbf{x}^{k+1})$, where $\mathbf{x}^{k+1}$ has been obtained with *one* iteration of steepest descent method.

(ii) Show the current point, direction vector, and new point on a sketch of the contour map of the function (draw, say, three contours of f).

(iii) Repeat above for one iteration of Newton's method.

P3.10. Given $f = x_1^2 + 25x_2^2$, a point $\mathbf{x}^0 = (5, 1)^T$, $f(\mathbf{x}^0) = 50$:

(i) Construct $f(\alpha)$ along the steepest descent direction.

(ii) Perform a line search (i.e., minimize $f(\alpha)$ with respect to α), and obtain a new point $\mathbf{x}^1$. Give value of $f(\mathbf{x}^1)$.

(iii) Provide a plot showing contour(s) of the function, steepest descent direction, $\mathbf{x}^0$ and $\mathbf{x}^1$.

P3.11. A cardboard box (see Fig. P3.11) is to be designed to have a volume of 1.67 ft^3. Determine the optimal values of l, w, and h so as to minimize the amount of cardboard material. (*Hint*: use the volume constraint equation to eliminate one of the variables.)

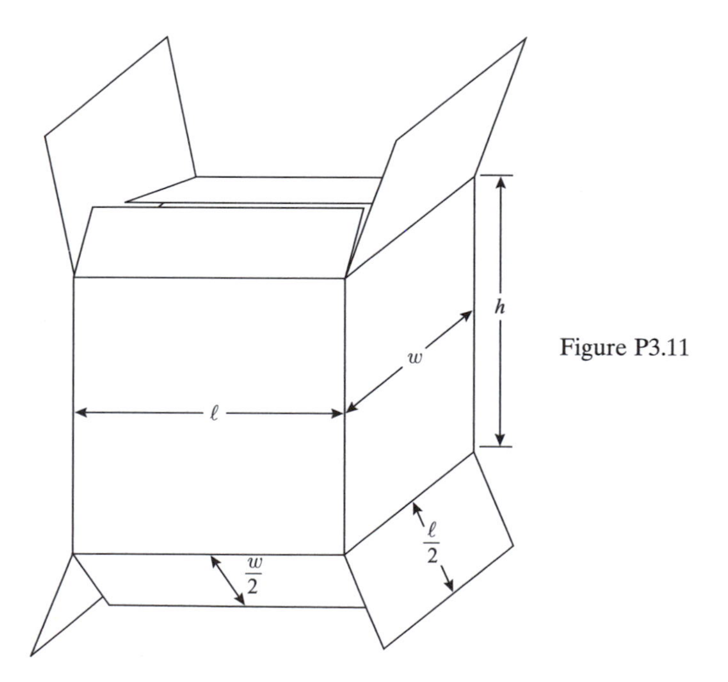

Figure P3.11

P3.12. Prove that if $\mathbf{A}$ is a positive definite matrix, then conjugate vectors – that is, vectors satisfying (3.25) – are linearly independent.

P3.13. A sequence $\{f(\mathbf{x}^k)\}$ is generated using an optimization program as: $f^k = [10.01, 4.01, 1.60, 0.639, 0.252\ldots]$ converging to $f^* = 0$. Determine whether the convergence rate is linear or nonlinear.

P3.14. [Courtesy: Dr. H.J. Sommer] A cylindrical coordinate robot is to be used for palletizing a rectangular area. Find the maximum rectangular area available within the annular footprint of the robot workspace (Fig. P3.14). Take $r_1 = 12''$ and $r_2 = 24''$.

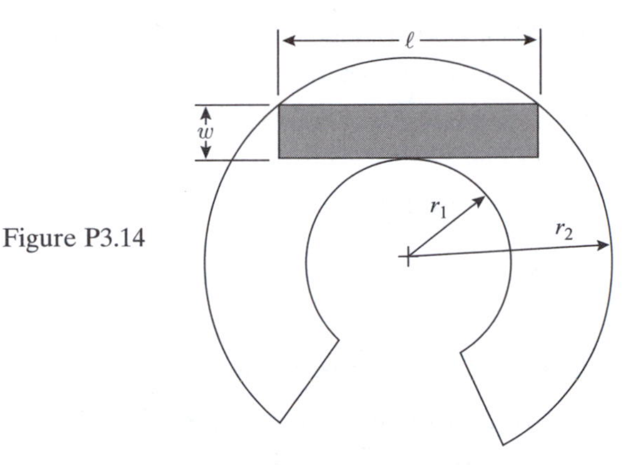

Figure P3.14

P3.15. (i) Construct linear and quadratic approximations to the function $f = x_1 x_2^2$ at the point $\mathbf{x}_0 = (1, 2)^T$. Plot the original function and its approximations (as in Example 1.21, Chapter 1).

(ii) For the function $f = x_1\, x_2$, determine expressions for $f(\alpha)$ along the line $x_1 = x_2$ and also along the line joining (0, 1) to (1, 0).

P3.16. Consider Example 3.1 in the text with $k_1 = k_2 = 1$ lb/in and $F_1 = 0$. Determine the vertical displacement Q_2 for $F_2 = 2.5$ lb, 3 lb, 6 lb, 10 lb. Plot Q_2 (vertical axis) versus F_2.

P3.17. Perform two iterations, by hand calculations, using the methods of steepest descent, Fletcher–Reeves, DFP, and Newton's for the quadratic function

$$f = (x_1 + 2x_2 - 6)^2 + (2x_1 + x_2 - 5)^2, \quad \mathbf{x}_0 = (0, 0)^T.$$

P3.18. It is desired to fit an equation of the form $f = a\, x_1^{c_1}\, x_2^{c_2}$, through some data obtained by a computer simulation, given below. Specifically, our task is to determine the coefficients *a*, c_1, and c_2 (which are our "variables") in the correlation given in the preceding text by minimizing the sum of square of the deviations between simulated and predicted values of *f*.

$f_{\text{simulated}}$	x_1	x_2
2.122	5.0	10.0
9.429	3.0	1.0
23.57	0.6	0.6
74.25	0.1	2.0
6.286	3.0	1.8

P3.19. Introduce scaling, $\mathbf{x} = \mathbf{T}\ \mathbf{y}$, into Program Steepest. Then, choose a $\mathbf{T}$ matrix to reduce the number of iterations needed to solve Rosenbrock's problem, starting from $(-1.2, 1)^{\mathrm{T}}$. (*Hint*: Use a diagonal $\mathbf{T}$ matrix. Include a listing of your modified code highlighting the changes to the code.)

P3.20. Implement the Polak–Ribiere algorithm (see (3.32)) and compare its performance with the Fletcher–Reeves algorithm for the problems in Example 3.12.

P3.21. Determine the function y(x) that leads to a stationary value of the functional

$$f = \int_0^2 \left[\left(\frac{\mathrm{d}y}{\mathrm{d}x} \right)^2 + 2y\frac{\mathrm{d}y}{\mathrm{d}x} + 4x^2 \right] \mathrm{d}x$$
$$y(0) = 1,\ y(2)\ \text{is free}$$

(*Hint*: discretize the function $y(x)$ as in Example 3.13).

P3.22. Implement the BFGS formula and compare with the DFP code. You may consider the test functions given in Example 3.12 and also Brook's function given by

$$f = x_1^2 \exp[1 - x_1^2 - 20.25(x_1 - x_2)^2]$$

with starting points $\mathbf{x}_0 = (1.9, 0.1)^{\mathrm{T}}$ and $\mathbf{x}_0 = (0.1, 0.1)^{\mathrm{T}}$.

P3.23. Compare the results in Table 3.1 with Matlab *fminunc* by providing analytical gradients as discussed in Section 3.11.

P3.24. Figure P3.24 shows a *rigid* plate supported by five springs and subjected to a load P as shown. Obtain the spring forces by minimizing the potential energy in the system which is given by $\Pi = \sum_{i=1}^{5} \frac{1}{2} k_i \delta_i^2 - P\delta$, where δ_i = vertical deflection of the ith spring and δ = the vertical deflection under the load. Assume only axial spring deflections and ignore any side loading on the springs caused by plate tipping. (*Hint*: the δ_i and the δ are related by the fact that the rigid plate remains flat).

P3.25. Implement Newton's with Hessian computed only once at the starting point $\mathbf{x}^0$ and compare with the regular Newtons's algorithm on two problems of your choice. (Choose problems from the literature with known solutions).

P3.26. Implement Levenberg–Marquardt algorithm discussed in the text on two least-squares problems of your choice and compare with FLREEV code. (Choose problems from the literature with known solutions).

P3.27. Repeat Example 3.13 and solve the Brachistochrone problem with friction – what happens to shape of path as μ increases in value? Give physical interpretation.

P3.28. Implement approximate line search based on sufficient decrease in Eq. (3.49) into STEEPEST, FLREEV, or DFP codes and resolve problems in Example 3.12. Study the effect of parameter β.

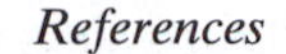

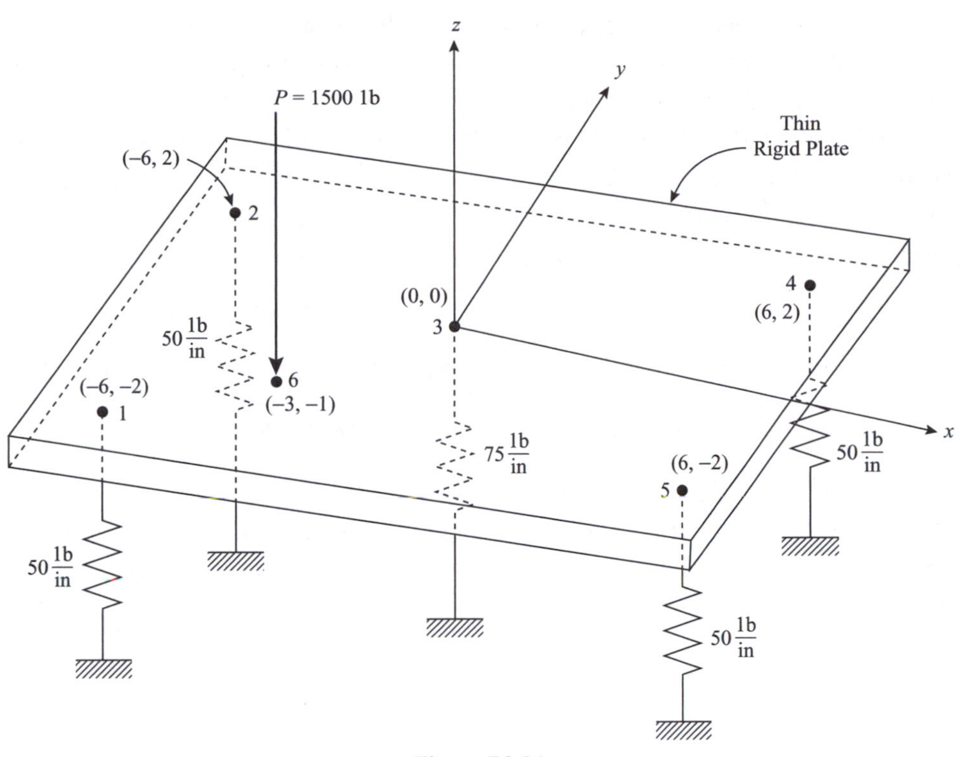

Figure P3.24

P3.29. Implement approximate line search based on both Eqs. (3.49) and (3.51) into STEEPEST, FLREEV, or DFP codes and resolve problems in Example 3.12.

REFERENCES

Armijo, L. Minimization of functions having Lipschitz-continuous first partial derivatives, *Pacific Journal of Mathematics*, **16**, 1–3, 1966.

Barzilai, J. and Borwein, M., Two-point step size gradient methods, *IMA Journal of Numerical Analysis*, 1988 **8**(1):141–148.

Cauchy, A., Methode generale pour la resolution des systemes d'equations simultanes, *C. R. Acad. Sci. Par.*, 25, 536–538, 1847.

Chandrupatla, T.R. and Belegundu, A.D. *Introduction to Finite Elements in engineering*. Prentice-Hall, Englewood Cliffs, NJ, 1991.

Curry, H., The method of steepest descent for nonlinear minimization problems, *Quarterly of Applied Mathematics*, **2**, 258–261, 1944.

Davidon, W.C., Variable metric method for minimization, Research and Development Report ANL-5990 (Revised), Argonne National Laboratory, U.S. Atomic Energy Commision, 1959.

Dennis, J.E. and Schnabel, R.B., *Numerical Methods for Unconstrained Optimization and Nonlinear Equations*, Prentice-Hall, Englewood-Cliffs, NJ, 1983.

Fletcher, R., *Practical Methods of Optimization*, 2nd Edition, Wiley, New York, 1987.

Fletcher, R. and Powell, M.J.D. A rapidly convergent descent method for minimization. *The Computer Journal*, **6**, 163–168, 1963.

Fletcher, R. and Reeves, C.M., Function minimization by conjugate gradients. *The Computer Journal*, **7**, 149–154, 1964.

Gill, P.E. and Murray, W., Quasi-Newton methods for unconstrained optimization. *Journal of the Institute of Mathematics and Its Applications*, **9**, 91–108, 1972.

Goldstein, A.A., On steepest descent, *SIAM Journal on Control*, **3**, 147–151, 1965.

Kardestuncer, H. (Ed.), *Finite Element Handbook*, McGraw-Hill, New York, 1987.

Luenberger, D.G., *Introduction to Linear and Nonlinear Programming*, Addison-Wesley, Reading, MA, 1965.

Polak, E., *Computational Methods in Optimization: A Unified Approach*, Academic Press, New York, 1971.

Pshenichny, B.N. and Danilin, Y.M., *Numerical Problems in Extremal Problems*, MIR Publishers, Moscow, 1978, 1982.

Wolfe, P., Convergence conditions for ascent methods, *SIAM Review*, **11**, 226–235, 1968.

4

Linear Programming

4.1 Introduction

Linear programming (*LP*) is the term used for defining a wide range of optimization problems, in which the objective function to be minimized or maximized is linear in the unknown variables and the constraints are a combination of linear equalities and inequalities. LP problems occur in many real-life economic situations where profits are to be maximized or costs minimized with constraint limits on resources. While the simplex method introduced in the following can be used for hand solution of LP problems, computer use becomes necessary even for a small number of variables. Problems involving diet decisions, transportation, production and manufacturing, product mix, engineering limit analysis in design, airline scheduling, and so on, are solved using computers. Linear programming also has applications in nonlinear programming (NLP). Successive linearization of a nonlinear problem leads to a sequence of LP problems that can be solved efficiently.

Practical understanding and geometric concepts of LP problems including computer solutions with EXCEL SOLVER and MATLAB, and output interpretation are presented in Sections 4.1–4.5. Thus, even if the focus is on nonlinear programming, the student is urged to understand these sections. Subsequently, algorithms based on Simplex (Tableau-, Revised-, Dual-) and Interior methods, and Sensitivity Analysis, are presented.

4.2 Linear Programming Problem

The general form of the linear programming problem has an objective function to be minimized or maximized, and a set of constraints.

$$\text{maximize} \qquad c_1x_1 + c_2x_2 + c_3x_3 + \cdots + c_nx_n$$

subject to

$$
\begin{aligned}
&a_{i1}x_1 + a_{i2}x_2 + a_{i3}x_3 + \cdots + a_{in}x_n \le b_i && \text{LE Constraints } (i = 1 \text{ to } \ell)\\
&a_{j1}x_1 + a_{j2}x_2 + a_{j3}x_3 + \cdots + a_{jn}x_n \ge b_j && \text{GE Constraints } (j = \ell + 1 \text{ to } \ell + r)\\
&a_{k1}x_1 + a_{k2}x_2 + a_{k3}x_3 + \cdots + a_{kn}x_n = b_k && \text{EQ Constraints } (k = \ell + r + 1 \text{ to } \ell + r + q)\\
&x_1 \ge 0, x_2 \ge 0, \ldots, x_n \ge 0 && (4.1)
\end{aligned}
$$

We note that the number of constraints is $m = \ell + r + q$, c_j and a_{ij} are constant coefficients, b_i are fixed real constants, which are adjusted to be nonnegative; x_j are the unknown variables to be determined. In engineering design problems, the x_j are referred to as *design variables*. The limits on j and i are $j = 1$ to n and $i = 1$ to m, respectively.

LP's are *convex* problems, which implies that a local maximum is indeed a global maximum. As discussed in Chapter 1, the constraints define a feasible region Ω and can be bounded, unbounded, or inconsistent (in which case, a solution does not exist).

4.3 Problem Illustrating Modeling, Solution, Solution Interpretation, and Lagrange Multipliers

Practical aspects of formulating, solving, and interpreting an LP is presented through an example. Modeling of different types of problems is discussed in Section 4.4.

Structural Design Problem

Consider the platform support system shown in Fig. 4.1 used for scaffolding. Cable 1 can support 120 lb, cable 2 can support 160 lb, and cables 3 and 4 can support 100 lb each. Determine the maximum total load that the system can support.

Formulation

Using free-body diagrams and static equilibrium equations, a weight of W acting at a from left support and b from the right support will cause reactions of $Wb/(a+b)$ and $Wa/(a+b)$ respectively. Applying this principle and simplifying, we obtain the constraints

$$W_2 \le 2S_{3,4} \qquad 4W_1 + 3W_2 \le 8S_2 \qquad 4W_1 + W_2 \le 8S_1$$

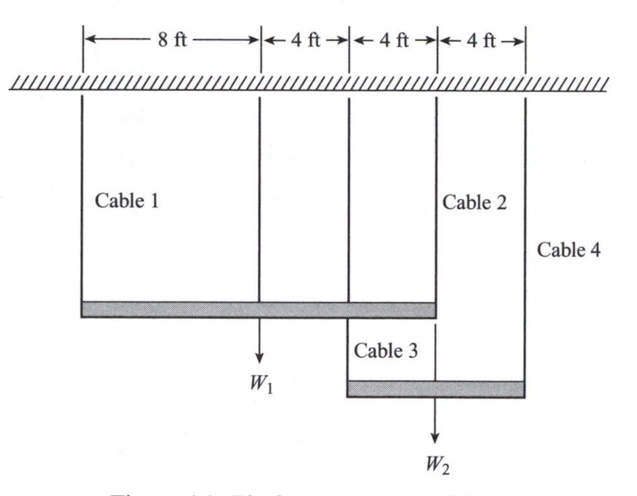

Figure 4.1. Platform support problem.

where S represents cable strength. Simplifying, we obtain

$$
\begin{aligned}
\text{maximize} \quad & f = W_1 + W_2 \\
\text{subject to} \quad & \\
& W_2 \le 200 \text{ (Cables 3 and 4)} \\
& 4W_1 + 3W_2 \le 1280 \quad \text{(Cable2)} \\
& 4W_1 + W_2 \le 960 \quad \text{(Cable1)} \\
& W_1 \ge 0, \quad W_2 \ge 0
\end{aligned}
\tag{4.2}
$$

We can model large-scale scaffolding problems using the same ideas. Graphical solution is given in the following as well as use of EXCEL SOLVER. Graphical solutions were discussed in Chapter 1. The variables are W_1 and W_2. First step is to identify the feasible region Ω. This can be done by plotting, in variable space with W_1 as the x-axis and W_2 as the y-axis, the line defined by each constraint, and then determining which side of this line is feasible. Thus, we plot lines corresponding to $W_2 = 200$, $4W_1 + 3W_2 = 1280$, $4W_1 + W_2 = 960$, and consider the first quadrant only. After obtaining Ω, plot contours of the objective function by plotting lines $W_1 + W_2 = c$ for various values of c. The optimum is defined by the highest value contour passing through Ω. From Fig. 4.2, it is clear that the optimum point lies at the intersection of the *active* (or *binding*) constraints "cable 2," and "cables 3 and 4." Thus, solving the two equations $W_2 = 200$ and $4W_1 + 3W_2 = 1280$ gives the optimum solution $W_1^* = 170$, $W_2^* = 200$, and objective $f^* = 370$.

Uniqueness of the optimum point is evident from Fig 4.2 since the objective function contour with value $f = 370$ passes thru a *vertex*. If this contour had been parallel to the constraint boundary, then any point on that boundary line, including the two vertices at its ends, would have the same optimum value of 370. Further, an

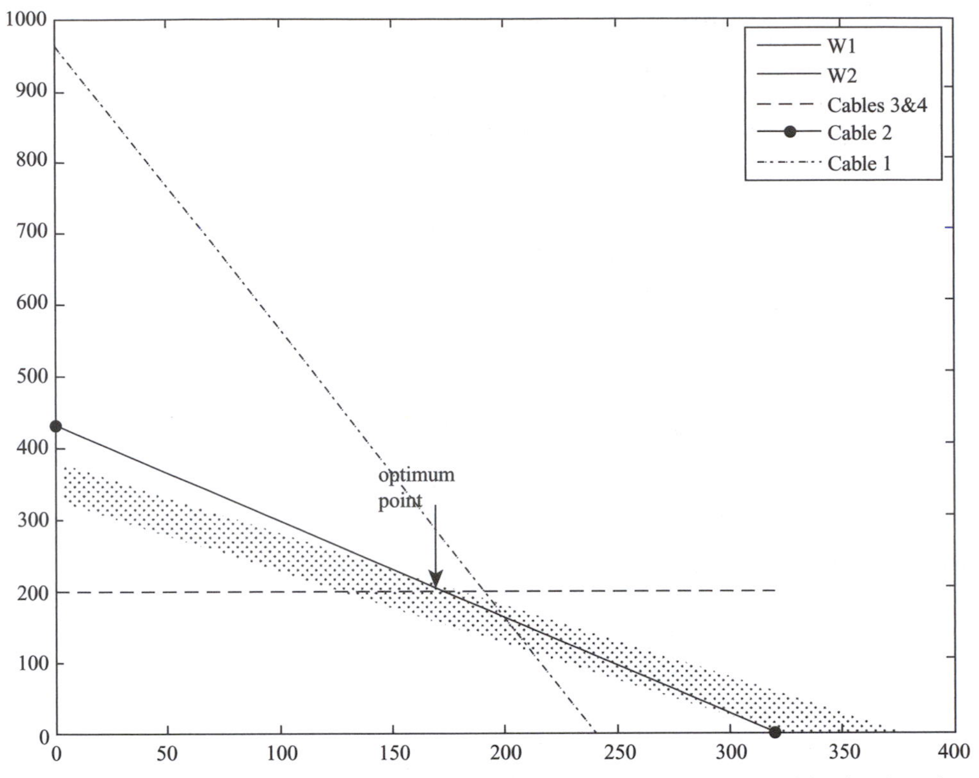

Figure 4.2. Plot of structural design problem showing feasible region and objective function contours (dotted lines).

LP problem is *convex*, which implies that there do not exist local optima. Geometrical concepts are described in more detail in Section 4.5.

Physical Interpretation of Optimum Solution

Constraints associated with cables 3,4 and cable 2 are active, which means that these cables are stressed to their respective limits. However, cable 1 has an 80 lb slack. Perhaps, this cable can be made of cheaper material with a lower strength, within limits, although a multitude of cables with differing strengths increases manufacturing cost and chance of making errors during assembly.

Computer Solution

Codes given in the attached CD-ROM, based on the theory discussed later in this chapter, may be used to obtain the solution to the above problem. In the following, we show how to use EXCEL SOLVER and MATLAB (*LINPROG* routine), respectively.

Excel Solver

First, the formulas for objective, and constraints and initial guesses for variables are defined on the Excel spreadsheet in cells A1-F2 as:

W1	W2	OBJ	Cable 3, 4	Cable 2	Cable 1
1	1	2	1	7	5

Then, SOLVER window is opened under Tools menu (if not visible, click "Add-ins"), and the optimization problem is defined as shown below:

```
Set Target Cell: $C$2
Equal to: ''Max''
By Changing Cells: $A$2:$B$2
Subject to the Constraints:
$A$2:$B$2 ≥ 0
$D$2 ≤ 200
$E$2 ≤ 1280
$F$2 ≤ 960
```

Solving (and also choosing additional answer sheets) we get

W1	W2	OBJ	Cable 3, 4	Cable 2	Cable 1
170	200	370	200	1280	880

Constraints

Cell	Name	Final value	Lagrange multiplier
$D$2	Cable 3, 4	200	0.25
$E$2	Cable 2	1280	0.25
$F$2	Cable 1	880	0

Significance of Lagrange Multipliers

For nondegenerate problems wherein active constraints have corresponding nonzero Lagrange multipliers, a Lagrange multiplier λ_i, associated with a constraint g_i, represents the sensitivity of the optimum objective function value to a relaxation in the constraint limit. Specifically, if $g_i(x) \le c_i$ is relaxed to $g_i(x) \le c_i + e_i$, then

$$\lambda_i = \left.\frac{df^*}{de_i}\right|_{e_i=0} \quad \text{for a maximization problem}$$
$$\lambda_i = -\left.\frac{df^*}{de_i}\right|_{e_i=0} \quad \text{for a minimization problem} \tag{4.3}$$

In this example, say, the strength limit of cables 3,4 is relaxed from current value of 200 to 202. Resolving the problem, we obtain $f^* = 370.5$. Thus, $df^*/de_i = (370.5 - 370)/2 = 0.25$ as obtained by SOLVER above. However, it is not necessary to resolve the problem to obtain λ_i. (Why is the Lagrange multiplier associated with cable 1 constraint equal to zero?)

More practical understanding of λ_i is obtained by writing the sensitivity relation $\lambda_i = \frac{df^*}{de_i}\big|_{e_i=0}$ as $\Delta f^* = \lambda_i \Delta e_i$, from which we can state that λ_i equals the improvement in objective for a *unit* change in resource *i*. λ_i are also known as *dual prices* (see Duality in Section 4.7) and as *shadow prices* in economics. We may say that λ_i reflects return on unit additional investment with respect to resource *i*. They are not prices as such, but can be used to set prices. For instance, if a constraint limits the amount of labor available to 40 hours per week, the shadow price λ_i will tell us how much we would be willing to pay for an additional hour of labor. If $\lambda_i = \$10$ for the labor constraint, we should pay no more than \$10 an hour for overtime. Labor costs of less than \$10/hour will increase the objective value, labor costs of more than \$10/hour will decrease the objective value, and labor costs of exactly \$10 will cause the objective function value to remain the same.

MATLAB (using the Optimization Toolbox)

```
function [] = LP()
close all; clear all;
f=[-1 -1]; %max f <--> min. -f
A=[0 1; 4 3; 4 1];
b=[200 1280 960]';
XLB=[0 0]; %lower bounds on variables
[X,FVAL,EXITFLAG,OUTPUT,LAMBDA] = LINPROG(f,A,b,[],[],XLB)
X =
170.0000
200.0000
FVAL =
-370.0000
```

Note: the objective has a negative value because $f_{new} = -f_{old}$ has been defined to convert maximization problem to a minimization problem as required by the routine.

4.4 Problem Modeling

Linear programming is one of the most widely used techniques in optimization. A wide variety of problems from various fields of engineering and economics fall into this category. In this section, we present some problems and their formulation. Detailed solution are left as exercises at the end of the chapter.

PM1: Diet Problem

Nutritional content data of a group of foods and the weekly need for an adult are given in the table shown below. Determine the lowest weekly cost for meeting the minimum requirements.

	Food	Proteins	Fats	Carbohydrates	Cost $ per 100g
1	Bread	8%	1%	55%	0.25
2	Butter	–	90%	–	0.5
3	Cheese	25%	36%	–	1.2
4	Cereal	12%	3%	75%	0.6
5	Diet Bar	8%	–	50%	1.5
	Weekly requirement (g)	550	600	2000	

Formulation

The table has all the information for the LP formulation. Let x_1, x_2, x_3, x_4, x_5 be the number of grams of each of the food items. The LP problem can be stated as

$$\begin{aligned} \text{Minimize} \quad & 0.25x_1 + 0.5x_2 + 1.2x_3 + 0.6x_4 + 1.5x_5 \\ \text{Subject to} \quad & 0.08x_1 + 0.25x_3 + 0.12x_4 + 0.08x_5 \geq 550 \\ & 0.01x_1 + 0.9x_2 + 0.36x_3 + 0.03x_4 \geq 600 \\ & 0.55x_1 + 0.75x_4 + 0.5x_5 \geq 2000 \\ & x_1 \geq 0, \quad x_2 \geq 0, \quad x_3 \geq 0, \quad x_4 \geq 0, \quad x_5 \geq 0 \end{aligned}$$

PM2: Alloy Manufacture

An alloy manufacturer is planning to produce 1000 kg of an alloy with 25% by weight of metal A and 75% by weight of metal B by combining five available alloys. The composition and prices of these alloys is shown in the table as follows.

Alloy	1	2	3	4	5
%A	10	15	20	30	40
%B	90	85	80	70	60
Available quantity kg	300	400	200	700	450
Price $/kg	6	10	18	24	30

Formulation

Let x_1, x_2, x_3, x_4, x_5 be the number of kg of each of the available alloys melted and mixed to form the needed amount of the required alloy. Then the problem can be formulated as

$$\begin{aligned} \text{Minimize} \quad & 6x_1 + 10x_2 + 18x_3 + 24x_4 + 30x_5 \\ \text{Subject to} \quad & 0.1x_1 + 0.15x_2 + 0.2x_3 + 0.3x_4 + 0.4x_5 = 250 \\ & x_1 + x_2 + x_3 + x_4 + x_5 = 1000 \\ & x_1 \le 300, \quad x_2 \le 400, \quad x_3 \le 200, \quad x_4 \le 700, \quad x_5 \le 450 \\ & x_1 \ge 0, \quad x_2 \ge 0, \quad x_3 \ge 0, \quad x_4 \ge 0, \quad x_5 \ge 0 \end{aligned}$$

The problem is now ready for solving using the simplex or other methods.

PM3: Refinery Problem

A refinery uses two different crude oils, a light crude costs $40 per barrel, and a heavy crude costs $30 per barrel, to produce gasoline, heating oil, jet fuel, and lube oil. The yield of these per barrel of each type of crude is given in the following table:

	Gasoline	Heating oil	Jet fuel	Lube oil
Light crude oil	0.4	0.2	0.3	0.1
Heavy crude oil	0.3	0.45	0.1	0.05

The demand is 8 million barrels of gasoline, 6 million barrels of heating oil, 7 million barrels of jet fuel, and 3 million barrels of lube oil. Determine the amounts of light crude and heavy crude to be purchased for minimum cost.

Formulation

Let x_1 and x_2 be the millions of barrels of light and heavy crude purchased, respectively. Then the problem can be put in the form

$$\begin{aligned}
\text{Minimize} \quad & 40x_1 + 30x_2 \\
\text{Subject to} \quad & 0.4x_1 + 0.3x_2 \geq 8 \\
& 0.2x_1 + 0.45x_2 \geq 6 \\
& 0.3x_1 + 0.1x_2 \geq 7 \\
& 0.1x_1 + 0.05x_2 \geq 3 \\
& x_1 \geq 0, \quad x_2 \geq 0
\end{aligned}$$

This is a simple formulation of petroleum refinery problem. Some of the real problems may have several more variables and more complex structure but they are still LP problems.

PM4: Agriculture

A vegetable farmer has the choice of producing tomatoes, green peppers, or cucumbers on his 200 acre farm. A total of 500 man-days of labor is available.

	Yield $/ acre	Labor man-days/acre
Tomatoes	450	6
Green Peppers	360	7
Cucumbers	400	5

Assuming fertilizer costs are same for each of the produce, determine the optimum crop combination.

Formulation

Let x_1, x_2, and x_3 be the acres of land for tomatoes, green peppers, and cucumbers respectively. The LP problem can be stated as

$$\begin{aligned}
\text{Maximize} \quad & 450x_1 + 360x_2 + 400x_3 \\
\text{Subject to} \quad & x_1 + x_2 + x_3 \leq 200 \\
& 6x_1 + 7x_2 + 5x_3 \leq 500 \\
& x_1 \geq 0, \quad x_2 \geq 0, \quad x_3 \geq 0
\end{aligned}$$

PM5: Trim Problem

A paper mill makes jumbo reels of width 1m. They received an order for 200 reels of width 260 mm, 400 reels of width 180 mm, and 300 reels of width 300 mm. These rolls are to be cut from the jumbo roll. The cutter blade combinations are such that one 300 mm width reel must be included in each cut. Determine the total number of jumbo reels and cutting combinations to minimize trim.

Formulation

First we prepare a table of all combinations with the amount of trim identified in each combination.

Combination	Number of 300 mm	Number of 260 mm	Number of 180 mm	Trim mm
1	3	0	0	100
2	2	1	0	140
3	2	0	2	40
4	1	2	1	0
5	1	1	2	80
6	1	0	3	160

Let x_i be the number of jumbo reels subjected to combination i. The problem is now easy to formulate. We make the first formulation realizing that the best cutting combination leads to least number of total jumbo reels.

Formulation 1 based on minimum total number:

$$\begin{aligned} \text{minimize} \quad & x_1 + x_2 + x_3 + x_4 + x_5 + x_6 \\ \text{subject to} \quad & 3x_1 + 2x_2 + 2x_3 + x_4 + x_5 + x_6 \geq 300 \\ & x_2 + 2x_4 + x_5 \geq 200 \\ & 2x_3 + x_4 + 2x_5 + 3x_6 \geq 400 \\ & x_1 \geq 0, \quad x_2 \geq 0, \quad x_3 \geq 0, \quad x_4 \geq 0, \quad x_5 \geq 0, \quad x_6 \geq 0 \end{aligned}$$

Formulation based on minimum trim will require that excess rolls produced in each size be treated as waste. The surplus variable of each constraint must be added to the cost function.

Formulation 2 based on minimum trim:

$$\begin{aligned} \text{minimize} \quad & 100x_1 + 140x_2 + 40x_3 + 80x_5 + 160x_6 + 300x_7 + 260x_8 + 180x_9 \\ \text{subject to} \quad & 3x_1 + 2x_2 + 2x_3 + x_4 + x_5 + x_6 - x_7 = 300 \\ & x_2 + 2x_4 + x_5 - x_8 = 200 \\ & 2x_3 + x_4 + 2x_5 + 3x_6 - x_9 = 450 \\ & x_1 \geq 0, \quad x_2 \geq 0, \quad x_3 \geq 0, \quad x_4 \geq 0, \quad x_5 \geq 0, \\ & x_6 \geq 0, \quad x_7 \geq 0, \quad x_8 \geq 0, \quad x_9 \geq 0 \end{aligned}$$

The equivalence of the two formulations can be accomplished easily. By eliminating x_7, x_8, and x_9 using the constraint equations, the objective function becomes $1000(x_1 + x_2 + \cdots + x_6)$.

In the aforementioned formulations, the solutions may have fractional values. We know that each of the variables x_i must be an integer. This may be solved later as an integer programming problem. The general strategy in integer programming problems is to start with a solution where this requirement is not imposed on the initial solution. Other constraints are then added, and the solution is obtained iteratively.

PM6: Straightness Evaluation in Precision Manufacture

In the straightness evaluation of a dimensional element, coordinate location along the edge x_i in mm and the coordinate measuring machine probe reading y_i in μm (micron meter $= 10^{-6}$m) have been recorded as shown.

	1	2	3	4	5
x_i mm	0	2.5	5	7.5	10
y_i μm	10	5	19	11	8

Straightness is defined as the smallest distance between two parallel lines between which the measured points are contained. Determine the straightness for the data shown.

Formulation

Let v be the interpolated value of the measured value given by $ax + b$. If v_i is the calculated value at point i, then the problem can be posed as a linear programming problem:

$$\begin{aligned} \text{Minimize} \quad & 2z \\ \text{subject to} \quad & z \geq v_i - y_i \\ & z \geq -v_i + y_i \quad i = 1 \text{ to } 5 \end{aligned}$$

The variables z, a, and b may be left as unrestricted in sign. Note that the final value of z is the maximum deviation from the minimum zone best fit straight line.

PM7: Machine Scheduling

A machine shop foreman wishes to schedule two types of parts, each of which has to undergo turning, milling, and grinding operations on three different machines.

The time per lot for the two parts available machine time and profit margins are provided in the table as follows.

Part	Turning	Milling	Grinding	Profit $ per lot
Part 1	12 hrs/lot	8 hrs/lot	15 hrs/lot	120
Part 2	6 hrs/lot	6.5 hrs/lot	10 hrs/lot	90
Machine time available hrs	60 hrs	75 hrs	140 hrs	

Schedule the lots for maximizing the profit.

Formulation

Let x_1 and x_2 be the number of lots of parts 1 and 2 to be produced, respectively. Then the LP problem can be formulated as

$$\begin{aligned} \text{Maximize} \quad & 120x_1 + 90x_2 \\ \text{Subject to} \quad & 12x_1 + 6x_2 \leq 60 \\ & 8x_1 + 6.5x_2 \leq 75 \\ & 15x_1 + 10x_2 \leq 140 \\ & x_1 \geq 0, x_2 \geq 0 \end{aligned}$$

Problems of complex nature can be formulated with a careful consideration. Some problems of great importance in transportation, assignment, and networking are of the linear programming category, but have special structures. These problems will be considered in greater detail in a later chapter.

4.5 Geometric Concepts: Hyperplanes, Halfspaces, Polytopes, Extreme Points

To fix some ideas, we consider a two-variable linear programming problem.

$$\begin{aligned} \text{maximize} \quad & f = 2x_1 + x_2 \\ \text{subject to} \quad & 2x_1 - x_2 \leq 8 \quad [1] \\ & x_1 + 2x_2 \leq 14 \quad [2] \\ & -x_1 + x_2 \leq 4 \quad [3] \\ & x_1 \geq 0, \quad x_2 \geq 0 \end{aligned} \tag{4.4}$$

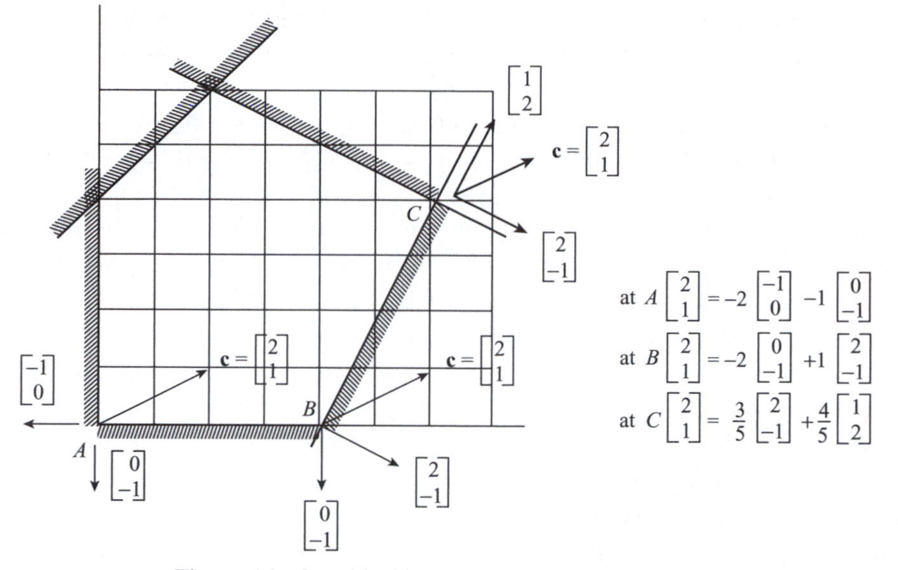

Figure 4.3. Graphical illustration of the LP problem.

The region enclosed by the constraints is illustrated in Fig. 4.3. Each constraint defines a line separating the space R^2 into two half spaces defining feasible and infeasible regions. Writing the objective function as $\mathbf{c}^{\mathrm{T}}\mathbf{x}$ results in $\mathrm{c} = [2, 1]^{\mathrm{T}}$. Vector $\mathbf{c}$ is also the gradient of f. Similarly, gradients of the constraints are $[2, -1]^{\mathrm{T}}$, $[1, 2]^{\mathrm{T}}$, $[-1, 1]^{\mathrm{T}}$, respectively, and $[-1, 0]^{\mathrm{T}}$, $[0, -1]^{\mathrm{T}}$ for the bounds recall from the earlier chapters that the gradient is normal or perpendicular to the boundary and points in the direction of increasing function value. Every point in the feasible half space satisfies the constraint. The common region of all the feasible half spaces defines the feasible region $ABCDE$ shown in the figure. Every point in the *feasible region* satisfies all the constraints of the problem.

In n dimensions (R^n), the surface where the constraint is satisfied with an equality is a *hyperplane*, which is $n - 1$ dimensional. The feasible region, which is a *polygon*, in two dimensions is a *polytope* in n-dimensions. This polytope is a *convex* region where a line segment joining any pair of points in the region lies in the region. Points A, B, C, D, and E are extreme points of the two-dimensional convex region. A point $\mathbf{x}$ is an *extreme point* of the convex set if there are no two distinct points $\mathbf{x}_1$ and $\mathbf{x}_2$ in the set for which $\mathbf{x} = \alpha\mathbf{x}_1 + (1 - \alpha)\mathbf{x}_2$ with $0 \leq \alpha \leq 1$. If the feasible region formed by LE constraints is nonempty, it defines a bounded polytope, called a *polyhedron*. The polytope formed by the feasible region is the *convex hull* of its extreme points. An extreme point occurs at the intersection of n-hyperplanes forming the boundary of the polytope.

With some geometric ideas fixed in mind, let us explore the geometric meaning of the optimum. The set of points satisfying $2x_1 + x_2 =$ constant, define a line with the increasing normal direction or the gradient $[2 \quad 1]^T$. To reach the optimum, we keep moving this plane along the gradient direction (increasing the value of the constant) until it passes through an extreme point. Moving beyond this leads to infeasibility. We note that the maximum of the problem illustrated in Fig. 4.3 is at the extreme point C where x_1 and x_2 take the values 6 and 4 and the maximum value of the objective function is 16. In n-dimensions $\mathbf{c}^T\mathbf{x} =$ constant represents a hyperplane with the gradient direction $\mathbf{c}$. If we can visualize this hyperplane with respect to the polytope defining the feasible region, we keep moving it in the direction of $\mathbf{c}$ until it becomes a supporting hyperplane. A *supporting hyperplane* is one that contains the polytope region in one of its closed half spaces and contains a boundary point of the polytope. From the two-dimensional problem posed above, it is easy to see that the supporting hyperplane with normal direction $\mathbf{c}$ passes through an extreme point. The maximum value occurs on the boundary, and there is always an extreme point with this maximum value.

In n-dimensions, a hyperplane is $(n-1)$ dimensional. Let us consider n-hyperplanes with linearly independent normal directions. The intersection of these n-hyperplanes is a point, which is zero-dimensional. The intersection of $(n-1)$ hyperplanes is a line, which is of dimension 1. In general, the intersection of k hyperplanes is $(n-k)$ dimensional for $k = 1$ to n. If a supporting hyperplane with normal direction $\mathbf{c}$ described in the aforementioned touches the polytope at the maximum on a facet formed by the intersection of k-hyperplanes, then all the points on that facet have the same value. There are multiple solutions in this case.

A constraint is said to be *active* or *binding* at a point if the constraint is satisfied with an equality, for instance, at C in Fig. 4.3, constraints [1] and [2] are active.

4.6 Standard form of an LP

Standard form of the LP is one in which all constraints are converted into the equality (EQ) type. This form is a starting point for discussing the simplex and interior methods discussed in the following. Computer codes usually do this conversion internally, as opposed to requiring the user to do so. LE type constraints are converted into equalities by adding nonnegative variable $x_i (i = 1 \text{ to } \ell)$, called *slack variables* on the left-hand side. GE type constraints are converted into equalities by subtracting nonnegative variables $x_j (j = \ell + 1 \text{ to } \ell + r)$, called *surplus variables*. Variables *unrestricted in sign*, also called "free variables," can be brought into the standard form by expressing each such variable as a difference of two nonnegative variables and substituting into the preceding relations.

Example 4.1

Consider the problem

$$\begin{aligned} \text{minimize} \quad & 3x_1 - 5x_2 \\ \text{subject to} \quad & x_1 + x_2 \leq -2 \\ & 4x_1 + x_2 \geq -5 \\ & x_1 \geq 0, \quad x_2 \text{ unrestricted in sign} \end{aligned}$$

To convert into standard form, we first multiply the constraints by –1 to make the right-hand sides positive, and note that minimize f is equivalent to maximize $-f$:

$$\begin{aligned} \text{maximize} \quad & -3x_1 + 5x_2 \\ \text{subject to} \quad & -x_1 - x_2 \geq 2 \\ & -4x_1 - x_2 \leq 5 \\ & x_1 \geq 0, \quad x_2 \text{ unrestricted in sign} \end{aligned}$$

We now introduce surplus and slack variables, S and s, respectively, and also express x_2 as $x_2 = y_1 - y_2$ where y_1 and y_2 are both nonnegative. We obtain

$$\begin{aligned} \text{maximize} \quad & -3x_1 + 5y_1 - 5y_2 \\ \text{subject to} \quad & -x_1 - y_1 + y_2 - S = 2 \\ & -4x_1 - y_1 + y_2 + s = 5 \\ & x_1, y_1, y_2, S, s \geq 0 \end{aligned}$$

The standard form of the LP problem Eq. (4.1) can be written as

$$\begin{aligned} \text{maximize} \quad & c_1x_1 + c_2x_3 + c_3x_3 + \cdots + c_nx_n \\ \text{subject to} \quad & a_{i1}x_1 + a_{i2}x_2 + a_{i3}x_3 + \cdots + a_{in}x_n + x_{n+i} = b_i \quad (i = 1 \text{ to } \ell) \\ & a_{j1}x_1 + a_{j2}x_2 + a_{j3}x_3 + \cdots + a_{jn}x_n - x_{n+j} = b_j \quad (j = \ell + 1 \text{ to } \ell + r) \\ & a_{k1}x_1 + a_{k2}x_2 + a_{k3}x_3 + \cdots + a_{kn}x_n = b_k \quad (k = \ell + r + 1 \text{ to } m) \\ & x_1 \geq 0, \quad x_2 \geq 0, \ldots, x_{n+\ell+r} \geq 0 \end{aligned} \tag{4.5}$$

By adding some zero coefficients in $\mathbf{c}$ and making it an $(n + \ell + r) \times 1$ vector, $\mathbf{x}$ an $(n + \ell + r) \times 1$ vector, $\mathbf{A}$ an $m \times (n + \ell + r)$ matrix, and $\mathbf{b}$ an $m \times 1$ vector,

We can write the above formulation in the form

$$\begin{aligned} \text{maximize} \quad & \mathbf{c}^{\mathrm{T}}\mathbf{x} \\ \text{subject to} \quad & \mathbf{A}\mathbf{x} = \mathbf{b} \\ & \mathbf{x} \geq \mathbf{0} \end{aligned} \tag{4.6}$$

4.7 The Simplex Method – Starting with LE (≤) Constraints

The simplex method provides a systematic algebraic procedure for moving from one extreme point to an adjacent one while improving the function value. The method was developed by George Dantzig in 1947 and has been widely used since then. The first step is to convert the LP problem into standard form as in (4.6), with $\mathbf{b} \geq 0$. By introducing slack variables, x_3, x_4, and x_5, the problem defined in (4.4) then becomes

$$\begin{aligned} \text{maximize} \quad & 2x_1 + x_2 \\ \text{subject to} \quad & 2x_1 - x_2 + x_3 = 8 && (1) \\ & x_1 + 2x_2 + x_4 = 14 && (2) \\ & -x_1 + x_2 + x_5 = 4 && (3) \\ & x1 \geq 0, \quad x2 \geq 0, \quad x3 \geq 0, \quad x4 \geq 0, \quad x5 \geq 0 \end{aligned}$$

Denoting $\mathbf{c}^{\mathrm{T}} = [2,\ 1,\ 0,\ 0,\ 0]$, $\mathbf{x}^{\mathrm{T}} = [x_1,\ x_2,\ x_3,\ x_4,\ x_5]$, $\mathbf{b}^{\mathrm{T}} = [8,\ 14,\ 4]$, and $\mathbf{A} = \begin{bmatrix} 2 & -1 & 1 & 0 & 0 \\ 1 & 2 & 0 & 1 & 0 \\ -1 & 1 & 0 & 0 & 1 \end{bmatrix}$, the problem can be stated in the form of (4.6). The three equality constraints are defined in terms of five variables. If these equations are linearly independent (**A** is of rank 3), a solution can be obtained if any two (= 5 – 3) variables are set equal to zero. This solution is called a *basic solution*. If **A** is an m row $n + m$ column matrix, where n is the original dimension of the problem and m is the number of constraints, a basic solution is obtained by setting n variables equal to zero. If all the variables of the basic solution are nonnegative, then the solution is a *basic feasible solution* (bfs). Every basic feasible solution is an extreme point of the feasible polytope.

The simplex method is initiated when:

(i) a basic feasible solution is obtained,
(ii) the objective is written only in terms of the nonbasic variables.

For the problem with LE constraints, the standard form given by (4.6) is canonical. If we set $x_1 = 0$ and $x_2 = 0$, then $x_3 = 8\ (= b_1)$, $x_4 = 14\ (= b_2)$,

$x_5 = 4\ (= b_3)$. The basic feasible solution is, thus, readily obtained: the slacks are basic, the original variables are nonbasic. Moreover, the cost row is in terms of the original or nonbasic variables.[1] This starting point is the corner A in Fig. 4.3. Now, since x_1 has a larger coefficient we decide to move along x_1 from A. We have, thus, decided that x_1 should turn into a basic variable or, in other words, we say x_1 enters the basis. In the simplex method, the first row is set as negative of the objective function row for maximization, and we look for the variable with the largest negative coefficient (explained in detail subsequently). The simplex method is initiated with the following tableau

First Tableau (Corresponds to Point A):

	x_1	x_2	x_3	x_4	x_5	rhs
f	−2	−1	0	0	0	0
x_3	[2]	−1	1	0	0	8
x_4	1	2	0	1	0	14
x_5	−1	1	0	0	1	4

As x_1 is increased, we look for the largest value that still retains feasibility. This is checked by performing the *ratio test*. For each constraint i with a positive coefficient a_{i1}, we evaluate the ratio b_i/a_{i1}. Among these ratios, we identify the smallest value. This row is now referred to as the pivot row and the column corresponding to the variable entering the basis is the pivot column. In the problem at hand, we compare 8/2 and 14/1. The first row is the pivot row. The third coefficient is negative, and this does not control the upper limit on x_1. The basic variable corresponding to this row is x_3. Pivot element is [2]. Thus, x_3 leaves the basis. That is, x_3 becomes zero, which, in view of the fact that it is the slack in constraint 1, means that we move to a point where the constraint becomes active. In Fig. 4.3, we move from A to B. Elementary row operations (ERO) as discussed in detail in Chapter 1 are now performed for pivot element at row 1 and column 1. The idea of the operation is to make the pivot location equal to 1 by dividing the entire equation by the pivot value, and making the other coefficients in that column equal to zero by adding an appropriate multiple of the row to each of the other constraint equations. The elementary row operation is performed on the row corresponding to the objective function also. The second tableau is as follows:

[1] Significance of cost row coefficients are explained further under the heading "Optimality Conditions for Problems with LE Constraints."

Second Tableau (Corresponds to Point B):

	x_1	x_2	x_3	x_4	x_5	rhs
f	0	−2	1	0	0	8
x_1	1	$-\frac{1}{2}$	$\frac{1}{2}$	0	0	4
x_4	0	$\left[\frac{5}{2}\right]$	$-\frac{1}{2}$	1	0	10
x_5	0	$\frac{1}{2}$	$\frac{1}{2}$	0	1	8

Now the most negative coefficient in the first row is −2 corresponding to variable x_2. Thus, x_2 enters the basis. The ratio test 10/(5/2) and 8/(1/2) shows that x_4 should leave the basis. Pivot element is 5/2. After performing the elementary row operations, the third tableau becomes

Third Tableau (Corresponds to Point C):

	x_1	x_2	x_3	x_4	x_5	rhs
f	0	0	$\frac{3}{5}$	$\frac{4}{5}$	0	16
x_1	1	0	$\frac{2}{5}$	$\frac{1}{5}$	0	6
x_2	0	1	$-\frac{1}{5}$	$\frac{2}{5}$	0	4
x_5	0	0	$\frac{3}{5}$	$-\frac{1}{5}$	1	6

Now all the coefficients in the first row are positive, and it yields the maximum value of the function $f = 16$. The solution is $x_1 = 6$, $x_2 = 4$, $x_3 = 0$, $x_4 = 0$, $x_5 = 6$.

In the program SIMPLEX, the first row is set as the last row. Further, the ratio test needs to be implemented with a small tolerance as: $\mathrm{B(I)/(A(I, ICV) + 1E - 10)}$, where ICV is the incoming variable.

Minimization

If $\mathbf{c}^{\mathrm{T}}\mathbf{x}$ is to be minimized, it is equivalent to maximization of $-\mathbf{c}^{\mathrm{T}}\mathbf{x}$. Taking the negative coefficients of the maximization problem, the first row has just the $\mathbf{c}$ values. The right-hand side will be obtained as negative of the function value. We need to apply a final change in sign for the function value.

A *degenerate basic feasible solution* is one in which one or more variables in the basis have a value of zero. In this situation, since nonbasic variables are at zero, bringing zero level basic variable into nonbasic set does not improve the function value. However, degenerate solutions do not pose problems in the working of the simplex method. Some aspects of recycling are discussed in the next section.

Steps Involved in the Simplex Method for LE Constraints

Step 1: Create the initial tableau in standard form, with negative of the objective coefficients for maximization problem (original coefficients for minimization).

Step 2: Choose the incoming variable j corresponding to the most negative objective row coefficient (pivot column). *If there is no negative coefficient, the optimum has been attained.*

Step 3: Perform the ratio test b_i/a_{ij} for positive coefficients[2] a_{ij} and determine the row i corresponding to the least ratio (pivot row). *If there is no positive coefficient in the column, the solution is unbounded.*

Step 4: Perform the elementary row operations to make the pivot equal to 1 and other coefficients in the column equal to 0, including the first row. Go to Step 2.

Optimality Conditions and Significance of Cost Row Coefficients

In the preceding text, slacks x_3, x_4, x_5 were chosen as basic to start with, point A in Fig. 4.3, which resulted in the objective $f = 2\,x_1 + x_2$ to be automatically expressed in terms of nonbasic variables. This is desired because we may then scan the coefficients to see which nonbasic variable, currently at zero value, should enter the basis so as to increase the objective function value. However, for the objective maximize $\mathbf{c}^{\mathrm{T}}\mathbf{x}$, it was stated earlier, without explanation that $-\mathbf{c}$ must be inserted in the cost row in the simplex tableau. The reasoning behind this has to do with optimality and is now explained. Once the initial tableau is defined, the steps above will lead to an optimum (or detection of infeasibility or unboundedness).

Recall from the earlier chapters that the gradient of a function is normal to its contour and points in the direction of increasing function value. Now, it is evident from Fig. 4.3 that point A is not an maximum. Observing the direction of $\mathbf{c}$ and the feasible region Ω, any "move" along the x_1- or x_2-axis will increase f while staying in Ω. More generally, since $\mathbf{c}$ is the gradient of f, any direction of travel that points into the half-space toward $\mathbf{c}$ will increase f. At point B, a move from B to C will improve the objective, at point E, a move from point E to D, and at point D a move from

[2] In a computer code, a small tolerance must be used to decide when a_{ij} is positive.

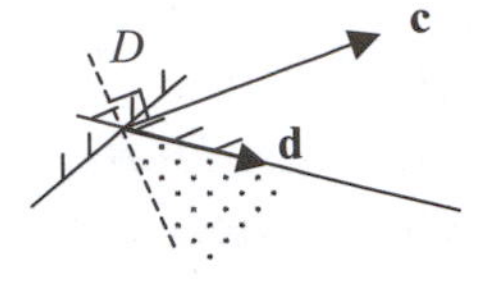

Figure 4.4. Nonoptimality at point D in Fig. 4.3.

D to C. At point C, there is no move that will improve the objective indicating that C is optimum. Figure 4.4 illustrates the situation at point D – any direction vector in the dotted cone will improve f while staying in Ω, although the simplex method will choose $\mathbf{d}$ as in the figure, and thus, move along the boundary, from vertex to vertex.

Karusch–Kuhn–Tucker or KKT conditions describe optimality or nonoptimality at a point generally so as to be applicable in N-dimensions. While these are explained in greater detail in Chapter 5, they are also discussed below to illuminate the choice and significance of cost row coefficients. For maximization, the cost gradient is expressed as a linear combination of *active* constraint gradients as

$$\mathbf{c} = \sum_{i \in I} \mu_i \mathbf{N}^i \tag{4.7}$$

where I = index set of active constraints at a point, and $\mathbf{N}^i$ is the gradient or normal of the ith active constraint (either $\mathbf{a}_i$ or $-\mathbf{e}_i$), μ_i are Lagrange multipliers. Equation (4.7) states that $\mathbf{c}$ lies in the *convex cone* formed by the gradients of the active constraints.

Optimality with LE ($\leq$) constraints: Equation (4.7) must be satisfied with all $\mu_i \geq 0, i \in I$, and the point must be feasible.

At point A, active constraint gradients are $\left[\begin{smallmatrix}-1\\0\end{smallmatrix}\right], \left[\begin{smallmatrix}0\\-1\end{smallmatrix}\right]$. Thus, $\mathbf{c} = \left[\begin{smallmatrix}2\\1\end{smallmatrix}\right]$ are expressed in terms of the constraint gradients as $\left[\begin{smallmatrix}2\\1\end{smallmatrix}\right] = -2\left[\begin{smallmatrix}-1\\0\end{smallmatrix}\right] - 1\left[\begin{smallmatrix}0\\-1\end{smallmatrix}\right]$, from which we have $\mu_1 = -2$, $\mu_2 = -1$. The cost row coefficients play the role of μ_i. Hence, as per KKT conditions, point A is not optimal.

At point D, we can write $\left[\begin{smallmatrix}2\\1\end{smallmatrix}\right] = \mu_1\left[\begin{smallmatrix}-1\\1\end{smallmatrix}\right] + \mu_2\left[\begin{smallmatrix}1\\2\end{smallmatrix}\right]$, which, upon solving, yields $\mu_1 = -1$, $\mu_2 = 1$, hence, not optimal.

At point C, we can write $\left[\begin{smallmatrix}2\\1\end{smallmatrix}\right] = \mu_1\left[\begin{smallmatrix}1\\2\end{smallmatrix}\right] + \mu_2\left[\begin{smallmatrix}2\\-1\end{smallmatrix}\right]$, which gives $\mu_1 = 0.8$, $\mu_2 = 0.6$, hence, optimal. Point C is a "KKT point."

[What are the Lagrange multipliers at points B and E? Relate multipliers at B with the simplex tableau earlier]

Geometric Interpretation of the Pivot Operation

The choice of the variable entering the basis, based on most negative reduced cost coefficient (step 2 earlier) can be interpreted as *a move based on maximum expected increase in objective function* (as we choose the most negative cost row coefficient to enter the basis), until point C is reached. Thus, at point A, since $\mathbf{c} = [2, 1]^{\mathrm{T}}$, the "slope" or gradient of the objective from A to B equals 2 while from A to E equals 1. Thus, based on information at point A only, the move from A to B corresponds to *maximum expected increase* in objective function, and the *active bound constraint* x_2 becomes inactive (i.e., x_2 becomes positive). However, the *actual* increase also depends on the distance from A to B or from A to E, which will be known only after the pivot operation.

Further, the choice of the variable that leaves the basis is essentially the determination of a variable that changes from a positive value to zero. This reduces to determination of which constraint becomes *active*. If the variable is an original variable, then the *bound* becomes active. If the variable corresponds to a slack, then the corresponding inequality constraint becomes active. The ratio test in step 3 earlier determines the *closest constraint that intersects the path* from the existing vertex to the next.

In summary, incoming variable (step 2) determines a *direction of movement* in **x**-space, while the outgoing variable (step 3) determines the *step size* to the boundary of the feasible polytope.

KKT Conditions

KKT conditions in Eq. (4.7) may be restated by associating zero value Lagrange multipliers with inactive constraints. That is, for $i \notin I$, implying $g_i < 0$, we require $\mu_i = 0$. This can be enforced by the *complementarity condition* $\mu_i \, g_i = 0$.

Problem

$$\begin{aligned} &\text{maximize} \quad \mathbf{c}^{\mathrm{T}}\mathbf{x} \\ &\text{subject to} \quad \mathbf{A}\mathbf{x} \leq \mathbf{b} \\ &\qquad\qquad\quad \mathbf{x} \geq \mathbf{0} \end{aligned} \tag{4.8}$$

KKT Conditions

$$\text{Feasibility} \qquad \mathbf{A}\mathbf{x} + \mathbf{y} = \mathbf{b} \tag{4.9}$$

$$\text{Optimality} \qquad \mathbf{c} = \mathbf{A}^{\mathrm{T}}\mathbf{v} - \mathbf{u} \tag{4.10}$$

$$\text{Complementarity} \qquad \mathbf{u}^{\mathrm{T}}\mathbf{x} + \mathbf{v}^{\mathrm{T}}\mathbf{y} = 0 \tag{4.11}$$

$$\text{Nonnegativity} \qquad \mathbf{x} \geq \mathbf{0}, \quad \mathbf{y} \geq \mathbf{0}, \quad \mathbf{u} \geq \mathbf{0}, \quad \mathbf{v} \geq \mathbf{0} \tag{4.12}$$

Note that feasibility $\mathbf{A}\mathbf{x} \leq \mathbf{b}$ has been expressed as $\mathbf{A}\mathbf{x} + \mathbf{y} = \mathbf{b}$, $\mathbf{y} \geq \mathbf{0}$. Direct solution of the KKT conditions Eqs. (4.9)–(4.12) is difficult owing to the complementarity conditions, or equivalently, owing to the task of determining the active set.

The KKT conditions define a complementary problem. The linear complementary problem (LCP) for quadratic programming (QP) has a very similar structure, as discussed in a later section.

4.8 Treatment of GE and EQ Constraints

GE and EQ constraints need some special consideration in the simplex initiation. We present the steps involved in the solution by considering a problem involving LE, GE, and EQ constraints.

$$\begin{aligned}
\text{maximize} \quad & x_1 + x_2 + 2x_3 \\
\text{subject to} \quad & 2x_1 + x_2 + 2x_3 \leq 8 \quad [1] \\
& x_1 + x_2 + \ \ x_3 \geq 2 \quad [2] \\
& -x_1 + x_2 + 2x_3 = 1 \quad [3] \\
& x_1 \geq 0, \quad x_2 \geq 0, \quad x_3 \geq 0
\end{aligned} \tag{4.13}$$

We first bring the problem into standard form by introducing a slack variable x_4 and a surplus variable x_5 as

$$\begin{aligned}
\text{maximize} \quad & x_1 + x_2 + 2x_3 + 0x_4 + 0x_5 \\
\text{subject to} \quad & 2x_1 + x_2 + 2x_3 + x_4 = 8 \quad (1) \\
& x_1 + x_2 + \ \ x_3 - x_5 = 2 \quad (2) \\
& -x_1 + x_2 + 2x_3 \qquad = 1 \quad (3) \\
& x_1 \geq 0, \quad x_2 \geq 0, \quad x_3 \geq 0, \quad x_4 \geq 0, \quad x_5 \geq 0
\end{aligned} \tag{4.14}$$

In the preceding standardized form, a basic feasible starting point is not readily available. That is, a basic feasible solution whereby two of the variables are set to zero and the constraints then furnish the three nonnegative basic variables is not apparent, as was the case with only LE constraints. Our first step in the initiation of the simplex method is to establish a basic feasible solution. Two different approaches to tackle such problems will now be discussed. The Two-Phase approach and the Big M method presented in the following need the introduction of artificial variables.

The Two-Phase Approach

In the two-phase approach, we introduce *artificial variables* x_6 and x_7 in Eqs. (2) and (3) in (4.14), and set up an auxiliary problem that will bring these artificial variables to zero. This auxiliary LP problem is the Phase I of the simplex method.

Phase I for Problems with Artificial Variables

One *artificial variable* for each constraint of the GE and EQ type are added to the standard form in (4.14), and the objective function is set as the minimization of the sum of the artificial variables.

$$\begin{aligned} \text{minimize} \quad & x_6 + x_7 \\ \text{subject to} \quad & 2x_1 + x_2 + 2x_3 + x_4 = 8 \quad (1) \\ & x_1 + x_2 + x_3 - x_5 + x_6 = 2 \quad (2) \\ & -x_1 + x_2 + 2x_3 + x_7 = 1 \quad (3) \\ & x_1 \geq 0, \quad x_2 \geq 0, \quad x_3 \geq 0, \quad x_4 \geq 0, \\ & x_5 \geq 0, \quad x_6 \geq 0, \quad x_7 \geq 0 \end{aligned} \tag{4.15}$$

This set is in canonical form with the slack and artificial variables in the basis, excepting that coefficients of x_6 and x_7 are not zero in the objective function. Recall that the objective function must be written only in terms of the nonbasic variables. Since this is a minimization problem, we write the first row coefficients same as mentioned previously.

Initial Tableau for Phase I

	x_1	x_2	x_3	x_4	x_5	x_6	x_7	rhs
f	0	0	0	0	0	1	1	0
x_4	2	1	2	1	0	0	0	8
x_6	1	1	1	0	−1	1	0	2
x_7	−1	1	2	0	0	0	1	1

To make the coefficients of x_6 and x_7 zero in the first row, the last two equations are to be subtracted from the first row. Equivalently, we can use the constraints (2) and (3) in (4.15) to write $f = x_6 + x_7 = (2 - x_1 - x_2 - x_3 + x_5) + (1 + x_1 - x_2 - 2x_3) = -2x_2 - 3x_3 + x_5 + 3$.

First Step Is To Bring the Equations to Canonical Form

	x_1	x_2	x_3	x_4	x_5	x_6	x_7	rhs
f	0	−2	−3	0	1	0	0	−3
x_4	2	1	2	1	0	0	0	8
x_6	1	1	1	0	−1	1	0	2
x_7	−1	1	[2]	0	0	0	1	1

Phase I is solved using the simplex method. The tableaus leading to the minimum, are given as follows. If the minimum objective of zero cannot be achieved, the original problem does not have a feasible solution.

	x_1	x_2	x_3	x_4	x_5	x_6	x_7	rhs
f	$-\frac{3}{2}$	$-\frac{1}{2}$	0	0	1	0	$\frac{3}{2}$	$-\frac{3}{2}$
x_4	3	0	0	1	0	0	−1	7
x_6	$[\frac{3}{2}]$	$\frac{1}{2}$	0	0	−1	1	$-\frac{1}{2}$	$\frac{3}{2}$
x_3	$-\frac{1}{2}$	$\frac{1}{2}$	1	0	0	0	$\frac{1}{2}$	$\frac{1}{2}$

	x_1	x_2	x_3	x_4	x_5	x_6	x_7	rhs
f	0	0	0	0	0	1	1	0
x_4	0	−1	0	1	2	−2	0	4
x_1	1	$\frac{1}{3}$	0	0	$-\frac{2}{3}$	$\frac{2}{3}$	$-\frac{1}{3}$	1
x_3	0	$\frac{2}{3}$	1	0	$-\frac{1}{3}$	$\frac{1}{3}$	$\frac{1}{3}$	1

This point with $x_1 = 1$ and $x_3 = 1$ and all other variables at zero correspond to the point A shown in Fig. 4.5. The minimum function value is zero. Also note that the artificial variables are not in the basis. Phase II can be initiated now.

Phase II

Since all artificial variables are nonbasic, the columns corresponding to those variables can be dropped. First, the original objective function (negative for maximization problem) is put in the first row, and elementary row operations are performed to get zero values in the entries corresponding to basic variables.

Initial Tableau with Objective Function Added and Artificial Variables Dropped

	x_1	x_2	x_3	x_4	x_5	rhs
f	−1	−1	−2	0	0	0
x_4	0	−1	0	1	2	4
x_1	1	$\frac{1}{3}$	0	0	$-\frac{2}{3}$	1
x_3	0	$\frac{2}{3}$	[1]	0	$-\frac{1}{3}$	1

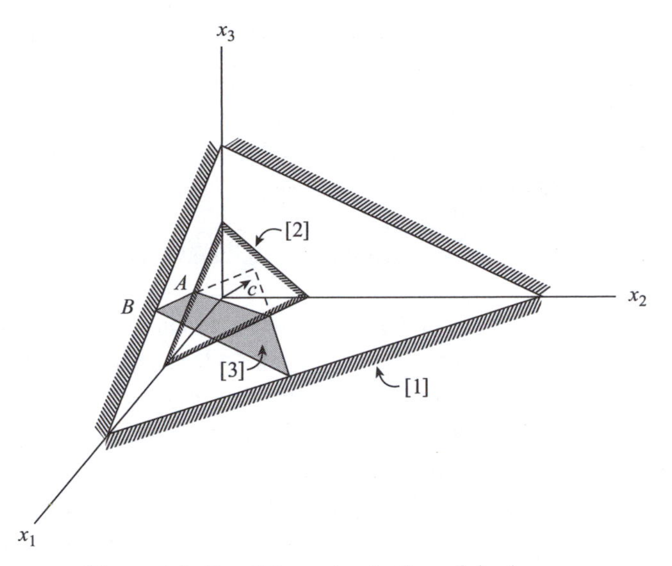

Figure 4.5. Feasible region in the original space.

Coefficients of Basic Variables in the First Row Made Zero

	x_1	x_2	x_3	x_4	x_5	rhs
f	0	$\frac{2}{3}$	0	0	$-\frac{4}{3}$	3
x_4	0	-1	0	1	[2]	4
x_1	1	$\frac{1}{3}$	0	0	$-\frac{2}{3}$	1
x_3	0	$\frac{2}{3}$	1	0	$-\frac{1}{3}$	1

The coefficients in the first row are nonnegative and thus, the solution is optimal.

Second Tableau

	x_1	x_2	x_3	x_4	x_5	rhs
f	0	0	0	$\frac{2}{3}$	0	$\frac{17}{3}$
x_5	0	$-\frac{1}{2}$	0	$\frac{1}{2}$	1	2
x_1	1	0	0	$\frac{1}{3}$	0	$\frac{7}{3}$
x_3	0	$\frac{1}{2}$	1	$\frac{1}{6}$	0	$\frac{5}{3}$

In general, there could exist artificial variables in the basis at zero value. First, redundant rows in the tableau must be detected and deleted. Second, the artificial variables must be pivoted out. See programs implementations.

The Big M Method

The Big M method is a penalty method where the two phases are combined into one. To the original objective function to be maximized, we add negative of a big number M times the sum of the artificial variables that must be nonnegative. The modified objective function is

$$\text{maximize } \mathbf{c}^{\mathrm{T}}\mathbf{x} - M(\text{sum of artificial variables}) \tag{4.16}$$

Thus, the, maximum must be achieved when the artificial variables attain zero values. First, the terms in the first row corresponding to the artificial variables are made zero by elementary row operations. The computer program SIMPLEX implements the Big M method. In the program, M is taken as ten times the sum of the absolute values of the objective function coefficients.

The output from the program for the problem presented in (4.14) above is given as follows.

Initial Tableau

4	2.00	1.00	2.00	1.00	0.00	0.00	0.00	8.00
6	1.00	1.00	1.00	0.00	−1.00	1.00	0.00	2.00
7	−1.00	1.00	2.00	0.00	0.00	0.00	1.00	1.00
OBJ	−1.00	−1.00	−2.00	0.00	0.00	40.00	40.00	0.00

Initial Tableau after Adjustment

4	2.00	1.00	2.00	1.00	0.00	0.00	0.00	8.00
6	1.00	1.00	1.00	0.00	−1.00	1.00	0.00	2.00
7	−1.00	1.00	2.00	0.00	0.00	0.00	1.00	1.00
OBJ	−1.00	−81.00	−122.00	0.00	40.00	0.00	0.00	−120.00

Iteration 1 Inc.Var 3 Pivot Row 3

4	3.00	0.00	0.00	1.00	0.00	0.00	−1.00	7.00
6	1.50	0.50	0.00	0.00	−1.00	1.00	−0.50	1.50
3	−0.50	0.50	1.00	0.00	0.00	0.00	0.50	0.50
OBJ	−62.00	−20.00	0.00	0.00	40.00	0.00	61.00	−59.00

Iteration 2 Inc.Var 1 Pivot Row 2

4	0.00	−1.00	0.00	1.00	2.00	−2.00	0.00	4.00
1	1.00	0.33	0.00	0.00	−0.67	0.67	−0.33	1.00
3	0.00	0.67	1.00	0.00	−0.33	0.33	0.33	1.00
OBJ	0.00	0.67	0.00	0.00	−1.33	41.33	40.33	3.00

Iteration 3 Inc.Var 5 Pivot Row 1

5	0.00	−0.50	0.00	0.50	1.00	−1.00	0.00	2.00
1	1.00	0.00	0.00	0.33	0.00	0.00	−0.33	2.33
3	0.00	0.50	1.00	0.17	0.00	0.00	0.33	1.67
OBJ	0.00	0.00	0.00	0.67	0.00	40.00	40.33	5.67

We note here that the number of iterations has not been reduced. In the Big M method, the constant M must be chosen carefully. If it is too small, inconsistent results may occur. If it is too large, rounding off will result in large errors.

4.9 Revised Simplex Method

In the conventional simplex approach, every element of the tableau is calculated at every iteration. This results in lot of unnecessary bookkeeping and wasted computational effort. The Revised Simplex method is a more efficient technique where only the necessary elements are evaluated at a given stage. The revised simplex method also makes it easy to do sensitivity analysis as discussed in Section 4.12.

Reviewing the simplex method, we note that at any stage we need to evaluate the reduced coefficients of the objective function. Then, having located the most negative element, we need to evaluate the corresponding column of $\mathbf{A}$ and the vector $\mathbf{b}$ for that iteration and then perform the ratio test to locate the pivot row. All these are done systematically in the revised simplex method. Consider the LP problem that has been set in the canonical form by using slack, surplus, and artificial variables as needed.

$$\begin{aligned} \text{Maximize} \quad & f \equiv \mathbf{c}^{\mathrm{T}}\mathbf{x} \\ \text{Subject to} \quad & \mathbf{Ax} = \mathbf{b} \\ & \mathbf{x} \geq \mathbf{0} \end{aligned} \tag{4.17}$$

In order to develop the steps involved, we partition $\mathbf{x}$ into two sets – basic variables $\mathbf{x}_\mathrm{B}$ and nonbasic variables $\mathbf{x}_\mathrm{N}$. This naturally partitions submatrices $\mathbf{B}$ and $\mathbf{N}$ such that $\mathbf{A} = [\mathbf{B} \quad \mathbf{N}]$. $\mathbf{c}$ is partitioned into $\mathbf{c}_\mathrm{B}$ and $\mathbf{c}_\mathrm{N}$. Thus, (4.17) can be written in the form

$$
\begin{aligned}
\text{Maximize} \quad & f = \mathbf{c}_\mathrm{B}^\mathrm{T}\mathbf{x}_\mathrm{B} + \mathbf{c}_\mathrm{N}^\mathrm{T}\mathbf{x}_\mathrm{N} \\
\text{Subject to} \quad & \mathbf{B}\mathbf{x}_\mathrm{B} + \mathbf{N}\mathbf{x}_\mathrm{N} = \mathbf{b} \\
& \mathbf{x} \geq \mathbf{0}
\end{aligned} \tag{4.18}
$$

If $\mathbf{A}$ is of size $m \times n$ and of rank m, the partitioned matrix $\mathbf{B}$ is of size $m \times m$ and has an inverse, say, $\mathbf{B}^{-1}$. We multiply the matrix equation $\mathbf{B}\mathbf{x}_\mathrm{B} + \mathbf{N}\mathbf{x}_\mathrm{N} = \mathbf{b}$ by $\mathbf{B}^{-1}$ to obtain

$$
\mathbf{x}_\mathrm{B} = \mathbf{B}^{-1}\mathbf{b} - \mathbf{B}^{-1}\mathbf{N}\mathbf{x}_\mathrm{N} \tag{4.19}
$$

We notice that at the start, the coefficients $\mathbf{c}_\mathrm{B}$ are each equal to zero, and $\mathbf{B}$ is the identity matrix $\mathbf{I}$ and its inverse $\mathbf{B}^{-1}$ is also $\mathbf{I}$. Thereafter, $\mathbf{B}$ is no longer $\mathbf{I}$. In the revised simplex method, only $\mathbf{B}^{-1}$ and $\mathbf{b}$ are stored and updated.

Substituting for $\mathbf{x}_\mathrm{B}$ from Eq. (4.19) into the objective function, we get

$$
f = \mathbf{c}_\mathrm{B}^\mathrm{T}\mathbf{B}^{-1}\mathbf{b} - \mathbf{r}_\mathrm{N}^\mathrm{T}\mathbf{x}_\mathrm{N} \tag{4.20}
$$

where $\mathbf{r}_\mathrm{N}$ is the set of reduced coefficients given by

$$
\mathbf{r}_\mathrm{N}^\mathrm{T} = \mathbf{c}_\mathrm{B}^\mathrm{T}\mathbf{B}^{-1}\mathbf{N} - \mathbf{c}_\mathrm{N}^\mathrm{T} \tag{4.21}
$$

Thus, the simplex tableau can be denoted as

	$\mathbf{x}^\mathrm{B}$	$\mathbf{x}^\mathrm{N}$	Current solution
Objective	$\mathbf{0}$	$\mathbf{r}_\mathrm{N}^\mathrm{T} = \mathbf{c}_\mathrm{B}^\mathrm{T}\,\mathbf{B}^{-1}\mathbf{N} - \mathbf{c}_\mathrm{N}^\mathrm{T}$	$\mathbf{c}_\mathrm{B}^\mathrm{T}\,\mathbf{B}^{-1}\mathbf{b}$
$\mathbf{x}^\mathrm{B}$	$\mathbf{I}$	$\mathbf{B}^{-1}\mathbf{N}$	$\mathbf{B}^{-1}\mathbf{b}$

At the beginning, since $\mathbf{c}_\mathrm{B} = \mathbf{0}$, $\mathbf{r}_\mathrm{N} = -\,\mathbf{c}_\mathrm{N}^\mathrm{T}$. We look for the most negative element of $\mathbf{r}_\mathrm{N}$ which is the variable entering the basis. If there is no negative element, the optimum is attained. Let us say that the entering variable (currently nonbasic) corresponds to column j of $\mathbf{A}$. Now the product of $\mathbf{B}^{-1}$ with column j of $\mathbf{A}$ gives the modified column $\mathbf{y}$ (obtained by elementary row operations in the conventional approach). Similarly, the product of $\mathbf{B}^{-1}$ with $\mathbf{b}$ gives the modified right-hand side $\mathbf{b}'$ $(=\mathbf{B}^{-1}\mathbf{b})$ for the current iteration. $\mathbf{b}'$ is the current basic solution to the problem. The ratio test is readily performed using elements of $\mathbf{b}'$ and $\mathbf{y}$ and the pivot row corresponding to the least ratio for positive column element is determined. The variable corresponding to this pivot row is the variable that exits from the basis.

The inverse of the new basis matrix may be obtained by direct inversion of the $\mathbf{B}$ matrix. However, computational savings are obtained using the following scheme, although it is advised to periodically invert $\mathbf{B}$ to avoid roundoff errors. If i is the index of the variable exiting the basis, it means that we are essentially replacing the column corresponding to ith variable in $\mathbf{B}$ with the column j of $\mathbf{A}$. Since we are having the inverse $\mathbf{B}^{-1}$ of $\mathbf{B}$ available at this stage, we only need to update the inverse when only one column of $\mathbf{B}$ is replaced. This updating is efficiently done by performing the following steps. Let $\mathbf{B}'$ be the new $\mathbf{B}$ with column i replaced by the column j of $\mathbf{A}$. Then, the product of $\mathbf{B}^{-1}$ and $\mathbf{B}'$ is readily written as

$$\mathbf{B}^{-1}\mathbf{B}' = \begin{bmatrix} 1 & 0 & \cdots & y_1 & 0 \\ 0 & 1 & \cdots & y_2 & 0 \\ \vdots & \vdots & \ddots & \vdots & \vdots \\ 0 & 0 & \cdots & y_i & 0 \\ 0 & 0 & \cdots & y_m & 1 \end{bmatrix} \tag{4.22}$$

The column i, which is $\mathbf{y}$, is readily available since this was already evaluated for the ratio test. Postmultiplying both sides of Eq. (4.22) by $\mathbf{B}'^{-1}$, we get

$$\begin{bmatrix} 1 & 0 & \cdots & y_1 & 0 \\ 0 & 1 & \cdots & y_2 & 0 \\ \vdots & \vdots & \ddots & \vdots & \vdots \\ 0 & 0 & \cdots & y_i & 0 \\ 0 & 0 & \cdots & y_m & 1 \end{bmatrix} \mathbf{B}'^{-1} = \mathbf{B}^{-1} \tag{4.23}$$

Solution of the above matrix set of equations is a simple back-substitution routine of Gauss elimination method. The steps are given in the following equation.

$$\mathbf{B}'^{-1} = \text{Divide row } i \text{ of } \mathbf{B}^{-1} \text{ by } y_i \text{ to obtain row } i \tag{4.24}$$

Then to obtain row j, subtract y_j times row i from row j ($j = 1$ to $n, j \neq i$)

Once the inverse of the updated partitioned basis matrix is obtained, the steps of evaluating the reduced coefficients $\mathbf{r}_\mathrm{N}$, determination of entering variable, evaluation of $\mathbf{y}$ and $\mathbf{b}'$, performing the ratio test and determination of variable exiting the basis are performed. The iterations are continued until all the reduced coefficients are nonnegative. In Eq. (4.20), $\mathbf{c}_\mathrm{B}^\mathrm{T}\,\mathbf{B}^{-1}\mathbf{b}$ gives the value of the function maximized. In the revised simplex method, we keep track of $\mathbf{B}^{-1}$, $\mathbf{y}$ and $\mathbf{b}'$, and the original $\mathbf{A}$, $\mathbf{b}$, and $\mathbf{c}$ are left intact.

In problems where the number of iterations is large, updating $\mathbf{B}^{-1}$ at each stage may result in the accumulation of round off errors. In such problems, it is recommended that $\mathbf{B}^{-1}$ be obtained by inverting $\mathbf{B}$.

Example 4.2

The revised simplex procedure will now be illustrated on a two-variable problem that was solved above using the tableau procedure. The problem is:

$$\begin{aligned} \text{maximize} \quad & 2x_1 + x_2 \\ \text{subject to} \quad & 2x_1 - x_2 \le 8 \quad [1] \\ & x_1 + 2x_2 \le 14 \quad [2] \\ & -x_1 + x_2 \le 4 \quad [3] \\ & x_1 \ge 0, x_2 \ge 0 \end{aligned}$$

$$\text{Note that } \mathbf{A} = \begin{bmatrix} 2 & -1 & 1 & 0 & 0 \\ 1 & 2 & 0 & 1 & 0 \\ -1 & 1 & 0 & 0 & 1 \end{bmatrix}, \quad \mathbf{b} = \begin{bmatrix} 8 \\ 14 \\ 4 \end{bmatrix}$$

Iteration 1

$\mathbf{x}_B = \{x_3, x_4, x_5\}$, $\mathbf{x}_{NB} = \{x_1, \ x_2\}$, $\mathbf{c}_B^T = [0, 0, 0]$, $\mathbf{c}_N^T = [2, 1]$, $\mathbf{B} = \begin{bmatrix} 1 & 0 & 0 \\ 0 & 1 & 0 \\ 0 & 0 & 1 \end{bmatrix}$, $\mathbf{r}_N^T = \mathbf{c}_B^T\mathbf{B}^{-1}\mathbf{N} - \mathbf{c}_N^T = [-2, -1] \equiv [x_1, x_2]$, $f = 0$. Based on the the most negative component of $\mathbf{r}_N$, we choose x_1 to be the incoming variable. The vector $\mathbf{y}$ is given by $\mathbf{y} = \mathbf{B}^{-1}\begin{bmatrix} 2 \\ 1 \\ -1 \end{bmatrix} = \begin{bmatrix} 2 \\ 1 \\ -1 \end{bmatrix}$. The ratio test involves $\mathbf{b}$ and $\mathbf{y}$: the smallest of 8/2 and 14/1 leads to x_3 being the variable to exit the basis.

Iteration 2

$\mathbf{x}_B = \{x_1, x_4, x_5\}$, $\mathbf{x}_{NB} = \{x_3, x_2\}$, $\mathbf{c}_B^T = [2, 0, 0]$, $\mathbf{c}_N^T = [0, 1]$. $\mathbf{B} = \begin{bmatrix} 2 & 0 & 0 \\ 1 & 1 & 0 \\ -1 & 0 & 1 \end{bmatrix}$, $\mathbf{B}^{-1} = \begin{bmatrix} 0.5 & 0 & 0 \\ -0.5 & 1 & 0 \\ 0.5 & 0 & 1 \end{bmatrix}$, $\mathbf{N} = \begin{bmatrix} 1 & -1 \\ 0 & 2 \\ 0 & 1 \end{bmatrix}$, $\mathbf{b} = \mathbf{B}^{-1}\mathbf{b} = \begin{bmatrix} 4 \\ 10 \\ 8 \end{bmatrix}$, $f = 8$. The new $r_{\mathbf{N}}^T = \mathbf{c}_B^T\mathbf{B}^{-1}\mathbf{N} - \mathbf{c}_N^T = [1, -2] \equiv [x_3, x_2]$. Thus, x_2 enters the basis. The $\mathbf{y}$ vector is determined from: $\mathbf{y} = \begin{bmatrix} 0.5 & 0 & 0 \\ -0.5 & 1 & 0 \\ 0.5 & 0 & 1 \end{bmatrix}\begin{bmatrix} -1 \\ 2 \\ 1 \end{bmatrix} = \begin{bmatrix} -0.5 \\ 2.5 \\ 0.5 \end{bmatrix} \equiv \begin{bmatrix} x_1 \\ x_4 \\ x_5 \end{bmatrix}$ The smaller of 10/2.5 and 8/0.5 identifies x_4 as the variable that exits the basis. The reader may observe that the numbers generated by the revised simplex procedure are identical to the tableau (regular simplex) procedure. The reader is encouraged to complete iteration 3 in this example and verify the optimum solution of $f = 16, x_1 = 6, x_2 = 4, x_3 = 0, x_4 = 0, x_5 = 6$.

Cycling is said to occur in the simplex method if the same set of variables are in the basis after several iterations. Though cycling occurs rarely, researchers have provided example problems where cycling occurs. Bland's rule for preventing cycling is simple, robust, and easy to implement.

Bland's Rule for Preventing Cycling

(a) Among the negative reduced coefficients, choose the negative coefficient with the lowest index to enter the basis. In other words, choose $j = \min$ (j such that $r_j < 0$).
(b) In case a tie occurs in the determination of which variable leaves the basis, select the one with the lowest index.

Revised simplex method with cycling prevention rule has been implemented in the program REVSIMC.

4.10 Duality in Linear Programming

Associated with every linear programming problem, here after referred to as *primal problem*, there is a corresponding *dual* linear programming problem. The dual LP problem is constructed with the same coefficients as that of the primal. If one is a maximization problem, the other is a minimization problem. Duality exists in problems of engineering and economics. An electric circuit design problem may be posed as one based on electric potential or its dual based on current flow. Problems in mechanics may be based on strain (displacement) or the dual based on stress (force). In resource allocation problems, if the objective of the primal has price per unit of product, the dual looks for price per unit of resource that the producer has to pay. The KKT conditions given in Eqs. (4.9) – (4.12) provide a binding relationship between the primal and dual.

Primal		Dual		
maximize	$\mathbf{c}^{\mathrm{T}}\mathbf{x}$	minimize	$\mathbf{b}^{\mathrm{T}}\mathbf{v}$	(4.25)
subject to	$\mathbf{A}\mathbf{x} \leq \mathbf{b}$	subject to	$\mathbf{A}^{\mathrm{T}}\mathbf{v} \geq \mathbf{c}$	
$\mathbf{x} \geq \mathbf{0}$		$\mathbf{v} \geq \mathbf{0}$		

KKT Conditions

$$\text{Feasibility} \quad \mathbf{A}\mathbf{x} + \mathbf{y} = \mathbf{b}, \quad \mathbf{A}^{\mathrm{T}}\mathbf{v} - \mathbf{u} = \mathbf{c} \tag{4.26}$$

$$\text{Optimality} \quad \mathbf{c} = \mathbf{A}^{\mathrm{T}}\mathbf{v} - \mathbf{u}, \quad \mathbf{b} = \mathbf{A}\mathbf{x} + \mathbf{y} \tag{4.27}$$

$$\text{Complementarity} \quad \mathbf{u}^T\mathbf{x} + \mathbf{v}^T\mathbf{y} = \mathbf{0} \tag{4.28}$$

$$\mathbf{x} \geq \mathbf{0}, \quad \mathbf{y} \geq \mathbf{0}, \quad \mathbf{u} \geq \mathbf{0}, \quad \mathbf{v} \geq \mathbf{0} \tag{4.29}$$

We note here that the KKT conditions of both the problems are essentially the same, except that the optimality conditions of one are the feasibility conditions of the other and vice versa. By interchanging the feasibility and optimality conditions above, one may view the problem as the primal or its dual. It is also easy to observe that *the dual of the dual is the primal.* We note that variables v_i and u_i are called Lagrange multipliers.

The dual formulation shown in (4.25) is the symmetric form. If the primal problem is in the standard form with $\mathbf{Ax} = \mathbf{b}$, then this set of conditions can be replaced by two inequalities $\mathbf{Ax} \leq \mathbf{b}$ and $-\mathbf{Ax} \leq -\mathbf{b}$. Thus,

Primal	Equivalent form	Dual
maximize $\mathbf{c}^T\mathbf{x}$	maximize $\mathbf{c}^T\mathbf{x}$	minimize $\mathbf{b}^T\mathbf{u} - \mathbf{b}^T\mathbf{v}$
subject to $\mathbf{Ax} = \mathbf{b}$	subject to $\mathbf{Ax} \leq \mathbf{b}$	subject to $\mathbf{A}^T\mathbf{u} - \mathbf{A}^T\mathbf{v} \geq \mathbf{c}$
$\mathbf{x} \geq \mathbf{0}$	$-\mathbf{Ax} \leq -\mathbf{b}$	$\mathbf{u} \geq \mathbf{0}$ $\mathbf{v} \geq \mathbf{0}$
	$\mathbf{x} \geq \mathbf{0}$	

In the above formulation, we partitioned the coefficient matrix as $\begin{bmatrix} \mathbf{A} \\ -\mathbf{A} \end{bmatrix}$, the right hand side of the primal as $\begin{bmatrix} \mathbf{b} \\ -\mathbf{b} \end{bmatrix}$ and the dual variable vector as $\begin{bmatrix} \mathbf{u} \\ \mathbf{v} \end{bmatrix}$. If we now set $\mathbf{w} = \mathbf{u} - \mathbf{v}$, then $\mathbf{w}$ is a vector with components, which are unrestricted in sign. The dual problem for equality constraints is given as follows.

Primal		Dual	
maximize	$\mathbf{c}^T\mathbf{x}$	minimize	$\mathbf{b}^T\mathbf{w}$
subject to	$\mathbf{Ax} = \mathbf{b}$	subject to	$\mathbf{A}^T\mathbf{w} \geq \mathbf{c}$
	$\mathbf{x} \geq \mathbf{0}$		

(4.30)

In general, every equality constraint in the primal results in a variable unrestricted in sign in the dual, and every variable unrestricted in sign in the primal results in a corresponding equality constraint in the dual. Using this idea, other unsymmetric forms of primal – dual relations can be developed.

For the symmetric form of the primal dual formulation given in (4.25), we have

$$\mathbf{c}^T\mathbf{x} \leq \mathbf{v}^T\mathbf{Ax} \leq \mathbf{v}^T\mathbf{b} \tag{4.31}$$

If the primal problem is a maximization problem, the objective function of the primal at a feasible point is less than or equal to the objective of the dual for some dual feasible point. The quantity $\mathbf{b}^T\mathbf{v} - \mathbf{c}^T\mathbf{x}$ is referred to as duality gap. The duality gap vanishes as the optima are approached. The duality theorem for linear programming is given in the following.

Duality Theorem for LP

If primal and dual problems have feasible solutions, then both have optimal solutions, and the maximum of the primal objective is equal to the minimum of the dual objective. If either problem has an unbounded objective, the other problem is infeasible.

If $\mathbf{x}_B$ is the basic solution at the optimum of the primal, and $\mathbf{v}$ is the solution of the dual, then $\mathbf{c}_B{}^T\mathbf{x}_B = \mathbf{v}^T\mathbf{b}$. We note here that the nonbasic variables $\mathbf{x}_N$ are at zero value. Substituting for $\mathbf{x}_B$ from Eq. 4.19, we get

$$\mathbf{v}^T\mathbf{b} = \mathbf{c}_B^T\mathbf{B}^{-1}\mathbf{b} \tag{4.32}$$

The dual solution $\mathbf{v}$ at the optimum is given by

$$\mathbf{v}^T = \mathbf{c}_B^T\mathbf{B}^{-1} \tag{4.33}$$

Example 4.3

Consider the dual problem corresponding to the primal in the structural design problem in Section 4.3. We have

$$\mathbf{c} = \begin{bmatrix} 1 \\ 1 \end{bmatrix}, \quad \mathbf{A} = \begin{bmatrix} 0 & 1 \\ 4 & 3 \\ 4 & 1 \end{bmatrix}, \quad \mathbf{b} = \begin{bmatrix} 200 \\ 1280 \\ 960 \end{bmatrix}.$$

Thus, the dual problem is, from (4.25),

$$\begin{aligned} \text{minimize} \quad & f_d = 200v_1 + 1280v_2 + 960v_3 \\ \text{subject to} \quad & 4v_2 + 4v_3 \geq 1 \\ & v_1 + 3v_2 + v_3 \geq 1 \\ & \mathbf{v} \geq \mathbf{0} \end{aligned}$$

The primal has 2 variables and 3 constraints, while the dual has 2 constraints and 3 variables. The solution is $\mathrm{v} = [0.25, 0.25, 0]^T$, which are indeed the Lagrange multipliers associated with the constraints in the primal. Moreover, the dual objective also equals 370. Further, any $\mathbf{v}$ that satisfies the constraints in the dual problem lead to an objective that is greater than 370. For example, $\mathbf{v} = [1, 1, 1]^T$ gives $f_d = 2440$. On the other hand, any $\mathbf{x}$ that satisfies the primal constraints gives a primal objective <370. At optimum, the duality gap vanishes. Significance of Lagrange multipliers, and hence of v_i, was discussed in Section 4.3.

4.11 The Dual Simplex Method

In some problems, basic feasible solution may not be readily available for the primal problem while the dual feasible solution (satisfying primal optimality) is available.

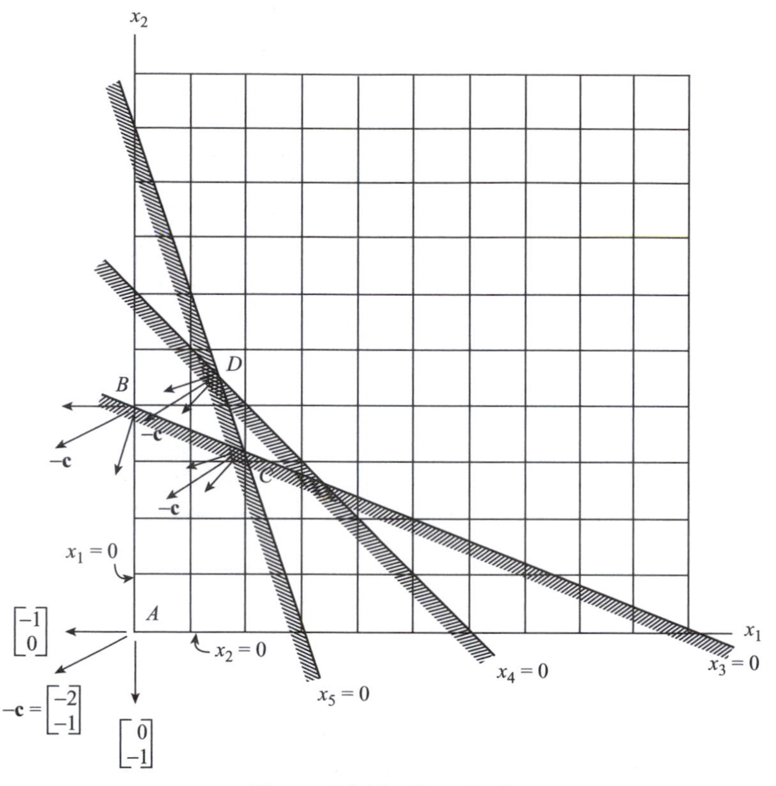

Figure 4.6. Dual method.

See example 4.5 given subsequently. The dual simplex method is a natural choice for these problems. The steps are very similar to the conventional simplex method discussed earlier. Here we maintain the dual feasibility, while achieving the primal feasibility. Thus, in direct contrast with the primal simplex procedure, the points generated by the dual simplex procedure are always optimal during the iterations, in that the cost coefficients are nonnegative; however, the points are infeasible (some basic $x_i < 0$). In the dual simplex method, the iterations terminate when we achieve a feasible point. The following example illustrates the dual simplex procedure. Consider the two variable problem (Fig. 4.6):

$$\begin{array}{ll} \text{minimize} & 2x_1 + x_2 \\ \text{subject to} & 2x_1 + 5x_2 \geq 20 \quad [1] \\ & x_1 + x_2 \geq 6 \quad [2] \\ & 3x_1 + x_2 \geq 9 \quad [3] \\ & x_1 \geq 0, x_2 \geq 0 \end{array} \tag{4.34}$$

The first step is to write the first tableau by introducing the surplus variables. We make the coefficients of the surplus variables positive by multiplying throughout by −1, and we do not require any artificial variables.

First Tableau (Corresponds to Point A)

	x_1	x_2	x_3	x_4	x_5	rhs
f	2	1	0	0	0	0
x_3	−2	[−5]	1	0	0	−20
x_4	−1	−1	0	1	0	−6
x_5	−3	−1	0	0	1	−9

Note that the first row coefficients are positive, which implies primal optimality or dual feasibility (see Fig. 4.6). However, the basic solution is infeasible since each of x_3, x_4, x_5 is negative.

Choice of the Pivot

(a) Choose the most negative basic variable to exit the basis, say, this corresponds to row i. If no pivot row can be located, the solution has been obtained. The pivot row for the above tableau is the second row corresponding to variable x_3.
(b) Choose the pivot column by performing the ratio test on $c_j/(-a_{ij})$ for only coefficients a_{ij}, which are negative, and non-basic, c_j that are positive. Choose the least of these corresponding to column j, which is the pivot column. If a pivot column cannot be obtained, the dual is unbounded. For the problem above the pivot column corresponds to the smaller of $\{2/(-(-2)), 1/(-(-5))\}$ or x_2. The pivot element is highlighted in the aforementioned.

Elementary row operations are performed for this pivot to make the number unity at the pivot, and rest of the coefficients in that column, zero. See point B in Fig. 4.6.

Second Tableau (Corresponds to Point B)

	x_1	x_2	x_3	x_4	x_5	rhs
f	1.6	0	0.2	0	0	−4
x_2	0.4	1	−0.2	0	0	4
x_4	−0.6	0	−0.2	1	0	−2
x_5	[−2.6]	0	−0.2	0	1	−5

The pivot row is the third row corresponding to x_4 and the pivot column is 1. The third tableau follows naturally.

Third Tableau (Corresponds to Point C)

	x_1	x_2	x_3	x_4	x_5	rhs
f	0	0	0.077	0	0.615	−7.077
x_2	0	1	−0.231	0	0.154	3.231
x_4	0	0	[−0.154]	1	−0.231	−0.846
x_1	1	0	0.077	0	−0.385	1.923

Fourth Tableau (Corresponds to Point D)

	x_1	x_2	x_3	x_4	x_5	rhs
f	0	0	0	0.5	0.5	−7.5
x_2	0	1	0	−1.5	0.5	4.5
x_3	0	0	1	−6.5	1.5	5.5
x_1	1	0	0	0.5	−0.5	1.5

Primal optimality has been achieved while maintaining the dual feasibility. The solution is $x_1 = 1.5$, $x_2 = 4.5$, and the function value is 7.5 (flipping the sign for the minimum problem).

In summary, the dual simplex method has the same steps as the primal simplex except that we decide the exiting variable first and then the entering variable. This method is attractive when all constraints are of the GE type. The method is useful when a GE type constraint is added to the current optimal tableau.

4.12 Sensitivity Analysis

We now have a clear idea of how to solve a linear programming problem using the simplex method. Having solved a problem, several questions arise. What if a coefficient in the objective function changes? What happens if the right-hand side of the constraints changes? What happens if a certain constraint coefficient is changed? What if a new constraint is added or a current constraint dropped? Sensitivity analysis deals with finding the answers to these questions. Also the coefficients in the formulation are never exact, and some statistical distribution may be associated with their values. Sensitivity analysis is done for some range of values of the coefficients. The technique is often referred to as *ranging*. The analysis becomes complex if several changes occur simultaneously. The standard practice in sensitivity analysis is to try variations in one or more entities while others are held at their current values. We now attempt to answer some questions posed in the preceding. The discussion follows the theoretical development ideas of the revised simplex method. This section further assumes that degeneracy is absent.

Change of the Objective Coefficients

We note that $\mathbf{c}_{\mathrm{B}}^{\mathrm{T}}\mathbf{B}^{-1}\mathbf{b}$ is the function value and $\mathbf{r}_{\mathrm{N}}^{\mathrm{T}} = \mathbf{c}_{\mathrm{B}}^{\mathrm{T}}\mathbf{B}^{-1}\mathbf{N} - \mathbf{c}_{\mathrm{N}}^{\mathrm{T}}$ is the vector of reduced coefficients. Or, if $\mathbf{v}$ is the optimal solution of the dual, then $\mathbf{v}^{\mathrm{T}} = \mathbf{c}_{\mathrm{B}}^{\mathrm{T}}\mathbf{B}^{-1}$. Thus, the dual optimal solution $\mathbf{v}$, which is obtained in the final simplex tableau, plays an important role in sensitivity analysis. As discussed earlier in detail, v_i is the change in the objective function for unit change in b_i, and is referred to as the *shadow price*. The results are summarized as follows.

$$\mathbf{v}^{\mathrm{T}} = \mathbf{c}_{\mathrm{B}}^{\mathrm{T}}\mathbf{B}^{-1} \tag{4.35}$$

$$f = \mathbf{v}^{\mathrm{T}}\mathbf{b} \tag{4.36}$$

The jth reduced coefficient is given by

$$r_j = \mathbf{v}^{\mathrm{T}}\mathbf{N}_{.j} - \mathbf{c}_j \tag{4.37}$$

where $\mathbf{N}_{.j}$ is the *jth* column of $\mathbf{N}$. If the objective coefficient corresponding to a basic variable is changed, it results in the change of the optimum value of the objective function. In addition, the reduced coefficients have to be calculated and checked if any of them are nonpositive. If a nonbasic coefficient is changed, there is no change in the function value – but we still need to check if any of the r_j is negative, in which case the current point is no more optimum.

Change of the Right-Hand Side of a Constraint

From the results presented in the revised simplex method, the change in the right-hand side results in the change in the basic solution $\mathbf{x}_{\mathrm{B}} = \mathbf{B}^{-1}\mathbf{b}$. If this basic solution is feasible for the altered $\mathbf{b}$, then this $\mathbf{x}_{\mathrm{B}}$ is optimal since change in $\mathbf{b}$ does not affect the reduced coefficients. No further steps are needed. The function value is given by $\mathbf{c}_{\mathrm{B}}^{\mathrm{T}}\mathbf{x}_{\mathrm{B}}$. If any of the basic variables become negative, we need to apply dual simplex steps in obtaining the optimum.

Change of Constraint Coefficients

We present the case of coefficient change by considering the change in a column of $\mathbf{A}$. Let us denote this column as $\mathbf{a}_j$. If x_j is a nonbasic variable, then we need to calculate only the reduced coefficient r_j and check if it is nonnegative.

$$r_j = \mathbf{v}^{\mathrm{T}}\mathbf{a}_j - \mathbf{c}_j \tag{4.38}$$

If r_j is negative, the simplex steps are performed to find the optimum. If x_j is a basic variable, it implies that an old $\mathbf{a}_j$ is replaced by a new $\mathbf{a}_j$. The product form of the inverse presented in Eqs. 4.23 and 4.24 can be used to update $\mathbf{B}^{-1}$. The reduced coefficients are calculated using Eqs 4.34–4.36 to see if further steps of the simplex method are needed. The function value is easily calculated.

Adding a New Variable

Adding a new variable can be treated as adding a new column to the nonbasic set. This is readily treated by using the discussion in the preceding section.

Adding and Dropping Inequality Constraints

Adding a new constraint or dropping an existing constraint poses some interesting challenges. We consider only inequality constraints that are brought to LE form. In the development of this treatment, we make use of the following matrix identities for updating the inverse when the dimension is increased by 1 or the dimension is collapsed by 1. Let $\begin{bmatrix} \mathbf{B} & \mathbf{c} \\ \mathbf{d} & e \end{bmatrix}$ and $\begin{bmatrix} \mathbf{P} & \mathbf{q} \\ \mathbf{r} & s \end{bmatrix}$ be two partitioned matrices with $\mathbf{c}$ and $\mathbf{q}$ column vectors, $\mathbf{d}$ and $\mathbf{r}$ row vectors, and e and s scalars, such that

$$\mathbf{B}' = \begin{bmatrix} \mathbf{B} & \mathbf{c} \\ \mathbf{d} & e \end{bmatrix} \quad \text{and} \quad \mathbf{B}'^{-1} = \begin{bmatrix} \mathbf{P} & \mathbf{q} \\ \mathbf{r} & s \end{bmatrix} \tag{4.39}$$

If $\mathbf{B}^{-1}$ is known and $\mathbf{c}$, $\mathbf{d}$, and e are added, $\mathbf{B}'^{-1}$ can be written using the relations

$$\begin{aligned} s &= \frac{1}{e - \mathbf{d}\mathbf{B}^{-1}\mathbf{c}} \\ \mathbf{q} &= -s\mathbf{B}^{-1}\mathbf{c} \\ \mathbf{r} &= -s\mathbf{d}\mathbf{B}^{-1} \\ \mathbf{P} &= \mathbf{B}^{-1} + \frac{\mathbf{q}\mathbf{r}}{s} \end{aligned} \tag{4.40}$$

The above relations correspond to increase of dimension by 1. These relations provide an alternative way of finding the inverse of a matrix. If the dimension is collapsed, then $\mathbf{P}$, $\mathbf{q}$, $\mathbf{r}$, and s are known and $\mathbf{B}^{-1}$ is to be determined. From the last relation in Eq. (4.40), we get

$$\mathbf{B}^{-1} = \mathbf{P} - \frac{\mathbf{q}\mathbf{r}}{s} \tag{4.41}$$

Addition of a constraint brings in a new row with an additional variable. The column corresponding to the new variable will have a 1 in that row and the rest of the elements in the column are zeroes. By rearranging the index set we can adjust the new basis matrix $\mathbf{B}$ to the form given in Eq. (4.39) so that $\mathbf{c} = \mathbf{0}$, $e = 1$, and

d corresponding to the variables in the basis. The updated inverse is constructed using Eq. (4.40) noting that $s = 1$, $\mathbf{q} = 0$, $\mathbf{r} = -\mathbf{dB}^{-1}$, $\mathbf{P} = \mathbf{B}^{-1}$. Since the objective coefficient for the new basic variable is zero, there is no change in the reduced coefficients. Thus, optimality condition is satisfied. Once this is done we need to check for updated $\mathbf{x}_B = \mathbf{B}^{-1}\mathbf{b}$, in which only the basic variable of the last row variable needs to be checked. If this is nonnegative, the basis is optimal, otherwise dual simplex iteration needs to be applied.

Dropping a constraint has to be treated with some care. If the variable corresponding to the constraint is *in the basis*, then it is not active. For LE type constraint, the original column corresponding to the basic variable has 1 at the constraint row and zeroes in other locations. Thus, $\mathbf{c} = \mathbf{0}$ and $e = 1$. From Eq. (4.40), we observe that $\mathbf{q} = \mathbf{0}$ and it follows from Eq. (4.41) that $\mathbf{B}^{-1} = \mathbf{P}$. There is no change in the partitioned inverse. Dropping of the row corresponding to the constraint and the column corresponding to the basic variable can be dropped. There is no change in the optimum point, the function value, and the reduced coefficients. This case is shown in Fig. 4.7a. If the variable corresponding to the constraint is *nonbasic* implying that the constraint is active, it leads to a two-step process. First, this variable has to be brought into basis with a move to an adjacent point, which is currently infeasible. This is done by performing the ratio test in the following way. If x_j is the variable to be brought into the basis, we perform the ratio test *for only negative* a_{ij} and i corresponding to the least $b_i/(-a_{ij})$ is chosen as the pivot row. This permits the optimum point to become infeasible with respect to the deleted constraint and keep feasibility with respect to the others. If there is no negative coefficient appearing in the column, the removal of this constraint leads to unbounded state. Once the pivot is identified, the operation is same as the interchange of column j with column i of the basis. Updating of the inverse is done using the technique suggested in the revised simplex method. Now that the constraint to be dropped is in the basis, the corresponding row and column can be deleted.

Example 4.4

Consider again the example

$$
\begin{aligned}
\text{maximize} \quad & 2x_1 + x_2 & \\
\text{subject to} \quad & 2x_1 - x_2 \leq 8 & [1] \\
& x_1 + 2x_2 \leq 14 & [2] \\
& -x_1 + x_2 \leq 4 & [3] \\
& x_1 \geq 0, x_2 \geq 0 &
\end{aligned}
$$

We know the optimum solution to be $\mathbf{x}_B = [x_1, x_2, x_5]$, $\mathbf{x}_{NB} = [x_3, x_4]$. If using the revised simplex method, we also know the current $\mathbf{B}^{-1}$ matrix.

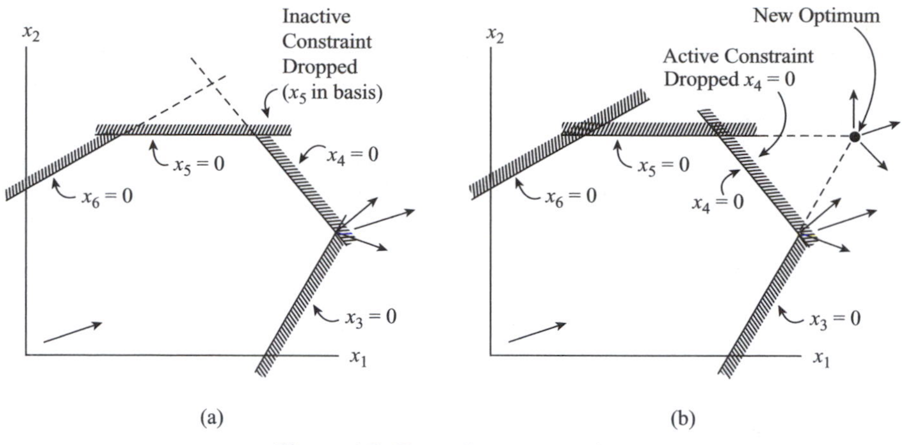

Figure 4.7. Dropping a constraint.

For the purposes of this example, we simply note that, at the optimum, $\mathbf{B} = \begin{bmatrix} 2 & -1 & 0 \\ 1 & 2 & 0 \\ -1 & 1 & 1 \end{bmatrix}$, $\mathbf{b} = \begin{bmatrix} 8 \\ 14 \\ 4 \end{bmatrix}$, $\mathbf{c}_\mathrm{B} = \begin{bmatrix} 2 \\ 1 \\ 0 \end{bmatrix}$, $\mathbf{N} = \begin{bmatrix} 1 & 0 \\ 0 & 1 \\ 0 & 0 \end{bmatrix}$ $\mathbf{c}_\mathrm{N} = [0, 0]^\mathrm{T}$, and $f = 16$.

Assume that we change the constraint limits to $\mathbf{b}^{\text{new}} = [10, 12, 4]^\mathrm{T}$. Then, the new basic solution is $\mathbf{x}_\mathrm{B}^{\text{new}} = \mathbf{B}^{-1}\,\mathbf{b}^{\text{new}} = [6.4, 2.8, 7.6]^\mathrm{T}$. Since $\mathbf{x}_\mathrm{B} \geq \mathbf{0}$, we can determine the new objective to be $f^{\text{new}} = \mathbf{c}_\mathrm{B}^\mathrm{T}\,\mathbf{x}_\mathrm{B}^{\text{new}} = 2(6.4) + 1(2.8) = 15.6$.

Now assume that we change the objective as $\mathbf{c}_\mathrm{B}^{\text{new}} = [2 + \Delta, 1, 0]^\mathrm{T}$ and wish to determine the limits on Δ, wherein the optimal basis does not change and where sensitivity may be used. We have $\mathbf{v}^\mathrm{T} = \mathbf{c}_\mathrm{B}^\mathrm{T}\,\mathbf{B}^{-1} = [0.6 + 0.4\Delta, 0.8 + 0.2\Delta, 0]$ and $f^{\,\text{opt}} = \mathbf{v}^\mathrm{T}\,\mathbf{b}$. The limits on Δ are obtained from requiring that $\mathbf{r} = \mathbf{v}^\mathrm{T}\mathbf{N} - \mathbf{c}_\mathbf{N} \geq \mathbf{0} : \Delta \geq -1.5$ and $\Delta \geq -4$. Thus, $\Delta \geq -1.5$. With this, sensitivity can be used to predict the solution to the above problem with the objective $f = (2 + \Delta)x_1 + x_2 : f^{\text{opt}} = 16 + 6\Delta$.

Example 4.5

Consider again Example 4.1.

(i) To what extent can the strength of cable 1 be reduced without changing the optimum basis?

At optimum, we know that cable 1 constraint is inactive. That is, force in cable 1 is less than its strength. Cable 1 strength may be reduced as long as $\mathbf{x}_\mathrm{B} \geq 0$. Thus,

$$\mathbf{x}_\mathbf{B} = \mathbf{B}^{-1}\mathbf{b} = \begin{bmatrix} 1 & 0 & 0 \\ 4 & 3 & 0 \\ 4 & 1 & 1 \end{bmatrix}^{-1} \begin{bmatrix} 200 \\ 1280 \\ 960 - \Delta \end{bmatrix} \geq \begin{bmatrix} 0 \\ 0 \\ 0 \end{bmatrix}$$

gives $\Delta \leq 80$. At $\Delta = 80$, cable 1 constraint becomes $4W_1 + W_2 \leq 880$ and is *active* at optimum. The slack corresponding to cable 1 is a basic variable with zero value, referred to as a degenerate solution (the reader is encouraged to plot the constraint set for this problem).

(ii) Assume the cable 1 constraint changes to $4W_1 + W_2 \leq 850$. Knowing the optimum to the problem with $4W_1 + W_2 \leq 960$, efficiently determine the new optimum.

We may interpret the solution to the original problem as "near-optimal" to the current perturbed problem. Thus, the task is one of driving the current optimal solution to a feasible solution. From the basis (revised simplex or tableau) *at optimum* corresponding to the original **b**, we have: basic variables $[x_1, x_2, x_5]$, nonbasic variables $[x_3, x_4]$, and

$$\mathbf{B} = \begin{bmatrix} 0 & 1 & 0 \\ 4 & 3 & 0 \\ 4 & 1 & 1 \end{bmatrix}, \quad \mathbf{b} = \begin{bmatrix} 200 \\ 1280 \\ 850 \end{bmatrix}, \quad \mathbf{c_B} = \begin{bmatrix} 1 \\ 1 \\ 0 \end{bmatrix}, \quad \mathbf{c_N} = \begin{bmatrix} 0 \\ 0 \end{bmatrix}, \quad \mathbf{N} = \begin{bmatrix} 1 & 0 \\ 0 & 1 \\ 0 & 0 \end{bmatrix}$$

Using Eqs. (4.34)–(4.36), the new values of variables and reduced cost coefficients are:

$$\mathbf{x_B} = \begin{bmatrix} 170 \\ 200 \\ -30 \end{bmatrix}, \quad \mathbf{r_N} = \begin{bmatrix} .25 \\ .25 \end{bmatrix}$$

The reduced cost coefficients remain positive, indicating "optimality", as expected since they are independent of the right-hand side **b**. On the other hand, the point is "infeasible" since basic variable $x_5 = -30$ (< 0). Thus, the dual simplex method is efficient here. A tableau-based dual simplex procedure was presented in Section 4.8, and will also be used in this example, although a matrix-based revised dual simplex is computationally more efficient. The starting step is the final primal simplex tableau corresponding to the original constraint $4W_1 + W_2 \leq 960$. It is left as an exercise to the reader to show that this tableau is:

0	0	2	−1	1	80	(x_5 basic)
0	1	1	0	0	200	(x_2 basic)
1	0	−0.75	0.25	0	170	(x_1 basic)
0	0	0.25	0.25	0	370	($\mathbf{r_N}$ row, $f = 370$)

We now change 80 to −30 to obtain a starting tableau for the dual simplex method:

0	0	2	−1	1	−30	(x_5 basic)
0	1	1	0	0	200	(x_2 basic)
1	0	−0.75	0.25	0	170	(x_1 basic)
0	0	0.25	0.25	0	370	($\mathbf{r}_N$ row, $f = 370$)

Thus, x_5 has to *leave* the basis. The only a_{ij} with a negative coefficient is column 4 in row 1. Thus, the pivot is as indicated in the preceding tableau, implying that x_4 *enters* the basis. This pivot operation is achieved by premultiplying the tableau by $\mathbf{T}$ (see Chapter 1) where

$$\mathbf{T} = \begin{bmatrix} -1 & 0 & 0 & 0 \\ 0 & 1 & 0 & 0 \\ .25 & 0 & 1 & 0 \\ .25 & 0 & 0 & 1 \end{bmatrix}$$

to obtain

0	0	−2	1	−1	30	(x_4 basic)
0	1	1	0	0	200	(x_2 basic)
1	0	−0.25	0	0.25	162.5	(x_1 basic)
0	0	0.75	0	0.25	362.5	($\mathbf{r}_N$ row, $f = 362.5$)

Thus, a single pivot operation yields the new optimum, $x_1 = 162.5$, $x_2 = 200$, $f = 162.5$. Cable 2 constraint is slack. It was not necessary to resolve the entire problem.

4.13 Interior Approach

Karmarkar's landmark paper published in 1984 introduced an interior algorithm that is polynomial in time [Karmarkar 1984]. That is, convergence is proved to depend on n^α as opposed to an exponential algorithm that depends on β^n, where n is the problem dimension. To see the advantage of polynomial convergence, note that $100^4 = 10^8$ while $2^{100} = 1.27 \times 10^{30}$. In theory, the simplex method is exponential while in practice it behaves polynomially. Since Karmarkar's paper, numerous papers have been published, and several conferences have been devoted to the interior methods (e.g., text by Roos et al. [1997]). Karmarkar's original algorithm requires a special simplex structure. We present here the affine scaling algorithm of Dikin published in 1967 [Dikin 1997]. This algorithm can be applied to linear programming problems in standard form; however, the polynomial complexity of this algorithm has not been proved.

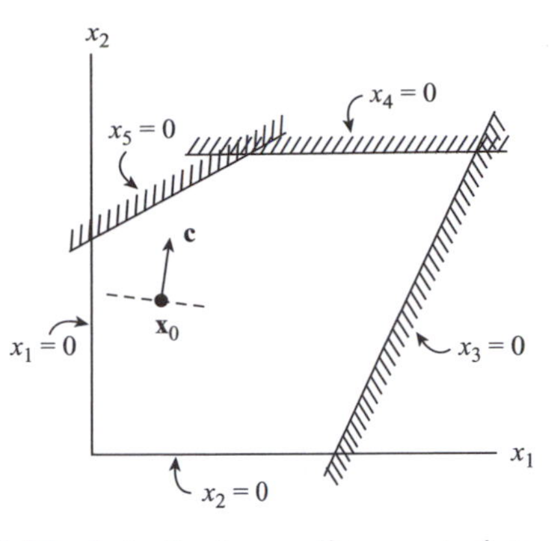

Figure 4.8. Interior point method.

We first introduce the slack and surplus variables into the inequality constraints and put the LP problem into the standard form

$$\begin{aligned} \text{maximize} \quad & \mathbf{c}^{\mathrm{T}}\mathbf{x} \\ \text{subject to} \quad & \mathbf{A}\mathbf{x} = \mathbf{b} \\ & \mathbf{x} \geq \mathbf{0} \end{aligned} \tag{4.42}$$

Let n be the number of variables in $\mathbf{x}$ and let m be the number of constraints corresponding to the number of rows of $\mathbf{A}$. The rows of $\mathbf{A}$ give the directions normal to the constraints. Thus, columns of $\mathbf{A}^{\mathrm{T}}$ are the normals to the constraints. $\mathbf{c}$ is the direction of steepest increase. Fig. 4.8 illustrates the polytope for a two-variable configuration.

In the interior point methods, we start from a strictly interior point, say, $\mathbf{x}^0$ such that all the components of $\mathbf{x}^0$ are strictly positive ($\mathbf{x}^0 > \mathbf{0}$). The constraint set and the interior point are shown in Fig. 4.8.

From this interior point, we proceed in some direction to increase the objective function for maximization. This direction must be obtained from the knowledge of direction $\mathbf{c}$ and the constraints in $\mathbf{A}$. Though $\mathbf{c}$ is the direction in which the function gets maximum increase, it may end up at the boundary and not proceed any further. We first use a scaling scheme and then a projection idea in obtaining a reasonable direction. There are various scaling schemes. Karmarkar uses a nonlinear scaling scheme and applies it to an equivalent problem. We present here an affine scaling scheme, which is not necessarily the most efficient, but one that is easy to follow and implement on a computer. If $\mathbf{x}^0 = [x_1, x_2, \ldots, x_n]^{\mathrm{T}}$ is the initial point, we define the diagonal scaling matrix $\mathbf{D}$ as

$$\mathbf{D} = \begin{bmatrix} x_1^0 & 0 & \cdots & 0 \\ 0 & x_2^0 & \cdots & 0 \\ \vdots & \vdots & \ddots & \vdots \\ 0 & 0 & \cdots & x_n^0 \end{bmatrix} \tag{4.43}$$

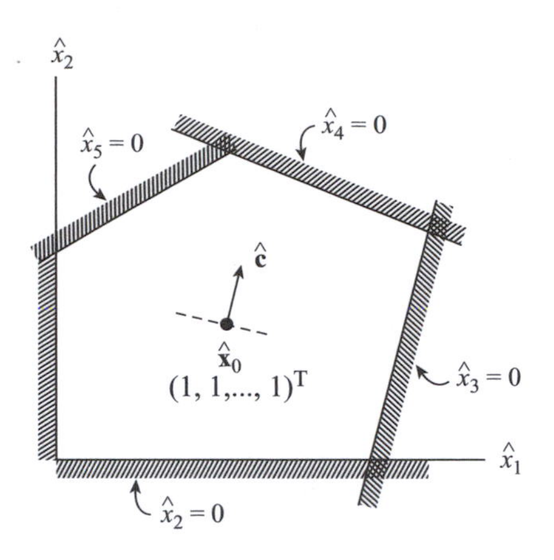

Figure 4.9. Scaled interior point.

and the transformed coordinates $\hat{\mathbf{x}}$ such that

$$\mathbf{x} = \mathbf{D}\hat{\mathbf{x}} \tag{4.44}$$

Thus, $\hat{\mathbf{x}}^0 = [1, 1, \ldots, 1]^T$ are the scaled coordinates corresponding to the current point $\mathbf{x}^0$. The transformed problem at the current point becomes

$$\begin{aligned} \text{max} \quad & \hat{\mathbf{c}}^T\hat{\mathbf{x}} \\ \text{subject to} \quad & \hat{\mathbf{A}}\hat{\mathbf{x}} = \mathbf{b} \\ & \hat{\mathbf{x}} \geq \mathbf{0} \end{aligned} \tag{4.45}$$

where $\hat{\mathbf{c}} = \mathbf{Dc}$ and $\hat{\mathbf{A}} = \mathbf{AD}$, and $\mathbf{D}\hat{\mathbf{x}} \geq \mathbf{0}$ resulted in $\hat{\mathbf{x}} \geq \mathbf{0}$.

The transformed problem is shown in Fig. 4.9.

Direction Finding

We turn our attention to the problem of finding the direction for the move. If $\Delta\hat{\mathbf{x}}$ is an increment from $\hat{\mathbf{x}}$, then we need

$$\hat{\mathbf{A}}\,(\hat{\mathbf{x}} + \Delta\hat{\mathbf{x}}) = \mathbf{b} \Rightarrow \hat{\mathbf{A}}\Delta\hat{\mathbf{x}} = \mathbf{0}$$

We note that the increment $\Delta\hat{\mathbf{x}}$ must be in the null space of $\hat{\mathbf{A}}$. We now look for the projection of $\hat{\mathbf{c}}$ onto the null space of $\hat{\mathbf{A}}$. Let V be the m-dimensional space spanned by the row vectors of $\hat{\mathbf{A}}$. Then the null space of $\hat{A}$ is the $n-m$-dimensional manifold formed by the intersection of the m-hyperplanes defining the constraints. The null space of $\hat{\mathbf{A}}$ denoted by U is orthogonal to V. If $\mathbf{v}$ is a vector in the space V, then it can be written as

$$\mathbf{v} = \hat{\mathbf{a}}_1^T y_1 + \hat{\mathbf{a}}_2^T y_2 + \cdots + = \hat{\mathbf{A}}^T\mathbf{y}$$

where $\hat{\mathbf{a}}_j$ is the jth row of $\hat{\mathbf{A}}$. A two-plane case in three dimensions is illustrated in Fig. 4.10.

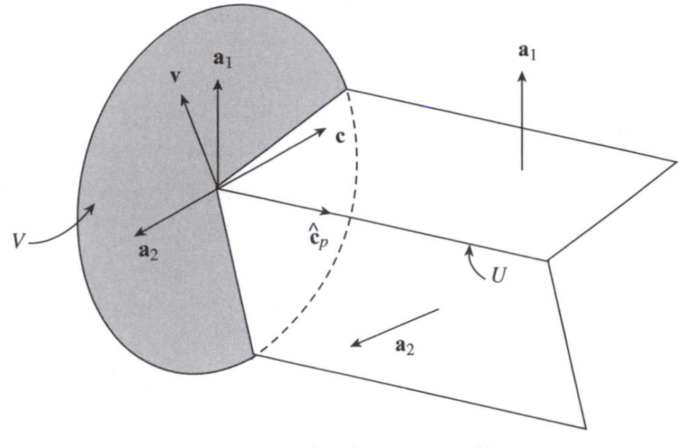

Figure 4.10. Projection onto null space.

If $\hat{\mathbf{c}}_{\mathrm{p}}$ is the projection of $\hat{\mathbf{c}}$ onto the null space of $\hat{\mathbf{A}}$, it can be written as

$$\hat{\mathbf{c}} = \mathbf{v} + \hat{\mathbf{c}}_{\mathrm{p}} = \hat{\mathbf{A}}^{\mathrm{T}}\mathbf{y} + \hat{\mathbf{c}}_{\mathrm{p}} \tag{4.46}$$

Multiplying throughout by $\hat{\mathbf{A}}$ and noting that $\hat{\mathbf{A}}\hat{\mathbf{c}}_{\mathrm{p}} = \mathbf{0}$, we get

$$\hat{\mathbf{A}}\hat{\mathbf{A}}^{\mathrm{T}}\mathbf{y} = \hat{\mathbf{A}}\hat{\mathbf{c}} \tag{4.47}$$

The above set of m equations can be solved to obtain $\mathbf{y}$. A usual representation of the projection vector $\hat{\mathbf{c}}_{\mathrm{p}}$ is written by introducing $\mathbf{y}$ from Eq. (4.47) into Eq. (4.46). Thus,

$$\hat{\mathbf{c}}_{\mathrm{p}} = \hat{\mathbf{c}} - \hat{\mathbf{A}}^{\mathrm{T}}\mathbf{y} = (\mathbf{I} - \hat{\mathbf{A}}^{\mathrm{T}}(\hat{\mathbf{A}}\hat{\mathbf{A}}^{\mathrm{T}})^{-1}\hat{\mathbf{A}})\hat{\mathbf{c}} \tag{4.48}$$

where $\mathbf{P} = \mathbf{I} - \hat{\mathbf{A}}^{\mathrm{T}}(\hat{\mathbf{A}}\hat{\mathbf{A}}^{\mathrm{T}})^{-1}\hat{\mathbf{A}}$ is called the projection operator[3]. Once $\hat{\mathbf{c}}_{\mathrm{p}}$ is obtained, we can move any length in that direction to increase the function without violating the constraints. However, we still need to keep $\hat{\mathbf{x}} > \mathbf{0}$ to remain in the interior. Since we move along along $\hat{\mathbf{c}}_{\mathrm{p}}$, the move is now controlled by the *most negative* component of $\hat{\mathbf{c}}_{\mathrm{p}}$ say $\hat{\mathbf{c}}_{\mathrm{p}}^{*}$. Note that $\hat{\mathbf{x}}$ has as value of 1. We then denote the maximum extension parameter σ as

$$\sigma = -\frac{1}{\hat{\mathbf{c}}_{\mathrm{p}}^{*}}$$

To keep in the interior, another parameter α in the open interval 0 to 1 is chosen. A value of 0.98 may be used. Values between 0.6 and 0.98 may be tried for a specific problem.

The increment $\Delta\hat{\mathbf{x}}$ is, thus, given by

$$\Delta\hat{\mathbf{x}} = \alpha\sigma\hat{\mathbf{c}}_{\mathrm{p}}$$

[3] We will come across this projection matrix when studying the Gradient Projection Method in Chapter 5.

Multiplying throughout by $\mathbf{D}$, we get the increment in the original coordinate system as

$$\Delta \mathbf{x} = \alpha \sigma \mathbf{D} \hat{\mathbf{c}}_{\mathrm{p}} \tag{4.49}$$

After the move is made to the new point, the process is repeated. The process is repeated until the function value converges to a maximum.

In the implementation of the interior method, an interior starting point satisfying the constraints is needed. This poses some problems when equality constraints are present. Two phase approach presented earlier can be used here. This has been implemented in the program LPINTER. This program works with the same data set as that for the earlier programs. The program uses the Gauss elimination approach for equation solving at each iteration. No attempt has been made to introduce more efficient updating schemes to exploit the special structure of the matrices.

4.14 Quadratic Programming (QP) and the Linear Complementary Problem (LCP)

There are many optimization problems that come under the quadratic programming category where the objective function is quadratic and the constraints are linear. Applications include least squares regression, frictionless contact problems in mechanics, and sequential QP approaches to solve the general NLP problem.

The application of the optimality conditions to a quadratic programming problem leads to a linear complementary problem (LCP). In fact, optimality conditions for an LP also lead to an LCP. LCPs can be solved using simplex-based methods developed by Wolfe and by Lemke, by interior point methods, or by treating it as an NLP problem and using techniques presented in Chapter 5. The approach by Lemke is presented in the following. Consider the quadratic programming problem

$$\begin{aligned} \text{minimize} \quad & f(\mathbf{x}) = \tfrac{1}{2}\mathbf{x}^{\mathrm{T}}\mathbf{Q}\mathbf{x} + \mathbf{c}^{\mathrm{T}}\mathbf{x} \\ \text{subject to} \quad & \mathbf{A}\mathbf{x} \le \mathbf{b} \\ & \mathbf{x} \ge \mathbf{0} \end{aligned} \tag{4.50}$$

The KKT optimality conditions for this problem are

$$\begin{aligned} \text{Optimality} \quad & \mathbf{Q}\mathbf{x} + \mathbf{c} + \mathbf{A}^{\mathrm{T}}\mathbf{v} - \mathbf{u} = \mathbf{0} \\ \text{Feasibility} \quad & \mathbf{A}\mathbf{x} + \mathbf{y} = \mathbf{b} \\ \text{Complementarity} \quad & \mathbf{x}^{\mathrm{T}}\mathbf{u} + \mathbf{v}^{\mathrm{T}}\mathbf{y} = 0 \\ & \mathbf{x} \ge \mathbf{0}, \quad \mathbf{y} \ge \mathbf{0}, \quad \mathbf{u} \ge \mathbf{0}, \quad \mathbf{v} \ge \mathbf{0} \end{aligned} \tag{4.51}$$

We denote

$$\mathbf{w} = \begin{bmatrix} \mathbf{u} \\ \mathbf{y} \end{bmatrix}, \quad \mathbf{z} = \begin{bmatrix} \mathbf{x} \\ \mathbf{v} \end{bmatrix}, \quad \mathbf{q} = \begin{bmatrix} \mathbf{c} \\ \mathbf{b} \end{bmatrix}, \quad \mathbf{M} = \begin{bmatrix} \mathbf{Q} & \mathbf{A}^{\mathrm{T}} \\ -\mathbf{A} & \mathbf{0} \end{bmatrix} \tag{4.52}$$

On rearranging the terms in Eq. 4.51 and introducing Eq. 4.52, we get the standard form for which an elegant solution is available.

$$\begin{aligned} \mathbf{w} - \mathbf{Mz} &= \mathbf{q} \\ \mathbf{w}^{\mathrm{T}}\mathbf{z} &= 0 \\ [\mathbf{w}, \mathbf{z}] &\geq \mathbf{0} \end{aligned} \tag{4.53}$$

From Wierstraas existence theorem (Chapter 1), we note that if the constraints in (4.53) define a bounded set Ω, a solution will exist (the $\leq$ ensures closedness of the set, and a quadratic objective function *is* continuous). However, numerical methods may not be able to determine a solution necessarily. If the Hessian $\mathbf{Q}$ is positive semidefinite, then the objective is convex and the problem is convex, which greatly increases the success rate of numerical methods. We assume here that $\mathbf{Q}$ is positive semidefinite, which makes $\mathbf{M}$ positive semidefinite. This together with the assumption of boundedness of the feasible set guarantees that Lemke's algorithm converges in finite number of steps (i.e., finite number of pivot operations). Further, for convex problems, any local minimum is also the global minimum. Of course, there are techniques, such as presented in Chapter 5 that can converge to a local minimum when $\mathbf{Q}$ is not positive semidefinite, although not in a finite number of steps.

Firstly, note that if all $q_i \geq 0$, then $\mathbf{w} = \mathbf{q}$, $\mathbf{z} = \mathbf{0}$ satisfies (4.53), and is hence the solution. Thus, it is understood that one or more $q_i < 0$. Lemke introduces artificial variable z_0 to set up a tableau for the set of equations

$$\mathbf{Iw} - \mathbf{Mz} - \mathbf{e}z_0 = \mathbf{q} \tag{4.54}$$

where $\mathbf{e} = [1\,1 \ldots 1]^{\mathrm{T}}$, and $\mathbf{I}$ is the identity matrix. In the above tableau, we start with an initial basic feasible solution $\mathbf{w} = \mathbf{q} + \mathbf{e}z_0$ and $\mathbf{z} = \mathbf{0}$. z_0 is chosen to be equal to the most negative component of $\mathbf{q}$. Let $q_s = \min q_i < 0$. This choice of z_0 ensures that the initial choice of $\mathbf{w}$ is nonnegative. The algorithm consists of an "initial step" followed by "typical steps". In the initial step, a pivot operation is performed to bring z_0 into the basis and let w_s leave the basis. Thereafter, the nonbasic variable that enters the basis is selected as the *complement* of the basic variable that just left the basis in the previous tableau. Thus, if w_r leaves the basis, then z_r enters in the next tableau or vice versa, which maintains the complementarity condition $w_r z_r = 0$.

Lemke's Algorithm for LCP

Step 1: z_0 enters basis, w_i corresponding to most negative q_i exits the basis. If there is no negative q_i, then solution is $z_0 = 0$, $\mathbf{w} = \mathbf{q}$, $\mathbf{z} = \mathbf{0}$. No further steps are necessary.

Step 2: Perform the elementary row operations (pivot operations) for column z_0 and row i. All elements of $\mathbf{q}$ are now nonnegative.

Step 3: Since basic variable in row i exited the basis, its complement, say, in column j enters the basis. (At first, iteration w_i exits and z_i enters basis.)

Step 4: Perform the ratio test for column j to find the least q_i/(positive row element i). The basic variable corresponding to row i now exits the basis. If there are no positive row elements, there is no solution to the LCP. We call this the *ray solution*, and terminate the calculations.

Step 5: Perform the elementary row operations to make the pivot 1 and other elements in the column zero. If the last operation results in the exit of the basic variable z_0, then the cycle is complete, STOP. If not go to Step 3.

The above steps have been implemented in the program QPLCP. Linear programming problems can also be solved using LCP formulation by setting $\mathbf{Q} = \mathbf{0}$.

Example 4.6

$$\begin{aligned} \text{minimize} \quad & x_1^2 + x_2^2 - 3x_1x_2 - 6x_1 + 5x_2 \\ \text{subject to} \quad & x_1 + x_2 \le 4 \\ & 3x_1 + 6x_2 \le 20 \\ & x_1 \ge 0, x_2 \ge 0 \end{aligned}$$

Solution:

For this problem, $\mathbf{Q} = \begin{bmatrix} 2 & -3 \\ -3 & 2 \end{bmatrix}$ $\mathbf{c} = \begin{bmatrix} -6 \\ 5 \end{bmatrix}$ $\mathbf{A} = \begin{bmatrix} 1 & 1 \\ 3 & 6 \end{bmatrix}$ $\mathbf{b} = \begin{bmatrix} 4 \\ 20 \end{bmatrix}$

Above data are input into the data statements in the program QPLCP. The tableaus are given below. The tableaus have been obtained by adding a few print statements in the included program.

Initial Tableau

	w1	w2	w3	w4	z1	z2	z3	z4	z0	q
w1	1.000	0.000	0.000	0.000	−2.000	3.000	−1.000	−3.000	−1.000	−6.000
w2	0.000	1.000	0.000	0.000	3.000	−2.000	−1.000	−6.000	−1.000	5.000
w3	0.000	0.000	1.000	0.000	1.000	1.000	0.000	0.000	−1.000	4.000
w4	0.000	0.000	0.000	1.000	3.000	6.000	0.000	0.000	−1.000	20.000

Tableau 1 (z_0 Enters Basis, $w_1 \equiv w_s$ Leaves)

z0	−1.000	0.000	0.000	0.000	2.000	−3.000	1.000	3.000	1.000	6.000
w2	−1.000	1.000	0.000	0.000	5.000	−5.000	0.000	−3.000	0.000	11.000
w3	−1.000	0.000	1.000	0.000	3.000	−2.000	1.000	3.000	0.000	10.000
w4	−1.000	0.000	0.000	1.000	5.000	3.000	1.000	3.000	0.000	26.000

Tableau 2 (z_1 Enters Basis, w_2 Leaves)

z0	−0.600	−0.400	0.000	0.000	0.000	−1.000	1.000	4.200	1.000	1.600
z1	−0.200	0.200	0.000	0.000	1.000	−1.000	0.000	−0.600	0.000	2.200
w3	−0.400	−0.600	1.000	0.000	0.000	1.000	1.000	4.800	0.000	3.400
w4	0.000	−1.000	0.000	1.000	0.000	8.000	1.000	6.000	0.000	15.000

Tableau 3 (z_2 Enters Basis, w_4 Leaves)

z0	−0.600	−0.525	0.000	0.125	0.000	0.000	1.125	4.950	1.000	3.475
z1	−0.200	0.075	0.000	0.125	1.000	0.000	0.125	0.150	0.000	4.075
w3	−0.400	−0.475	1.000	−0.125	0.000	0.000	0.875	4.050	0.000	1.525
z2	0.000	−0.125	0.000	0.125	0.000	1.000	0.125	0.750	0.000	1.875

Tableau 4 (z_4 Enters Basis, w_3 Leaves)

z0	−0.111	0.056	−1.222	0.278	0.000	0.000	0.056	0.000	1.000	1.611
z1	−0.185	0.093	−0.037	0.130	1.000	0.000	0.093	0.000	0.000	4.019
z4	−0.099	−0.117	0.247	−0.031	0.000	0.000	0.216	1.000	0.000	0.377
z2	0.074	−0.037	−0.185	0.148	0.000	1.000	−0.037	0.000	0.000	1.593

Tableau 5 (z_3 Enters Basis, z_4 Leaves)

z0	−0.086	0.086	−1.286	0.286	0.000	0.000	0.000	−0.257	1.000	1.514
z1	−0.143	0.143	−0.143	0.143	1.000	0.000	0.000	−0.429	0.000	3.857
z3	−0.457	−0.543	1.143	−0.143	0.000	0.000	1.000	4.629	0.000	1.743
z2	0.057	−0.057	−0.143	0.143	0.000	1.000	0.000	0.171	0.000	1.657

Tableau 6 (w_4 Enters Basis, z_0 Leaves)

w4	−0.300	0.300	−4.500	1.000	0.000	0.000	0.000	−0.900	3.500	5.300
z1	−0.100	0.100	0.500	0.000	1.000	0.000	0.000	−0.300	−0.500	3.100
z3	−0.500	−0.500	0.500	0.000	0.000	0.000	1.000	4.500	0.500	2.500
z2	0.100	−0.100	0.500	0.000	0.000	1.000	0.000	0.300	−0.500	0.900

The solution is obtained as $x_1 = 3.1$, $x_2 = 0.9$, and the minimum value of the function is -12.05. Plotting of this problem is shown in Fig. E4.6.

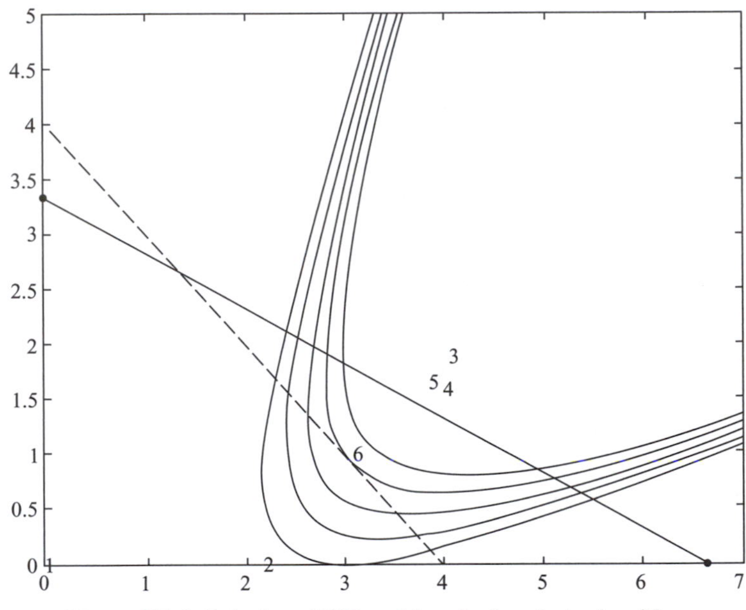

Figure E4.6. Solution of QP problem by Lemke's algorithm.

Treatment of Equality Constraints

The LCP formulation and its implementation in the program is provided for inequality constraints, which are put in the LE form. If there is an equality constraint, it can be treated by two inequality constraints. If $\ell_1 = b_1$ is the equality constraint, where ℓ_1 is a linear or nonlinear form, it can be replaced by two constraints $\ell_1 \leq b_1$ and $\ell_1 \geq b_1$. It is easy to check that a system of equality constraints

$$\begin{aligned} \ell_1 &= b_1 \\ \ell_2 &= b_2 \\ \ell_3 &= b_3 \\ \ell_4 &= b_4 \end{aligned} \tag{4.55}$$

is equivalent to

$$\begin{aligned} \ell_1 &\le b_1 \\ \ell_2 &\le b_2 \\ \ell_3 &\le b_3 \\ \ell_4 &\le b_4 \\ \ell_1 + \ell_2 + \ell_3 + \ell_4 &\ge b_1 + b_2 + b_3 + b_4 \end{aligned} \tag{4.56}$$

A system of n equalities can be treated by $n + 1$ inequalities.

COMPUTER PROGRAMS

SIMPLEX, REVSIMC, LPINTER, QPLCP, and EXCEL SOLVER, MATLAB

PROBLEMS

P4.1. Fill in the blank:
When a slack variable equals zero, the corresponding constraint is ___________

P4.2. Do all the calculations displayed in the structural design problem, Section 4.3.

P4.3. In the structural design problem in Section 4.3, Fig. 4.1, if the horizontal location s of load W_1 is a third variable, along with W_1 and W_2, is the problem still an LP? Justify.

P4.4. Since the optimum, if it exists, is at a vertex of the feasible polytope, is it possible to evaluate the objective at each vertex and determine the optimum? Discuss.

P4.5. Solve[4] the diet problem PM1 modeled in Section 4.4.

P4.6. Solve the alloy manufacturing problem PM2 modeled in Section 4.4.

P4.7. Solve the refinery problem PM3 modeled in Section 4.4.

P4.8. Solve the agriculture problem PM4 modeled in Section 4.4.

P4.9. Solve the trim problem PM5 modeled in Section 4.4.

P4.10. Solve the straightness evaluation problem PM6 modeled in Section 4.4.

P4.11. Solve the machine scheduling problem PM7 modeled in Section 4.4.

[4] All solutions must include a discussion of active constraints, output interpretations.

P4.12. A set of points belonging to set 1 and another set of points belonging to set 2, are to be separated or "classified" by a straight line. That is, a line $y - ax - b = 0$ must be determined, which separates the two sets of points. Thus, we will have $y - ax - b > 0$ for each point in set 1 and $y - ax - b < 0$ for points in set 2. Formulate this as an LP problem. Give formulation, solution, plot showing line, and points. More general problems of this type are discussed in [S. Boyd and L. Vandenberghe, Convex Optimization]. Use SOLVER.

Hint: Use 3 design variables, namely, a, b and (an additional variable) t, where t is the "gap" between the positive values for set 1 and negative values for set 2. Maximize t.

x_i	y_i
set 1	
1	1
1	2
0.8	3.5
2	1.8
2	2.7
3	3.5
set 2	
1.5	0.3
2.5	0.75
3	0.1
3.8	1
3.7	1.7
4	0
4	0.6
4.8	2.7
5	0.7
5.2	1

P4.13. Subcontracting dyeing operations in textile company: ABC textile mills manufacture fabrics with different colors. The factory has a high quality dyeing unit but with limited capacity. It is, thus, necessary for ABC mills to subcontract some of its dyeing operations – but the logistics manager wishes to know which fabrics are to be subcontracted and the corresponding quantity. The problem, ignoring time scheduling and other complications, can be formulated as follows.

Let n_f = number of different colored fabrics, x_i = quantity of fabric i dyed in-house, and y_i = quantity of fabric i, that is, subcontracted out for dyeing. Let P_i = unit profit, c_i = unit cost to dye in-house, s_i = unit cost to dye as charged by the subcontractor, and d_i = amount of fabric that needs to be dyed in a given time period (i.e., the demand). Further, let r = total amount of fabric that can be dyed in-house (of all types) and R = total fabric quantity that can be dyed by the subcontractor. A suitable objective function is the maximization of total revenue = total profit – total cost. This leads to the problem

$$\begin{aligned}
\text{maximize} \quad & f = \sum (P_i - c_i)x_i + \sum (P_i - s_i)y_i \\
\text{subject to} \quad & x_i + y_i = d_i \quad i = 1, \ldots, n_f \\
& \sum x_i \le r \\
& \sum y_i \le R \\
& \text{all} \quad x_i, y_i \ge 0.
\end{aligned}$$

The "design variables" are the quantities x_i and y_i, and the total number of variables $n = 2\,n_f$. The solution of course depends on the data. Typically, it is more expensive to subcontract, especially the high-end dyeing operations for the more profitable fabrics. However, owing to limited resources in-house, there is no option but to subcontract some amount. For the data below:

$$n_f = 4, \quad r = 50, \quad R = 1000\,(\text{unlimited}), \quad d_1 = 25, \quad d_2 = 45, \quad d_3 = 50, \quad d_4 = 60$$

$$(P_i - c_i) = \{10.0, 6.5, 6.0, 5.0\} \quad \text{and} \quad (P_i - s_i) = \{-2.0, 5.5, 5.0, 4.0\}$$

determine the net revenue, and the amount of each fabric dyed in-house versus dyed outside.

P4.14. The furnace used in a cogeneration plant is capable of burning coal, oil, or gas, simultaneously. Taking efficiency into consideration, total input rate of 4000 kW equivalent of fuel must be burnt. Regulations require that the sulphur content of the emissions is limited to 2.5%. The heating capacities, emission values and the costs are given as follows

Fuel Type	Sulphur Emission %	Cost $ per 1000 kg	Heating value kW/(kg/s)
Gas	0.12	55	61,000
Oil	0.45	41	45,000
Coal	2.8	28	38,000

Determine the minimum cost per hour and the corresponding fuel burning rates in thousands of kg per hour for the problem.

P4.15. Minimum weight design problem based on plastic collapse for the configuration shown in Fig. P4.15 is formulated as

$$\begin{aligned} \text{minimize} \quad & 8M_b + 8M_c \\ \text{subject to} \quad & |M_3| \leq M_b \\ & |M_4| \leq M_b \\ & |M_4| \leq M_c \\ & |M_5| \leq M_c \\ & |2M_3 - 2M_4 + M_5 - 1120| \leq M_c \\ & |2M_3 - M_4 - 800| \leq M_b \\ & |2M_3 - M_4 - 800| \leq M_c \\ & M_b \geq 0,\ M_c \geq 0 \\ & M_3,\ M_4,\ M_5 \text{ unrestricted in sign} \end{aligned}$$

Determine the plastic moment capacities M_b, M_c, and the plastic hinge moments M_3, M_4, M_4 for the above formulation.

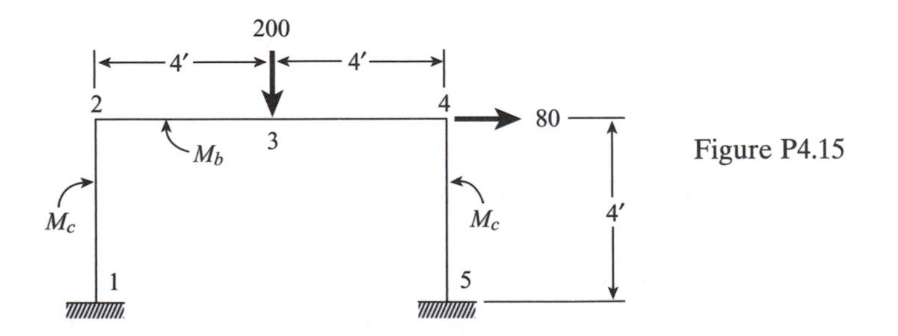

Figure P4.15

P4.16. A 30 V source is used to charge the three batteries of 6V, 12V, and 24V, as shown in Fig. P4.16. The settings are to be chosen such that the power to the charging batteries is maximized while satisfying the flow constraints. The problem is set as

$$\begin{aligned} \text{maximize} \quad & 6I_2 + 12I_3 + 24I_4 \\ \text{subject to} \quad & I_1 = I_2 + I_3 + I_4 \\ & I_1 \leq 8 \\ & I_2 \leq 6 \\ & I_3 \leq 4 \\ & I_4 \leq 2 \\ & I_1, I_2, I_3, I_4 \geq 0 \end{aligned}$$

Determine the currents for maximizing the power.

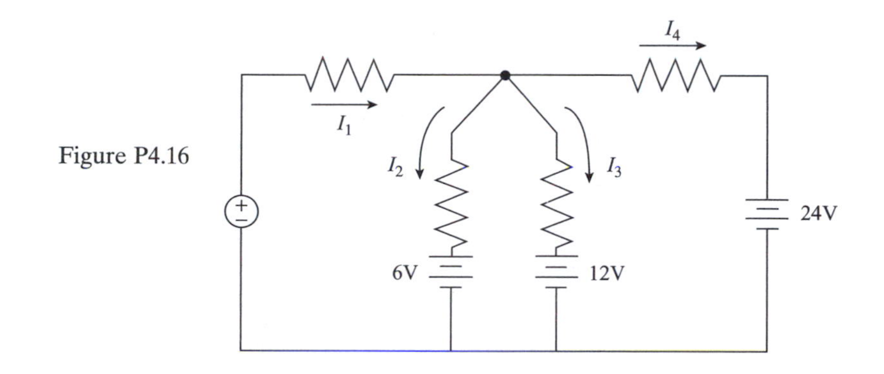

Figure P4.16

P4.17. Solve the problem

$$\begin{aligned} \text{maximize} \quad & x_1^{\alpha_1} x_2^{\alpha_2} \cdots x_n^{\alpha_n} \\ \text{subject to} \quad & x_1^{\beta_{i1}} x_2^{\beta_{i2}} \cdots x_n^{\beta_{in}} \leq b_i \quad i = 1 \text{ to } m \\ & x_j \geq 1 \quad j = 1 \text{ to } n \end{aligned}$$

Also develop a strategy if the limits on x_j are of the form $x_j \geq l_j$. (*Hint*: Take logarithm of the objective function and the constraints and change variables.)

P4.18. A parallel flow heat exchanger of a given length is to be designed. The conducting tubes, all of same diameter, are enclosed in an outer shell (Fig. P4.18). The area occupied by the tubes in the shell should not exceed 0.25 m^2. The diameter of the conducting tube is to be larger than 60 mm. Determine the number of tubes and the diameter for the largest surface area.

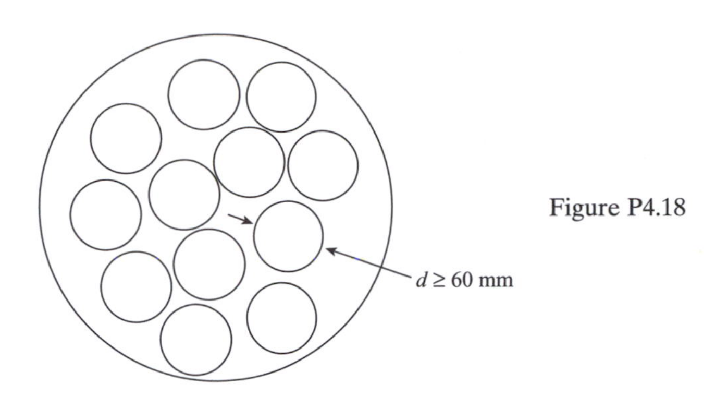

Figure P4.18

P4.19. Plot the feasible region for the structural design problem, Section 4.3, with cable 1 constraint $4W_1 + W_2 \leq 960$ replaced by $4W_1 + W_2 \leq 880$. Comment on the plot.

P4.20. In the LP in (4.4), use the simplex method and show tableau calculations to pivot from point A to point E.

P4.21. Solve structural design problem, Section 4.3, using the revised simplex procedure. Show calculations.

P4.22. Formulate and solve the dual for any one of the primal problems PM1-PM7, Section 4.4, and give physical interpretation to the dual variables.

P4.23. Given the primal LP:

$$\begin{aligned} \text{minimize} \quad & -x_1 - x_2 \\ \text{subject to} \quad & 0 \leq x_1 \leq 6 \\ & 0 \leq x_2 \leq 5 \\ & x_1 + 3x_2 \leq 18 \\ & 2x_1 + x_2 \leq 14 \end{aligned}$$

Formulate and solve the dual LP. Recover the primal solution. Show all tableau calculations.

P4.24. Prove that if $\mathbf{Q}$ is positive semidefinite, then $\mathbf{M}$ is positive semidefinite, in Eq. (4.52).

$$\textit{Hint}\text{: consider } \mathbf{y} = \begin{bmatrix} \mathbf{y}_1 \\ \mathbf{y}_2 \end{bmatrix} \text{ and } \mathbf{y}^{\mathrm{T}}\mathbf{M}\mathbf{y}$$

P4.25. Determine the optimal portfolio $\mathbf{x}$ (x_i represents investment of stock i as a % of the total) by solving the QP using QPLCP (Lemke) and EXCEL SOLVER computer programs:

$$\begin{aligned} \text{maximize} \quad & \mathbf{c}^{\mathrm{T}}\mathbf{x} - \frac{1}{\lambda}\mathbf{x}^{\mathrm{T}}\,\mathbf{A}\mathbf{x} \\ \text{subject to} \quad & 0 \le x_i \le 0.35, \quad i = 1, \ldots, n \\ & \sum_{i=1}^{n} x_i = 1 \end{aligned}$$

where $\lambda = 10$, $n = 3$, $\mathbf{c} = [2.0, 1.0, 0.5]^{\mathrm{T}}$, $\mathbf{A} = \begin{bmatrix} 32 & 3.2 & .016 \\ 3.2 & 8 & 0.2 \\ 0.016 & 0.2 & 0.02 \end{bmatrix}$

P4.26. In Fig. P4.26, determine the displacements Q_1 and Q_2 by solving the following QP:

$$\begin{aligned} \text{minimize} \quad & \Pi = \tfrac{1}{2}\mathbf{Q}^{\mathrm{T}}\mathbf{K}\mathbf{Q} - \mathbf{Q}^{\mathrm{T}}\mathbf{F} \\ \text{subject to} \quad & Q_2 \le 1.2 \end{aligned}$$

where $\mathbf{F} = [P, 0]^{\mathrm{T}}$, $\mathbf{Q} = [Q_1, Q_2]^{\mathrm{T}}$, $\mathbf{K} = \dfrac{10^5}{3}\left[\begin{smallmatrix} 2 & -1 \\ -1 & 1 \end{smallmatrix}\right]$, $P = 60 \times 10^3\,\mathrm{N}$

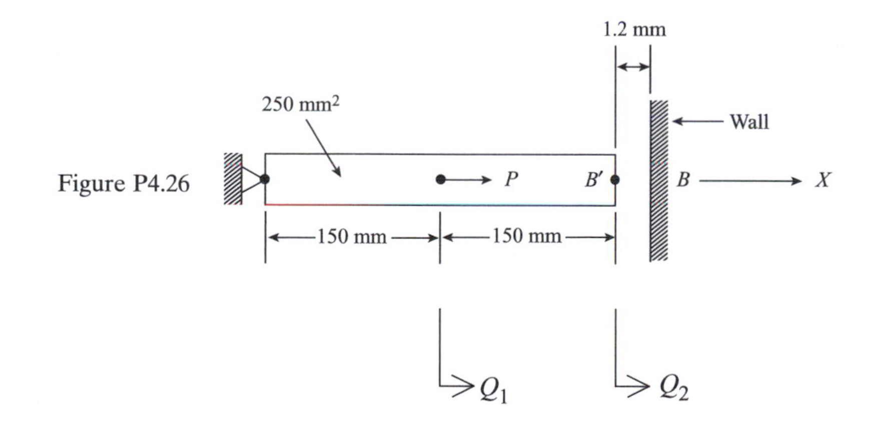

Figure P4.26

P4.27. Try solving Problem PM3 (refinery problem) in the text using Program LPINTER. Why does it fail ? Add the constraints $x_1 \le 1000$ and $x_2 \le 1000$. Now why does it succeed?

REFERENCES

Dantzig, G.B., Linear programming – the story about how it began: some legends, a little about its historical significance, and comments about where its many mathematical programming extensions may be headed, *Operations Research*, **50** (1), 42–47, 2002.

Dantzig, G.B., *Linear Programming and Extensions*, Princeton University Press, Princeton, NJ, 1963.

Dikin, I.I., Iterative solution of problems of linear and quadratic programming (in Russian), *Doklady Academiia Nauk USSR*, **174**, 747–748. English translation, *Soviet Mathematics Doklady*, **8**, 674–675, 1967.

Karmarkar, N., A new polynomial time algorithm for linear programming, *Combinatorica*, **4**, 373–395, 1984.

Lemke, C.E., Bimatrix equilibrium points and mathematical programming, *Management Science*, **11**, 681–689, 1965.

Murty, K.G., *Linear Programming*, Wiley, New York, 1983.

Roos, C., Terlaky, T., and Vial, J.P., *Theory and Algorithms for Linear Optimization: An Interior Point Approach*, Wiley, New York, 1997.

5

Constrained Minimization

5.1 Introduction

The majority of engineering problems involve *constrained* minimization – that is, the task is to minimize a function subject to constraints. A very common instance of a constrained optimization problem arises in finding the minimum weight design of a structure subject to constraints on stress and deflection. Important concepts pertaining to linear constrained problems were discussed in Sections 4.1–4.5 in Chapter 4 including active or binding constraints, Lagrange multipliers, computer solutions, and geometric concepts. These concepts are also relevant to nonlinear problems. The numerical techniques presented here directly tackle the nonlinear constraints, most of which call an LP solver within the iterative loop. In contrast, penalty function techniques transform the constrained problem into a sequence of unconstrained problems as discussed in Chapter 6. In this chapter, we first present graphical solution for two variable problems and solution using EXCEL SOLVER and MATLAB. Subsequently, formulating problems in "standard NLP" form is discussed followed by optimality conditions, geometric concepts, and convexity. Four gradient-based numerical methods applicable to problems with differentiable functions are presented in detail:

1. Rosen's Gradient Projection method for nonlinear objective and linear constraints
2. Zoutendijk's Method of Feasible Directions
3. The Generalized Reduced Gradient method
4. Sequential Quadratic Programming method

Each of these methods is accompanied by a computer program in the disk at the end of the book. The reader can learn the theory and application of optimization with the help of the software.

Constrained problems may be expressed in the following general *nonlinear programming* (NLP) form:

$$\begin{array}{lll} \text{minimize} & f(\mathbf{x}) & \\ \text{subject to} & g_i(\mathbf{x}) \le 0 & i = 1, \ldots, m \\ \text{and} & h_j(\mathbf{x}) = 0 & j = 1, \ldots, \ell \end{array} \tag{5.1}$$

where $\mathbf{x} = (x_1, x_2, \ldots, x_n)^\mathrm{T}$ is a column vector of n real valued design variables. Here, f is the *objective* or *cost* function, g's are *inequality constraints* and h's are *equality constraints*. The notation $\mathbf{x}^0$ for starting point, $\mathbf{x}^*$ for optimum and $\mathbf{x}^k$ for the (current) point at the kth iteration will be generally used. A point $\mathbf{x}$ that satisfies all inequality and equality constraints is called *feasible*. Also, an inequality constraint $g_i(\mathbf{x}) \le 0$ is said to be *active* at a feasible point $\mathbf{x}$ if $g_i(\mathbf{x}) = 0$ and *inactive* if $g_i(\mathbf{x}) < 0$. We consider an equality constraint $h_i(\mathbf{x}) = 0$ as *active* at any feasible point. Conceptually, problem (5.1) can be expressed in the form

$$\begin{array}{ll} \text{minimize} & f(\mathbf{x}) \\ \text{subject to} & \mathbf{x} \in \Omega \end{array} \tag{5.2}$$

where Ω is the feasible region defined by all the constraints as $\Omega = \{\mathbf{x} : \mathbf{g} \le \mathbf{0}, \mathbf{h} = \mathbf{0}\}$. Looking at the problem in the form (5.2), we see that a point $\mathbf{x}^* = [x_1^*, x_2^*, \ldots, x_n^*]^\mathrm{T}$ is a global minimum if $f(\mathbf{x}^*) \le f(\mathbf{x})$ for all $\mathbf{x} \in \Omega$.

Practical aspects of formulating, solving and interpreting an NLP is presented through an example as follows. Graphical solution is explained in Section 5.2. Use of EXCEL SOLVER and MATLAB are discussed in Section 5.3. Formulation of problems to fit the standard form in (5.1) is discussed in Section 5.4.

Example 5.1 (Truss Problem)

The following simple example, obtained from Fox [1971], illustrates use of nonlinear programming in design. Consider the two-bar planar truss in Fig. E5.1.

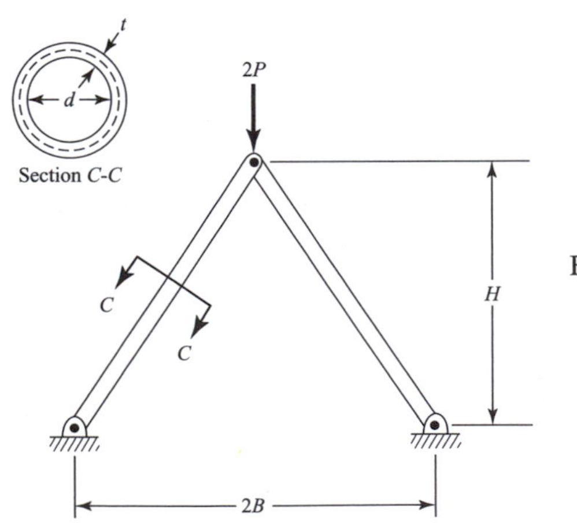

Figure E5.1. Two-bar truss.

The members are thin-walled tubes of steel, pinned together with a downward load of magnitude $2P$ applied as shown. We will assume that the wall thickness of the tube is fixed at some value t and that the half-span is fixed at some value B. The design problem is to select $d =$ the mean diameter of the tube, and $H =$ height of the truss.

The stress in the members is given by

$$\sigma = \frac{P}{\pi t}\frac{(B^2 + H^2)^{1/2}}{Hd} \tag{a}$$

Thus, a constraint expressing the requirement that the stress be less than an allowable limit, σ_{all}, can be written as

$$\frac{P}{\sigma_{\text{all}}\,\pi t}\frac{(B^2 + H^2)^{1/2}}{Hd} - 1 \le 0 \tag{b}$$

The normalized manner in which the above constraint is expressed is important for numerical solution as the magnitudes of diverse types of constraints will be roughly in the range [0, 1].

Since the members are in compression, we would also like to safeguard against buckling. Euler's formula for critical buckling load is

$$\sigma_{\text{cr}} = \frac{\pi^2 E(d^2 + t^2)}{8(B^2 + H^2)} \tag{c}$$

Using (a) and (c), the constraint $\sigma \le \sigma_{\text{cr}}$ can be written in normalized form as

$$\frac{8P(B^2 + H^2)^{1.5}}{\pi^3 Et\,Hd(d^2 + t^2)} - 1 \le 0 \tag{d}$$

Of course, we must impose $H, d \ge 0$. The objective function is the weight of the truss:

$$f = 2\rho\pi dt(B^2 + H^2)^{1/2} \tag{e}$$

where $\rho =$ weight density of the material. Our problem is to minimize f subject to the two inequality constraints in (b) and (d). Note that after obtaining a solution (H^*, d^*), we must ensure that the tubular cross-sections are indeed thin-walled (i.e., $d/t \gg 1$). For data $\rho = 0.3$ lb/in^3, $P = 33{,}000$ lb, $B = 30$ in., $t = 0.1$, $E = 30 \times 10^6$ psi, $\sigma_{\text{all}} = 100{,}000$ psi we find, using a numerical procedure, that the optimum solution is $H = 20$ in., $d = 1.9$. Both buckling and yield stress constraints are active. However, by changing the problem parameters (such as the value of B or yield stress limit), a different optimum point will result and importantly, the active set at the optimum may be different. We can impose lower and upper limits on the design variables or add a deflection constraint. It should be realized that for certain data, no feasible solution may exist.

The use of optimization techniques frees the design engineer from the drudgery of doing detailed calculations. Furthermore, the designer can, by studying the active constraints at the optimum design, determine if there are things that can be done at a higher judgemental level to further reduce cost, for example, by changing the material, changing the truss topology, adding constraints, etc.

5.2 Graphical Solution of Two-Variable Problems

Two-variable problems ($n = 2$) can be visualized graphically. The procedure as given in Section 1.5, Example 1.12, is as follows. Let $\mathbf{x} = [x_1, x_2]^{\mathrm{T}}$ and the problem be of the form: minimize $f(\mathbf{x})$, subject to $g_i\ (\mathbf{x}) \leq 0$, $i = 1, \ldots, m$. First, define the $x_1 - x_2$ plane. Then, for each g_i, plot the curve obtained by setting $g_i(x_1, x_2) = 0$. This can be done by choosing various values for x_1 and obtaining corresponding values of x_2 by solving the equation. Identify which side of the curve is feasible – that is, on which side does $g_i(\mathbf{x}) < 0$. The intersection of the feasible regions of all these m curves will define the feasible region. Now, plot contours of the objective function by plotting the curves $f(\mathbf{x}) = c$, where, say, c = 10, 20, 30, etc. Contours of f will represent a family of parallel or nonintersecting curves. Identify the lowest contour within the feasible region and this will define the optimum point. In some cases, the minimum contour may coincide with a constraint curve – in this case, the entire curve segment within the feasible region is optimum (there is an infinity of designs that are optimum).

If an equality constraint $h_1(x_1, x_2) = 0$ is also present in the problem, then we have to look for an optimum *on* the curve $h_i(\mathbf{x}) = 0$ and within the feasible region defined by the g_i. Note: two equality constraints will define a single point – f will not play a role.

Example 5.2

Let us consider the problem presented earlier. The graph of this two-variable problem is shown in Fig. E5.2. The optimum values of d and H agree with the numerically obtained result in Example 5.1. First, H was discretized from 1.0 to 50.0. Equation (b) was solved for the corresponding value of d – in closed form. The graph (obtained as a "scatter plot" with the EXCEL spreadsheet) of d versus H is the graph of the equation $g_1 = 0$. The same basic technique was used to plot $g_2 = 0$, except that finding a d for a given H involves finding the root of a cubic equation, which was also done numerically.

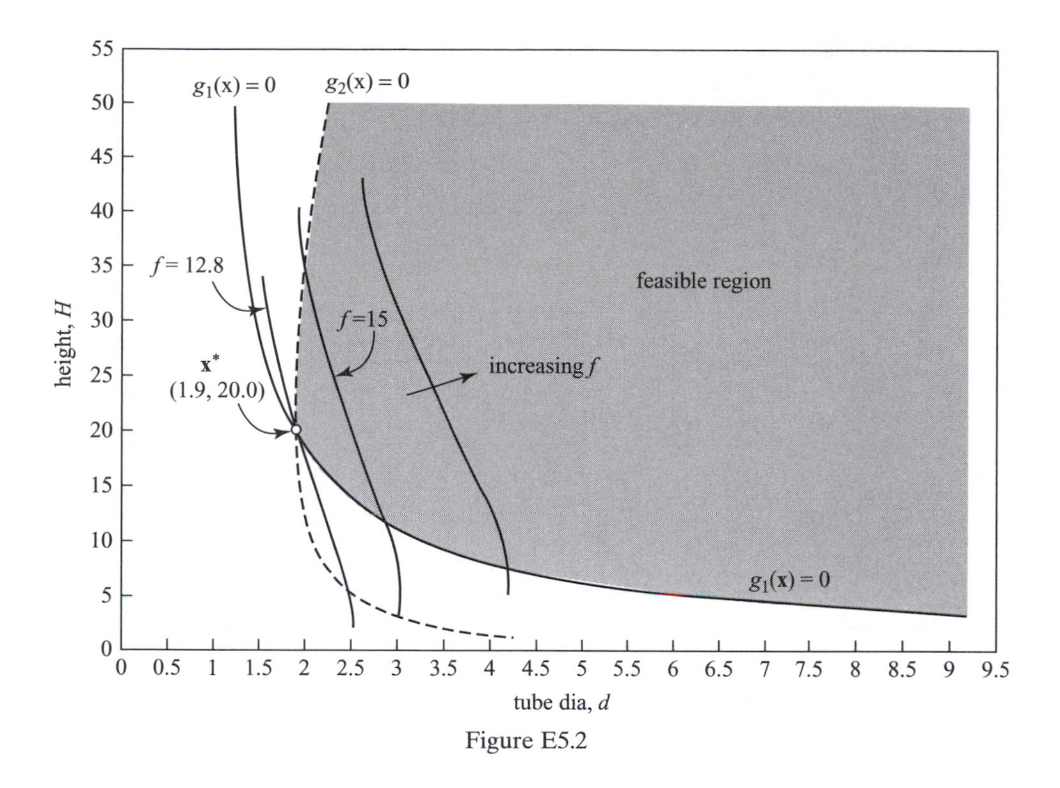

Figure E5.2

5.3 Use of EXCEL SOLVER and MATLAB

While details of numerical techniques along with in-house codes are presented subsequently, use of SOLVER and MATLAB, respectively, are demonstrated in the following. Consider the truss example given in Example 5.1.

SOLVER

As in Section 4.3, Chapter 4, first, the formulas for objective and constraints, initial guesses for variables, and constants are defined on the EXCEL spreadsheet in cells A1-G7. Then, SOLVER window is opened under Tools menu (if not visible, click "Add-ins"), and the optimization problem is defined and solved.

Set Target Cell:___, Equal to: "Min," By Changing Cells:___
Subject to the Constraints: (variables) ≥ 0, (constraints g_1, g_2) ≤ 0
Use initial guess: $d = 0.5$, $H = 5$.

Solving (and also choosing additional answer sheets) we get:

Target Cell (Min)

Cell	Name	Original value	Final value
$C$2	OBJ = f	2.866434354	12.81260005

Adjustable Cells

Cell	Name	Original value	Final value
$A$2	dia, d	0.5	1.878356407
$B$2	height, H	5	20.23690921

Constraints

Cell	Name	Cell value	Formula	Status	Slack
$D$2	stress con	3.81829E-08	$D$2<=0	Binding	0
$E$2	buckling con	2.17391E-07	$E$2<=0	Binding	0
$A$2	dia, d	1.878356407	$A$2>=0	Not Binding	1.878356407
$B$2	height, H	20.23690921	$B$2>=0	Not Binding	20.23690921

Further, the Sensitivity Report (at optimum) provides the Lagrange multipliers:

Cell	Name	Final value	Lagrange multiplier
$D$2	stress con	3.81829E-08	−5.614014237
$E$2	buckling con	2.17391E-07	−2.404064294

MATLAB (Using the Optimization Toolbox)

Main M-file: 'fmincon1.m':

```
function [] = LP()
close all; clear all;
XLB=[0 0];  %lower bounds on variables
X = [ 0.5  5];  %starting guess
A=[];B=[];Aeq=[];Beq=[];XUB=[5 100];
[X,FOPT,EXITFLAG,OUTPUT,LAMBDA] = fmincon('GETFUN',X,A,B,Aeq,
Beq,XLB,XUB,...
                                            'NONLCON')
LAMBDA.ineqnonlin
```

File: GETFUN.m

```
function [f] = GETFUN(X)
pi=3.1415927;
rho = 0.3; t = 0.1; B = 30;
f = 2*pi*rho*X(1)*t*sqrt(B^2+X(2)^2);
```

File; NONLCON.m

```
function [c, ceq] = NONLCON(X)
pi=3.1415927;
rho = 0.3; t = 0.1; B = 30; E = 30e6; sall = 1e5; P=33e3;
c(1) = P*sqrt(B^2+X(2)^2)/(sall*pi*t*X(2)*X(1)) - 1;
c(2) = 8*P*(B^2+X(2)^2)^1.5/(pi^3*E*t*X(2)*X(1)*(X(1)^2+t^2)) - 1;
ceq=[];
```

Executing the main m-file produces the output:

```
X =     1.8784    20.2369
FOPT =      12.8126
EXITFLAG =        1
OUTPUT =
          iterations: 9
           funcCount: 30
LAMBDA =
    5.6140
    2.4041
```

5.4 Formulation of Problems in Standard NLP Form

Sometimes, problems have to be modified so as to be of the form in (5.1) as illustrated in the following. Examples are distributed in this chapter, which illustrate (i)–(iv).

(i) Maximize $f \Leftrightarrow$ minimize $-f$

(ii) "Minimax" problems are of the form

$$\text{minimize : maximum } \{f_1, f_2, \ldots, f_p\}$$

These can be expressed as

$$\begin{aligned} &\text{minimize} \quad f = x_{n+1} \\ &\text{subject to} \quad f_i(\mathbf{x}) - x_{n+1} \le 0, \quad i = 1, \ldots, p \end{aligned}$$

where x_{n+1} is a newly introduced variable in the problem. Methods which solve (5.1) can be applied to the preceding. There are special-purpose methods that have also been developed for the minimax problem.

(iii) *Pointwise* constraint functions that have to be satisfied for every value of a parameter, say $\theta \in [0, 1]$, can be handled by discretizing the parameter as θ_i, $i = 1, \ldots, n_d$ where n_d = number of discretization points. Often, θ is time. Consider

$$\psi(\mathbf{x}, \theta) \leq 0, \qquad \text{for all} \quad \theta \in [\theta_{\min}, \theta_{\max}] \tag{a}$$

This can be written as

$$\psi(\mathbf{x}, \theta_i) \leq 0, \quad i = 1, \ldots, n_d \tag{b}$$

Alternative techniques to handle pointwise constraints exist in the literature; however, discretization as in (b) coupled with optimizer that uses an "active set strategy" is a robust approach. This has been applied in Example 5.20 at the end of this chapter.

(iv) Often, in regression analysis, which involves fitting a model equation to data, we have to minimize error which can be defined with various norms. Thus, we come across

$$f = \text{minimize} \sum_{i=1}^{n} |\psi_i(\mathbf{x})| \tag{d}$$

where ψ_i represents an error between model and data. As in $y = |x|$, we know that the function in (d) is not differentiable at points where $\psi_i = 0$. This is overcome by posing the problem in the form:

$$\begin{aligned} &\text{minimize} \quad \sum_{i=1}^{n} (p_i + n_i) \\ &\text{subject to} \quad \psi_i = p_i - n_i \\ &\text{and} \quad p_i \geq 0, \quad n_i \geq 0 \end{aligned}$$

where p_i and n_i, $i = 1, \ldots, n$, are additional variables representing the positive and negative errors, respectively.

Lastly, consider the "min-max" error objective:

$$\underset{\boldsymbol{x}}{\text{minimize}} \; \underset{i}{\text{maximize}} \; |\psi_i(\mathbf{x})|,$$

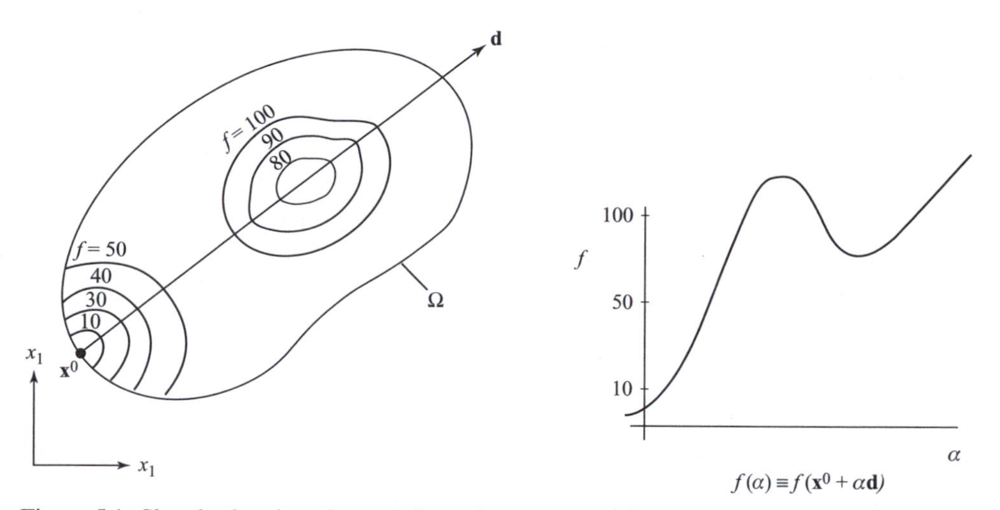

Figure 5.1. Sketch showing that gradient does not vanish when minimum occurs on the boundary.

which can be posed in differentiable form as

$$\begin{aligned} &\text{minimize} \quad Z \\ &\text{subject to} \quad -Z \le \psi_i(\mathbf{x}) \le Z \end{aligned}$$

where Z is an additional variable. As Z decreases, the error is squeezed down from both below and above. This is also the technique presented in (ii) aforementioned.

5.5 Necessary Conditions for Optimality

Recall from Chapter 3 that for $\mathbf{x}^*$ to be an *unconstrained* local minimum of the function f, it is necessary that $\nabla f(\mathbf{x}^*) = \mathbf{0}$. We found this central concept of "zero slope" to also help us solve problems by hand calculations and useful as a stopping criterion in algorithms. We would now like to identify the necessary conditions for optimality of Problem (5.1) or (5.2). Figure 5.1 shows that the slope need not equal zero if the optimum lies on the boundary. If the optimum lies in the interior, where all constraints are inactive, then the zero slope condition holds true.

Optimality conditions will be presented in two steps. First, only equality constraints will be considered via the method of Lagrange multipliers. This method has made a great impact in mechanics. It was used in conjunction with the Calculus of Variations to analytically solve problems, such as the Brachistochrone Problem: to find a plane curve along which a particle descends in the shortest possible time, starting from a given location A and arriving at a location B which is below A. See Lanczos [1970] for further examples.

Equality Constrained Problem – The Method of Lagrange Multipliers

The necessary conditions for optimality that are given in the following are for any *stationary* point of the function f that corresponds to a minimum, maximum or a saddle point (inflection point). It is analogous to the zero-slope condition for an unconstrained function. We may generalize the method for ℓ constraints as follows. Consider

$$\text{minimize} \quad f(\mathbf{x}) \tag{5.3a}$$

$$\text{subject to} \quad h_j(\mathbf{x}) = 0, \quad j = 1, \ldots, \ell \tag{5.3b}$$

Our first thought would be to eliminate one of the x_i – for example, x_n – from the constraint equation (5.3b), expressing it in terms of the other x_i. Then our function f would depend only on the $n-1$ unconstrained variables, for which we know the optimality conditions. This method is entirely justified and sometimes advisable. But frequently the elimination is a rather cumbersome procedure. Moreover, the condition (5.3b) may be symmetric in the variables $x_1, x_2, \ldots, x_n$, and there would be no reason why one of the variables should be artificially designated as dependent, the others as independent variables. Lagrange devised a beautiful method that preserves the symmetry of the variables without elimination. First, the Lagrangian function is formed as

$$L(\mathbf{x}) = f(\mathbf{x}) + \sum_{j=1}^{\ell} \lambda_j h_j(\mathbf{x}) \tag{5.4}$$

where λ_j is a scalar multiplier associated with constraint h_j. If $\mathbf{x}^*$ is a stationary point (minimum or maximum or a point of inflection), then it is necessary that the following conditions be satisfied:

$$\frac{\partial L}{\partial x_i} = \frac{\partial f}{\partial x_i} + \sum_{j=1}^{\ell} \lambda_j \frac{\partial h_j}{\partial x_i} = 0, \quad i = 1, \ldots, n \tag{5.5a}$$

$$h_j(x) = 0, \quad j = 1, \ldots, \ell \tag{5.5b}$$

Equations (5.5) can also be written as $\nabla_{\mathbf{x}} L = 0$ and $\mathbf{h} = 0$. Optimality conditions in (5.5) can be used to obtain candidate solutions or to verify if a given design point $\mathbf{x}^*$ is a minimum point. However, if we plan to use them to determine $\mathbf{x}^*$ and λ, then we should determine all candidate points that satisfy (5.5) and then compare the function values at these points – and pick the minimum point, if this is what we desire. "Sufficiency" conditions, which can also help us to pick a minimum from the candidates, will be discussed in Section 5.6.

Example 5.3
Consider

$$\begin{array}{ll} \text{minimize} & 2x_1 + x_2 \\ \text{subject to} & x_1^2 + x_2^2 - 1 = 0 \end{array}$$

We have $L = 2\,x_1 + x_2 + \lambda\,(x_1^2 + x_2^2 - 1)$. The optimality conditions are

$$\frac{\partial L}{\partial x_1} = 2 + 2\lambda x_1 = 0, \quad x_1 = -\frac{1}{\lambda}$$

$$\frac{\partial L}{\partial x_2} = 1 + 2\lambda x_2 = 0, \quad x_2 = -\frac{1}{2\lambda}$$

Upon substituting for $\mathbf{x}(\lambda)$ into the constraint equation $h = 0$, we get $\lambda = \pm\sqrt{5}/2$. For each root, we obtain

$$\lambda = \frac{\sqrt{5}}{2} : x_1^* = -\frac{2}{\sqrt{5}}, \quad x_2^* = -\frac{1}{\sqrt{5}} \Rightarrow \text{see point } A \text{ in Fig. E5.3}$$

$$\lambda = -\frac{\sqrt{5}}{2} : x_1{}^* = \frac{2}{\sqrt{5}}, \quad x_2{}^* = \frac{1}{\sqrt{5}} \Rightarrow \text{see point } B \text{ in Fig. E5.3}$$

Evidently, the minimum point is A. The necessary conditions only identified the two possible candidates. Sufficiency conditions, discussed in Section 5.6, will identify the proper minimum among the candidates (see Example 5.8)

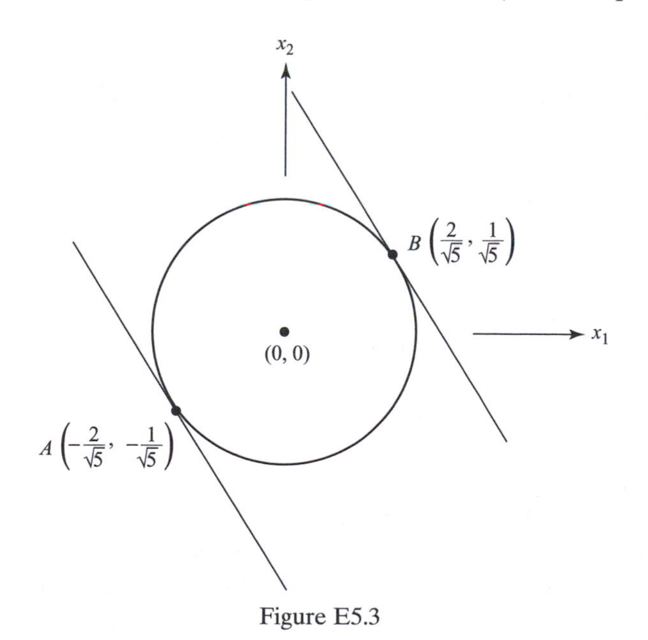

Figure E5.3

Inequality Constrained Problem – Concept of Descent and Feasible Directions and Optimality (KKT) Conditions

We will focus on understanding the necessary conditions for optimality for the inequality constrained problem:

$$\text{minimize} \quad f(x) \tag{5.6a}$$

$$\text{subject to} \quad g_i(\mathbf{x}) \le 0, \quad i = 1, \ldots, m \tag{5.6b}$$

Both the discussion as well as the development of numerical methods later in this chapter require knowing what a *descent direction* and a *feasible direction* means. A "direction" is a vector in design variable space or the space of $x_1, x_2, \ldots, x_n$. We consider only differentiable functions here – that is, functions whose derivative is continuous.

Descent Direction

The vector **d** is a *descent direction* at a point $\mathbf{x}_k$ if

$$\nabla f^{\mathrm{T}} \mathbf{d} < 0 \tag{5.7}$$

where the gradient ∇f is evaluated at $\mathbf{x}_k$. This condition ensures that for sufficiently small $\alpha > 0$, the inequality

$$f(\mathbf{x}_k + \alpha \mathbf{d}) < f(\mathbf{x}_k)$$

should hold. This means that we can reduce f by changing the design along the vector **d**. Recall from Chapter 3 that ∇f points in the direction of increasing f; thus, **d** is a descent direction if it is pointing in the opposite half-space from ∇f (Fig. 5.2).

Example 5.4
Given: $f = x_1 x_2$, $\mathbf{x}^0 = [1, 3]^{\mathrm{T}}$, $\mathbf{d} = [1, 1]^{\mathrm{T}}$. Is **d** a descent direction?

Solution: $\nabla f(\mathbf{x}^0) = [3, 1]^{\mathrm{T}}$. The dot product $\nabla f(\mathbf{x}^0)\ \mathbf{d} = 4 > 0$. Thus, **d** is *not* a descent direction.

Feasible Direction

Now consider a feasible point $\mathbf{x}_k$. That is, $\mathbf{x}_k \in \Omega$, where Ω denotes the feasible region defined by the collection of all points that satisfy (5.6b). We say that **d** is a feasible direction at $\mathbf{x}_k$ if there is a $\bar{\alpha} > 0$ such that $(\mathbf{x}_k + \alpha\ \mathbf{d}) \in \Omega$ for all α, $0 \le \alpha \le \bar{\alpha}$. Now, we will develop a more practical definition of a feasible direction using the

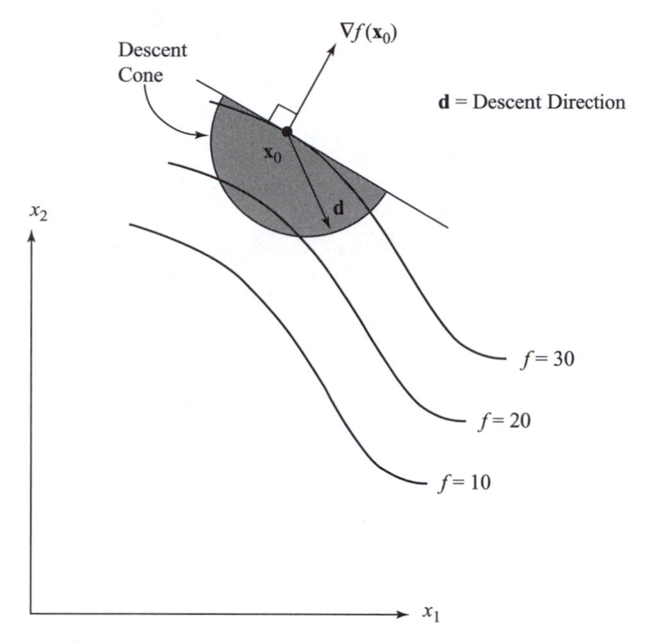

Figure 5.2. A descent cone and a descent direction.

fact that $g_i(\mathbf{x})$ are differentiable functions. At a feasible point $\mathbf{x}_k$, let one or more inequality constraints be active at $\mathbf{x}_k$ – that is, one or more $g_i = 0$. Let us refer to the active set I by

$$I = \{i : g_i(x_k) = 0, \quad i = 1, \ldots, m\} \tag{5.8}$$

Regular Point

Let $\mathbf{x}^k$ be a feasible point. Then $\mathbf{x}^k$ is said to be a *regular point* if the gradient vectors $\nabla g_i(\mathbf{x}_k))$, $i \in I$, are linearly independent.

With the assumption that $\mathbf{x}^k$ is a *regular point*, we can say that a vector $\mathbf{d}$ is a *feasible direction* if

$$\nabla g_i^{\mathrm{T}} \mathbf{d} < \mathbf{0} \quad \text{for each} \quad i \in I, \tag{5.9}$$

which will ensure that for sufficiently small $\alpha > 0$, $(\mathbf{x}_k + \alpha\, \mathbf{d})$ will be feasible or $g_i(\mathbf{x}_k + \alpha \mathbf{d}) < 0, i = 1, \ldots, m$. In fact, for each $i \in I$, (5.9) defines a half-space, and the intersection of all these half-spaces forms a "feasible cone" within which $\mathbf{d}$ should lie. Figure 5.3 illustrates this.

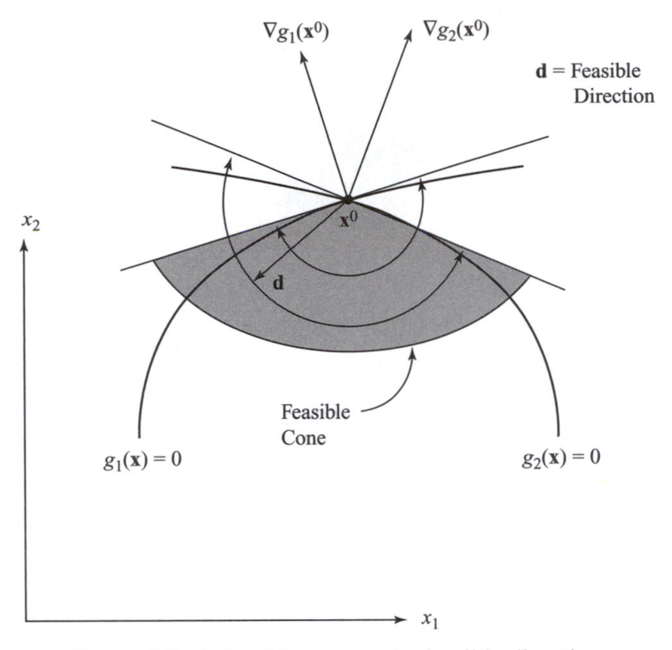

Figure 5.3. A feasible cone and a feasible direction.

Example 5.5

Consider:

$$\begin{array}{ll} \text{minimize} & f = -(x_1 + x_2) \\ \text{s.t.} & g_1 \equiv x_1^2 + 4x_2^2 - 1 \leq 0, \quad g_2 \equiv -x_1 \leq 0, \quad g_3 \equiv -x_2 \leq 0 \\ \text{point} & \mathbf{x}^0 = \left(\frac{1}{\sqrt{5}}, \frac{1}{\sqrt{5}}\right)^{\mathrm{T}} \end{array}$$

Identify the feasible and/or descent properties of each of the following direction vectors: $\mathbf{d}^1 = (1, 0)^{\mathrm{T}}$, $\mathbf{d}^2 = (1, -0.5)^{\mathrm{T}}$, $\mathbf{d}^3 = (0, -1)^{\mathrm{T}}$.

To check descent, we need to use (5.7), with $\nabla f(\mathbf{x}^0) = [-1, -1]^{\mathrm{T}}$, which gives: $\mathbf{d}^1$ is a descent vector, $\mathbf{d}^2$ is a descent vector, $\mathbf{d}^3$ is not a descent vector.

To check feasibility of $\mathbf{d}^i$, we need to use (5.9), with active set $I = \{1\}$. That is, only g_1 is active at $\mathbf{x}^0$. Further, $\nabla g_1(\mathbf{x}^0) = \frac{1}{\sqrt{5}}(2, 8)^{\mathrm{T}}$. Evaluating the dot products, we conclude: $\mathbf{d}^1$ is not a feasible vector, $\mathbf{d}^2$ is a feasible vector, $\mathbf{d}^3$ is a feasible vector.

Thus, only $\mathbf{d}^2$ is both descent and feasible. Figure E5.5 illustrates this problem.

Having understood the concept of a descent direction and a feasible direction, the optimality conditions for the inequality constrained problem (5.6) can now be derived. The reader is urged to understand the derivation as important

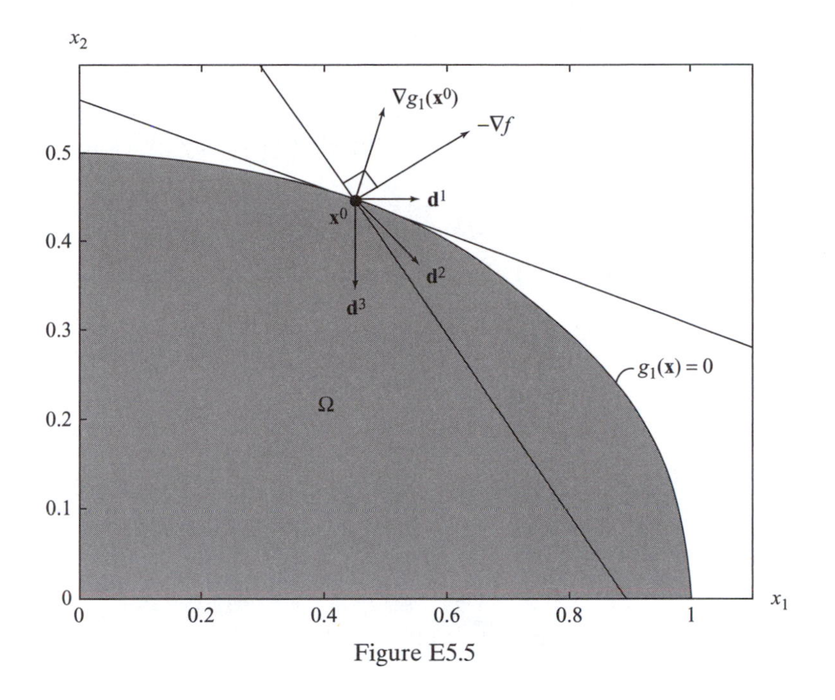

Figure E5.5

concepts are embedded in it, which also help to understand numerical methods, in particular, the method of feasible directions.

Consider, first the case when there are no active constraints at a current point $\mathbf{x}_k$. That is, we are currently at an "interior" point with respect to the constraints. Now, we can reduce the objective function still further by simply travelling in the direction of steepest descent $-\nabla f(\mathbf{x}_k)$. Upon determining the step size along this direction, we will thus obtain a new point with a lesser value of *f*. Evidently, we cannot repeat this procedure if we arrive at an interior point where $\nabla f = \mathbf{0}$. Thus, we see that if $\mathbf{x}^*$ is a solution to (5.6), and lies in the interior, then it is necessary that $\nabla f(\mathbf{x}^*)) = 0$. This is the optimality condition in the absence of active constraints that, of course, agrees with that derived in Chapter 3 on unconstrained problems.

However, in almost all engineering problems, some constraints *will* be active at the solution to (5.6). These active constraints can include lower or upper limits on variables, stress limits, etc. We saw this to be the case in Example 5.1. Assume that the active set at a current point $\mathbf{x}_k$ is defined by *I* as in (5.8) and the regular point assumption is satisfied. Now, if we can find a direction $\mathbf{d}$ that is both descent and feasible, then we can travel in this direction to reduce *f* while satisfying the constraints – thus, our current point $\mathbf{x}_k$ cannot be a minimum! This leads us to conclude that in order for a point $\mathbf{x}^*$ to be a solution to (5.6), it is necessary that there does not exist any vector **d** at the point $\mathbf{x}^*$ that is both a descent direction *and* a feasible direction. That is, the intersection of the descent–feasible cone should be "empty." This statement can be expressed in an elegant and useful way by making use of the following lemma due to Farkas.

Farkas Lemma

Given the vectors $\mathbf{a}^i$, $i = 1, 2, \ldots, t$ and **g**, there is no vector **d** satisfying the conditions

$$\mathbf{g}^{\mathrm{T}}\mathbf{d} < 0 \tag{5.10a}$$

and

$$\mathbf{a}^{i\mathrm{T}}\mathbf{d} \geq 0, \quad i = 1, 2, \ldots, t \tag{5.10b}$$

if and only if **g** can be written in the form

$$\mathbf{g} = \sum_{i=1}^{t} \mu_i \, \mathbf{a}^i, \quad \mu_i \geq 0 \tag{5.11}$$

If we associate **g** with $\nabla f(\mathbf{x}^*)$ and each $\mathbf{a}^i$ with the negative of an active constraint gradient, $-\nabla g_i\,(\mathbf{x}^*)$, $i \in I$, and denote t as the number of active constraints, we immediately see that the a vector **d** which is *both* descent and feasible cannot exist if $\nabla f = \sum_{i \in I} \mu_i(-\nabla g_i)$, $\mu_i \geq 0$. The scalars μ_i are called Lagrange multipliers. The aforementioned condition may be written as

$$-\nabla f = \sum_{i \in I} \mu_i \nabla g_i, \quad \mu_i \geq 0 \tag{5.12}$$

The geometric significance of this is that a vector **d** cannot be both descent and feasible, that is, satisfy both (5.7) and (5.9), if $-\nabla f$ lies in the *convex cone spanned by the vectors* ∇g_i, $i \in I$. In the case when there is just a single active constraint, (5.12) reduces to $-\nabla f = \nabla g_i$: the negative cost gradient aligns itself with the constraint gradient; there is no way in which we can travel in a direction that reduces f without making $g > 0$ – this is indeed necessary if you are currently at an optimum point. Figure 5.4 shows various situations, reinforces the preceding discussion in a pictorial manner.

> The reader may wish to read Section 4.7 and Fig. 4.3 now to see the connection between the reduced cost coefficients in the simplex method for LP and KKT conditions.

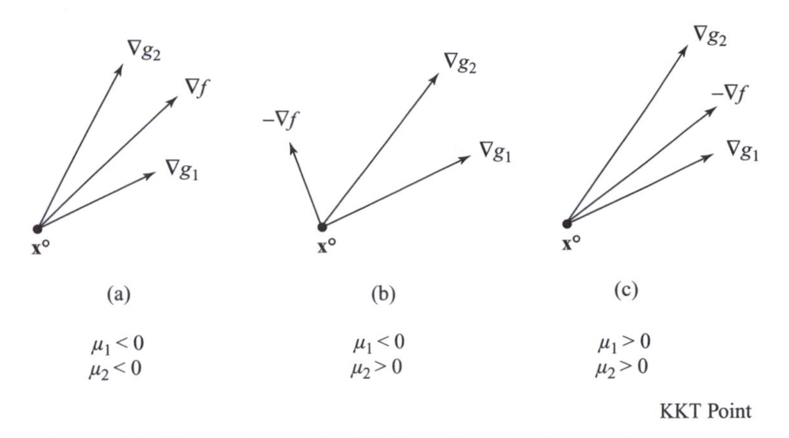

Figure 5.4. Illustration of the KKT condition, $\nabla f = \mu_1 \nabla g_1 + \mu_2 \nabla g_2$.

The optimality conditions in (5.12) do not refer to constraints that are inactive ($g_j < 0$). It is usual to include the inactive constraints as well by assigning a zero Lagrange multiplier to each of them. Thus, for each inactive constraint g_j, we introduce a scalar $\mu_j \geq 0$ and impose the condition that $\mu_j\, g_j = 0$.

We may now state the necessary conditions for optimality (making the assumption of a regular point as discussed before): A point $\mathbf{x}^*$ is a local minimum of f subject to the conditions $g_i \leq 0, i = 1, \ldots, m$ only if

$$\frac{\partial L}{\partial x_p} = \frac{\partial f}{\partial x_p} + \sum_{i=1}^{m} \mu_i \frac{\partial g_i}{\partial x_p} = 0, \quad p = 1, \ldots, n \tag{5.13a}$$

$$\mu_i \geq 0, \quad i = 1, \ldots, m \tag{5.13b}$$

$$\mu_i g_i = 0, \quad i = 1, \ldots, m \tag{5.13c}$$

$$g_i \leq 0, \quad i = 1, \ldots, m \tag{5.13d}$$

where the Lagrangian L is defined as $L = f + \sum_{i=1}^{m} \mu_i g_i$. The optimality conditions above can be compactly written as

$$\nabla L = \mathbf{0}, \quad \boldsymbol{\mu}^{\mathrm{T}} \mathbf{g} = 0, \quad \boldsymbol{\mu} \geq 0, \quad \mathbf{g} \leq \mathbf{0} \tag{5.14}$$

These necessary conditions for optimality are called Karush–Kuhn–Tucker (KKT) conditions and a point satisfying these is referred to as a KKT point [Karush 1939; Kuhn and Tucker 1951].

A problem that we will run into on several future occasions is discussed in the following example:

Example 5.6
Consider:

$$\begin{aligned} &\text{minimize} \quad f(\mathbf{x}) \\ &\text{subject to} \quad \mathbf{x} \geq 0 \end{aligned}$$

If $\mathbf{x}^*$ is a solution to the above problem, then it should satisfy the optimality conditions:

$$\left.\frac{\partial f}{\partial x_i}\right|_{\mathbf{x}^*} = 0, \quad \text{if } x_i^* > 0$$

$$\left.\frac{\partial f}{\partial x_i}\right|_{\mathbf{x}^*} \geq 0, \quad \text{if } x_i^* = 0$$

It is left as an exercise for the reader to see the equivalence between the preceding conditions and the KKT conditions. Figure E5.6 illustrates the optimality conditions.

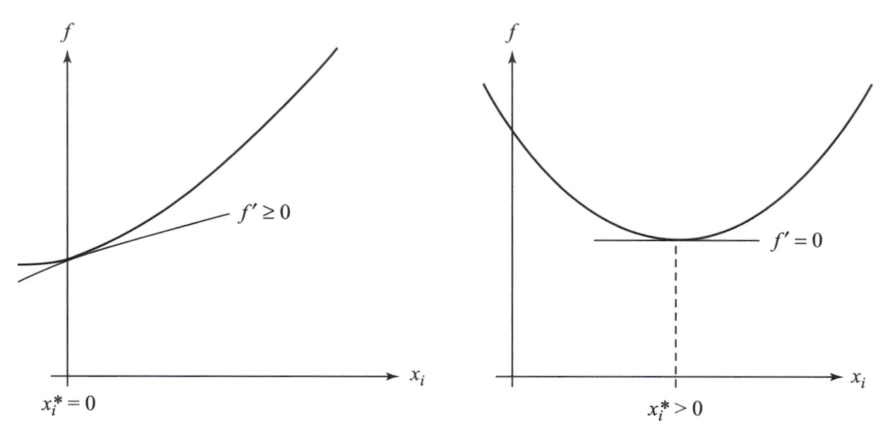

Figure E5.6

Example 5.7
For the problem

$$\begin{aligned} \text{minimize} \quad & f = (x_1 - 3)^2 + (x_2 - 3)^2 \\ \text{subject to} \quad & 2x_1 + x_2 - 2 \leq 0 && \equiv g_1 \\ & -x_1 \leq 0 && \equiv g_2 \\ & -x_2 \leq 0 && \equiv g_3 \end{aligned}$$

(i) Write down the KKT conditions and solve for the KKT point(s)
(ii) Graph the problem and check whether your KKT point is a solution to the problem
(iii) Sketch the descent–feasible "cone" at the point (1, 0) – that is, sketch the intersection of the descent and feasible cones.
(iv) What is the descent–feasible cone at the KKT point obtained in (i)?

We have $L = f + \mu_1(2x_1 + x_2 - 2) - \mu_2 x_1 - \mu_3 x_2$ and the KKT conditions:

$$2(x_1 - 3) + 2\mu_1 - \mu_2 = 0$$
$$2(x_2 - 3) + \mu_1 - \mu_3 = 0$$
$$\mu_1(2x_1 + x_2 - 2) = 0, \quad \mu_2 x_1 = 0, \quad \mu_3 x_2 = 0$$
$$\mu_1, \mu_2, \mu_3 \geq 0$$
$$2x_1 + x_2 \leq 2, \quad x_1 \geq 0, \quad x_2 \geq 0$$

Owing to the "switching" nature of the conditions $\mu_i\, g_i = 0$, we need to guess which constraints are active, and then proceed with solving the KKT conditions. We have the following guesses or cases:

$$g_1 = 0, \quad \mu_2 = 0, \quad \mu_3 = 0 \quad \text{(only 1st constraint active)}$$
$$g_1 = 0, \quad g_2 = 0, \quad \mu_3 = 0 \quad \text{(1st and 2nd constraint active)}$$
$$g_1 = 0, \quad \mu_2 = 0, \quad g_3 = 0 \qquad \ldots$$
$$g_1 = 0, \quad g_2 = 0, \quad g_3 = 0$$
$$\mu_1 = 0, \quad \mu_2 = 0, \quad \mu_3 = 0$$
$$\mu_1 = 0, \quad g_2 = 0, \quad \mu_3 = 0$$
$$\mu_1 = 0, \quad \mu_2 = 0, \quad g_3 = 0$$
$$\mu_1 = 0, \quad g_2 = 0, \quad g_3 = 0$$

In general, there will be 2^m cases – thus, this approach of guessing is only for illustration of the KKT conditions and is not a solution technique for large m. Now, let us see what happens with Case 2: $g_1 = 0, g_2 = 0, \mu_3 = 0$. We get, upon using the KKT conditions:

$$x_1 = 0, \quad x_2 = 2, \quad \mu_1 = 2, \quad \mu_2 = -2, \quad \mu_3 = 0, \quad g_1 = 0, \quad g_2 = 0, \quad g_3 < 0$$

Since μ_2 is negative, we have violated the KKT conditions and, hence, our guess is incorrect. The correct guess is:

Case 1: $g_1 = 0, \mu_2 = 0, \mu_3 = 0$. This leads to: $x_1 = 3 - \mu_1, x_2 = 3 - 0.5\,\mu_1$. Substituting this into $g_1 = 0$ gives $\mu_1 = 2$. We can then recover, $x_1 = 0.2, x_2 = 1.6$. Thus, we have the KKT point

$$x_1 = 0.2, \quad x_2 = 1.6, \quad \mu_1 = 2, \quad \mu_2 = 0, \quad \mu_3 = 0, \quad g_1 = 0, \quad g_2 < 0, \quad g_3 < 0$$

The descent–feasible cone at $\mathbf{x}_0 = (1,0)^T$ is shown in Fig. E5.7. On the other hand, at the point $\mathbf{x}^* = (0.2, 1.6)^T$, we have the negative cost gradient to be $-\nabla f = (5.6, 2.8)^T = 2.8\ (2, 1)^T$ and the active constraint gradient to be

$\nabla g_1 = (2, 1)^T$. Since these two vectors are aligned with one another, there is no descent–feasible cone (or the cone is empty).

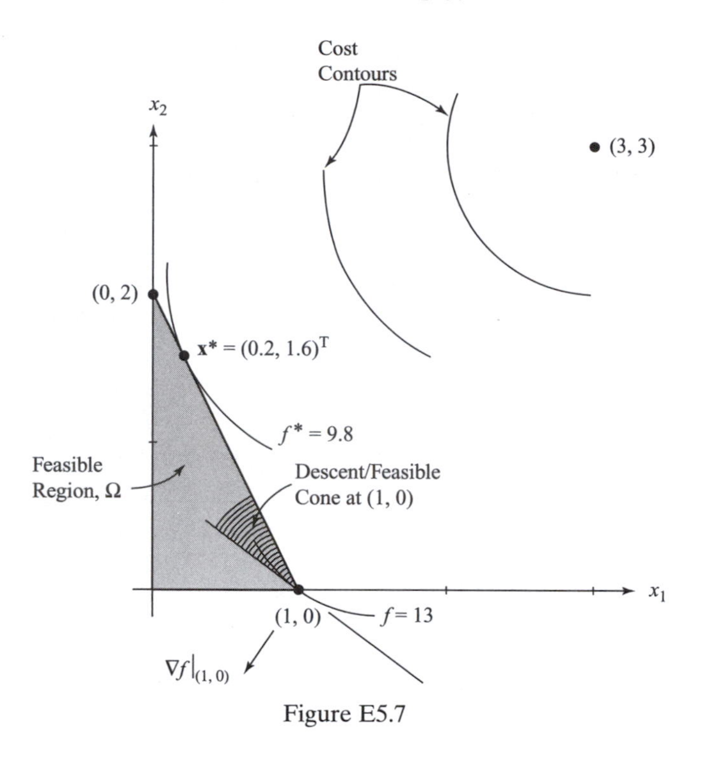

Figure E5.7

Optimality (KKT) Conditions With Both Equality And Inequality Constraints

In view of (5.5) and (5.13), we can now state the optimality conditions for the general problem in (5.1). We assume that **x*** is a regular point (gradients of active inequalities *and* of all the equality constraints are linearly independent). Then, **x*** is a local minimum of f subject to the constraints in (5.1) only if

$$\text{Optimality: } \frac{\partial L}{\partial x_p} = \frac{\partial f}{\partial x_p} + \sum_{i=1}^{m} \mu_i \frac{\partial g_i}{\partial x_p} + \sum_{j=1}^{\ell} \lambda_j \frac{\partial h_j}{\partial x_p} = 0, \quad p = 1, \ldots, n \tag{5.15a}$$

$$\text{Nonnegativity: } \mu_i \geq 0, \quad i = 1, \ldots, m \tag{5.15b}$$

$$\text{Complementarity: } \mu_i g_i = 0, \quad i = 1, \ldots, m \tag{5.15c}$$

$$\text{Feasibility: } g_i \leq 0, \quad i = 1, \ldots, m \quad \text{and} \quad h_j = 0,\ j = 1, \ldots, \ell \tag{5.15d}$$

Note that the Lagrange multipliers λ_j associated with the equality constraints are unrestricted in sign – they can be positive or negative; only the multipliers associated with the inequalities have to be nonnegative. Physical significance of nonnegative

Lagrange multipliers was explained in Section 4.3 and is also explained in Section 5.8. Finally, the KKT conditions can be expressed in matrix form as

$$\nabla L = 0, \quad \boldsymbol{\mu}^{\mathrm{T}}\mathbf{g} = \mathbf{0}, \quad \boldsymbol{\mu} \geq \mathbf{0}, \quad \mathbf{g} \leq \mathbf{0}, \quad \mathbf{h} = \mathbf{0}$$

where $L = f + \boldsymbol{\mu}^{\mathrm{T}}\mathbf{g} + \boldsymbol{\lambda}^{\mathrm{T}}\mathbf{h}$.

5.6 Sufficient Conditions for Optimality

The KKT conditions discussed earlier are *necessary* conditions: if a design $\mathbf{x}^*$ is a local minimum to the problem in (5.1), then it should satisfy (5.15). Conversely, if a point $\mathbf{x}^*$ satisfies (5.15), it can be a local minimum or a saddle point (a saddle point is neither a minimum nor a maximum). A set of *sufficient* conditions will now be stated which, if satisfied, will ensure that the KKT point in question is a strict local minimum.

Let *f*, $\mathbf{g}$, $\mathbf{h}$ be twice continuously differentiable functions. Then, the point $\mathbf{x}^*$ is a strict local minimum to Problem (5.1) if there exist Lagrange multipliers $\boldsymbol{\mu}$ and $\boldsymbol{\lambda}$, such that

- KKT necessary conditions in (5.15) are satisfied at $\mathbf{x}^*$
  and
- the Hessian matrix

$$\nabla^2 L(\mathbf{x}^*) = \nabla^2 f(\mathbf{x}^*) + \sum_{i=1}^{m} \mu_i \nabla^2 g_i(\mathbf{x}^*) + \sum_{j=1}^{\ell} \lambda_j \nabla^2 h_j(\mathbf{x}^*) \tag{5.16}$$

is positive definite on a subspace of R^n as defined by the condition:

$\mathbf{y}^{\mathrm{T}}\nabla^2 L(\mathbf{x}^*)\mathbf{y} > 0$ for every vector $\mathbf{y} \neq 0$ which satisfies:
$\nabla h_j(\mathbf{x}^*)^{\mathrm{T}}\mathbf{y} = 0,\ j = 1, \ldots, \ell$ and $\nabla g_i(\mathbf{x}^*)^{\mathrm{T}}\mathbf{y} = 0$ for all i for which $g_i(\mathbf{x}^*) = 0$ with $\mu_i > 0$ (5.17)

To understand (5.17), consider the following two remarks:

(1) To understand what $\mathbf{y}$ is, consider a constraint $h(\mathbf{x}) = 0$. This can be graphed as a surface in x-space. Then, $\nabla h(\mathbf{x}^*)$ is normal to the tangent plane to the surface at $\mathbf{x}^*$. Thus, if $\nabla h(\mathbf{x}^*)^{\mathrm{T}}\ \mathbf{y} = 0$, it means that $\mathbf{y}$ is perpendicular to ∇h and is, therefore, in the tangent plane. The collection of all such $\mathbf{y}$'s will define the entire tangent plane. This tangent plane is obviously a subspace of the entire space R^n. Figure 5.5 illustrates a few tangent planes. Thus, the KKT sufficient conditions only require $\nabla^2 L\ (\mathbf{x}^*)$ to be positive definite on the tangent space M defined by $M = \{\mathbf{y}: \nabla h_j(\mathbf{x}^*)^{\mathrm{T}}\mathbf{y} = 0,\ j = 1, \ldots, \ell$ and $\nabla g_i(\mathbf{x}^*)^{\mathrm{T}}\mathbf{y} = 0$ for all i for which $g_i(\mathbf{x}^*) = 0$ with $\mu_i > 0\}$

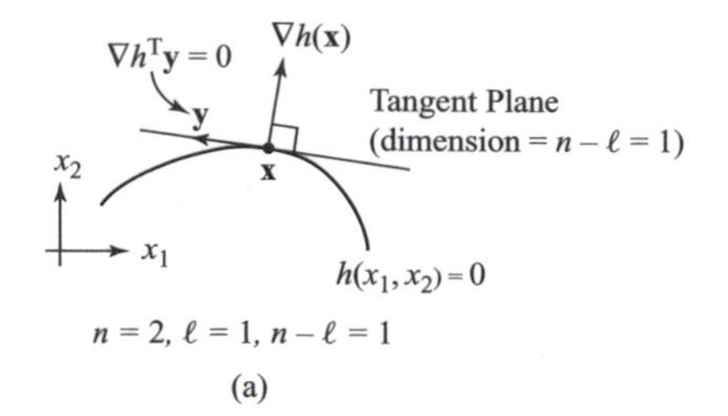

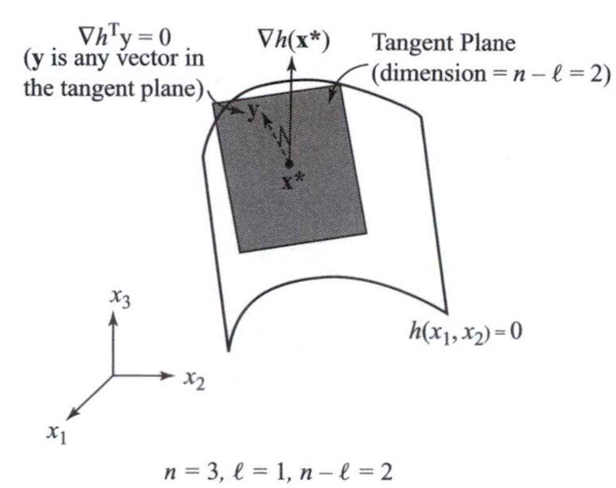

Figure 5.5. Tangent planes.

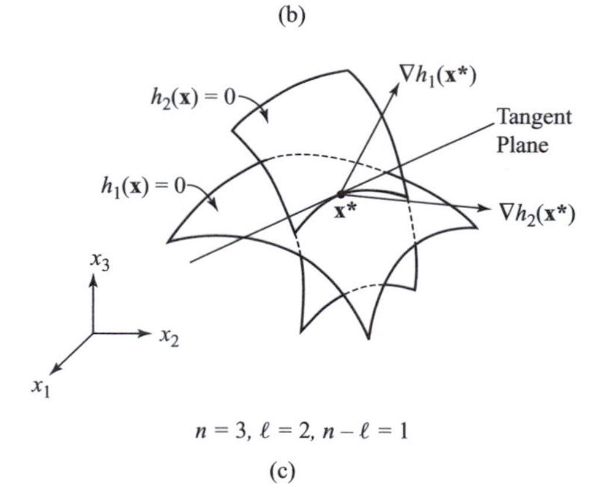

(2) If $\mathbf{y}^{\mathrm{T}}\ \nabla^2 L(\mathbf{x}^*)\ \mathbf{y} > 0$ for *every* vector $\mathbf{y} \in R^n$, then it means that the Hessian matrix $\nabla^2 L(\mathbf{x}^*)$ is positive definite (on the whole space R^n). If this is the case, then it will also be positive definite on a tangent subspace of R^n – however, in the sufficient conditions, we are not requiring this stronger requirement that $\nabla^2 L(\mathbf{x}^*)$ be positive definite.

Example 5.8

We now verify the sufficient conditions in Example 5.3:

$$\text{At } A\colon \nabla^2 L = \begin{bmatrix} 2\lambda & 0 \\ 0 & 2\lambda \end{bmatrix} = \sqrt{5}\,\mathbf{I}, \quad \nabla h(\mathbf{x}^*) = [-8, -2]^{\mathrm{T}}/\sqrt{5}$$

$\nabla h(\mathbf{x}^*)^{\mathrm{T}}\mathbf{y} = 0$ gives $\mathbf{y} = (1, -2)^{\mathrm{T}}$, and thus $\mathbf{y}^{\mathrm{T}}\ \nabla^2 L(\mathbf{x}^*)\ \mathbf{y} = 5\sqrt{5} > 0$, so the Hessian is positive definite on the tangent subspace, and thus point A is a strict local minimum. In fact, we note that the Hessian $\nabla^2 L(\mathbf{x}^*)$ is positive definite (on the whole space R^2) implying that the Hessian is positive definite on the subspace. Further,

At B: $\nabla^2 L = -\sqrt{5}\,\mathbf{I}$, $\nabla h(\mathbf{x}^*)^{\mathrm{T}}\ \mathbf{y} = 0$ gives $\mathbf{y} = (1, -2)^{\mathrm{T}}$, and thus $\mathbf{y}^{\mathrm{T}}\ \nabla^2 L(\mathbf{x}^*)$ $\mathbf{y} = -\,5\sqrt{5}$, and it is not a minimum.

Example 5.9

Consider the two-variable problem in Example 5.1. The solution, obtained graphically and numerically, is $\mathbf{x}^* \equiv [d, H] = [1.9, 20.0]$ in. Let us verify the KKT necessary conditions as well as the sufficient conditions at this point. At $\mathbf{x}^*$, we have

$$\nabla f = \begin{bmatrix} 6.80 \\ 0.20 \end{bmatrix}, \quad \nabla g_1 = \begin{bmatrix} -0.52 \\ -0.03 \end{bmatrix}, \quad \nabla g_2 = \begin{bmatrix} -1.49 \\ 0.00 \end{bmatrix}$$

These gradient vectors can be obtained by analytical differentiation or by using program GRADIENT given in Chapter 1. Using Eq. (5.13a), we obtain

$$\mu_1 = 6.667 \quad \text{and} \quad \mu_2 = 2.237$$

Thus, since μ's are nonnegative, we have shown $\mathbf{x}^*$ to be a KKT point.

To verify the sufficient conditions, we need to evaluate Hessian matrices. The calculations in the following have been verified using program HESSIAN given in Chapter 1. We have, at $\mathbf{x}^*$:

$$\nabla^2 f = \begin{bmatrix} \dfrac{\partial^2 f}{\partial d^2} & \dfrac{\partial^2 f}{\partial d \partial H} \\ \dfrac{\partial^2 f}{\partial d \partial H} & \dfrac{\partial^2 f}{\partial H^2} \end{bmatrix} = \begin{bmatrix} 0.011 & 0.103 \\ 0.103 & 0.007 \end{bmatrix},$$

$$\nabla^2 g_1 = \begin{bmatrix} 0.552 & 0.018 \\ 0.018 & 0.004 \end{bmatrix} \quad \nabla^2 g_2 = \begin{bmatrix} 3.202 & 0.006 \\ 0.006 & 0.003 \end{bmatrix}$$

Thus,

$$\nabla^2 L = \nabla^2 f + \mu_1 \nabla^2 g_1 + \mu_2 \nabla^2 g_2 = \begin{bmatrix} 10.854 & 0.236 \\ 0.236 & 0.040 \end{bmatrix}$$

Condition (5.17) requires us to find all $\mathbf{y}$'s such that $\nabla g_1^{\mathrm{T}} y = 0$ *and* $\nabla g_2^{\mathrm{T}}\mathbf{y} = 0$. This leads to

$$-0.52 y_1 - 0.03 y_2 = 0$$
$$-1.49 y_1 = 0$$

This gives $\mathbf{y} = \mathbf{0}$. Thus, since there does not exist a nonzero $\mathbf{y}$, (5.17) is satisfied by default and, hence, $\mathbf{x}^*$ is a strict local minimum to the problem. We did not even have to evaluate $\nabla^2 L$ and also, the observation that $\nabla^2 L$ is positive definite (a stronger property than required in the conditions) does not enter into the picture.

5.7 Convexity

We have studied convexity in connection with unconstrained problems in Chapter 3. Convexity is a beautiful property – we do not have distinct local minima, a local minimum is indeed the global minimum, and a point satisfying the necessary conditions of optimality is our solution. We have seen that the problem: *minimize* $f(\mathbf{x})$, $\mathbf{x} \in R^n$ is convex if the function f is convex. Note that the space R^n is convex. Here, we need to study the convexity of problems of the form in (5.1). This study is greatly simplified if we rewrite the problem as in (5.2):

$$\begin{array}{ll} \text{minimize} & f(\mathbf{x}) \\ \text{subject to} & \mathbf{x} \in \Omega \end{array}$$

where Ω is the feasible region defined by all the constraints as $\Omega = \{\mathbf{x}: \mathbf{g} \leq \mathbf{0}, \mathbf{h} = \mathbf{0}\}$.

For the problem to be convex, we require:

(i) the set Ω to be convex, and
(ii) the function f to be convex over the set Ω.

While it is relatively simple to prove, we will simply state here the important result:

> Ω is a convex set if all the inequalities $\mathbf{g}$ are convex functions and if all equality constraints are linear.

However, if $\mathbf{g}$ are not convex functions, then the set Ω may still turn out to be a convex set. The reason an equality constraint h_j has to be linear if it is to be convex is due to the fact that if two points lie on the curve, then the entire line segment joining the two points can lie on the curve only if the curve is a straight line. Thus, the problem in (5.1) is convex if all g_i are convex functions, all h_j are linear, and if f is a convex function.

The following are the key aspects of a convex problem:

(1) Any relative minimum is also a global minimum.
(2) The KKT necessary conditions are also sufficient – that is, if $\mathbf{x}^*$ satisfies (5.15), then $\mathbf{x}^*$ is a relative (and global) minimum to problem (5.1).

A nonconvex feasible region Ω can lead to distinct local minima as illustrated in Fig. 5.6.

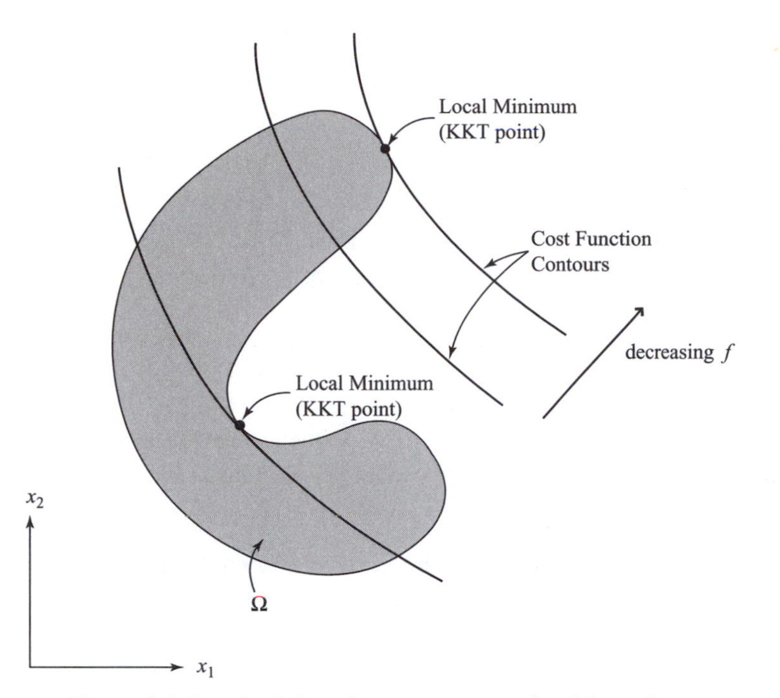

Figure 5.6. Local minima due to nonconvex feasible regions.

Example 5.10
Is the feasible region Ω defined by the three constraints as follows a convex set?

$$g_1 \equiv x_1 x_2 \geq 5$$
$$g_2 \equiv 0 < x_1 \leq 10$$
$$g_2 \equiv 0 < x_2 \leq 10$$

Obviously, g_2 and g_3 are linear and define convex regions. If we write g_1 as $5 - x_1\, x_2 \leq 0$, and apply the Hessian test, we find $\nabla^2 g_1 = \left[\begin{smallmatrix} 0 & -1 \\ -1 & 0 \end{smallmatrix}\right]$, which is not positive semidefinite since its eigenvalues are -1, 1. While g_1 is not a convex function, the constraint $g_1 \leq 0$ does indeed define a convex set which is evident from a graph of the function. To prove this, we use the basic definition of a convex set: given any two points in the set, the entire line segment joining the two points must be in the set. Let point A $= (x_1^a, x_2^a)$ and point B $= (x_1^b, x_2^b)$. Since A and B are assumed to be in the feasible region, we have

$$x_1^a, x_2^a \geq 5 \quad \text{and} \quad x_1^b x_2^b \geq 5$$

Points on the line segment joining A and B are given by

$$x_1 = \theta x_1^a + (1-\theta)x_1^b, \quad x_2 = \theta x_2^a + (1-\theta)x_2^b, \quad 0 \leq \theta \leq 1$$

Consider

$$\begin{aligned} 5 - x_1 x_2 &= 5 - \left[\theta x_1^a + (1-\theta) x_1^b\right] \cdot \left[\theta x_2^a + (1-\theta) x_2^b\right] \\ &= 5 - \left[\theta^2 x_1^a x_2^a + (1-\theta)^2 x_1^b x_2^b + \theta(1-\theta)\left(x_1^b x_2^a + x_1^a x_2^b\right)\right] \\ &\le 5 - \left[5\theta^2 + 5(1-\theta)^2 + \theta(1-\theta)\left(x_1^b x_2^a + x_1^a x_2^b\right)\right] \end{aligned}$$

Since $x_1^b \ge \frac{5}{x_2^b}$ and $x_2^a \ge \frac{5}{x_1^a}$, we have $x_1^b x_2^a + x_1^a x_2^b \ge x_1^a x_2^b + \frac{25}{x_1^a x_2^b} \ge 10$, the latter coming from the inequality $(\sqrt{y} - \frac{5}{\sqrt{y}})^2 \ge 0$. Thus, the preceding expression is

$$\begin{aligned} &\le 5 - [5\theta^2 + 5(1-\theta)^2 + 10\,\theta(1-\theta)] \\ &= 0 \end{aligned}$$

Thus, $5 - x_1\, x_2 \le 0$ showing that $g_1 \le 0$ defines a convex set. Since the intersection of convex sets is a convex set, we conclude that Ω is a convex set.

5.8 Sensitivity of Optimum Solution to Problem Parameters

Let p be a parameter where $f = f(\mathbf{x}, p)$, $\mathbf{g} = \mathbf{g}(\mathbf{x}, p)$. Consider the problem

$$\begin{aligned} \text{minimize} \quad & f(\mathbf{x}, p) \\ \text{subject to} \quad & g_i(\mathbf{x}, p) \le 0 \quad i = 1, \ldots, m \\ & h_j(\mathbf{x}, p) = 0 \quad j = 1, \ldots, \ell \end{aligned} \tag{5.18}$$

Suppose that for $p = p_0$ there is a local minimum $\mathbf{x}^*$ to problem (5.18). We will assume that

(i) there exist Lagrange multipliers $\boldsymbol{\mu}, \boldsymbol{\lambda} \ge \mathbf{0}$ that together with $\mathbf{x}^*$ satisfy the sufficient conditions for a strict local minimum
(ii) $\mathbf{x}^*$ is a regular point
(iii) for every active constraint g_i, $i \in I$, $\mu_i > 0$. That is, there are no "degenerate" active inequality constraints

These assumptions allow us to conclude that for values of p in a small interval containing p_0, the solution to (5.18) denoted as $\mathbf{x}^*(p)$, depends continuously on p, with $\mathbf{x}^*(p_0) = \mathbf{x}^*$.

Differentiating $f(\mathbf{x}^*, p)$, g_i with $i \in I$ and h_j with respect to p, we obtain

$$\mathrm{d}f/\mathrm{d}p = \nabla f^{\mathrm{T}} \mathrm{d}\mathbf{x}^*/\mathrm{d}p + \partial f/\partial p$$

$$\mathrm{d}g_i/\mathrm{d}p = \nabla g_i{}^{\mathrm{T}} \mathrm{d}\mathbf{x}^*/\mathrm{d}p + \partial g_i/\partial p = 0, \; \mathrm{d}h_j/\mathrm{d}p = \nabla h_j^{\mathrm{T}} \mathrm{d}\mathbf{x}^*/\mathrm{d}p + \partial h_j/\partial p = 0$$

where ∇ means the column vector d/d$\mathbf{x}$ as always. From the KKT conditions, we have $\nabla f = -\,[\nabla \mathbf{g}]\,\boldsymbol{\mu} - [\nabla \mathbf{h}\boldsymbol{\lambda}\,\boldsymbol{\lambda}$. From these, we readily obtain the sensitivity result

$$\frac{df(\mathbf{x}^*, p)}{dp} = \frac{\partial f(\mathbf{x}^*, p)}{\partial p} + \sum_{i \in I} \mu_i \frac{\partial g_i(\mathbf{x}^*, p)}{\partial p} + \sum_{i \in I} \lambda_j \frac{\partial h_j(\mathbf{x}^*, p)}{\partial p} \tag{5.19}$$

Significance of Lagrange Multipliers

The significance of the Lagrange multipliers that we have come across in our treatment of optimality conditions can be shown from the previous result. If we consider the original constraints $g_i(\mathbf{x}) \le 0$ and $h_j(\mathbf{x}) = 0$ in the form

$$g_i(\mathbf{x}) \le c_i \quad \text{and} \quad h_j(\mathbf{x}) = d_j$$

then substituting $p = -c_i$ or $p = -d_j$ into (5.19), and noting that $\partial f/\partial p = 0$, we obtain

$$\left.\frac{\partial f(\mathbf{x}^*(\mathbf{c},\mathbf{d}))}{\partial c_i}\right|_{\mathbf{0},\mathbf{0}} = -\mu_i \qquad \left.\frac{\partial f(\mathbf{x}^*(\mathbf{c},\mathbf{d}))}{\partial d_j}\right|_{\mathbf{0},\mathbf{0}} = -\lambda_j \tag{5.20}$$

Equations (5.20) imply that the negative of the Lagrange multiplier is a sensitivity of the optimum cost with respect to the constraint limit. The above result tells us the following:

(a) With $\mu_i > 0$ for an active inequality, $\partial f/\partial c_i < 0$. This means that the optimum cost will *reduce* as c_i is increased from its zero value. This is to be expected as increasing the value of c_i implies a "relaxation" of the constraint, resulting in an enlarged feasible region and, consequently, a lesser value of the optimum value of f. The nonnegativity restrictions on μ_i in the KKT conditions now make physical sense.

(b) Let f^* be the optimum cost at $\mathbf{x}^*$. The new value of f^* at $\mathbf{x}^*(\mathbf{c},\mathbf{d})$ can be predicted, to a first-order approximation for small values of $(\mathbf{c},\mathbf{d})$ from

$$f^*(\mathbf{c},\mathbf{d}) - f^* \approx -\sum \mu_i c_i - \sum \lambda_j d_j \tag{5.21}$$

Here, changes in $\mathbf{c}$, $\mathbf{d}$ should not change the active set.

Example 5.11.

Consider the simple problem in Example 5.5:

$$\begin{aligned} &\text{minimize} && f = (x_1-3)^2 + (x_2-3)^2 && \\ &\text{subject to} && 2x_1 + x_2 - 2 \le 0 && \equiv g_1 \\ &&& -x_1 \le 0 && \equiv g_2 \\ &&& -x_2 \le 0 && \equiv g_3 \end{aligned}$$

The solution is $x_1^* = 0.2$, $x_2^* = 1.6$, $\mu_1 = 2$, $\mu_2 = 0$, $\mu_3 = 0$ and the corresponding value of $f^* = 9.8$. Now, assume that $g_1 \equiv 2x_1 + x_2 - 2 \le c$. The new optimum is approximately (see (5.21)): $f^*_{\text{new}} \approx 9.8 - 2\,c$. Comparison with the exact solution is presented below for various values of c. Of course, small changes in the limits of the inactive constraints g_2 and g_3 will not influence the solution.

c	f^* as per (5.21)	f^*_{exact}	Error
0.005	9.79	9.786	0.04 %
0.010	9.78	9.772	0.08 %
0.040	9.72	9.688	0.33 %

Example 5.12

Consider the truss problem in Example 5.1, with optimum solution including Lagrange multipliers given in Section 5.3. What is the (post-optimal) sensitivity of f^* and $\mathbf{x}^*$ to Young's modulus E? Substituting $p \equiv E$, $\mu_1 = 5.614$, $\mu_2 = 2.404$, and

$$\frac{\partial f(\mathbf{x}^*, E)}{\partial E} = 0, \quad \frac{\partial g_1(\mathbf{x}^*, E)}{\partial E} = 0, \quad \frac{\partial g_2(\mathbf{x}^*, E)}{\partial E} = \frac{8P(B^2 + H^2)^{1.5}}{\pi^3 t Hd(d^2 + t^2)} \frac{(-1)}{E^2}\bigg|_{(H^*,d^*)}$$

into Eq. (5.19) yields $\frac{df(\mathbf{x}^*, p)}{dE} = (2.404)(-3.33E - 08) = -8.01E - 08$. This result may be verified using forward differences, which involves resolving optimization problem with a suitably small perturbed value for E.

To obtain post-optimal sensitivity for the variables $\mathbf{x}^*$, we need to use equations leading to Eq. (5.19):

$$\begin{bmatrix} \frac{\partial g_1}{\partial d} & \frac{\partial g_1}{\partial H} \\ \frac{\partial g_2}{\partial d} & \frac{\partial g_2}{\partial H} \end{bmatrix} \begin{pmatrix} \frac{dd}{dE} \\ \frac{dH}{dE} \end{pmatrix} = - \begin{pmatrix} \frac{dg_1}{dE} \\ \frac{dg_2}{dE} \end{pmatrix}, \quad \text{which results in}$$

$$\begin{bmatrix} -0.532 & -0.033956 \\ -1.593 & -0.00305 \end{bmatrix} \begin{pmatrix} \frac{dd}{dE} \\ \frac{dH}{dE} \end{pmatrix} = - \begin{pmatrix} 0 \\ -3.33E \quad -08 \end{pmatrix}, \quad \text{giving}$$

$$\begin{pmatrix} \frac{dd^*}{dE} \\ \frac{dH^*}{dE} \end{pmatrix} = 10^{-6} \begin{pmatrix} -0.0216 \\ 0.3376 \end{pmatrix}$$

5.9 Rosen's Gradient Projection Method for Linear Constraints

The next few sections will focus on numerical methods for solving constrained problems. We will present Rosen's gradient projection method in this section, followed by Zoutendijk's method of feasible directions, the generalized reduced gradient method and sequential QP. Rosen's Gradient Projection Method will be presented for linear constraints [Rosen 1960]. While Rosen did publish an extension of his method for nonlinear constraints [Rosen 1961], it is very difficult to implement and other methods are preferred. However, it can be effective in special cases, such as in problems where there is only a single nonlinear constraint. Consider problems that can be expressed in the form

$$\begin{aligned} &\text{minimize} && f(\mathbf{x}) \\ &\text{subject to} && \mathbf{a}^i\mathbf{x} - b_i \le 0 \quad i = 1, \ldots, m \\ & && \mathbf{a}^i\mathbf{x} - b_i = 0 \quad i = m+1, \ldots, m+\ell \end{aligned} \tag{5.22}$$

Let t = number of active constraints, as consisting of all the equalities and the active inequalities:

$$I(\mathbf{x}) = \{m+1, \ldots, m+\ell\} \cup \{j : \mathbf{a}^j\mathbf{x} = \mathrm{b}_j, \quad j = 1, \ldots, m\} \tag{5.23a}$$

In actual computation, the active set is implemented as

$$I(\mathbf{x}) = \{m+1, \ldots, m+l\} \cup \{j : \mathbf{a}^j\mathbf{x} + \varepsilon - \mathrm{b}_j \ge 0, \quad j = 1, \ldots, m\} \tag{5.23b}$$

Here, ε is an user-defined scalar or "constraint thickness" parameter that allows constraints near the limit to be included in the active set. For example, let $m = 5$, and $\mathbf{x}_k$ be the current value of $\mathbf{x}$, and let $\mathbf{g} \equiv \mathbf{a}^j\,\mathbf{x}_k - \mathrm{b}_j = \{-0.1, -0.0003, -0.9, 0.0, -0.015\}$. With $\varepsilon = 0.01$, the active inequalities are {2, 4}. For purposes of simplicity in presentation, we will assume that the constraints are rearranged so that the first t constraints are active.

Now, assume that $\mathbf{x}_k$ is the current design. We require $\mathbf{x}_k$ to be feasible – that is, $\mathbf{x}_k$ satisfies the constraints in (5.22). To determine such an $\mathbf{x}_k$, we may use the Phase I simplex routine discussed in Chapter 4. Our task is to determine a direction vector $\mathbf{d}$ followed by a step length along this direction, which will give us a new and better design point. We repeat this iterative process until we arrive at a KKT point. If we are at an "interior" point, then the active set is empty (no equalities exist and no active inequalities at $\mathbf{x}_k$), and the steepest descent direction is the obvious choice: $\mathbf{d} = -\nabla f(\mathbf{x}_k)$. Proceeding along this, we are bound to hit the boundary as most realistic problems will have the minimum on the boundary. Thus, in general, we may assume that some constraints are active at our current point $\mathbf{x}_k$.

As in the discussion immediately following (5.17), we can introduce the "tangent plane" as $M = \{\mathbf{y} : \mathbf{a}^i\,\mathbf{y} = 0,\, i \in I(\mathbf{x}_k)\}$. The tangent plane can be described by $n - t$ independent parameters (t = number of active constraints). For instance, if $n = 3$ and $t = 2$, we have a line as our tangent plane (see Fig. 5.5c). If we denote a $t \times n$-dimensional matrix $\mathbf{B}$ to consist of rows of the gradient vectors of the active constraints, or equivalently,

$$\mathbf{B}^{\mathrm{T}} = \{\mathbf{a}^{1\mathrm{T}} | \mathbf{a}^{2\mathrm{T}} \ldots | \mathbf{a}^{t\mathrm{T}}\}$$

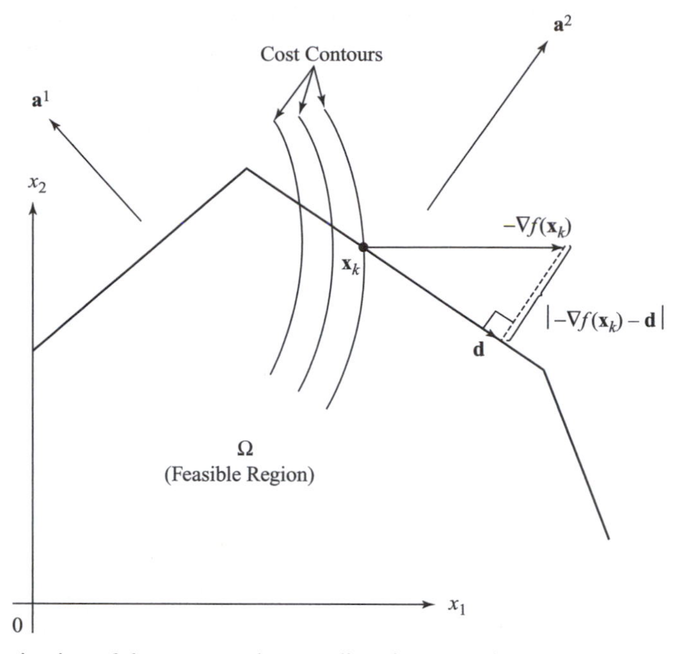

Figure 5.7. Projection of the steepest descent direction onto the tangent plane: $\mathbf{d} = [\mathbf{P}]\,[-\nabla f]$.

then we can conveniently define $M = \{\mathbf{y} : \mathbf{B}\,\mathbf{y} = \mathbf{0}\}$. Rosen's idea was to determine the direction vector by *projecting the steepest descent direction onto the tangent plane*. Referring to Fig. 5.7, we are seeking a vector $\mathbf{d}$ that satisfies $\mathbf{B}\,\mathbf{y} = \mathbf{0}$ and that makes the length of the "perpendicular" $|-\nabla f(\mathbf{x}_k) - \mathbf{d}\,|$ a minimum.

Thus, $\mathbf{d}$ can be obtained as the solution of

$$\begin{aligned} &\text{minimize} \quad (-\nabla f(\mathbf{x}_k) - \mathbf{d})^{\mathrm{T}}(-\nabla f(\mathbf{x}_k) - \mathbf{d}) \\ &\text{subject to} \quad \mathbf{B}\,\mathbf{d} = \mathbf{0} \end{aligned} \tag{5.24}$$

Defining the Lagrangian $L = (\nabla f + \mathbf{d})^{\mathrm{T}}\,(\nabla f + \mathbf{d}) + \boldsymbol{\beta}^{\mathrm{T}}\,\mathbf{B}\,\mathbf{d}$, we have the optimality conditions

$$\frac{\partial L^{\mathrm{T}}}{\partial \mathbf{d}} = (\nabla f^{\mathrm{T}} + \mathbf{d}) + \mathbf{B}^{\mathrm{T}}\boldsymbol{\beta} = \mathbf{0} \tag{5.25}$$

Premultiplying the preceding equation by $\mathbf{B}$ and using $\mathbf{B}\,\mathbf{d} = \mathbf{0}$, we get

$$\mathbf{B}\mathbf{B}^{\mathrm{T}}\boldsymbol{\beta} = -\mathbf{B}\nabla f \tag{5.26}$$

Here we make the assumption that the current point is a regular point. This means that the rows of $\mathbf{B}$ are linearly independent or that the rank of $\mathbf{B}$ equals t. If there is dependency, then we have to "pre-process" our mathematical model by deleting the redundant constraints. Luenberger has discussed some general approaches to this problem [Luenberger 1973]. With the previous assumption, we can show that the $(t \times t)$ matrix $\mathbf{B}\,\mathbf{B}^{\mathrm{T}}$ has rank t and is nonsingular. Thus, we can solve the matrix equations (5.26) for the Lagrange multipliers $\boldsymbol{\beta}$. The direction vector $\mathbf{d}$ can be recovered from (5.25):

$$\mathbf{d} = -\nabla f - \mathbf{B}^{\mathrm{T}}\boldsymbol{\beta} \tag{5.27}$$

Computationally, the direction vector $\mathbf{d}$ is obtained by solving (5.26) followed by (5.27). However, we can introduce a "projection matrix" $\mathbf{P}$ by combining the two equations:

$$\begin{aligned}\mathbf{d} &= -\nabla f + \mathbf{B}^{\mathrm{T}}[\mathbf{B}\mathbf{B}^{\mathrm{T}}]^{-1}\mathbf{B}\nabla f \\ &= \mathbf{P}(-\nabla f)\end{aligned}$$

where

$$\mathbf{P} = [\mathbf{I} - \mathbf{B}^{\mathrm{T}}[\mathbf{B}\mathbf{B}^{\mathrm{T}}]^{-1}\mathbf{B}] \tag{5.28}$$

Given any vector $\mathbf{z}$, $\mathbf{Pz}$ is the projection of $\mathbf{z}$ onto the tangent plane; in particular, $\mathbf{P}\ (-\nabla f)$ projects the steepest descent direction $-\nabla f$ onto the tangent plane. The vector $\mathbf{d}$, if not zero, can be shown to be a descent direction. Also, $\mathbf{P}\,\mathbf{P} = \mathbf{P}$.

We must consider the possibility when the projected direction $\mathbf{d} = \mathbf{0}$.

Case (i): If all Lagrange multipliers associated with inequality constraints are ≥ 0, then this fact together with (5.27) that yields $-\nabla f = \mathbf{B}^{\mathrm{T}}\ \boldsymbol{\beta}$ implies that the current point is a KKT point and we terminate the iterative process.

Case (ii): If at least one β_i (associated with the inequalities) is negative, then we can drop this constraint and move in a direction that reduces f. In the method, we delete the row corresponding to the most negative β_i (associated with an inequality) from $\mathbf{B}$ and resolve (5.26) and (5.27) for a new $\mathbf{d}$. Figure 5.8 illustrates two simple situations where case (ii) occurs.

Now, having determined a direction $\mathbf{d}$, our next task is to determine the step size. As depicted in Fig. 5.9, the first task is to evaluate α_{U} = step to the nearest intersecting boundary. This can be readily determined from the requirement that

$$\mathbf{a}^{i\mathrm{T}}(\mathbf{x}_k + \alpha\mathbf{d}) - \mathbf{b}_i \leq 0, \quad i \notin I(\mathbf{x}_k)$$

Next, we evaluate the slope of f at $\alpha = \alpha_{\mathrm{U}}$ (Fig. 5.8). The slope is evaluated as

$$f' = \mathrm{d}f/\mathrm{d}\alpha = \nabla f(\mathbf{x}_k + \alpha_{\mathrm{U}}\mathbf{d})^{\mathrm{T}}\mathbf{d} \tag{5.29}$$

If $f' < 0$, then evidently our step is $\alpha_0 = \alpha_{\mathrm{U}}$. If $f' > 0$, then it means that the minimum of f is in the interval $[0, \alpha_{\mathrm{U}}]$. The minimum α_0 is determined in the program ROSEN using a simple bisection strategy : the interval is bisected between points having $f' < 0$ and $f' > 0$ until the interval is small enough.

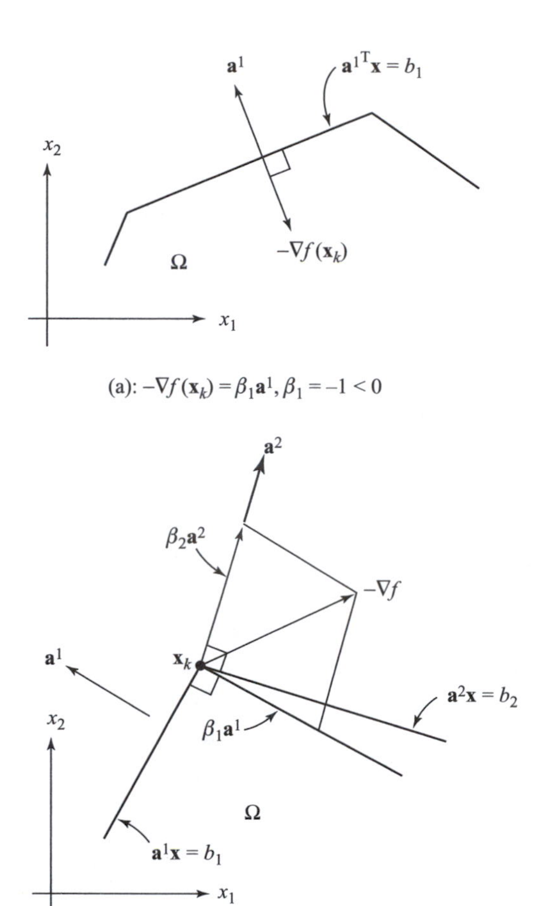

Figure 5.8. Examples of projected direction $\mathbf{d} = 0$ with some $\beta_i < 0$.

A typical iteration of Rosen's gradient projection method for linear constraints may be summarized as follows:

1. Determine a feasible starting point $\mathbf{x}_0$
2. Determine the active set and form the matrix $\mathbf{B}$.
3. Solve (5.26) for β and evaluate $\mathbf{d}$ from (5.27).
4. If $\mathbf{d} \neq \mathbf{0}$, obtain step α_k as discussed above. Update the current point $\mathbf{x}_k$ as $\mathbf{x}_k + \alpha_k\, \mathbf{d}$ and return to step (2).
5. If $\mathbf{d} = \mathbf{0}$,
   (a) If $\beta_j \geq 0$ for all j corresponding to active inequalities, stop, $\mathbf{x}_k$ satisfies KKT conditions.
   (b) Otherwise, delete the row from $\mathbf{B}$ corresponding to the most negative component of $\boldsymbol{\beta}$ (associated with an inequality) and return to step 3.

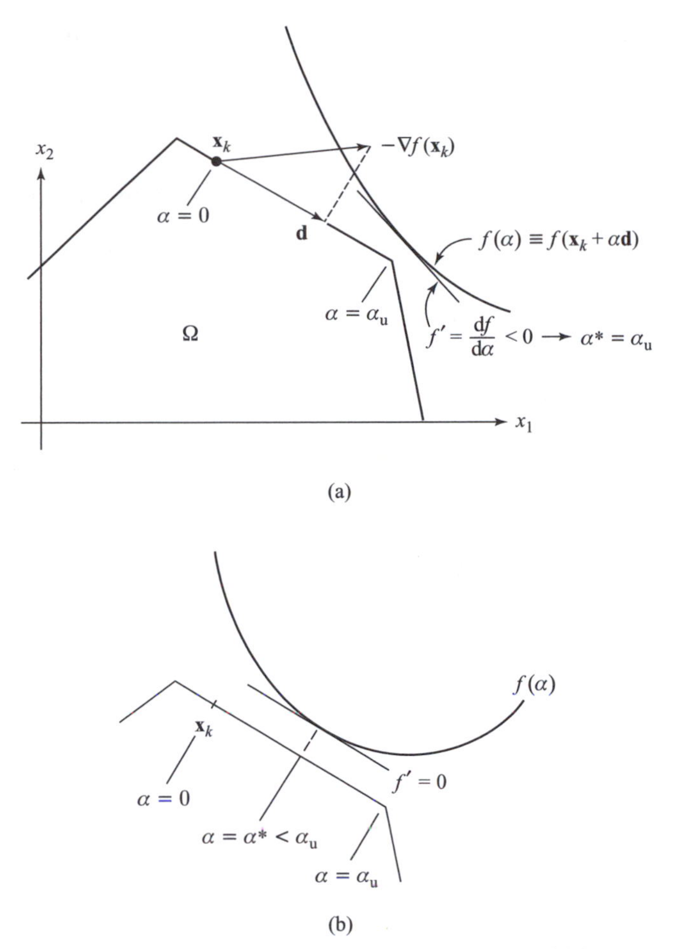

Figure 5.9. Step size calculations in program ROSEN.

Example 5.13

Perform one iteration using the gradient projection method.

$$\begin{aligned} \text{Minimize} \quad & f(\mathbf{x}) = 0.01x_1^2 + x_2^2 - 100 \\ \text{subject to} \quad & g_1 = 2 - x_1 \le 0 \\ & g_2 = 10 - 10x_1 + x_2 \le 0 \\ & -50 \le x_1, \quad x_2 \le 50 \end{aligned}$$

With a starting point $\mathbf{x}_0 = (2, 5)$ that satisfies the constraints, we have the active set $\{g_1\}$, $\nabla f = (0.04,\ 10)^T$, $\mathbf{d}_0 = -\ [\mathbf{P}]\ \nabla f = (0,\ -10)^T$. Thus, the normalized $\mathbf{d}_0 = (0, -1)^T$. The maximum step to the nearest constraint boundary is $\alpha_U =$ 55.0. Since the slope $f' = 100 > 0$, we have to minimize $f(\mathbf{x}_0 + \alpha\ \mathbf{d}_0)$ – this gives the step $\alpha_0 = 5.0$. The new design point is $\mathbf{x}_1 = \mathbf{x}_0 + \alpha\, \mathbf{d}_0 = (2, 0)^{\mathrm{T}}$. At the next iteration, we get $\mathbf{d} = (0, 0)^{\mathrm{T}}$, which means that the point obtained is the optimum.

5.10 Zoutendijk's Method of Feasible Directions (Nonlinear Constraints)

In 1960, Zoutendijk developed the Method of Feasible Directions [Zoutendijk 1960], which is still considered as one of the most *robust* methods for optimization. The method is geared to solving problems with inequality constraints, where the feasible region Ω has an "interior." Nonlinear equality constraints may be tackled (but not very effectively) by introducing penalty functions as discussed at the end of this section. Linear equality constraints may be readily incorporated, but this is left as an exercise for the reader. Thus, consider problems that can be expressed in the form

$$\begin{aligned} &\text{minimize} \quad f(\mathbf{x}) \\ &\text{subject to} \quad g_i(\mathbf{x}) \le 0 \quad i = 1, \ldots, m \end{aligned} \tag{5.30}$$

We assume that functions f and g_i have continuous derivatives, but we do not require them to be convex. Recall the definition and meaning of a "descent" direction ($\nabla f^{\mathrm{T}} \mathbf{d} < 0$) and a "feasible direction" ($\nabla g_i{}^{\mathrm{T}}\, \mathbf{d} < 0$ holds for each $i \in I$) – see discussion on KKT optimality conditions immediately following (5.6). Assume that the current design point $\mathbf{x}_k$ is feasible. Zoutendijk's method of feasible directions is based on the following idea: if we have a procedure to determine a direction $\mathbf{d}$ that is both descent and feasible, then a line search along $\mathbf{d}$ will yield an improved design, that is, a feasible point with a lesser value of f. Zoutendijk used the term "usable" instead of "descent," but we will continue to use "descent" here.

We define the active set I as

$$I = \{j : g_j(\mathbf{x}_k) + \varepsilon \ge 0, \quad j = 1, \ldots, m\} \tag{5.31}$$

See the comment following (5.23b) for explanation of the constraint thickness parameter ε. We denote $\nabla f = \nabla f(\mathbf{x}_k)$ and $\nabla g_j = \nabla g_j(\mathbf{x}_k)$, and we introduce an artificial variable α as

$$\alpha = \max\{\nabla f^{\mathrm{T}} \mathbf{d},\ \nabla g_j^{\mathrm{T}} \mathbf{d} \text{ for each } j \in I\} \tag{5.32}$$

Now, if $\alpha < 0$, then it means that $\nabla f^{\mathrm{T}} \mathbf{d} < 0$ *and* $\nabla g_j{}^{\mathrm{T}} \mathbf{d} < 0$ holds for each $j \in I$ which means that $\mathbf{d}$ is a descent–feasible direction. If this is not clear, consider the argument when $\alpha = \max\ \{y_1, y_2, \ldots, y_n\}$; if $\alpha < 0$, then all y_i's have to negative (<0) since the maximum of these y_i is negative. Thus, to obtain a descent–feasible direction, our aim would be to try and reduce α till it becomes a negative number.

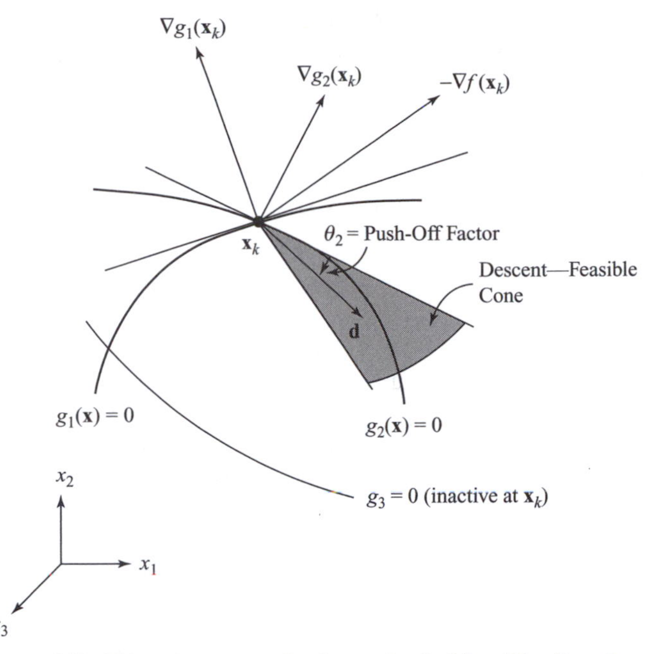

Figure 5.10. Direction vector in the method of feasible directions.

We can achieve this by minimizing α as much as possible. We can formulate the following "subproblem" to determine a descent–feasible direction:

$$\text{minimize } \alpha, \text{ subject to } \alpha = \max\{\nabla f^{\mathrm{T}}\mathbf{d}, \nabla g_j^{\mathrm{T}}\mathbf{d} \text{ for each } j \in I\}$$

which is equivalent to

$$\begin{aligned} &\text{minimize} && \alpha \\ &\text{subject to} && \nabla f^{\mathrm{T}}\mathbf{d} \le \alpha \\ &&& \nabla g_j^{\mathrm{T}}\mathrm{d} \le \alpha \quad \text{for each } j \in I \\ &&& -1 \le \mathrm{d}_i \le 1 \quad i = 1, \ldots, n \end{aligned} \tag{5.33}$$

Here, there are $n + 1$ variables in the subproblem: $\mathbf{d}$ and α. The constraint that the magnitude of d_i must be less than unity has been imposed to ensure that the solution is bounded. A negative α means that the dot product (in hyperspace) between the constraint normal and the direction vector is negative, which in turn means that the angle between these two is obtuse. In fact, Zoutendijk also introduced "push-off" factors θ_j to control this angle: $\nabla g_j^{\mathrm{T}}\mathbf{d} \le \theta_j\alpha$, for each $j \in I$. If $\theta_j = 0$, then it is possible for $\nabla g_j^{\mathrm{T}}\mathbf{d}^* = 0$, that is, for $\mathbf{d}^*$ to be tangent to the constraint. This is permissible only for linear constraints. In general, we require $\theta_j > 0$. See Fig. 5.10. By default, we use $\theta_j = 1$ in program ZOUTEN. Also, we note that at the solution to (5.33), $\alpha^* \leq 0$ since $\mathbf{d} = \mathbf{0}$ is a feasible solution with $\alpha = 0$. With the substitution of variables:

$$s_i = d_i + 1, \quad i = 1, \ldots, n \quad \text{and} \quad \beta = -\alpha$$

the sub-problem in (5.33) can be written as:

Direction Finding Subproblem

$$\begin{aligned} \text{minimize} \quad & -\beta \\ \text{subject to} \quad & \nabla f^{\mathrm{T}} \mathbf{s} + \beta \leq c_0 \\ & \nabla g_j^{\mathrm{T}} \mathbf{s} + \theta_j \beta \leq c_j \quad \text{for each } j \in I \\ & s_i \leq 2 \quad i = 1, \ldots, n \\ & s_i \geq 0, \quad \beta \geq 0 \end{aligned} \tag{5.34}$$

where $c_0 = \sum_{i=1}^{N} \partial f / \partial x_i$, $c_j = \sum_{i=1}^{N} \partial g_j / \partial x_i$. Problem (5.34) is an LP problem of the form: minimize $\mathbf{c}^{\mathrm{T}}\mathbf{x}$ subject to $\mathbf{A}\ \mathbf{x} \leq \mathbf{b}$, $\mathbf{x} \geq \mathbf{0}$, where $\mathbf{x}^{\mathrm{T}} = (\mathbf{s}^{\mathrm{T}}, \beta)$. The Simplex method, which is based on the Big-M technique, which is used to solve for $d_i{}^* = s_i{}^* - 1$ and β^*. See Chapter 4 for LP solution techniques. The gradients ∇f and ∇g_j used in (5.34) are normalized as unit vectors by dividing by their length.

The next step is to check for convergence. From (5.32), it is evident that $\beta^* > 0$ or $\alpha^* < 0$ is equivalent to $\nabla f^{\mathrm{T}} \mathbf{d}^* < 0$ *and* $\nabla g_j^T \mathbf{d}^* < 0$ for each active constraint. Thus, $\beta^* > 0$ means that $\mathbf{d}^*$ is a descent–feasible direction. Now, $\beta^* = 0$ means that no descent–feasible direction exists. From Farkas Lemma discussed earlier, we conclude that the current design point is a KKT point and we have convergence. In the program, the convergence check is implemented as: If $|\beta^*| < tol$ OR $\sqrt{\mathbf{d}^{*\mathrm{T}}\, \mathbf{d}^*} < tol$, stop, where *tol* is a tolerance, say, 10^{-6}. If $\beta^* > 0$, then we have a descent–feasible direction and *line search* has to be performed. This is discussed next.

Line Search

The step-size problem is depicted in Fig. 5.11. We should realize that the step-size problem is a constrained one-dimensional search and can be expressed as: minimize $f(\alpha) \equiv f(\mathbf{x}_k + \alpha\ \mathbf{d})$, subject to $g_i(\alpha) \equiv g_i\ (\mathbf{x}_k + \alpha\ \mathbf{d}) \leq 0$, $i = 1, \ldots, m$. The technique that is used in program ZOUTEN to solve this problem is now discussed.

Our first step is to determine α_{U} = step to the nearest intersecting boundary. However, since the constraints are nonlinear, we cannot determine this analytically as was done in Rosen's gradient projection method discussed in Section 5.9. We first determine a step limit based on just the lower and upper limits on $\mathbf{x}$ – in engineering

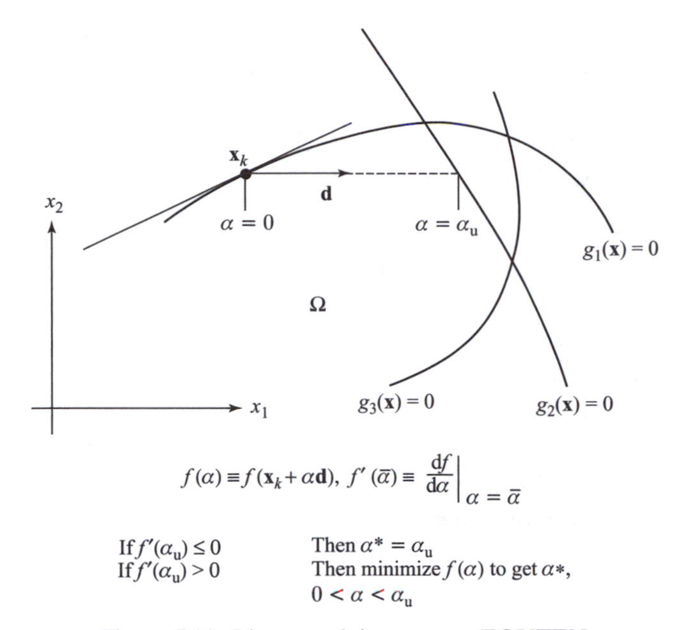

Figure 5.11. Line search in program ZOUTEN.

problems, for instance, a negative x_i may be meaningless at which g_j need not even be defined. The maximum step based on bounds only, α_{U1}, is determined from

$$x_i^L \le x_i^k + \alpha d_i^* \le x_i^U \tag{5.35}$$

where $\mathbf{x}_k$ = current design point, $\mathbf{x}^L$ = lower bound vector, $\mathbf{x}^U$ = upper bound vector.

Now, we evaluate all constraints g_j, $j = 1, \ldots, m$, at $\alpha = \alpha_{U1}$, that is, at the point $(\mathbf{x}_k + \alpha_{U1}\mathbf{d}^*)$. If $g_{max} = \max\{g_j\} \le 0$, then we set $\alpha_U = \alpha_{U1}$. Otherwise, we have to find the step α_U, $0 < \alpha_U < \alpha_{U1}$ to the nearest intersecting constraint. This is implemented in the program ZOUTEN using a simple but robust *bisection* scheme: The interval containing α_U is $[0, \alpha_{U1}]$; g_{max} is evaluated at $\alpha_{U1}/2$. If $g_{max} > 0$, then the new interval containing α_U is $[0, \alpha_{U1}/2]$ whereas if $g_{max} < 0$, then the new interval is $[\alpha_{U1}/2, \alpha_{U1}]$. This process is repeated until a point α_U is obtained at which

$$-g_{low} \le g_{max} \le 0 \tag{5.36}$$

where g_{low} is a tolerance set by the user, say 10^{-4}. Also, a limit on the number of bisections is set, at the end of which we define α_U = lower limit of the interval (which is feasible).

Having determined the step limit α_U, the procedure to determine α is as in Section 5.9: the slope of f at $\alpha = \alpha_U$ is evaluated from Eq. (5.29), (Fig. 5.11). If $f' < 0$, then our step is $\alpha_0 = \alpha_U$. If $f' > 0$, then it means that the minimum of f is in the interval $[0, \alpha_U]$. The minimum α_k is determined using a bisection strategy: the interval is bisected between points having $f' < 0$ and $f' > 0$ until the interval is small enough.

A typical iteration of Zoutendijk's Method of Feasible Directions may be summarized as follows.

1. Determine a feasible starting point $\mathbf{x}_0$ which satisfies $g_j \leq 0,\ j=1,\ldots,m$. See the end of this section for discussion on how to obtain a feasible point. Also, specify parameters ε and g_{low}.
2. Determine the active set from (5.31).
3. Solve the LP problem (5.34) for$(\beta^*, \mathbf{d}^*)$.
   (a) If $\beta^* = 0$, then the current point $\mathbf{x}_0$ is a KKT point; stop.
   (b) Otherwise, perform line search: determine maximum step α_U and, hence, optimum step size α_k as discussed above. Update the current point $\mathbf{x}_k$ as $\mathbf{x}_k + \alpha_k \mathbf{d}^*$ and return to step (2).

Comment on the Constraint Thickness Parameter ε

When ε is very large, a constraint that is well satisfied will become active. In this situation, we will enforce $\nabla g_j^{\mathrm{T}} \mathbf{d} < 0$ in the direction-finding subproblem in 5.34. However, we may then not be able to get sufficiently close to the boundary to home-in on the exact minimum. Thus, if we choose a large ε to begin with, we may benefit from resolving the problem with a smaller ε using the old optimum as the new starting point. On the other hand, if ε is too small to begin with, then constraints that are nearly active can get discarded only to reenter the active set at the next iteration. This can lead to oscillations.

Example 5.14

$$\begin{aligned}
\text{minimize} \quad & f = -(x_1 + 2x_2) \\
\text{s.t.} \quad & g_1 \equiv x_1^2 + 6x_2^2 - 1 \leq 0 \\
& g_2 \equiv -x_1 \leq 0 \\
& g_3 \equiv -x_2 \leq 0
\end{aligned}$$

Consider the problem P1. At the point $\mathbf{x}^0 = (1, 0)^{\mathrm{T}}$,

(i) Sketch the feasible region Ω along with objective function contours. On this sketch, show the usable–feasible cone

(ii) Write down the LP subproblem that will enable us to determine an usable–feasible direction $\mathbf{d}^0$. Use EXCEL SOLVER or MATLAB or a Chapter 4 in-house code to solve this LP to obtain $\mathbf{d}^0$, and show $\mathbf{d}^0$ on your sketch.

(iii) Write down expressions for $f(\alpha)$ and $\mathbf{g}(\alpha)$, where $f(\alpha) \equiv f(\mathbf{x}^0 + \alpha\, \mathbf{d}^0)$, etc. Then solve the line search problem: {Minimize $f(\alpha)$ subject to $\mathbf{g}(\alpha) \leq \mathbf{0}$} (you may use SOLVER again if you like or use your sketch), to obtain step size α_0. Obtain the new point $\mathbf{x}^1 = \mathbf{x}^0 + \alpha_0\, \mathbf{d}^0$. Evaluate f_1 and compare this to initial value f_0

For sketch in (i), see Fig. E5.14. Current value of objective is $f_0 = -1$. At $\mathbf{x}^0 = (1, 0)^{\mathrm{T}}$, the active set is $I = \{1, 3\}$. Thus, the LP subproblem is

$$\begin{aligned} &\text{minimize} && \beta \\ &\text{subject to} && -d_1 - 2d_2 \leq \beta \\ &&& 2d_1 \leq \beta \\ &&& -d_2 \leq \beta \\ &&& -1 \leq d_1 \leq 1, \quad -1 \leq d_2 \leq 1 \end{aligned}$$

The solution is $\mathbf{d}^0 = (-0.5, 1)^{\mathrm{T}}$, $\beta = -1$. Negative value of β indicates that an usable–feasible direction has been found. Direction $\mathbf{d}^0$ is shown in the figure. It is clear that the step size is governed by the constraint g_1 as the objective function is linear. Thus, we have

$$x(\alpha) = \begin{pmatrix} 1 - 0.5\alpha \\ \alpha \end{pmatrix}.$$

Setting $g_1(\mathbf{x}(\alpha)) = 0$, yields $\alpha_0 = 0.16$, from which the new point is $\mathbf{x}^1 = [0.92, 0.16]^{\mathrm{T}}$ with $f_1 = -1.24 < f_0$.

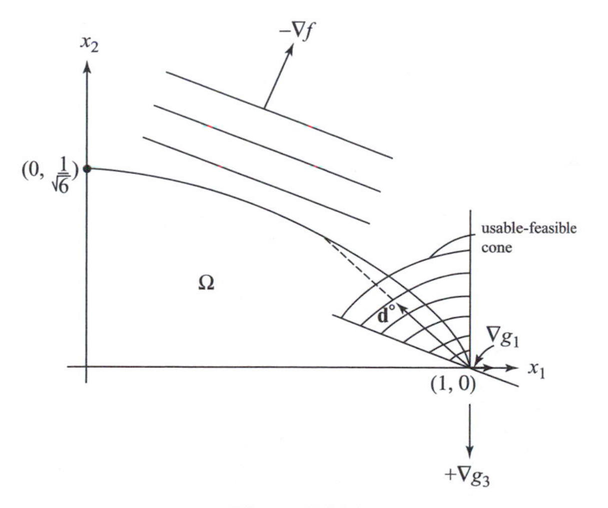

Figure E5.14

Obtaining a Feasible Starting Point, x_0

The method of feasible directions presented earlier needs a feasible starting point: a point that satisfies

$$g_j(x_0) \leq 0 \quad j = 1, \ldots, m \tag{5.37}$$

A simple yet effective approach to obtaining such an $\mathbf{x}_0$ is to minimize the objective function

$$\begin{aligned} &\text{minimize} \quad \hat{f} = \sum_{j=1}^{m} \max(0, g_j + \varepsilon)^2 \\ &\text{subject to} \quad \mathbf{x}^{\mathrm{L}} \leq \mathbf{x} \leq \mathbf{x}^{\mathrm{U}} \end{aligned} \tag{5.38}$$

where ε is a small positive number set by the user. The only constraints in this problem are any lower and upper bounds on the variables defined by the user. The use of $\varepsilon > 0$ is to provide a point that is slightly "interior," with $\mathbf{g} < \mathbf{0}$. We recommend use of program ZOUTEN to solve (5.38) – numerical results has shown this to be a very effective approach to obtaining a feasible starting point. Note that

$$\nabla \hat{f} = 2 \sum_{j=1}^{m} \max(0, g_j + \varepsilon) \nabla g_j^{\mathrm{T}}$$

Handling Equality Constraints

The method of feasible directions is applicable only to problems where the feasible region has an interior. Thus, nonlinear equality constraints cannot be handled. However, it is possible to approximately account for these by using penalty functions [Vanderplaats 1984]. Thus, problems in the form (5.1) are handled as

$$\begin{aligned} &\text{minimize} \quad f - r \sum h_i \\ &\text{subject to} \quad g_i \geq 0 \qquad i = 1, \ldots, m \\ &\qquad\qquad\quad\;\; h_j \leq 0 \qquad j = 1, \ldots, \ell \end{aligned} \tag{5.39}$$

The idea is to require h to be negative through the inequality constraints and then to penalize the objective function when h indeed becomes negative in value. This way, each successive point $\mathbf{x}$ will hover near the boundary of the equalities. The penalty parameter r is user-defined.

Another approach is to convert the equality into two inequalities. For example, material balances in chemical processes involve equations of the form:

$$x_3 = g(x_1, x_2)$$

This can be approximated as, say,

$$\tfrac{9}{10}x_3 \leq g(x_1, x_2) \leq \tfrac{10}{9}x_3$$

The reader should experiment with these techniques. Also, the next two methods in this chapter can explicitly account for equality constraints.

Quadratic Normalization

In (5.33), the normalization constraint used to ensure that **d** is bounded is:

$$-1 \leq d_i \leq 1 \quad i = 1, \ldots, n \tag{5.40}$$

However, Zoutendijk proposed a somewhat superior normalization constraint

$$\mathbf{d}^{\mathrm{T}}\mathbf{d} \leq 1 \tag{5.41}$$

based on the grounds that (5.33) represents a hypercube that can bias affect the direction since the corners are a little farther away than the sides of the cube. The hypersphere constraint in (5.41) does not suffer from this weakness. Thus, the direction-finding subproblem becomes

$$\begin{aligned} &\text{minimize} && \alpha \\ &\text{subject to} && \nabla f^{\mathrm{T}}\mathbf{d} \leq \alpha \\ &&& \nabla g_j^{\mathrm{T}}\mathbf{d} \leq \theta_j\alpha \quad \text{for each } j \in I \\ &&& \tfrac{1}{2}\mathbf{d}^{\mathrm{T}}\mathbf{d} \leq 1 \end{aligned} \tag{5.42}$$

A factor of ½ has been introduced for convenience. By introducing

$$\mathbf{y} = \{\mathbf{d}^{\mathrm{T}}, \alpha\}^{\mathrm{T}}$$

$$\mathbf{p} = \{0, \ldots, 0, 1\}^{\mathrm{T}}$$

$$\mathbf{A} = \begin{bmatrix} \nabla g_1 & -\theta_1 \\ \nabla g_2 & -\theta_2 \\ \ldots & \ldots \\ \nabla g_t & -\theta_t \\ \nabla f & -1 \end{bmatrix}$$

where **A** consists of the constraint gradients (assumed to be numbered from 1 to t for notational simplicity), we can write (5.42) as

$$\begin{aligned} &\text{minimize} \quad \mathbf{p}^{\mathrm{T}}\mathbf{y} \\ &\text{subjectto} \quad \mathbf{A}\mathbf{y} \leq \mathbf{0} \\ &\qquad\qquad\quad \tfrac{1}{2}\mathbf{d}^{\mathrm{T}}\mathbf{d} \leq 1 \end{aligned} \tag{5.43}$$

The dual problem in (5.43) can be written as (see Chapter 6 for more on duality):

$$\begin{aligned} &\text{minimize} \quad \tfrac{1}{2}\boldsymbol{\mu}^{\mathrm{T}}\{\mathbf{A}\mathbf{A}^{\mathrm{T}}\}\boldsymbol{\mu} + \boldsymbol{\mu}^{\mathrm{T}}(\mathbf{A}\mathbf{p}) \\ &\text{subject to} \quad \boldsymbol{\mu} \geq \mathbf{0} \end{aligned}$$

The quadratic programming (QP) sub-problem earlier has simple constraints, and can be solved very efficiently (for the variables $\boldsymbol{\mu}$) by a modified conjugate gradient method. Once $\boldsymbol{\mu}$ is obtained, **d** is recovered from the equation

$$\mathbf{y} = -\mathbf{p} - \mathbf{A}^{\mathrm{T}}\boldsymbol{\mu} \tag{5.45}$$

Example 5.15

Example 5.1 is solved using the method of feasible directions. The starting point is $\mathbf{x}_0 = (5, 50)$, which is *feasible*. Table 5.1 presents the iterations, and the path taken by the method are shown in Fig. E5.15. At the end of iteration 6, the number of function evaluations was NFV = 86.

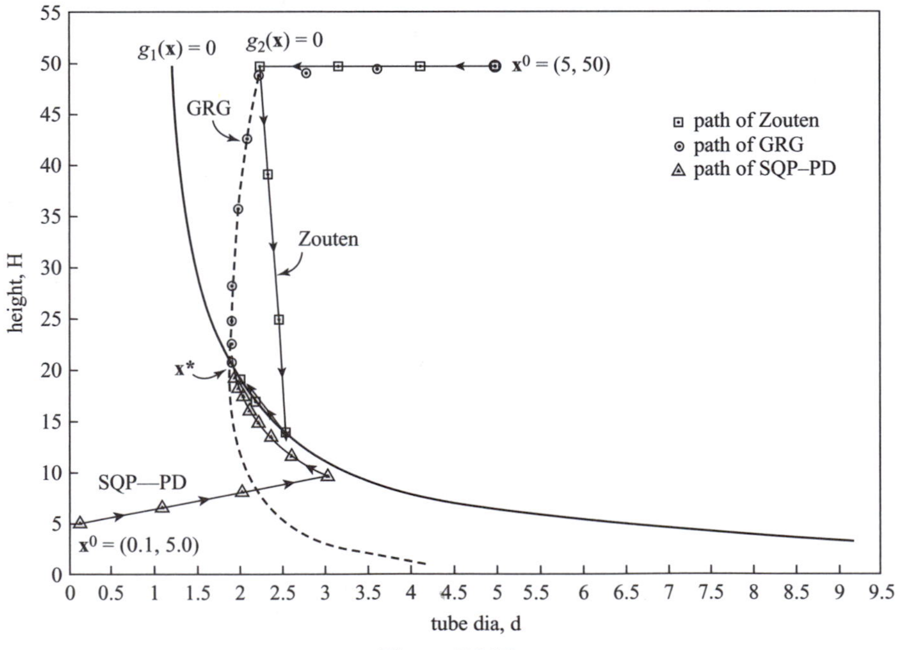

Figure E5.15

Table 5.1. *Solution of Example 5.1 using program ZOUTEN.*

Iteration	f	x_1	x_2	Active set
1	55.0	5.0	50.0	–
2	24.5	2.24	49.8	2
3	15.6	2.50	13.9	1
4	12.9	1.88	20.8	2
5	12.9	1.89	20.0	1
6	12.8	1.88	20.3	1,2

User Subroutine for Program ZOUTEN

The solution of a new problem using the in-house program needs the following modifications:

(A) Need to define NDV and NCON (= n, m, respectively) under the appropriate "CASE" (or Problem#)
(B) In Subroutine USER, the name of the subroutine corresponding to the Problem# needs to be given.
(C) An user subroutine must be provided, as per the format given below, which supplies function and gradient information. The reader must note that gradients of only *active* constraints need to be computed in this method. For large problems, such as those in Chapter 12, this aspect becomes critical from the viewpoint of computational time.

USER SUBROUTINE FORMAT
SUB TESTPROB (INF)

INF = −1 -> *first call*
  Provide $\mathbf{x}_0$ = starting point, $\mathbf{x}^L$ and $\mathbf{x}^U$ = bounds
INF = 1 -> *function call*
  At the given $\mathbf{x}$, provide $f(\mathbf{x})$ and $g_i(\mathbf{x})$ for all $i = 1, \ldots, m$
INF = 2 -> *call for cost function gradient only*
  At the given $\mathbf{x}$, provide $\partial f/\partial x_i$, $i = 1, \ldots, n$ in the array "DF(i)"
INF = 3 -> *call for cost gradient and active constraint gradients*

At the given $\mathbf{x}$, provide $\partial f/\partial x_i$, $i = 1, \ldots, n$ in the array "DF(i)"
Also, for the *NC* active constraints whose constraint numbers are

$$NACT(1), NACT(2), \ldots, NACT(NC)$$

provide the gradients $\partial g_j/\partial x_i$, $i = 1, \ldots, n$ for each active j. Provide this information in the (NC, n) – dimensional array **A** as

$$A(1, i) = \frac{\partial g(NACT(1))}{\partial x_i}, \quad i = 1, \ldots, n$$

$$\mathbf{A}(NC, i) = \frac{\partial g(NACT(NC))}{\partial x_i}, \quad i = 1, \ldots, n$$

5.11 The Generalized Reduced Gradient Method (Nonlinear Constraints)

The generalized reduced gradient (GRG) method is well suited to handle nonlinear *equality* constraints. Excel Solver is based on a GRG algorithm. In GRG, the optimization problem should be expressed in the form

$$\begin{array}{lll} \text{minimize} & f(\mathbf{x}) & \\ \text{subject to} & h_j(\mathbf{x}) = 0, & j = 1, \ldots, \ell \\ \text{and bounds} & \mathbf{x}^{\mathrm{L}} \le \mathbf{x} \le \mathbf{x}^{\mathrm{U}} & \end{array} \tag{5.46}$$

where **x** is $(n \times 1)$. If there is an inequality constraint $g_i \le 0$ in the problem, then this can be converted to an equality as required, through the addition of a slack variable as

$$g_i(\mathbf{x}) + x_{n+1} = 0, \quad 0 \le x_{n+1} \le \mathrm{U} \text{ (a large number, in practice)} \tag{5.47}$$

The starting point x_0 must be chosen to satisfy $\mathbf{h}(\mathbf{x}) = \mathbf{0}$,

$$\mathbf{x}^{\mathrm{L}} \le \mathbf{x}_0 \le \mathbf{x}^{\mathrm{U}}$$

The first step in GRG is to partition **x** into *independent* variables **z** and *dependent* variables **y** as

$$\mathbf{X} = \begin{Bmatrix} \mathbf{Y} \\ \mathbf{Z} \end{Bmatrix} \begin{array}{l} \ell \text{ dependent variables} \\ n - \ell \text{ independent variables} \end{array} \tag{5.48}$$

The **y**-variables are chosen so as to be strictly off the bounds, that is

$$\mathbf{y}^{\mathrm{L}} < \mathbf{y} < \mathbf{y}^{\mathrm{U}} \tag{5.49}$$

Analogous to the partitioning of variables in the revised simplex method for linear programming (Chapter 4), **y** and **z** are also referred to as “basic” and “non-basic” variables, respectively. In (5.48), the basic variables are indicated as being the first ℓ components of **x** only for simplicity of notation. We assume that each point $\mathbf{x}_k$ in the iterative process is a regular point (Section 5.5). This implies that the $(n \times \ell)$

gradient matrix $[\nabla \mathbf{h}(\mathbf{x}^k)]$ has ℓ linearly independent columns or rank $[\nabla \mathbf{h}] = \ell$. Let the $(\ell \times \ell)$ square matrix $\mathbf{B}$, and the $\ell \times (n - \ell)$ matrix $\mathbf{C}$ be defined from the partition

$$[\nabla_{\mathbf{Y}} \mathbf{h}^{\mathrm{T}}, \nabla_{\mathbf{Z}} \mathbf{h}^{\mathrm{T}}] = [\mathbf{B}, \mathbf{C}] \tag{5.50}$$

Our regular point assumption implies that there for a certain choice of dependent variables $\mathbf{y}$, the matrix $\mathbf{B}$ will be nonsingular. Thus, we will choose $\mathbf{y}$ to satisfy (5.49) *and* such that $\mathbf{B}$ is nonsingular – a technique for this is given subsequently. The nonsingularity of $\mathbf{B}$ allows us to make use of the Implicit Function Theorem. The theorem states that there is a small neighborhood of $\mathbf{x}_k$ such that for $\mathbf{z}$ in this neighborhood, $\mathbf{y} = \mathbf{y}(\mathbf{z})$ is a differentiable function of $\mathbf{z}$ with $\mathbf{h}\,(\mathbf{y}(\mathbf{z}), \mathbf{z}) = \mathbf{0}$. Thus, we can treat f as an implicit function of $\mathbf{z}$ as $f = f\,(\mathbf{y}(\mathbf{z}), \mathbf{z})$. This means that we can think of f as dependent only on $\mathbf{z}$. The gradient of f in the $(n-\ell)$-dimensional $\mathbf{z}$-space is called the *reduced gradient* and is given by

$$\mathbf{R}^{\mathrm{T}} \equiv \frac{df}{d\mathbf{z}} = \frac{\partial f}{\partial \mathbf{z}} + \frac{\partial f}{\partial \mathbf{y}} \frac{\partial \mathbf{y}}{\partial \mathbf{z}} \tag{5.51}$$

where $\frac{\partial f}{\partial \mathbf{z}} = [\frac{\partial f}{\partial z_1}, \frac{\partial f}{\partial z_2}, \ldots, \frac{\partial f}{\partial z_{n-\ell}}]$, etc. Differentiating $\mathbf{h}\,(\mathbf{y}(\mathbf{z}), \mathbf{z}) = \mathbf{0}$ we get

$$\mathbf{B} \frac{\partial \mathbf{y}}{\partial \mathbf{z}} + \mathbf{C} = \mathbf{0} \tag{5.52}$$

Combining the earlier two equations, we get

$$\mathbf{R}^{\mathrm{T}} = \frac{\partial f}{\partial \mathbf{z}} - \frac{\partial f}{\partial \mathbf{y}} \mathbf{B}^{-1} \mathbf{C} \tag{5.53}$$

A direction vector $\mathbf{d}$, an $(n \times 1)$ column vector, is now determined as follows. Again, we partition $\mathbf{d}$ as

$$\mathbf{d} = \begin{bmatrix} \mathbf{d}_{\mathrm{y}} \\ \mathbf{d}_{\mathrm{Z}} \end{bmatrix} \tag{5.54}$$

We choose the direction $\mathbf{d}_{\mathbf{Z}}$ to be the steepest descent direction in $\mathbf{z}$-space, with components of $\mathbf{z}$ on their boundary held fixed if the movement would violate the bound. Thus, we have

$$(\mathbf{d}_{\mathbf{Z}})_i = \begin{cases} -R_i \\ 0 \quad if\ Z_i = Z_i^L\ and\ R_i > 0 \\ 0 \quad if\ Z_i = Z_i^U\ and\ R_i < 0 \end{cases} \tag{5.55}$$

Of course, in practice, we implement the check $Z_i < Z_i^{\mathrm{L}} + \varepsilon_x$ rather than $Z_i = Z_i^{\mathrm{L}}$ where ε_x is a small tolerance (see program GRG). Next, $\mathbf{d}_{\mathbf{y}}$ is obtained from

$$\mathbf{d}_{\mathbf{y}} = -\mathbf{B}^{-1} \mathbf{C}\, \mathbf{d}_{\mathbf{Z}} \tag{5.56}$$

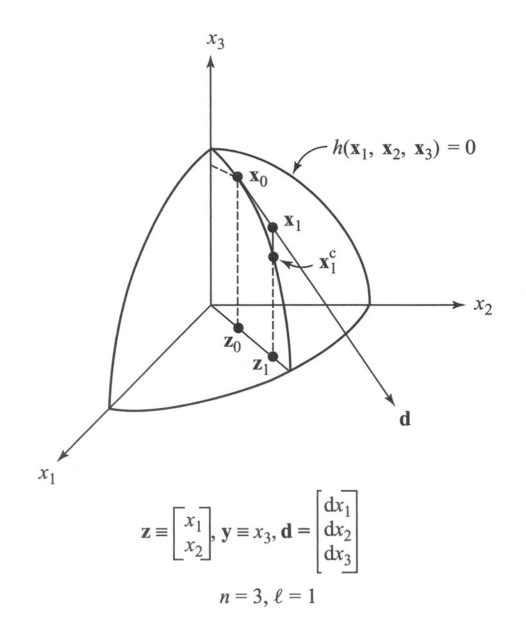

Figure 5.12. The GRG method with nonlinear constraints.

If the direction vector $\mathbf{d} = \mathbf{0}$, then the current point is a KKT point and the iterations are terminated. The maximum step size α_U along $\mathbf{d}$ is then determined from

$$\mathbf{x}^L \leq \mathbf{x}_k + \alpha \mathbf{d} \leq \mathbf{x}^U \tag{5.57}$$

We are assured an $\alpha_U > 0$ in view of $\mathbf{y}$ satisfying (5.49) and also since bounds on $\mathbf{z}$ are accounted for in (5.55). As with other methods discussed in this chapter, a simple strategy to determine the step size α is as follows. We check the slope of f at $\alpha = \alpha_U$ using (5.29). If $f' < 0$, then evidently our step is $\alpha_k = \alpha_U$. If $f' > 0$, then it means that the minimum of f is in the interval $[0, \alpha_U]$. The minimum α_k is determined using a bisection strategy: the interval is bisected between points having $f' < 0$ and $f' > 0$ until the interval is small enough. The new point is then obtained as

$$\mathbf{x}_{k+1} = \mathbf{x}_k + \alpha_k \mathbf{d}$$

If the constraints $\mathbf{h}$ in (5.46) were all linear, then $\mathbf{x}_1$ is indeed a new and improved point. We can reset $\mathbf{x}_k = \mathbf{x}_{k+1}$ and repeat the earlier procedure. However, with nonlinear constraints, $\mathbf{x}_{k+1}$ may not be a feasible point as shown in Fig. 5.12. When $\mathbf{x}_{k+1}$ is infeasible, or $\max_j \{|h_j(\mathbf{x}_{k+1})| >\} \varepsilon_G$ where ε_G is a tolerance, then we adjust the basic or dependent variables to "return" to the feasible region while keeping the independent variables fixed. Let $\mathbf{x}_{k+1} = \{\mathbf{y}_{k+1}, \mathbf{z}_{k+1}\}$. Thus, our problem is to solve

$\mathbf{h}(\mathbf{y}, \mathbf{z}_{k+1}) = \mathbf{0}$ by adjusting $\mathbf{y}$ only. With the starting point $\mathbf{y}_{k+1}^{(0)} \equiv \mathbf{y}_{k+1}$, Newton's iterations are executed for $r = 0, 1, 2, \ldots,$ as

$$\mathbf{J}\Delta\mathbf{y} = -\mathbf{h}\left(\mathbf{y}_{k+1}^{(r)}, \mathbf{z}_{k+1}\right) \tag{5.58}$$

followed by

$$\mathbf{y}_{k+1}^{(r+1)} = \mathbf{y}_{k+1}^{(r)} + \Delta\mathbf{y} \tag{5.59}$$

Above, the $\mathbf{J} = \frac{\partial \mathbf{h}\left(\mathbf{y}_{k+1}^{(r)}, \mathbf{z}_{k+1}\right)}{\partial \mathbf{y}}$ is a $(\ell \times \ell)$ Jacobian matrix. The iterations are continued till a feasible point, $\mathbf{x}_{k+1}^{c}$ is obtained.

In the correction process just described, there are three pitfalls which have to be safeguarded against:

(i) Equations (5.58–5.59) yield a feasible design, $\mathbf{x}_{k+1}^{c}$ but the value of f at this point is higher than the value at $\mathbf{x}_k$ (Fig. 5.13a).
(ii) The bounds on $\mathbf{y}$ are violated during the Newton's iterations (Fig. 5.13b).
(iii) The constraint violations start to increase (Fig. 5.13c).

In all these cases, it is necessary to reduce the step along the $\mathbf{d}$ vector in $\mathbf{x}_{k+1} = \mathbf{x}_k + \alpha_k \mathbf{d}$ as

$$\alpha_k(\text{new}) = \alpha_k/2 \tag{5.60}$$

A point $\mathbf{x}_{k+1}$ closer to $\mathbf{x}_k$ is obtained and the Newton iterations are reexecuted to obtain a new feasible point $\mathbf{x}_{k+1}^{c}$ with $f\left(\mathbf{x}_{k+1}^{c}\right) < f\left(\mathbf{x}_k\right)$. We then reset $\mathbf{x}_k = \mathbf{x}_{k+1}^{c}$ and repeat the process starting from the choice of dependent and independent variables, evaluating the reduced gradient, etc.

Selection of Dependent Variables

At the start of each iteration in the GRG method, we have a point $\mathbf{x}_k$ which satisfies the constraints in (5.47). Our task is to partition $\mathbf{x}$ into dependent (basic) variables $\mathbf{y}$ and independent variables $\mathbf{z}$. The gradient matrix is also partitioned as

$\mathbf{A} = [\nabla \mathbf{h}]^{\mathrm{T}} = [\mathbf{B}, \mathbf{C}]$. The variables $\mathbf{y}$ must satisfy two requirements:

(i) $\mathbf{y}^{\mathrm{L}} < \mathbf{y} < \mathbf{y}^{\mathrm{U}}$ and
(ii) $\mathbf{B} = \left[\frac{\partial \mathbf{h}}{\partial \mathbf{y}}\right]$ is nonsingular.

For illustration, consider $\mathbf{A}$ as

$$\mathbf{A} = \begin{bmatrix} x_1 & x_2 & x_3 & x_4 & x_5 \\ 1 & -1 & \mathbf{2} & 0 & 1 \\ 1 & 1 & -3 & 0 & 0 \\ -2 & -1 & 1 & 0 & 5 \end{bmatrix}$$

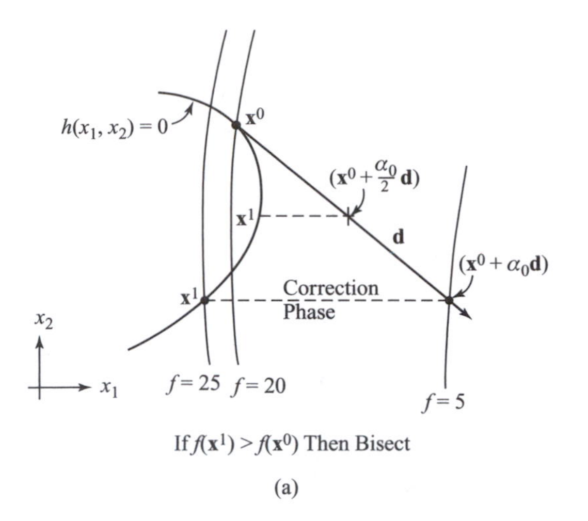

(a)

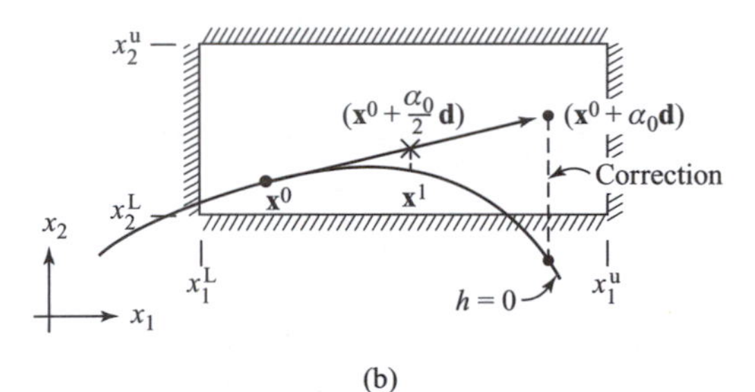

(b)

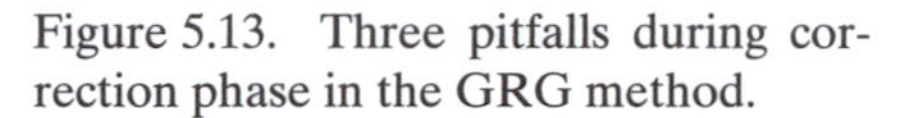

Figure 5.13. Three pitfalls during correction phase in the GRG method.

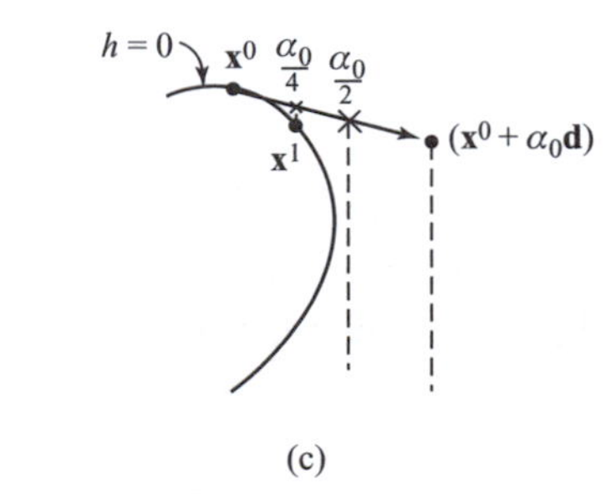

(c)

In this example, x_4 cannot be a basic variable since **B** will then consist of a column of zeroes that will make it singular. Also, a variable that is at its lower or upper bound cannot be considered as a basic variable. A general approach for computer implementation is given in [Vanderplaats 1984]. This is explained below and has been implemented in Program GRG. The idea is to perform Gaussian elimination operations on matrix **A** using pivot search. We first search for the largest entry in the first row, the "pivot" that equals 2, and identify the variable x_3. If this variable is not at its lower or upper bound (we will assume this to be the case for simplicity), then we choose it to be a basic variable. We will place the column corresponding to x_3 in the first column, again for simplicity. The **A** matrix now appears as

$$\mathbf{A} = \begin{bmatrix} x_3 & x_2 & x_1 & x_4 & x_5 \\ \mathbf{2} & -1 & 1 & 0 & 1 \\ -3 & 1 & 1 & 0 & 0 \\ 1 & -1 & -2 & 0 & 5 \end{bmatrix}$$

We perform row operations on rows 2 and 3 to insert zeroes in column 1. This is done by the row operation: Row II – Row I. $(\frac{-3}{2})$, and Row III – Row I. $(\frac{1}{2})$. We get the matrix on the left as follows:

$$\mathbf{A} = \begin{bmatrix} x_3 & x_2 & x_1 & x_4 & x_5 \\ 2 & -1 & 1 & 0 & 1 \\ 0 & -0.5 & \mathbf{2.5} & 0 & 1.5 \\ 0 & -0.5 & -2.5 & 0 & 4.5 \end{bmatrix} \Rightarrow \mathbf{A} = \begin{bmatrix} x_3 & x_1 & x_2 & x_4 & x_5 \\ 2 & 1 & -1 & 0 & 1 \\ 0 & \mathbf{2.5} & -0.5 & 0 & 1.5 \\ 0 & -2.5 & -0.5 & 0 & 4.5 \end{bmatrix}$$

\-------/ \---/

B C

We now scan the second row for the largest element in the second row, pivot = 2.5 and hence identify x_1. We place the x_1-column in the second column of **A** to get the matrix on the right shown above. Now, to introduce a zero below the pivot, we perform the operation: Row III – Row II. (–1). This gives (matrix on the left as follows):

$$\mathbf{A} = \begin{bmatrix} x_3 & x_1 & x_2 & x_4 & x_5 \\ 2 & 1 & -1 & 0 & 1 \\ 0 & 2.5 & -0.5 & 0 & 1.5 \\ 0 & 0 & -1 & 0 & \mathbf{6} \end{bmatrix} \Rightarrow \mathbf{A} = \begin{bmatrix} x_3 & x_1 & x_5 & x_4 & x_2 \\ 2 & 1 & 1 & 0 & -1 \\ 0 & 2.5 & 1.5 & 0 & -0.5 \\ \underbrace{0 \quad 0 \quad 6}_{\mathbf{B}} & & & \underbrace{0 \quad -1}_{\mathbf{C}} & \end{bmatrix}$$

Scanning the last row, we see that the pivot (= 6) corresponds to x_5. Thus, our third column corresponds to x_5. The matrices **B, C** are shown above on the right.

We, thus, obtain $\mathbf{y} = [x_3, x_1, x_5]$ and $\mathbf{z} = [x_4, x_2]$. Also, notice that we have decomposed **B** as an upper-triangular matrix. This can be taken advantage of in evaluating the reduced gradient $\mathbf{R} = \frac{\partial f}{\partial \mathbf{z}} - \frac{\partial f}{\partial \mathbf{y}} \mathbf{B}^{-1}\mathbf{C}$. If we let $\mathbf{B}^{-1}\mathbf{C} = \mathbf{\Lambda}$ or equivalently $\mathbf{B\Lambda} = \mathbf{C}$, then we can obtain $\mathbf{\Lambda}$ by reduction of the columns of **C** followed by back-substitution, noting that **B** has already been decomposed. Then we have $\mathbf{R} = \frac{\partial f}{\partial \mathbf{z}} - \frac{\partial f}{\partial \mathbf{y}} \mathbf{\Lambda}$. The reader can see these steps in Program GRG.

Table 5.2. *Solution of Example 5.1 using program GRG.*

Iteration	f	x_1	x_2	x_3	x_4
1	55.0	5.0	50.0	0.755	0.91
2	25.6	2.50	49.82	0.51	0.29
3	24.5	2.34	49.81	0.47	0.12
4	13.9	2.24	49.80	0.45	0.00
5	13.2	1.89	24.95	0.13	0.00
9	12.8	1.88	20.24	0.00	0.00

Summary of the GRG Method

1. Choose a starting point $\mathbf{x}_0$ which satisfies $\mathbf{h}(\mathbf{x}_0) = \mathbf{0}$, with $\mathbf{x}^L \le \mathbf{x}_0 \le \mathbf{x}^U$.
2. Perform pivoted Gaussian elimination and determine the basic and nonbasic variables and also evaluate the reduced gradient $\mathbf{R}$.
3. Determine the direction vector $\mathbf{d}$ from (5.55) and (5.56). If $\mathbf{d} = \mathbf{0}$, stop; the current point is a KKT point.
4. Determine the step-size α_k and determine $\mathbf{x}_{k+1}$.
5. If $\mathbf{x}_{k+1}$ is feasible, then set $\mathbf{x}_k = \mathbf{x}_{k+1}$ and go to step 2. Otherwise, perform Newton iterations to return to the feasible surface as per (5.58–5.59) to obtain a corrected point $\mathbf{x}^c_{k+1}$. If $f(\mathbf{x}^c_{k+1}) < f(\mathbf{x}_k)$, then set $\mathbf{x}_k = \mathbf{x}^c_{k+1}$ and go to step 2. If $f(\mathbf{x}^c_{k+1}) > f(\mathbf{x}_k)$, then set $\alpha_k = \alpha_k/2$, obtain a new $\mathbf{x}_{k+1} = \mathbf{x}_k + \alpha_k\, \mathbf{d}$ and go to the beginning of this step.

Example 5.1 is solved using the GRG method. Slack variables, x_3 and x_4, have been added to g_1 and g_2, respectively, as is required by the GRG method. Thus, $n = 4, m = 2$. The starting point is $\mathbf{x}_0 = (5, 50, 0.755, 0.91)$ which is feasible. Table 5.2 presents the iterations, and the path taken by the method are shown in Fig. E5.15. At the end of iteration 9, the number of function evaluations was NFV = 18.

Example 5.16 (Hand Calculation)

Consider the problem

$$\begin{array}{ll} \text{minimize} & f = -x_1 - x_2 \\ \text{subject to} & h = -x_1 + 2x_2 - 1 + x_3 = 0 \quad (x_3 \text{ is a slack variable}) \\ \text{and} & x_1, x_2, x_3 \ge 0 \end{array}$$

At Point "A" in Fig. E5.16, we have $\mathbf{x} = [0, 0.5, 0]^{\mathrm{T}}$. Thus, we choose our dependent variable as $y = x_2$ since it is off its bound, and independent variables as $\mathbf{z} = [x_1, x_3]$. Thus, $\mathbf{B} = [2]$, $\mathbf{C} = [-1, 1]$, $\partial f/\partial y = -1$, $\partial f/\partial \mathbf{z} = [-1, 0]$, and from Eq. (5.53), we get the reduced gradient as $\mathbf{R}^{\mathrm{T}} = [-1, 0] - [-1]\ [2]^{-1}\ [-1, 1] = [-1.5, 0.5]$. From (5.62), we have $\mathbf{d}_{\mathbf{z}} = [1.5, 0]^{\mathrm{T}}$. From (5.56), we have $\mathbf{d}_{\mathbf{y}} = -\ [2]^{-1}\ [-1, 1]\ [1.5, 0]^{\mathrm{T}} = 0.75$. Thus, $\mathbf{d} = [1.5, 0.75,$

0] = [dx_1, dx_2, dx_3]. We see that the direction vector is simply a move along the constraint boundary from A.

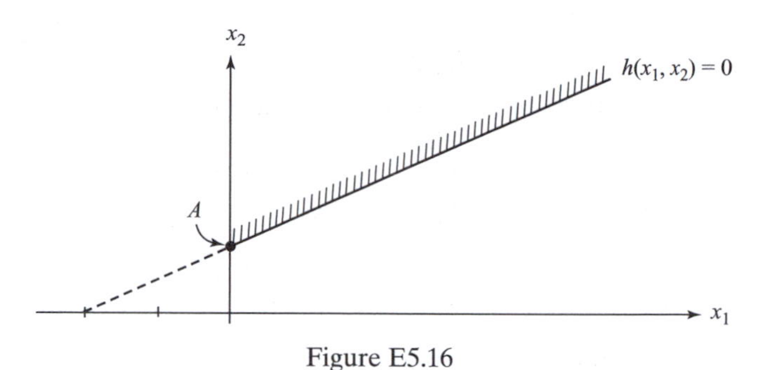

Figure E5.16

Example 5.17

One iteration of GRG will be performed on problem in Example 5.14, at $\mathbf{x}^0 = (1, 0)^T$, $f_0 = -1$. We first need to define upper bounds on variables, and reformulate the inequality constraint as an equality:

$$\begin{aligned} &\text{minimize} \quad f = -(x_1 + 2x_2) \\ &\text{subject to} \quad h \equiv x_1^2 + 6x_2^2 - 1 + x_3 = 0 \\ &\qquad\qquad 0 \le x_i \le 1.5, \quad i = 1, 2, 3 \end{aligned}$$

We have x_1 = dependent variable "y_1" and x_2, x_3 = independent variables "z_1, z_2." The following matrices are then readily determined as

$$\mathbf{R}^T = \{-2 \quad 0\} - \{-1\}\{2\}^{-1}\{0 \quad 1\} = \{-2 \quad 0.5\}$$

$$\mathbf{B} = 2x_1 = 2$$

$$\mathbf{C} = \{12x_2, \quad 1\} = \{0 \quad 1\}$$

$$d_2 = 2, \quad d_3 = 0$$

$$d_1 = -[2]^{-1}[0 \quad 1][2 \quad 0]^T = 0$$

$$\alpha_U = 0.75$$

which gives the new point as $[1, 1.5, 0]^T$. See Fig. E5.17. Since $|h| > 0$ at the new point, a correction phase needs to be initiated to return to $h(\mathbf{x}) = 0$. Eqs. (5.58)–(5.59) are implemented in the following MATLAB code written specifically for this example as shown here. Note that since the objective is linear here, descent is guaranteed along the reduced gradient direction, and hence its value need not be checked after correction. The steps are indicated in Fig. E5.17.

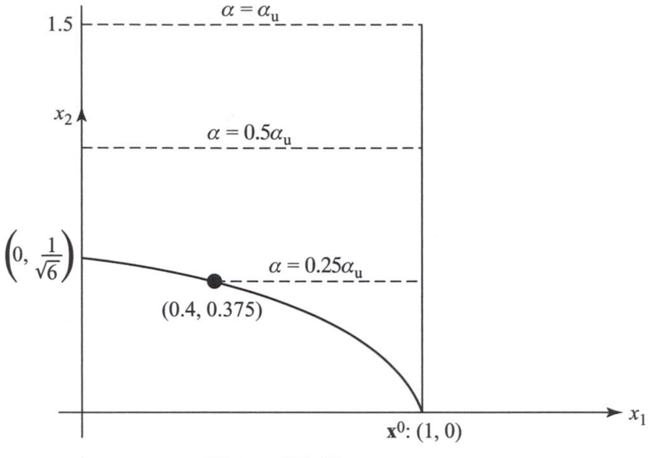

Figure E5.17

```
function [] = GRGcorrection()
close all; clear all;
x1=1; x2=1.5; x3=0; d2 = 2; alp = 0.75;  neval=-1;
eps=.001; itermax=5;
[h,neval] = getcon(x1,x2,x3,neval);
while (abs(h)>eps | alp<1e-2)
x2=0 + alp*d2;
iter=0; ifl=0;
while (iter <= itermax)
    iter=iter+1;
    J = 2*x1;
    [h,neval] = getcon(x1,x2,x3,neval);
    if (abs(h) <= eps)
        ifl=1;
        break
    end
    dx1 = -h/J;
    x1=x1+dx1;
end
if (ifl == 1)
    break
elseif (ifl == 0)
    alp=alp/2;
end
```

```
end
x1, x2, x3, h,neval

function [h,neval] = getcon(x1,x2,x3,neval)
neval = neval+1;
h = x1^2 + 6*x2^2 -1 + x3;
```

The final point after the correction phase is $x_1 = 0.3957$, $x_2 = 0.375$, $x_3 = 0$, $h = 3\text{e}{-4}$, with 14 function evaluations. At this point, $f = -1.1457$.

User Subroutine for Program GRG

The user subroutine is similar in format to that for Program ZOUTEN discussed in the previous section. However, note that the user must add slack or surplus variables to formulate all constraints (not including bounds on variables) as *equalities*. There is no concept of active constraints here. Thus, when supplying constraint gradients, provide $\mathbf{A}$ $(j, i) = \partial g_j / \partial x_i$, $i = 1, \ldots, n$ for each j.

5.12 Sequential Quadratic Programming (SQP)

Sequential quadratic programming (SQP) methods have received a lot of attention in recent years owing to their superior rate of convergence (see Han [1976], Powell [1978], and Schittkowski [1983]). While SQP methods may be interpreted as Newton's method applied to the solution of the KKT conditions, this aspect will not be elaborated on. The method given in the following was first published by Pshenichny in 1970 in Russian and later in a book by Pshenichny and Danilin in 1978 [Pshenichny and Danilin 1982, 2nd printing]. They termed their SQP method as "linearization method" and proved convergence under certain conditions. Many variations of SQP methods exist in the literature. In fact, Matlab "fmincon" optimizer is based on a SQP algorithm. A code is provided based on the algorithm shown below.

The sequential QP method discussed here has several attractions: the starting point can be infeasible, gradients of only active constraints are needed, and equality constraints can be handled in addition to the inequalities. As with all gradient methods, there are two tasks: direction-finding or "where to go" in the design space, and step-size selection or "how far to go." A new and improved design point is then obtained as $\mathbf{x}_{k+1} = \mathbf{x}_k + \alpha_k\, \mathbf{d}_k$. In this method, the direction vector at a given point $\mathbf{x}_k$ is determined from solving a QP subproblem.

QP Problem for Direction Finding

$$\begin{aligned} &\text{minimize} && \tfrac{1}{2}\,\mathbf{d}^{\mathrm{T}}\mathbf{d} + \nabla f^{\mathrm{T}}\,\mathbf{d} && \text{(5.61a)}\\ &\text{subject to} && \nabla g_i^{\mathrm{T}}\,\mathbf{d} + g_i^k \le 0 \quad i \in I_1 && \text{(5.61b)}\\ & && \nabla h_i^{\mathrm{T}}\,\mathbf{d} + h_i^k = 0 \quad i \in I_2 && \text{(5.61c)}\\ & && \mathbf{x}^{\mathrm{L}} \le \mathbf{d} + \mathbf{x}_k \le \mathbf{x}^{\mathrm{U}} && \text{(5.61d)} \end{aligned}$$

Note that d_i, $i = 1, \ldots, n$ are variables in the aforementioned QP. We will denote the solution to the above QP as $\mathbf{d}_k$. The gradient vectors ∇f, ∇g, and ∇h are evaluated at $\mathbf{x}_k$. Other SQP methods use $\frac{1}{2}\,\mathbf{d}^{\mathrm{T}}\,\mathbf{B}\,\mathbf{d}$ in (5.61a) where $\mathbf{B}$ = a positive definite matrix approximation of the Hessian of the Lagrangian that is updated using, say, the BFGS formula in Chapter 3 with ∇L replacing ∇f. However, in the linearization method, $\mathbf{B} = \mathbf{I}$ = Identity matrix.

Various techniques exist to solve the QP. In the code provided in the text, Dikin's interior method for LP problems in Chapter 4 is extended to solve the QP.

Definition of Active Constraints

First, let us understand the constraints in the preceding QP. Constraints (5.61b) and (5.61c) are simply linearized versions of the original nonlinear constraints in (1). The sets I_1 and I_2 are "active sets" – not all constraints in the original problem enter into the QP subproblem. Specifically, the active sets are defined as

$$I_1 = \{j : g_j(\mathbf{x}_k) \ge \mathrm{V}(\mathbf{x}_k) - \delta, \quad j = 1, \ldots, m\} \tag{5.62a}$$

$$I_2 = \{j : |h_j(\mathbf{x}_k)| \ge \mathrm{V}(\mathbf{x}_k) - \delta, \quad j = 1, \ldots, m\} \tag{5.62b}$$

where δ is a small number, which is specified by the user, and V represents the maximum violation as defined by

$$\mathrm{V}(\mathbf{x}) = \max\{0, |h_j(\mathbf{x})|, g_i(\mathbf{x})\}, \quad j = 1, \ldots, \ell, \quad i = 1, \ldots, m \tag{5.63}$$

For illustration, at the current point, let the values of six inequality constraints ($\le$ type) be

$$\mathbf{g} = \{-0.90, 0.00, 1.60, 1.30, 0.05, 1.62\}$$

We see that the maximum violation is $\mathrm{V}(\mathbf{x}_k) = 1.62$. With $\delta = 0.1$, we see that the active set consists of constraints that are greater than or equal to 1.52 in value. Thus,

$I_1 = \{3, 6\}$. Further, particularly in engineering problems, bounds on variables have to be accounted for as in (5.61d) – no active set strategy is used for this.

Geometrical Significance of the Direction Vector

Now, we consider the objective function in (5.61a). We will present a geometric picture of the solution to the QP, which is also of practical value. Assume that we have solved the QP (discussed below). At the solution $\mathbf{d}_k$, only a subset of the constraints in (5.61) will be active: let us denote this active set as

$$\bar{I} = \{I_2\} \cup \{i : \nabla g_i^T \mathbf{d}_k + g_i^k = 0 \quad i \in I_1\} \cup \{\text{active bounds in (5.61d}\}$$

Thus, the solution $\mathbf{d}_k$ must satisfy the KKT conditions to the problem

$$\text{minimize} \quad \tfrac{1}{2}\mathbf{d}^{\mathrm{T}}\mathbf{d} + \nabla f^{\mathrm{T}}\mathbf{d} \tag{5.64a}$$

$$\text{subject to} \quad \{\mathbf{A}\}\mathbf{d} + \widehat{g} = \mathbf{0} \tag{5.64b}$$

where the rows of the matrix $\mathbf{A}$ are the gradients of the constraints in $\bar{I}$ and $\widehat{g}$ = value of the constraints in the set. We assume that rows of $\mathbf{A}$ are linearly independent. We have $L = \frac{1}{2}\,\mathbf{d}^{\mathrm{T}}\,\mathbf{d} + \nabla f^{\mathrm{T}}\,\mathbf{d} + \mu^{\mathrm{T}}\,(\mathbf{A}\,\mathbf{d} + \widehat{g})$. Setting $\partial L/\partial \mathbf{d} = \mathbf{0}$, we get $\mathbf{d} = -(\nabla f + \mathbf{A}^{\mathrm{T}}\,\mu)$. Substituting this into (5.64b) we obtain $\mu = [\mathbf{A}\mathbf{A}^{\mathrm{T}}]^{-1}\,(-\mathbf{A}\nabla f + \mathbf{A}\,\widehat{g})$, which allows us to write

$$\mathbf{d} \equiv \mathbf{d}^1 + \mathbf{d}^2 \tag{5.65}$$

where

$$\mathbf{d}^1 = -[\mathbf{P}]\nabla f \tag{5.66a}$$

$$\mathbf{d}^2 = -\mathbf{A}^{\mathrm{T}}\{\mathbf{A}\mathbf{A}^{\mathrm{T}}\}^{-1}\widehat{g} \tag{5.66b}$$

where $[\mathbf{P}] = \mathbf{I} - \mathbf{A}^{\mathrm{T}}\,[\mathbf{A}\mathbf{A}^{\mathrm{T}}]^{-1}\,\mathbf{A}$ is a projection matrix. Interestingly, $\mathbf{d}^1$ is the projection of the steepest descent direction onto the tangent plane to the active constraints at $\mathbf{x}_k$ (we discussed this in connection with Rosen's gradient projection method in Section 5.9). The vector $\mathbf{d}^2$ is a "correction step" that points directly toward the constraint region. That is, $\mathbf{d}^2$ will reduce V in (5.63). It is left as an exercise to show that $\mathbf{d}^1$ and $\mathbf{d}^2$ are orthogonal or $\mathbf{d}^{1\mathrm{T}}\,\mathbf{d}^2 = 0$. If there are no constraint violations, then $\mathbf{d}^2$ will equal zero, and $\mathbf{d}_k$ will simply be the projected steepest descent direction. Figure 5.14 shows the components $\mathbf{d}^1$ and $\mathbf{d}^2$. Thus, we see that the direction vector $\mathbf{d}_k$ is a combination of a cost reduction step, $\mathbf{d}^1$, and a constraint correction step, $\mathbf{d}^2$. We will use this interpretation to scale the cost function f shown below. Finally, the QP in (5.61) can be interpreted as finding the smallest distance $\mathbf{d}^{\mathrm{T}}\mathbf{d}$ with a penalty on the accompanying increase in f while satisfying the linearized constraints.

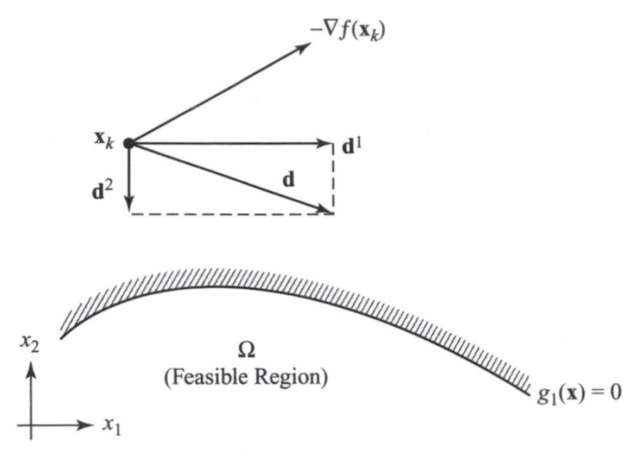

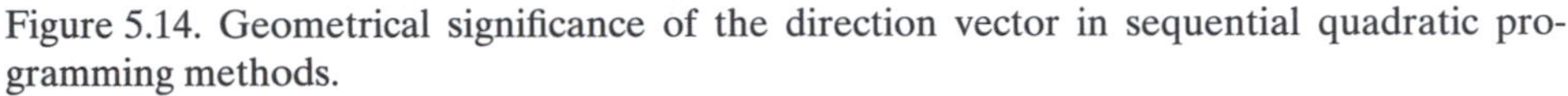
Figure 5.14. Geometrical significance of the direction vector in sequential quadratic programming methods.

Line Search

In the method of feasible directions and in the GRG method, the step-size α along a direction vector was determined by minimizing the cost function f. We cannot use f alone in SQP methods because, while the $\mathbf{d}^1$ component of $\mathbf{d}$ reduces f, the $\mathbf{d}^2$ component will, in general, increase the value of f.

In fact, we have, to a first-order approximation:

$\mathbf{d}^1$ = cost reduction step:	will reduce f	no change in V
$\mathbf{d}^2$ = constraint correction step:	can increase f	will reduce V

From these observations, we can expect the "exact penalty function"

$$\theta(\mathbf{x}) = f(\mathbf{x}) + RV(\mathbf{x}) \tag{5.67}$$

to reduce along $\mathbf{d}$ provided the penalty parameter R is sufficiently large. That is, if R is sufficiently large, we can be assured that θ will reduce in value and, hence, can be used as a "descent function" for determining the step size. It can be shown that R should be chosen such that

$$\mathrm{R} \geq \sum_i |\mu_i|, \quad i \in I_1 \cup I_2 \tag{5.68}$$

where μ_i are the Lagrange multipliers associated with the constraints in the QP in (5.61). In practice, we compute $\mathrm{C} = \sum_i |\mu_i|$ at each iteration and increase R if necessary based on

$$\text{IF} R < \mathrm{C} \quad \text{THEN} \quad R = 2^*\mathrm{C}$$

The value of R is never reduced – this fact is used in proving convergence of the linearization method.

Having identified a descent function, namely, θ, the procedure to find the step α is based on an approximate line search strategy as follows. The step α_0 is chosen equal to $(0.5)^J$ where J is the first of the integers $q = 0, 1, 2, \ldots$, for which the following inequality holds:

$$\theta(\mathbf{x}_k + (0.5)^q \mathbf{d}_k) \leq \theta(\mathbf{x}_k) - \gamma(0.5)^q \|\mathbf{d}\|^2 \tag{5.69}$$

Once the step α_0 is determined, a new point is obtained from $\mathbf{x}_{k+1} = \mathbf{x}_k + \alpha_k \mathbf{d}_k$. We reset $\mathbf{x}_k = \mathbf{x}_{k+1}$ and again solve the QP in (5.61) for a new direction vector. Note that if the starting point $\mathbf{x}_0$ for SQP-PD is highly feasible (that is, well in the interior of Ω), then the algorithm will take a steepest descent step and obtain the step size based on an approximate strategy as per (5.69). This may result in inefficiency compared to an infeasible starting point.

Stopping Criteria

If $\mathbf{d}_k = \mathbf{0}$ is the solution to the QP in (5.61), then we can show that $\mathbf{x}_k$ is a KKT point. Thus, $\|\mathbf{d}\| < \mathrm{TOL_{KKT}}$ is used for convergence. In addition, the iterative process is terminated if there is no change in f for three consecutive iterations *with* $V \leq \mathrm{TOL_G}$. The need for checking constraint violations is due to the fact that SQP methods iterate through the infeasible region.

Algorithm

1. Choose a starting point $\mathbf{x}_0$ which satisfies the bounds $\mathbf{x}^L \leq \mathbf{x}_0 \leq \mathbf{x}^U$. Select scalars δ, R, γ (typically, 0.05, 100., 0.1, respectively)
2. Determine the active set, evaluate gradients of active constraints and solve the QP in (5.61) to obtain $\mathbf{d}_k$
3. Check for convergence as discussed earlier.
4. Determine the step size α_k from (5.69). Update the design point as $\mathbf{x}_{k+1} = \mathbf{x}_k + \alpha_k \mathbf{d}_k$, set $\mathbf{x}_{k+1} = \mathbf{x}_k$, and return to step (1).

Cost Scaling

It is evident from Fig. 5.14 that if ∇f is very large in magnitude compared to the gradients of the constraints, then the direction vector will point in the direction of the tangent to the constraints and not directly toward the boundary. At the same time, the line search in (5.69) will yield a very small step α. Thus, convergence will not be obtained. This situation is special to SQP methods. Scaling f and/or variables can remedy this difficulty so that the cost gradient is of similar magnitude to the constraint gradient.

Example 5.18 (hand calculation)

Two iterations of SQP will be performed on Example 5.14. At $\mathbf{x}^0 = (1, 0)^T$, $f_0 = -1$, maximum violation $V = 0$. The active set is $\{1, 3\}$, and QP in (5.61) reduces to

$$\begin{aligned} \text{minimize} \quad & \tfrac{1}{2}(d_1^2 + d_2^2) - d_1 - 2d_2 \\ \text{subject to} \quad & 2d_1 \le 0 \\ & -d_2 \le 0 \\ & 0 \le d_1 + 1 \\ & 0 \le d_2 \end{aligned}$$

The solution obtained using MATLAB "fmincon" is: $\mathbf{d} = (0, 2)^T$, as shown in Fig. E5.18, with Lagrange multipliers (1, 0, 0, 0). Now, step size is determined from (5.69). Assume parameters $R = 5$ and $\gamma = 0.1$. There is no need to increase R based on (5.68). Note that $f = -x_1 - 2\,x_2$ and $V \equiv g_1 = x_1^2 + 6\,x_2^2$ since this is the only violated constraint along **d**. Referring to the left- and right-sides of (5.69) as LHS and RHS, respectively, a spreadsheet calculation gives the solution $q = 3$ with associated $\mathbf{x^1} = (1, 0.25)^T, f_1 = -1.5, V = +0.375$ (i.e., infeasible). This step is shown in the figure.

q	LHS	RHS
1	27	3.8
2	5.5	3.9
3	0.375	3.95

A second iteration will now be performed from the point $\mathbf{x}^1$. The direction-finding QP is

$$\begin{aligned} \text{minimize} \quad & \tfrac{1}{2}(d_1^2 + d_2^2) - d_1 - 2d_2 \\ \text{subject to} \quad & 1.5 + 2d_1 + 3d_2 \le 0 \\ & 0 \le d_1 + 1 \\ & 0 \le d_2 + 0.25 \end{aligned}$$

The solution is: $\mathbf{d} = (-0.4615, -0.1923)^T$, as shown in Fig. E5.18, with Lagrange multipliers (0.738, 0, 0). Now, step size is again determined from (5.69), with

$R = 5$ and $\gamma = 0.1$. A spreadsheet calculation gives the solution $q = 0$ with associated $\mathbf{x^2} = (0.5385, 0.0577)^{\mathrm{T}}$, $f_2 = -0.6539$, $V = -0.69$ (i.e., feasible). This step is shown in the figure.

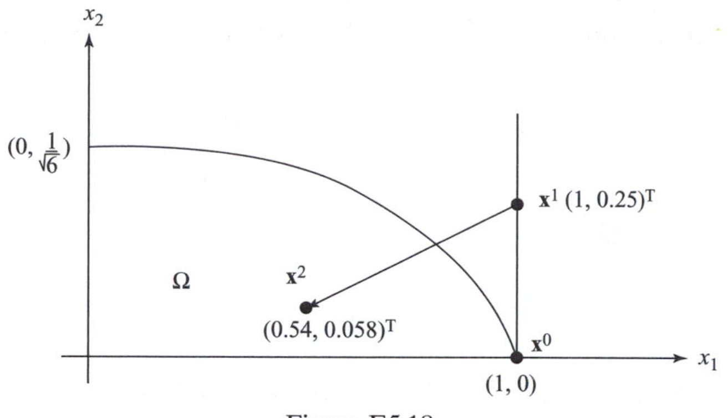

Figure E5.18

User Subroutine for Program SQP-PD: The format for the user subroutine is identical to that for Program ZOUTEN given earlier. Here also, an active set strategy is used to compute constraint gradients.

Example 5.19

Example 5.1 is solved using the sequential quadratic programming method. The starting point is $\mathbf{x}_0 = (0.1, 5.0)$, which is *infeasible*. Table 5.3 presents the iterations, and the path taken by the method are shown in Fig. E5.15. At the end of iteration 6, the number of function evaluations was NFV = 66. Lagrange multipliers printed by the program are 5.6 and 2.4.

5.13 Features and Capabilities of Methods Presented in this Chapter

Four methods have been presented: Rosen's Gradient Projection Method ("ROSEN"), Zoutendijk's Method of Feasible Directions ("ZOUTEN"),

Table 5.3. *Solution of Example 5.1 using program SQP.*

Iteration	f	x_1	x_2	Maximum violation
1	0.6	0.1	5.0	7983.0
2	0.9	0.15	5.05	3269.0
3	1.2	0.21	5.11	1349.0
7	4.0	0.70	5.57	40.5
11	10.9	1.88	6.72	1.56
34	12.8	1.88	20.2	7.8×10^{-10}

Table 5.4. *Features of the gradient methods.*

	ROSEN	ZOUTEN	GRG	SQP
Must the starting point be feasible with respect to the constraints?	Yes	Yes	Yes	No
Can nonlinear constraints be handled?	No	Yes	Yes	Yes
Can equality constraints be handled?	Yes	No[1]	Yes	Yes
Is an active set strategy used for the constraints?	Yes	Yes	No[2]	Yes
Are all points feasible during the iterative process?	Yes	Yes	No[3]	No
Do all functions have to be differentiable?	Yes	Yes	Yes	Yes

[1] Except approximately through a penalty function approach as discussed.
[2] Constraints may be reduced in number by physical understanding of the problem.
[3] However, a correction process is initiated the moment an infeasible point is generated.

Generalized Reduced Gradient Method ("GRG"), and a Sequential Quadratic Programming Method (SQP). Penalty function and duality based methods are discussed in the next chapter. Features of these methods are summarized in Table 5.4.

Computationally Expensive Functions and Large Scale Problems

When interfacing optimizers to simulation codes, each function evaluation can take considerable amount of computing time. In this case, it is desirable to avoid function calls during line search. Instead, approximations are used particularly in the line search phase. This is discussed further in Chapter 12. Further, with large number of constraint functions, an active set strategy is highly desirable as this reduces gradient computations for direction finding. Penalty function methods are competitive for large-scale problems (Chapter 6).

Example 5.20 – Dynamic Impact Absorber with Pointwise (i.e., time-dependent) Constraints

This example illustrates two formulation techniques discussed in Section 5.4. The first is on how a min-max problem is converted to standard NLP form using an artificial variable as

$$\underset{\mathbf{x}}{\text{minimize}} \left\{ \begin{matrix} \text{maximum}\, f(t) \\ 0 \le t \le T \end{matrix} \right\}$$

can be expressed as

$$\begin{aligned} &\text{minimize} && x_{n+1} \\ &\text{subject to} && f(t) \le x_{n+1}, \quad t \in [0, T] \end{aligned}$$

The second technique involves handling a *pointwise* constraint as seen earlier using a discretization procedure. Specifically, the above constraint $f(t) \le x_{n+1}$, $t \in [0, T]$ is formulated as

$$f(t_i) - x_{n+1} \le 0, \quad t_i \in [0, T]$$

where t_i is a discretized time value in the interval. Usually, $[0, T]$ is discretized as $[0, \Delta t, 2\Delta t, 3\Delta t, \ldots, n_d \Delta t = T]$. Thus, the number of constraints depends on the number of discretization points.

The absorber optimization problem is as follows. A mass M impacts a fixed barrier with a given velocity v_0 (Figure E5.20a). The "cushion" between the mass and the barrier is a spring-damper system. Design variables relate to the spring and damper coefficients. The objective is to minimize maximum acceleration within a simulation time of 12 seconds while limiting maximum displacement ("excursion") of the mass. The mass M may represent a vehicle, an instrument, etc.

Figure 5.20a

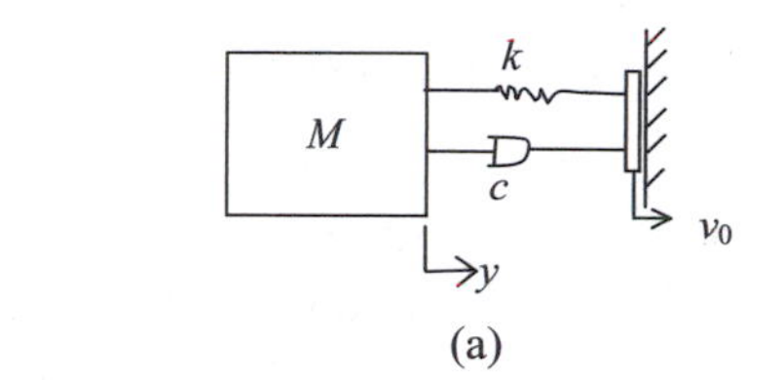

The equations of motion are ordinary differential equations:

$$M\ddot{y} + c\dot{y} + ky = 0$$
$$y(0) = 0, \quad \dot{y}(0) = v_0$$

Preceding equations can be expressed in first-order form (suitable for MATLAB) by defining $z_1(t) = y(t)$ and $z_2(t) = \dot{y}(t)$ as

$$\begin{bmatrix} 1 & 0 \\ 0 & M \end{bmatrix} \begin{bmatrix} \dot{z}_1 \\ \dot{z}_2 \end{bmatrix} = \begin{bmatrix} z_2 \\ -cz_2 - kz_1 \end{bmatrix}$$
$$\begin{bmatrix} z_1(0) \\ z_2(0) \end{bmatrix} = \begin{bmatrix} 0 \\ v_0 \end{bmatrix} \tag{1}$$

Design variables are scalars k and c. The objective is to minimize the magnitude of maximum acceleration of the mass, so an artificial variable α is introduced as the objective to be minimized, which represents an upper bound on the acceleration at any time t. Thus, there are three design variables in all.

$$\text{minimize} \quad \alpha \tag{2}$$

$$\text{subject to} \quad \frac{1}{M}|(cz_2 + kz_1)| - \alpha \le 0, \; 0 \le t \le T \tag{3}$$

and a limit on maximum displacement of the mass:

$$z_1 - z_1^{\max} \le 0, \quad 0 \le t \le T \tag{4}$$

and bounds on c and k.

Data

$M=1$, $v_0=1$, $z_1^{\max}=1$, $T=12$ s. A discretized grid of 0.15 s between points was used to enforce constraints in (3) and (4), which translates into $81+81=162$ constraints. It is important that the time grid includes $t=0$ s, as the maximum acceleration may occur at $t=0$. If not, an additional constraint may be added to the aforementioned as $(\frac{cv_0}{M}) - \alpha \le 0$. Denoting design variable vector as $\mathbf{x} = [k, c, \alpha]$, lower bounds on $\mathbf{x}$ are [0, 0, 0], upper bounds are [5., 5., 100.], initial guess are [0.1, 0.1, 10.].

Results

At the starting design, $y_{\max} \equiv \max\limits_{0 \le t \le T} |y(t)| = 2.522$ that exceeds $z_1^{\max} = 1$. That is, the initial design is infeasible. At optimum (verified from a few different starting designs), we have $y_{\max} = 1$, and maximum absolute acceleration $\ddot{y}_{\max} = 0.5205$ ($= \alpha^*$, the objective). Studying the output, we see that active constraints are the displacement limit on the mass at $t = 2.1$ s, and the peak acceleration magnitude occurs at $t = 0.45$ s. A plot of $y(t)$ versus t shows this (Fig. E5.20b)

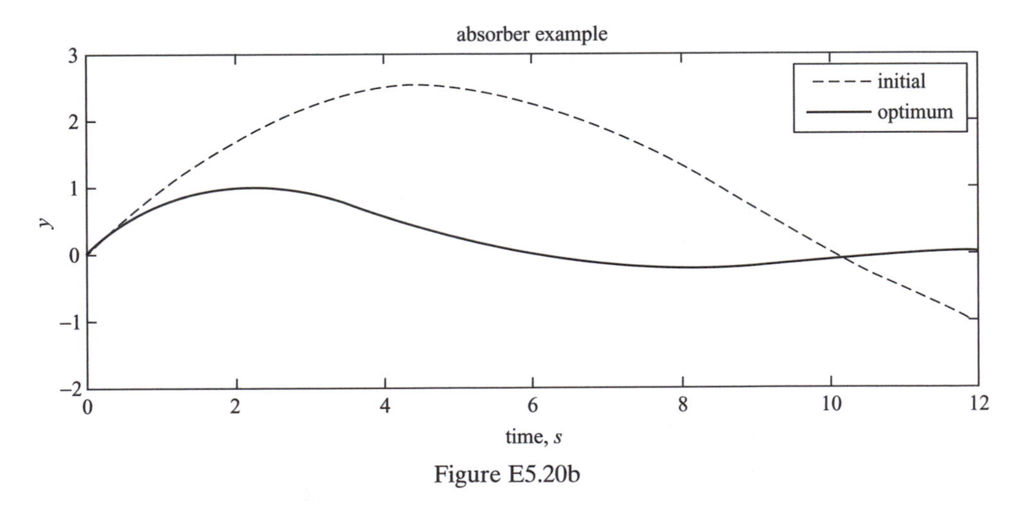

Figure E5.20b

Matlab code to solve the problem is given as follows; "fmincon" is used as the optimizer (see Section 5.3) and "ode45" is used to solve the differential equations in (1).

It is noted that the Matlab optimizer "fmincon" computes gradients by divided differences (divided differences was explained in Chapter 1). This implies additional calls to the constraint subroutine that integrates the differential equations. Analytical derivatives can be used instead (Chapter 12). However, "fmincon" used SQP with active set strategy, and since only three constraints were active (with 0.15 s discretization of time variable), constraint derivative computing time is not an issue here. Note that Method of Feasible Directions (MFD) also uses an active set strategy. If using the generalized Reduced Gradient (GRG) method, then a search for critical points (in time) where the constraints are maximum may be done and constraints imposed at these times, since GRG does not use an active set.

```
function [] = absorber()
close all; clear all;
global M v0 tau pnu omega z1max
M=1; v0=1; tau=12; pnu=1; omega=1; z1max=1;
XLB=[0 0 0];                    %lower bounds on variables
XUB=[5.  5.  100.];             %upper bounds
X = [ 0.1 0.1 10.];             %starting guess
A=[];B=[];Aeq=[];Beq=[];         %linear constraint info
disp('initial response maxima')
[ymax, ydotmax, yddotmax] = responseinfo(X)
disp('initial constraint values')
[c, ceq] = ABSCON(X)
%
[XOPT,FOPT,EXITFLAG,OUTPUT,LAMBDA] = fmincon(@ABSFUN,X,A,B,Aeq,Beq,
                                             XLB,... XUB,@ABSCON)
 L1 = length(LAMBDA.ineqnonlin)
 for i=1:L1
     if abs(LAMBDA.ineqnonlin(i)) > 0
         i, LAMBDA.ineqnonlin(i)
     end
 end
disp('final response maxima')
[ymax, ydotmax, yddotmax] = responseinfo(XOPT)
disp('final constraint values')
[c, ceq] = ABSCON(XOPT)

function [f] = ABSFUN(X)
global M v0 tau pnu omega z1max
f = X(3);
```

```
function [c, ceq] = ABSCON(X)
global M v0 tau pnu omega z1max
ceq=[];
b1 = X(1); b2=X(2); b3=X(3);
tspan = [0 tau];
y0 = [0 v0];
sol = ode45(@odefun,tspan,y0,[],b1,b2);
npoints = 81;
t = linspace(0,tau,npoints);
y = deval(sol,t,1); ydot = deval(sol,t,2);
%
c = (b2*(abs(ydot).^omega).*sign(ydot) + b1*(abs(y).^pnu).*sign(y))/M -
b3;
c(npoints+1) = b2*v0^omega/M - b3;
cc = abs(y)-z1max;
c(npoints+2:2*npoints+1) = cc;

function [f] = odefun(t,y,b1,b2)
global M v0 tau pnu omega z1max
f(1,1) = y(2);
f(2,1) = -b2*abs(y(2))^omega*sign(y(2)) - b1*abs(y(1))^pnu*sign(y(1));
f(2,1)=f(2,1)/M;

function [ymax, ydotmax, yddotmax] = responseinfo(X)
global M v0 tau pnu omega z1max
b1=X(1); b2=X(2);
tspan = [0 tau];
y0 = [0 v0];
sol = ode45(@odefun,tspan,y0,[],b1,b2);
npoints = 80;
t = linspace(0,tau,npoints);
y = deval(sol,t,1); ydot = deval(sol,t,2);
yddot = -b2*(abs(ydot).^omega).*sign(ydot) - b1*(abs(y).^pnu).*sign(y);
ymax = max(abs(y)); ydotmax = max(abs(ydot)); yddotmax = max(abs(yddot));
%plot(t,yddot)
plot(t,y)
hold
```

COMPUTER PROGRAMS

ROSEN, ZOUTEN, GRG, SQP

PROBLEMS

> When using the computer programs: (i) use different starting points, (ii) quote the values of design variables, objective function, active constraints at both the initial point and at the final optimum, and (iii) try to understand the solution using graphical or other means

P5.1. Consider the problem

$$\begin{aligned} \text{minimize} \quad & f = 0.01x_1^2 + x_2^2 \\ \text{subject to} \quad & g_1 = 25 - x_1 x_2 \le 0 \\ & g_2 = 2 - x_1 \le 0 \\ & x_1 \ge 0, x_2 \ge 0 \end{aligned}$$

(i) Obtain the solution graphically [Answer: $\mathbf{x}^* = (15.81, 1.58), f^* = 5$]

(ii) Obtain the solution using KKT conditions as in Example 5.4 (i.e., considering the various "cases").

(iii) Verify the sufficient conditions for optimality at $\mathbf{x}^*$ (see Section 5.6)

P5.2. Is the following problem convex?

$$\begin{aligned} \text{minimize} \quad & f = (x-5)^2 \\ \text{subject to} \quad & x \in \Omega \end{aligned}$$

where $\Omega = \{x : 1 \le x \le 3\} \cup \{x : 3.5 \le x \le 8\}$

P5.3. Is the problem P5.1 above convex?

P5.4. Consider the cost function $f = x_1 + x_2$ and the constraints

$$x_2 \le 150, \quad x_1 + x_2 \le 450, \quad 4x_1 + x_2 \le 900, \quad x_1 \ge 0, \quad x_2 \ge 0$$

Draw the feasible region in $x_1 - x_2$ space. Show the "descent cone" and the "feasible cone" on your figure at the point $\mathbf{x} = (225, 0)^{\mathrm{T}}$. Show also the descent–feasible cone (the intersection of the aforementioned two cones).

P5.5. Consider:

$$\begin{aligned} \text{minimize} \quad & \pi DH + \pi D^2/2 & \\ \text{subject to} \quad & 1500 - \pi D^2 H \le 0 & g_1 \\ & 4.5 - D \le 0 & g_2 \\ & D - 12 \le 0 & g_3 \\ & 9 - H \le 0 & g_4 \\ & H - 18 \le 0 & g_5 \end{aligned}$$

(i) Write down the KKT conditions for the above problem

(ii) Solve the KKT conditions for the following "cases":

(a) Constraints g_1 and g_2 active, rest inactive. Do you have a KKT point?

(b) Constraints g_1 and g_4 active, rest inactive. Do you have a KKT point?

P5.6. Consider the problem: min $f(\mathbf{x})$, subject to $\mathbf{x}^T\mathbf{x} \leq 1$. Under what conditions can you say that a solution *exists* for this specific problem? [see Chapter 1].

P5.7. Sketch the usable–feasible cone in the following figure, at points "A" and "B," respectively. The problem is: {min. f : $\mathbf{x} \in \Omega$}. Contour is tangent at point B.

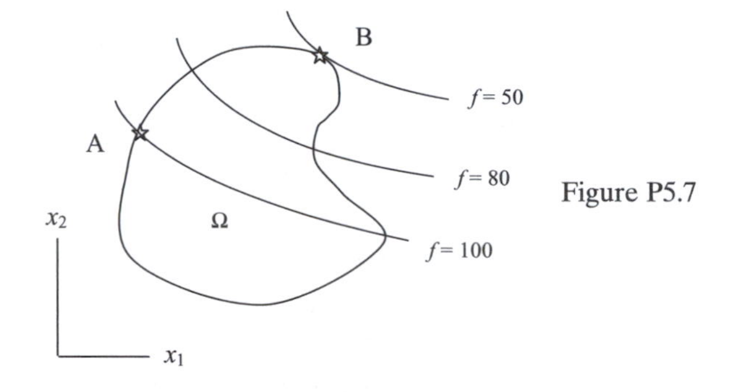

Figure P5.7

P5.8. Consider the problem:

$$\begin{aligned}
\text{minimize} \quad & f = -2x_1 - x_2 \\
\text{subject to} \quad & g_1 \equiv x_2 - 1 \leq 0 \\
& g_2 \equiv 4x_1 + 6x_2 - 7 \leq 0 \\
& g_3 \equiv 10x_1 + 12x_2 - 15 \leq 0 \\
& g_4 \equiv x_1 + 0.25x_2^2 - 1 \leq 0 \\
& g_5 \equiv -x_1 \leq 0 \\
& g_6 \equiv -x_2 \leq 0
\end{aligned}$$

Assume the active set is $\{1, 2\}$. That is $g_1 = g_2 = 0$, rest < 0. Are KKT conditions satisfied with this assumption ?

P5.9. The first constraint in P5.1 is changed to: $g_1 = 26 - x_1\, x_2 \leq 0$. Use the Lagrange multipliers and Eq. (5.21) to estimate the new value of f^*. Compare this with the exact optimum.

P5.10. Verify the Lagrange multipliers for Example 5.1 given in Example 5.12 using Eq. (5.20) with divided differences (that is, relax the constraint by a small amount, resolve for f^*).

P5.11. It is required to solve

$$\begin{aligned}
\text{minimize} \quad & f = \tfrac{1}{2}\mathbf{V}^T\mathbf{B}\mathbf{V} \\
\text{subject to} \quad & V_\beta = 1
\end{aligned}$$

where $\mathbf{V} = (n \times 1)$ vector of nodal velocities in a structural model, $\mathbf{B}$ is a square symmetric matrix of rank $(n - 1)$ and β is some particular node number. Use the method of Lagrange multipliers, and develop a solution procedure to solve this problem. Also, state the requirements that $\mathbf{B}$ must satisfy for f to have a strict local minimum.

P5.12. Use the method of Lagrange multipliers and check for points that satisfy the necessary and sufficient conditions for the problem:

$$\begin{aligned} &\text{minimize} && f = x_1^2 + x_2^2 - 3x_1x_2 \\ &\text{subject to} && x_1^2 + x_2^2 = 6 \end{aligned}$$

P5.13. The minimum weight design problem based on plastic collapse (see P4.15 in Chapter 4 on LP with accompanying figure is more realistically posed as: for the

$$\begin{aligned} &\text{minimize} && 8M_b^{1.15} + 8M_c^{1.15} \\ &\text{subject to} && |M_3| \le M_b \\ &&& |M_4| \le M_b \\ &&& |M_4| \le M_c \\ &&& |M_5| \le M_c \\ &&& |2M_3 - 2M_4 + M_5 - 1120| \le M_c \\ &&& |2M_3 - M_4 - 800| \le M_b \\ &&& |2M_3 - M_4 - 800| \le M_c \\ &&& M_b \ge 0, \quad M_c \ge 0 \\ &&& M_3, M_4, M_5 \text{ unrestricted in sign} \end{aligned}$$

Determine the plastic moment capacities M_b, M_c, and the plastic hinge moments M_3, M_4, M_4 for the above formulation. Compare your solution with the LP solution to P4.11 (Program ROSEN may be used).

P5.14. (Betts Problem): Solve

$$\begin{aligned} &&& n = 3, \quad m = 0, \quad \ell = 2 \\ &\text{minimize} && f = (x_1 - 1)^2 + (x_1 - x_2)^2 + (x_3 - 1)^2 + (x_4 - 1)^4 + (x_5 - 1)^6 \\ &&& h_1 = x_4x_1^2 + \sin(x_4 - x_5) - 2\sqrt{2} = 0 \\ &&& h_2 = x_2 + x_3^4x_4^2 - 8 - \sqrt{2} = 0 \\ &&& -10 \le x_i \le 10, \quad \mathbf{x}_0 = (0, 0, 0, 0). \end{aligned}$$

Note: You may use programs GRG or SQP [Answer: $\mathbf{x}^* = (1.166, 1.182, 1.380, 1.506, 0.610)$, $f^* = 0.24$].

P5.15. For the two constraints in Example 5.1 in the text, use Program GRADIENT in Chapter 1 to compare your analytical derivatives with numerical derivatives.

P5.16. For the two constraints in Example 5.1 in the text, use Matlab to compare your analytical derivatives with numerical derivatives.

P5.17. (Rosen–Suzuki Test Problem) Solve

$$
\begin{aligned}
& n = 4, \quad m = 3, \quad \ell = 0 \\
\text{minimize} \quad & f = x_1^2 + x_2^2 + 2x_3^2 - x_4^2 - 5x_1 - 5x_2 - 21x_3 + 7x_4 + 100 \\
& g_1 = x_1^2 + x_2^2 + x_3^2 + x_4^2 + x_1 - x_2 + x_3 - x_4 - 8 \le 0 \\
& g_2 = x_1^2 + 2x_2^2 + x_3^2 + 2x_4^2 - x_1 - x_4 - 10 \le 0 \\
& g_3 = 2x_1^2 + x_2^2 + x_3^2 + 2x_1 - x_2 - x_4 - 5 \le 0 \\
& -100 \le x_i \le 100, \quad \mathbf{x}_0 = (0, 0, 0, 0)
\end{aligned}
$$

Note: you may use programs ZOUTEN, GRG or SQP.

P5.18. Solve Example 5.1 with all data remaining the same, except: $B = 15$ in.

P5.19. Consider

$$
\begin{aligned}
\text{minimize} \quad & f = (x_1 - x_2)^2 + (x_2 + x_3 - 2)^2 + (x_4 - 1)^2 + (x_5 - 1)^2 \\
\text{subject to} \quad & h_1 = x_1 + 3x_2 = 0 \\
& h_2 = x_3 + x_4 - 2x_5 = 0 \\
& h_3 = x_2 - x_5 = 0 \\
& -10 \le \mathbf{x} \le 10.
\end{aligned}
$$

(i) Do one iteration by hand calculations using the GRG method, starting from $\mathbf{x}_0 = (0,0,0,0,2)^{\mathrm{T}}$. Verify your calculations with that obtained using Program GRG.

(ii) Proceed, using Program GRG, to obtain the optimum solution ($f^* = 4.09$).

P5.20. Repeat P5.19 using SQP method instead of GRG method (for hand calculations, you may use an LP code for the direction finding subproblem).

P5.21. If the sequential QP method, if the solution to the QP in (5.70) is $\mathbf{d} = \mathbf{0}$, show that the current point is a KKT point.

P5.22. (From Box [1966]) Solve, using Programs GRG, ZOUTEN and SQP-PD:

$$
\begin{aligned}
\text{maximize} \quad & f = x_1 x_2 x_3 \\
\text{subject to} \quad & -g_i = x_i \ge 0 \quad i = 1, 2, 3 \\
& -g_i + 3 = 42 - x_i \ge 0 \quad i = 1, 2, 3 \\
& -g_7 = x_1 + 2x_2 + 2x_3 \ge 0 \\
& -g_8 = 72 - (x_1 + 2x_2 + 2x_3) \ge 0 \\
& 0 \le \mathbf{x} \le 100
\end{aligned}
$$

starting with $\mathbf{x}_0 = (10,10,10)^{\mathrm{T}}$. [Answer: $\mathbf{x}^* = (24,12,12)$, $f^* = 3{,}456$]. Verify KKT conditions at the optimum.

P5.23. This problem involves fitting a model to observed data. Consider the model:

$$y = a_1 + a_2 e^{a_3 x},$$

where a_1, a_2, a_3 are our design variables. A set of six observed (x, y) values are given as follows:

Independent variable, x	Dependent variable, y
−5	127
−3	150
−1	379
1	426
2	445
5	420

Determine the best fit using the following strategies:

(a) Least absolute value: minimize $\sum_{i=1}^{6} |y_i - (a_1 + a_2 e^{a_3 x_i})|$

(b) Min-max error: $\underset{\mathbf{a}}{\text{minimize}}\ \underset{i}{\text{maximize}}\ |y_i - (a_1 + a_2 e^{a_3 x_i})|$

For Part (a), read Section 5.4 to see how the above functions can be formulated in standard nonlinear programming form. For Part (b), introduce an artificial variable and then formulate it as a standard NLP problem. Discuss your solution by each of the formulations. That is, compare the average error, maximum error, and provide plots.

P5.24. State, with justification, whether True or False:

(i) A point $\mathbf{x}^*$ that satisfies the KKT conditions for the problem : minimize f subject to $g_i \leq 0$, will also satisfy the KKT conditions for the problem: maximize f subject to $g_i \leq 0$.

(ii) The intersection of the descent cone and the feasible cone at a KKT point is empty.

(iii) The NLP problem

$$\begin{array}{ll} \text{minimize} & f(\mathbf{x}) \\ \text{subject to} & g_i(\mathbf{x}) \leq 0 \quad i = 1, \ldots, m \end{array}$$

is convex if f and all $\{g_i\}$ are convex functions.

(iv) Rosen's gradient projection method can find the global minimum to the optimization problem.

(v) For a convex problem, the KKT conditions are also sufficient in concluding that the point is a global minimum.

(vi) Zoutendijk's method of feasible directions, in theory, can find a local minimum even for nonconvex problems.

(vii) The set $\Omega = \{(x_1, x_2): x_1^2 + x_2 = 1\}$ is a convex set.

P5.25. In truss Example 5.1, compute (postoptimal) sensitivity df^*/dB and df^*/dt (see Example 5.12 for steps)

P5.26. In truss Example 5.1, compute (postoptimal) sensitivity $d\mathbf{x}^*/dB$ and $d\mathbf{x}^*/dt$ (see Example 5.12 for steps)

P5.27. Do a second iteration of the Method of Feasible Directions by hand calculations in Example 5.14.

P5.28. Do a second iteration of the Generalized Reduced Gradient method by hand calculations on Example 5.17 (i.e., a reduced gradient step followed by a constraint correction step).

P5.29. Do one iteration of GRG by hand calculations on Example 5.3, using point B as starting point.

P5.30. Do two additional iterations of the Sequential Quadratic Programming method by hand calculations on Example 5.18.

P5.31. Redo Example 5.20 (linear impact absorber problem) with varying number of discretization points for t.

P5.32 Provide a graphical solution to Example 5.20 (linear impact absorber problem) with a plot of contours of objective and constraint functions in $k - c$ space.

P5.33. Redo Example 5.20 (linear impact absorber problem) with a nonlinear spring and absorber. Specifically, replace the differential equations in Example 5.20 with

$$M\ddot{y} + c|\dot{y}|^{\alpha} sign(\dot{y}) + k|y|^{\beta} sign(y) = 0$$
$$y(0) = 0, \quad \dot{y}(0) = v_0$$

and optimize with the same objective and constraints with:

(i) $\alpha = 2, \beta = 2$.
(ii) with α and β as additional design variables (i.e., in addition to k and c). Use lower bounds equal to 1.0. Can you improve over the linear impact absorber?

P5.34. [From D. L. Bartel, A. I. Krauter, ASME Paper 70-WA/DE-5]. A simplified model of two railroad cars engaging with each other is shown in Fig. P5.34a. Spring and damper constants k_P, M_P and M_a are given and it is required to choose the absorber parameters k_a and c_a so that the initial energy is dissipated as rapidly as possible. Data: $W_a = W_P =$ 2,000 lb, $k_P =$ 24,000 lb/ft, velocity $V_a = -2$ mph $= -2.933$ ft/s, $V_P = 0$, $X_P(0) = X_a(0) = 0$. Optimize for $\varepsilon = 0.01$, $\varepsilon = 0.02$.

Hint: Let denote $E(t) =$ fraction of energy remaining in the system at time t; note that $E \leq 1$. Our goal is to reduce the value of E to, say 0.02 (2%), in minimum time T. Thus, the optimization problem is

$$\begin{aligned} &\text{minimize} && T \\ &\text{subject to} && E = \varepsilon \end{aligned}$$

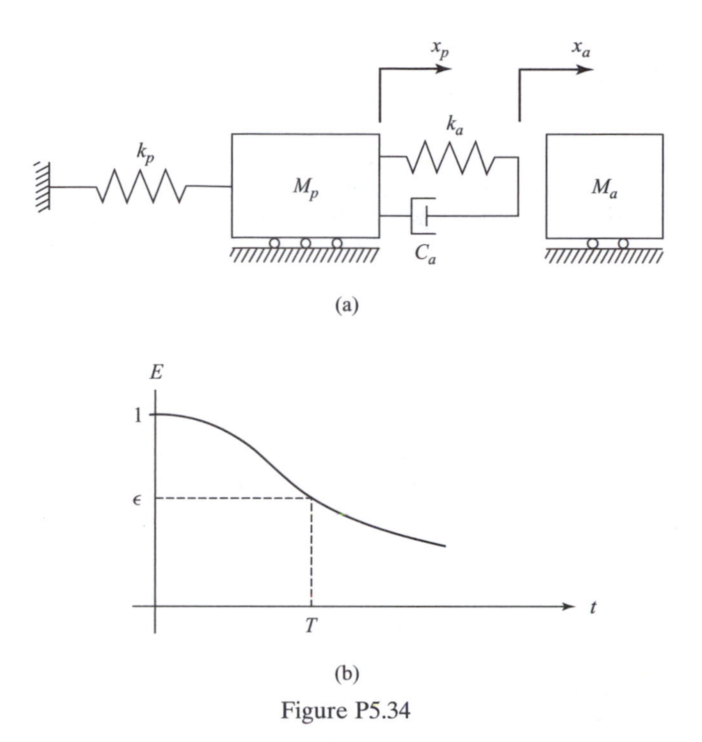

Figure P5.34

Note that E is a monotonically decreasing function of time (Fig. P5.34b). All springs and dampers are linear and there are no external forces on the system. The equations of motion, are:

$$M_P \ddot{X}_P + c_a(\dot{X}_P - \dot{X}_a) + k_P X_P + k_a(X_P - X_a) = 0$$
$$M_a \ddot{X}_a - c_a(\dot{X}_P - \dot{X}_a) - k_a(X_P - X_a) = 0$$

which are solved with the initial conditions. Further, the expression for E is

$$E(t) = \frac{k_P X_P^2 + M_P \dot{X}_P^2 + k_a(X_P - X_a)^2 + M_a \dot{X}_a^2}{k_P X_P(0)^2 + M_P \dot{X}_P(0)^2 + k_a(X_P(0) - X_a(0))^2 + M_a \dot{X}_a(0)^2}$$

Use Matlab for this problem.

P5.35. Research an algorithm based on Sequential Linear Programming (SLP) and develop a computer code.

REFERENCES

Arora, J.S., *Introduction to Optimum Design*, McGraw-Hill, New York 1989.

Betts, J.T., An accelerated multiplier method for nonlinear programming, *Journal of Optimization Theory and Applications*, **21** (2), pp. 137–174, 1977.

Box, M.J., A comparison of several current optimization methods and the use of transformations in constrained problems, *Computer Journal*, **9**, 67–77, 1966.

Fox, R.L., Computers in optimization and design, ASME Paper 74-DE-31. See also: *Optimization Techniques for Engineering Design*, Addison-Wesley, Reading, MA, 1971.

Haftka, R.T., *Elements of Structural Optimization*, 3rd edition, Kluwer Academic Publishers, Dordrecht, 1992.

Han, S.-P., Superlinearly convergent variable metric algorithms for general nonlinear programming problems, *Mathematical Programming*, **11**, 263–282, 1976.

Haug, E.J. and Arora, J.S., *Applied Optimal Design*, Wiley, New York, 1979.

Karush, W., Minima of functions of several variables with inequalities as side conditions, MS Thesis, Department of Mathematics, University of Chicago, Chicago, IL, 1939.

Kuhn, H.W. and Tucker, A.W., "Non-linear programming," in J. Neyman (Ed.), *Proceedings of the Second Berkeley Symposium on Mathematical Statistics and Probability*, University of California Press, Berkeley, 1951, pp. 481–493.

Lanczos, C., *The Variational Principles of Mechanics*, 4th Edition, University of Toronto Press, Toronto, 1970.

Lasdon, L., Plummer, J., and Warren, A.D., "Nonlinear programming," in M. Avriel and B. Golany (Eds.), *Mathematical Programming for Industrial Engineers*, Marcel Dekker, New York, 1996.

Luenberger, D.G., *Introduction to Linear and Nonlinear Programming*, Addison-Wesley, Reading, MA, 1973.

Powell, M.J.D., "A fast algorithm for nonlinearly constrained optimization calculations," in G.A. Watson (ed.), *Numerical Analysis, Lecture Notes in Mathematics*, Vol. 630, Springer, Berlin, 1978.

Pshenichny, B.N. and Danilin, Y.M., *Numerical Methods in Extremal Problems*, 2nd Printing, Mir Publishers, Moscow, 1982.

Rao, S.S., *Engineering Optimization*, 3rd Edition, Wiley, New York, 1996.

Rosen, J., The gradient projection method for nonlinear programming: I. Linear constraints, *Journal of the Society for Industrial and Applied Mathematics*, **8**, 181–217, 1960.

Rosen, J., The gradient projection method for nonlinear programming: II. Non-Linear constraints, *Journal of the Society for Industrial and Applied Mathematics*, **9**, 514–532, 1961.

Schittkowski, K., On the convergence of a sequential quadratic programming method with an augmented Lagrangian line search function, *Optimization, Mathematische Operationsforschung und Statistik*, **14**, 197–216, 1983.

Siddall, J.N., *Analytical Decision-Making in Engineering Design*, Prentice-Hall, Englewood Cliffs, NJ, 1972.

Vanderplaats, G.N., *Numerical Optimization Techniques for Engineering Design: With Applications*, McGraw-Hill, New York, 1984.

Zoutendijk, G., *Methods of Feasible Directions*, Elsevier, New York, 1960.

6

Penalty Functions, Duality, and Geometric Programming

6.1 Introduction

In this chapter, we present the following methods: (1) Penalty functions – interior and exterior, (2) Duality concepts and their application to separable problems, (3) the Augmented Lagrangian method, and (4) Geometric Programming. In the first three methods, a sequence of *unconstrained* minimizations is performed. Thus, simple algorithms based on conjugate directions or quasi-Newton updates can be used at each step, which makes these methods attractive to implement. However, each unconstrained minimization can itself be a big hurdle unless the problem has some special feature such as involving quadratic functions or separability, which we can exploit. In the absence of any special features, the methods in Chapter 5 are generally more efficient. On the other hand, the methods in this chapter are robust even for large problems and require very few algorithmic parameters to tune.

Again, we consider the general nonlinear programming (NLP) problem

$$\begin{array}{lll} \text{minimize} & f(\mathbf{x}) & \\ \text{subject to} & g_i(\mathbf{x}) \le 0 & i = 1, \ldots, m \\ \text{and} & h_j(\mathbf{x}) = 0 & j = 1, \ldots, \ell \end{array} \tag{6.1}$$

where $\mathbf{x} = (x_1, x_2, \ldots, x_n)^{\mathrm{T}}$ is a column vector of n real-valued design variables. In the preceding, f is the *objective* or *cost* function, g are *inequality constraints* and h are *equality constraints*. We denote the feasible region Ω as: $\Omega = \{\mathbf{x} : \mathbf{g} \le \mathbf{0}, \mathbf{h} = \mathbf{0}\}$.

6.2 Exterior Penalty Functions

To illustrate the concept, consider a simple problem involving only one variable, x:

$$\begin{array}{ll} \text{minimize} & f(x) \\ \text{subject to} & g(x) \equiv x - 5 \le 0 \end{array}$$

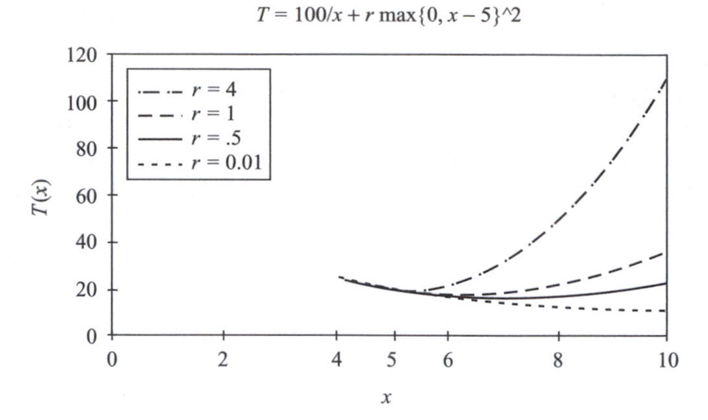

Figure 6.1. Example of quadratic penalty function.

where $f(x)$ is a monotonically decreasing function of x as depicted in Fig. 6.1 ($x^* = 5$ is evidently the solution). We will now define a penalty function $P = [\max(0, g(x))]^2$, or

$$P(x) = \max(0, x - 5)^2$$

We see that the preceding "quadratic loss" function has a positive value only when the constraint in (b) is violated. Now, let us define a composite function

$$T(x) = f(x) + r\,P(x)$$

where $r > 0$ is called a "penalty parameter." The parameter r controls the degree of penalty for violating the constraint. Figure 6.1 shows a sketch of the function $T(x)$ for various values of r for a sample problem.

From the figure, we may make a key observation: the *unconstrained* minimization of $T(x)$, for increasing values of r, results in a sequence of points that converges to the minimum of $x^* = 5$ from the *outside* of the feasible region. Since convergence is from the outside of the feasible region, we call these "exterior point" methods. The idea is that if x strays too far from the feasible region, the penalty term $r_i P(x)$ becomes large when r_i is large. As $r_i \to \infty$, the tendency will be to draw the unconstrained minima towards the feasible region so as to reduce the value of the penalty term. That is, a large penalty is attached to being infeasible.

Further, we have $\frac{dT}{dx} = \frac{df}{dx} + 2r\,\max(0, x - 5)$. Comparing this with the KKT optimality condition $\frac{dL}{dx} = \frac{df}{dx} + \mu\frac{dg}{dx}$, we obtain a Lagrange multiplier *estimate* as

$$\mu \approx 2r\,\max(0, x - 5)$$

However, there is a catch to the simplicity of this method. Firstly, the initial choice of r is not easy. Secondly, for large r, the function $T(x)$ is very nonlinear. Finding its minimum directly from *any* starting point x^0 is very difficult, unless $T(x)$ is of simple structure (such as being a quadratic function). Thus, the procedure is to minimize $T(x)$ with, say, $r_1 = 10$ and obtain a point x^1. We use x^1 as a starting guess for the minimum of $T(x)$ with $r_2 = 100$ and obtain x^2. We then use x^2 as a starting point to obtain the minimum of $T(x)$ with $r_3 = 1000$, and so on. In actual computation, we stop the process when the constraint violation is small and when changes in f with the resulting x^i are insignificant. Theoretically, we arrive at x^* in the limit as r tends to infinity.

Procedure

1. For some $r_1 > 0$, find an unconstrained local minimum of $T(\mathbf{x}, r_1)$. Denote the solution by $\mathbf{x}(r_1)$.
2. Choose $r_2 > r_1$ and using $\mathbf{x}(r_1)$ as a starting guess, minimize $T(\mathbf{x}, r_2)$. Denote the solution as $\mathbf{x}(r_2)$.
3. Proceed in this fashion, minimizing $T(\mathbf{x}, r_i)$ for a strictly monotonically increasing sequence $\{r_i\}$.

For the general problem in (6.1), the quadratic penalty function P is developed as

$$P(\mathbf{x}) = \sum_{i=1}^{m} \{\max[0,\ g_i(\mathbf{x})]\}^2 + \sum_{j=1}^{k} [h_j(\mathbf{x})]^2 \tag{6.2}$$

and the composite function $\psi(\mathbf{x})$ is given by

$$T(\mathbf{x}) = f(\mathbf{x}) + r\,P(\mathbf{x}) \tag{6.3}$$

and the Lagrange multiplier estimates are

$$\mu_i \approx 2r\max[0, g_i], \quad \lambda_j \approx 2rh_j \tag{6.4}$$

Example 6.1

Consider the following example from the classical reference on penalty function methods [Fiacco and McCormick 1990]

$$\begin{aligned} \text{minimize} \quad & f = -x_1 x_2 \\ \text{subject to} \quad & g_1 = x_1 + x_2^2 - 1 \le 0 \\ & g_2 = -x_1 - x^2 \le 0 \end{aligned}$$

The composite function ψ is given by (see Eq. 6.2):

$$T(\mathbf{x}, r) = -x_1 x_2 + r\left\{\max\left[0, x_1 + x_2^2 - 1\right]\right\}^2 + r\{\max[0, -x_1 - x_2]\}^2$$

The minimum of T for various values of r leads to points $\mathbf{x}(r)$ as given in Table E6.1. The path of $\mathbf{x}(r)$ to the optimum is shown in Fig. E6.1

Table E6.1. *Solution using quadratic penalty function.*

Iteration	Starting point, $\mathbf{x}^0$	r	$\mathbf{x}^*$
0	(1, 5)	1	(−0.212, 1.163)
1	(−0.212, 1.163)	10	(0.686, 0.586)
2	(0.686, 0.586)	100	(0.669, 0.578)
3	(0.669, 0.578)	1000	(0.667, 0.577)

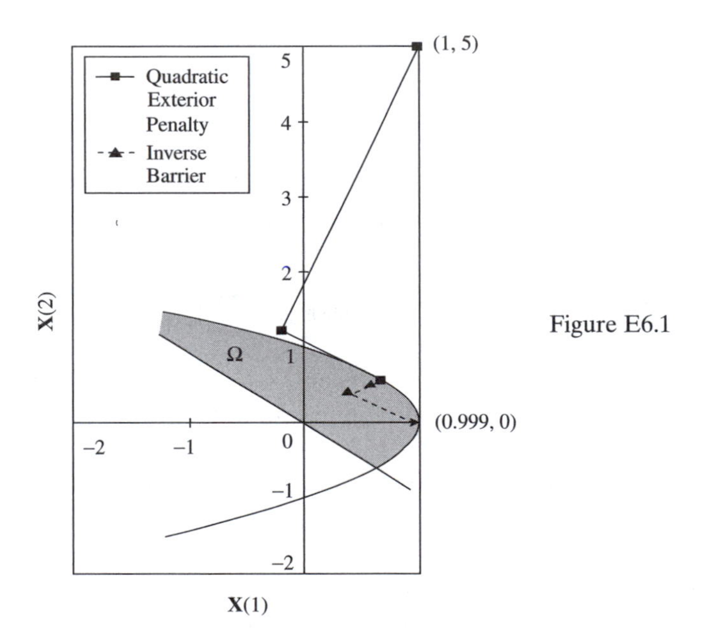

Figure E6.1

A second example will now be given wherein the functions have simple structure for which a single choice of the penalty parameter r provides a good enough answer. That is, no iteration is involved. Specifically, in this example, $f(\mathbf{x})$ is a quadratic function and the equality constraints are linear functions. The function $T(\mathbf{x})$ is then a quadratic function, and the optimality conditions result in a set of simultaneous equations.

Example 6.2
Consider the spring system shown in Fig. E6.2.

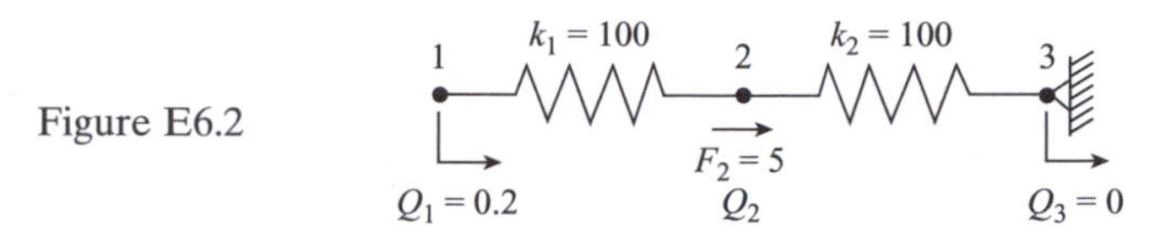

Figure E6.2

The equilibrium state is determined by solving:

$$\begin{aligned} \text{minimize} \quad & \pi = \tfrac{1}{2}100(Q_2 - Q_1)^2 + \tfrac{1}{2}100(Q_3 - Q_2)^2 - 5Q_2 \\ \text{subject to} \quad & Q_1 = 0.2 \\ & Q_3 = 0 \end{aligned}$$

where π represents the potential energy in a spring system and Q_i are displacements of the nodes in the system. In view of (6.2), our task reduces to solving the unconstrained problem

$$\text{minimize } T = \tfrac{1}{2}100(Q_2 - Q_1)^2 + \tfrac{1}{2}100(Q_3 - Q_2)^2 - 5\,Q_2 + \tfrac{1}{2}C(Q_1 - 0.2)^2 + \tfrac{1}{2}C\,Q_3^2$$

where the factor of ½ in front of the penalty parameter C has been added for convenience. Let us choose $C = 10^5$. Our solution, obtained by setting $\frac{\partial T}{\partial Q_i} = 0$, is

$$\mathbf{Q} = [0.1999251, 0.125025, 1.249001\text{E} - 04]$$

Note that the solution Q_1 is almost its specified value of 0.2. In fact, the difference is important in calculating Lagrange multiplier estimates as:

$$\begin{aligned} \lambda_1 &\approx (2)(\tfrac{1}{2})(10^5)\max(0, 0.1999251 - 0.2) = -7.490814 \\ \lambda_2 &\approx 10^5 \max(0, 1.249001\text{E} - 04) = 12.49001 \end{aligned}$$

This Lagrange multiplier estimates represent the negative reaction forces exerted on nodes 1 and 3 to enforce the specified displacements $Q_1 = 0.2$ and $Q_3 = 0$, respectively. In this example, C can be interpreted as the stiffness of a very stiff spring attached to node 1.

The above example is but a special case of the problem in finite element analysis where the goal is to minimize potential energy ($\pi = ½\,\mathbf{Q}^{\mathrm{T}}\mathbf{K}\mathbf{Q} - \mathbf{Q}^{\mathrm{T}}\mathbf{F}$) (see Fig. 3.7). Constraints are linear and take the form $a_1 Q_I + a_2 Q_J = a_0$. The constraints represent specified displacements as in the preceding example, or inclined rollers, press fits, rigid connections, etc. Matrices **K** and **F** refer to structural stiffness and force, respectively. Use of the quadratic exterior penalty function leads to the solution of

$$\hat{\mathbf{K}}\,\mathbf{Q} = \hat{\mathbf{F}}$$

where $\hat{\mathbf{K}}$ and $\hat{\mathbf{F}}$ are formed simply by modifying the original **K** and **F** matrices in Ith and Jth row and column locations as

$$\begin{bmatrix} K_{I,I} & K_{I,J} \\ K_{J,I} & K_{J,J} \end{bmatrix} \Rightarrow \begin{bmatrix} K_{I,I} + Ca_1^2 & K_{I,J} + Ca_1 a_2 \\ K_{J,I} + Ca_1 a_2 & K_{J,J} + Ca_2^2 \end{bmatrix} \text{ and } \begin{bmatrix} F_I \\ F_J \end{bmatrix} \Rightarrow \begin{bmatrix} F_I + Ca_0 a_1 \\ F_J + C\, a_0\, a_2 \end{bmatrix}$$

The penalty approach is easy to implement in programs, and is especially suited for classroom use [Chandrupatla and Belegundu 1997]. Finally, by expanding the preceding matrix equation, we can see that C must be chosen so as to be much larger than any element of **K**. Typically, $C = 10^5 \cdot \max_{I,J} |K_{I,J}|$. In this case, a single choice of the penalty parameter r and subsequent minimization of T provide a fairly accurate answer.

Convergence Aspects

Penalty functions are not restricted to be in the form (6.2). They can be of any form as long as they satisfy: (i) $P(\mathbf{x})$ is a continuous function, and (ii) $P(\mathbf{x}) = 0$ if $\mathbf{x} \in \Omega$ (that is, if $\mathbf{x}$ is feasible), and $P(\mathbf{x}) > 0$ otherwise. Importantly, for convergence, it is not necessary that the function f, g_i, and h_j be differentiable – we can use a zero-order method such as Hooke–Jeeves to minimize $T(\mathbf{x}, r)$ with $r = r_i$. Zero-order methods are discussed in Chapter 7. *The main drawback in penalty-function methods is that convergence is guaranteed only when the function* $T(\mathbf{x}, r)$ *is minimized exactly*. This is almost impossible for general large-scale problems that do not possess any special structure. Furthermore, the Hessian of $T(\mathbf{x}, r)$ becomes increasingly ill-conditioned as r_i increases [Luenberger 1973]. From Chapter 3, we know that this implies poor rate of convergence for gradient methods, such as the method of Fletcher–Reeves.

Finally, we should not use second-order gradient methods (e.g., pure Newton's method) with the quadratic loss penalty function with inequality constraints, since the Hessian is discontinuous. To see this clearly, consider the example illustrated in Fig. 6.1: minimize $f(x)$, subject to $g \equiv x - 5 \leq 0$, with $f(x)$ being a monotonically decreasing function of x. At the optimum $x^* = 5$, the gradient of $P(x)$ is $\frac{dP}{dx} = 2r\ \max(0,\ x - 5)$. Regardless of whether we approach x^* from the left or from the right, the value of dP/dx at x^* is zero. So, $P(x)$ is first-order differentiable ($P \in C^1$). However, $\frac{d^2P}{dx^2} = 0$ when approaching x^* from the left, while $\frac{d^2P}{dx^2} = 2r$ when approaching from the right. Thus, the penalty function is not second-order differentiable at the optimum.

Lower and Upper Limits

In some engineering problems, lower and upper limits and certain other geometry constraints (usually linear) must be satisfied at all times during the iterative process – not just in the limit of the process. In such cases, we can include these explicitly during minimization of T and exclude them from the penalty function as

$$\begin{aligned} &\text{minimize} \quad T(\mathbf{x}, \mathbf{r}_i) \\ &\text{subject to} \quad \mathbf{x}^{\mathrm{L}} \leq \mathbf{x} \leq \mathrm{x}^{\mathrm{U}} \end{aligned} \tag{6.5}$$

6.3 Interior Penalty Functions

In interior point or "barrier" methods, we approach the optimum from the interior of the feasible region. Only inequality constraints are permitted in this class of methods. Thus, we consider the feasible region to be defined by $\Omega = \{\boldsymbol{x} : g_i(\boldsymbol{x}) \leq 0, i = 1, \ldots, m\}$. However, Ω can be nonconvex. Interior penalty functions (or barrier functions), $B(\mathbf{x})$, are defined with the properties: (i) B is continuous, (ii) $B(\mathbf{x}) \geq 0$, (iii) $B(\mathbf{x}) \to \infty$ as $\mathbf{x}$ approaches the boundary of Ω. Two popular choices are:

$$\text{Inverse Barrier Function:} \quad B(\mathbf{x}) = -\sum_{i=1}^{m} \frac{1}{g_i(\mathbf{x})} \tag{6.6}$$

$$\text{Log Barrier Function:} \quad B(\mathbf{x}) = -\sum_{i=1}^{m} \log[-g_i(\mathbf{x})] \tag{6.7}$$

In (6.7), we should define $\log(-g_i) = 0$ when $g_i(\mathbf{x}) < -1$ to ensure that $B \geq 0$. This is not a problem when constraints are expressed in normalized form, in which case $g < -1$ implies a highly inactive constraint, which will not play a role in determining the optimum. For instance, we should express the constraint $x \leq 1000$ as $g \equiv x / 1000 - 1 \leq 0$. The T-function is defined as

$$T(\mathbf{x}, r) = f(\mathbf{x}) + \frac{1}{r} B(\mathbf{x}) \tag{6.8}$$

Starting from a point $\mathbf{x}^0 \in \Omega$, and a $r > 0$, the minimum of T will, in general, provide us with a new point, which is in Ω, since the value of T becomes infinite on the boundary. As r is increased, the weighting of the penalty term $\frac{1}{r} B(\mathbf{x})$ is decreased. This allows further reduction in $f(\mathbf{x})$ while maintaining feasibility.

Let us consider the problem: minimize $f(x)$, subject to $g \equiv x - 5 \leq 0$, where $f(x)$ is a monotonically decreasing function of x. Figure 6.2 shows a sketch of a sample function $T(x)$ for increasing values of r. As r increases, the function T matches the

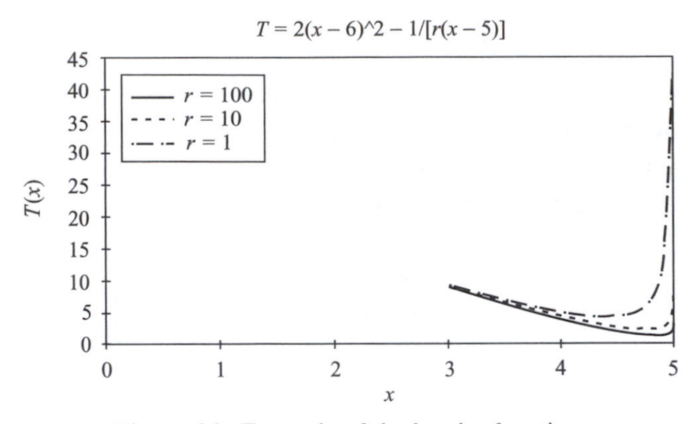

Figure 6.2. Example of the barrier function.

original cost function *f* more closely except near the boundary where *g* approaches zero.

We also see that the unconstrained minimization of $T(x)$, for increasing values of r, results in a sequence of points that converges to the minimum of $x^* = 5$ from the *inside* of the feasible region. Hence, the name "interior point" methods.

Example 6.3

The problem in Example 6.1 is solved using the inverse barrier method. The path of $\mathbf{x}(r)$ to the optimum is shown in Fig. E6.1.

Computational Aspects with Interior Point Methods

During minimization of $T(\mathbf{x}, r_i)$, we have to ensure that the minimum point does not fall outside Ω. If this happens, then the algorithm fails. Here, we use the method of Fletcher-Reeves to minimize T. The line search problem involves minimization of $T(\mathbf{x}(\alpha), r_i)$, where $\mathbf{x}(\alpha) = \mathbf{x}_k + \alpha \mathbf{d}_k$. To ensure feasibility, we perform the line search in the interval $0 < \alpha < \alpha_{\mathrm{U}}$, where α_{U} is the step to the nearest boundary defined by $g_{\max} = -10^{-6}$, where $g_{\max}$ is the maximum of $\{g_1, g_2, \ldots, g_m\}$. More specifically, in the computer program, we accept as α_{U} any point for which $-10^{-4} < g_{\max} < -10^{-6}$.

The user has to supply an initial feasible point $\mathbf{x}^0$. As noted in Chapter 5, this may be obtained by using the exterior point method on the problem

$$\text{minimize } \hat{f} = \sum_{j=1}^{m} \max(0, g_j + \varepsilon)^2$$

where $\varepsilon > 0$ is a small positive scalar.

Finally, there are various extensions of penalty function methods, such as mixed interior-exterior methods [Fiacco and McCormick 1990] and extended interior point methods [Haftka and Gurdal 1992]. Also, the classical interior point penalty function methods presented here will give the reader the background to pursue the newly developed "path following" interior point methods that hold great promise for engineering problems.

6.4 Duality

We have used duality in linear programming in Chapter 4. In the next two sections, and also in Chapter 12, on structural optimization, we will develop optimization techniques that use duality concepts to advantage. In this section, we present the basic concepts. We will assume only local convexity in the following treatment of duality. However, if the problem is known to be convex (that is, f and g_i are convex and h_j are linear), then the results are still valid except that the stronger "global" operations can replace the weaker "local" operations.

Firstly, we assume that all functions are twice continuously differentiable and that a solution $\mathbf{x}^*$, $\boldsymbol{\mu}^*$, $\boldsymbol{\lambda}^*$ of the "primal" problem in (6.1) is a regular point that satisfies the KKT necessary conditions as given in the previous chapter. Further, in the sufficient conditions for a minimum to problem (6.1), we required only that Hessian of the Lagrangian ($\nabla^2 L(\mathbf{x}^*, \boldsymbol{\mu}^*, \boldsymbol{\lambda}^*)$) be positive definite on the tangent subspace (Chapter 5). Now, to introduce duality, we make the *stronger* assumption that the Hessian of the Lagrangian $\nabla^2 L\ (\mathbf{x}^*)$ is positive definite. This is a stronger assumption because we require positive definiteness of the Hessian at $\mathbf{x}^*$ on the whole space R^n and not just on a subspace. This is equivalent to requiring L to be locally convex at $\mathbf{x}^*$. Under these assumptions, we can state the following *equivalent* theorems. The reader can see the proofs in Wolfe [1961] and Mangasarian [1969], for example.

(1) $\mathbf{x}^*$, together with $\boldsymbol{\mu}^*$ and $\boldsymbol{\lambda}^*$, is a solution to the primal problem (6.1)
(2) $\mathbf{x}^*$, $\boldsymbol{\mu}^*$, $\boldsymbol{\lambda}^*$ is a saddle point of the Lagrangian function $L(\mathbf{x}, \boldsymbol{\mu}, \boldsymbol{\lambda})$. That is,

$$L(\mathbf{x}^*, \boldsymbol{\mu}, \boldsymbol{\lambda}) \leq L(\mathbf{x}^*, \boldsymbol{\mu}^*, \boldsymbol{\lambda}^*) \leq L(\mathbf{x}, \boldsymbol{\mu}^*, \lambda^*) \tag{6.9}$$

in the neighborhood of $\mathbf{x}^*$, $\boldsymbol{\mu}^*$, $\boldsymbol{\lambda}^*$. A saddle function is shown in Fig. 6.3.

(3) $\mathbf{x}^*$, $\boldsymbol{\mu}^*$, $\boldsymbol{\lambda}^*$ solves the dual problem

$$\text{maximize} \quad L = f(\mathbf{x}) + \boldsymbol{\mu}^{\mathrm{T}}\mathbf{g}(\mathbf{x}) + \boldsymbol{\lambda}^{\mathrm{T}}\mathbf{h}(\mathbf{x}) \tag{6.10a}$$

$$\text{subject to} \quad \nabla L = \nabla f(\mathbf{x}) + [\nabla g(\mathbf{x})]\boldsymbol{\mu} + [\nabla \mathbf{h}(\mathbf{x})]\boldsymbol{\lambda} = \mathbf{0} \tag{6.10b}$$

$$\boldsymbol{\mu} \geq \mathbf{0} \tag{6.10c}$$

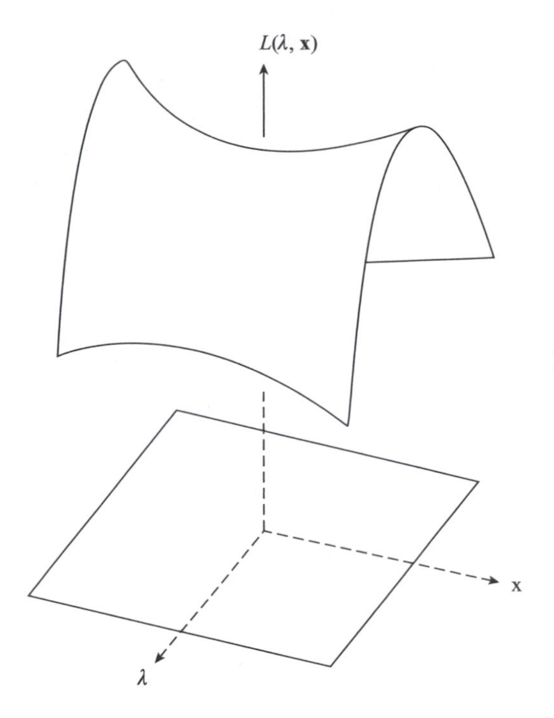

Figure 6.3. Saddle point condition.

where both $\mathbf{x}$ and Lagrange multipliers $\boldsymbol{\mu}$, $\boldsymbol{\lambda}$ are variables in the preceding problem. Further, the two extrema are equal: $L(\mathbf{x}^*, \boldsymbol{\mu}^*, \boldsymbol{\lambda}^*) = f(\mathbf{x}^*)$.

The following two corollaries are also true:

(a) Any $\mathbf{x}$ that satisfies the constraints in (6.1) is called primal-feasible and any $\mathbf{x}$, $\boldsymbol{\mu}$, $\boldsymbol{\lambda}$ that satisfies (6.10b–c) is called dual-feasible. Obviously, we know that any primal-feasible point provides us with a upper bound on the optimum cost:

$$f(\mathbf{x}^*) \leq f(\mathbf{x}) \quad \text{for any } \mathbf{x} \text{ that is primal-feasible.}$$

It can be shown that any dual-feasible point provides a lower bound on the optimum cost as

$$L(\mathbf{x}, \boldsymbol{\mu}, \boldsymbol{\lambda}) \leq f(\mathbf{x}^*) \quad \text{for any } \mathbf{x}, \boldsymbol{\mu}, \boldsymbol{\lambda} \text{ that is dual-feasible.} \tag{6.11}$$

It is important to realize that the lower bound is only on the local minimum of f in the neighborhood.

(b) If the dual objective is unbounded, then the primal has no feasible solution.

The Dual Function $\Phi\,(\mu, \lambda)$

The condition in (6.10b), namely, $\nabla L = \nabla f(\mathbf{x}) + \nabla \mathbf{g}(\mathbf{x})\,\boldsymbol{\mu} + \nabla \boldsymbol{h}(\mathbf{x})\boldsymbol{\lambda} = \mathbf{0}$ can be viewed as equations involving two sets of variables, $\mathbf{x}$ and $[\boldsymbol{\mu}, \boldsymbol{\lambda}]$. The Jacobian of the preceding system with respect to the $\mathbf{x}$ variables is $\nabla^2 L$. Since this matrix is nonsingular at $\mathbf{x}^*$ (by our assumption that it is positive definite), we can invoke the Implicit Function Theorem that states that in a small neighborhood of $[\mathbf{x}^*, \boldsymbol{\mu}^*, \boldsymbol{\lambda}^*]$, such that for $\boldsymbol{\mu}, \boldsymbol{\lambda}$ in this neighborhood, $\mathbf{x} = \mathbf{x}(\boldsymbol{\mu}, \boldsymbol{\lambda})$ is a differentiable function of $\boldsymbol{\mu}, \boldsymbol{\lambda}$ with $\nabla L\,(\mathbf{x}(\boldsymbol{\mu}, \boldsymbol{\lambda}), \boldsymbol{\mu}, \boldsymbol{\lambda}) = \mathbf{0}$. This allows us to treat L as an implicit function of $\boldsymbol{\mu}, \boldsymbol{\lambda}$ as $L = L(\mathbf{x}(\boldsymbol{\mu}, \boldsymbol{\lambda}), \boldsymbol{\mu}, \boldsymbol{\lambda})$, which we will denote as the "dual function" $\phi\,(\boldsymbol{\mu}, \boldsymbol{\lambda})$. Further, since $\nabla^2 L$ will remain positive definite near $\mathbf{x}^*$, we can define $\phi\,(\boldsymbol{\mu}, \boldsymbol{\lambda})$ from the minimization problem

$$\varphi(\mu, \lambda) = \underset{\mathbf{x}}{\text{minimum}} \;\; f(\mathbf{x}) + \boldsymbol{\mu}^{\mathrm{T}}\mathbf{g}(\mathbf{x}) + \boldsymbol{\lambda}^{\mathrm{T}}\mathbf{h}(\mathbf{x}) \tag{6.12}$$

where $\boldsymbol{\mu} \geq \mathbf{0}$, $\boldsymbol{\lambda}$ are near $\boldsymbol{\mu}^*, \boldsymbol{\lambda}^*$, and where the minimization is carried out with respect to $\mathbf{x}$ near $\mathbf{x}^*$.

Example 6.4

Consider Example 5.4:

$$\begin{aligned} &\text{minimize} && f = (x_1 - 3)^2 + (x_2 - 3)^2 \\ &\text{subject to} && 2x_1 + x_2 - 2 \leq 0 \quad \equiv g_1 \\ &&& -x_1 \leq 0 \quad \equiv g_2 \\ &&& -x_2 \leq 0 \quad \equiv g_3 \end{aligned}$$

We have $L = (x_1 - 3)^2 + (x_2 - 3)^2 + \mu_1(2x_1 + x_2 - 2) - \mu_2\, x_1 - \mu_3\, x_2$. It is evident that $\nabla^2 L$ is positive definite. Setting $\partial L/\partial x_1 = 0$ and $\partial L/\partial x_2 = 0$, we obtain

$$x_1 = \frac{\mu_2}{2} - \mu_1 + 3, \quad x_2 = \frac{\mu_3}{2} - \frac{\mu_1}{2} + 3 \tag{a}$$

Substituting into L, we obtain the dual function

$$\varphi\,(\mu) = -\frac{5}{4}\,\mu_1^2 - \frac{\mu_2^2}{4} - \frac{\mu_3^2}{4} + \mu_1\,\mu_2 + \frac{\mu_1\,\mu_3}{2} + 7\,\mu_1 - 3\,\mu_2 - 3\,\mu_3 \tag{b}$$

Gradient and Hessian of $\phi(\mu, \lambda)$ in Dual Space

The gradient of the dual function ϕ is

$$\frac{d\varphi}{d\mu_j} = g_j[\mathbf{x}(\boldsymbol{\mu}, \boldsymbol{\lambda})] \tag{6.13a}$$

$$\frac{d\varphi}{d\lambda_j} = h_j[\mathbf{x}(\boldsymbol{\mu}, \boldsymbol{\lambda})] \tag{6.13b}$$

The proof follows from: $\frac{d\varphi}{d\mu_j} = \frac{dL(\mathbf{x}(\boldsymbol{\mu},\boldsymbol{\lambda}),\boldsymbol{\mu},\boldsymbol{\lambda})}{d\mu_j} = \frac{\partial L}{\partial \mu_j} + \nabla L \,.\, \frac{d\mathbf{x}}{d\mu_j} = \frac{\partial L}{\partial \mu_j} = g_j$ in view of the fact that $\nabla L = \mathbf{0}$ by the very construction of ϕ.

The Hessian of ϕ in μ, λ – space is given by

$$\nabla_{\mathrm{D}}^2 \varphi = -\mathbf{A}[\nabla^2 L(\mathbf{x}^*)]^{-1}\mathbf{A}^{\mathrm{T}} \tag{6.13c}$$

where the rows of $\mathbf{A}$ are the gradient vectors of the constraints. Since we assume that $\mathbf{x}^*$ is a regular point, $\mathbf{A}$ has full row rank and, hence, the dual Hessian is negative definite.

The Dual Problem in Terms of the Dual Function

The dual problem in (6.10) can be written in an alternative form using the dual function as

$$\boxed{\underset{\boldsymbol{\mu} \geq 0,\, \boldsymbol{\lambda}}{\textbf{maximize}}\ \varphi(\boldsymbol{\mu},\boldsymbol{\lambda})} \tag{6.14}$$

Computationally, the maximization problem in (6.14) can be carried out using a gradient method like Fletcher–Reeves, BFGS, or Newton's method.

Example 6.5

Let us continue with Example 6.4. We will first illustrate (6.10). We have,

$$\begin{aligned}
\frac{\partial \varphi}{\partial \mu_1} &= -\frac{5}{2}\,\mu_1 + \mu_2 + \frac{\mu_3}{2} + 7\\
\frac{\partial \varphi}{\partial \mu_2} &= -\frac{\mu_2}{2} + \mu_1 - 3\\
\frac{\partial \varphi}{\partial \mu_3} &= -\frac{\mu_3}{2} + \frac{\mu_1}{2} - 3
\end{aligned}$$

In view of $x_1 = \frac{\mu_2}{2} - \mu_1 + 3$, $\quad x_2 = \frac{\mu_3}{2} - \frac{\mu_1}{2} + 3$, we have $\frac{\partial \varphi}{\partial \mu_i} = g_i$. Moreover, the dual problem is

$$\max\ \varphi(\mu) = -\frac{5}{4}\,\mu_1^2 - \frac{\mu_2^2}{4} - \frac{\mu_3^2}{4} + \mu_1\,\mu_2 + \frac{\mu_1\,\mu_3}{2} + 7\,\mu_1 - 3\,\mu_2 - 3\,\mu_3$$

with the restriction that $\mu_i \geq 0, i = 1, 2, 3$. The solution is $\mu_1^* = 2.8$, $\mu_2^* = \mu_3^* = 0$. Correspondingly, we obtain $x_1(\mu^*) = 0.2$ and $x_2(\mu^*) = 1.6$.

Duality Applied to Separable Problems

A problem of the form

$$\begin{aligned} \text{minimize} \quad & f = \sum_{i=1}^{q} f_i\,(x_i) \\ \text{subject to} \quad & g = \sum_{i=1}^{q} g_i(x_i) \le 0 \\ & h = \sum_{i=1}^{q} h_i\,(x_i) = 0 \end{aligned} \tag{6.15}$$

is called "separable" because the expression for each function can be written as a sum of functions, each function in the summation depending only on one variable. For example,

$$\begin{aligned} \text{minimize} \quad & f = x_1 + e^{x_2} \\ \text{subject to} \quad & \sin x_1 + x_2^3 \le 2 \\ & -x_1 \le 0 \\ & -x_2 \le 0 \end{aligned}$$

is a separable problem. There can be several g's and h's in (6.15), but this generality has been suppressed for notational simplicity. However, there are several variables x_i in (6.15). Dual methods are particularly powerful on separable problems. To see why, we write the dual function given by (6.12) for problem (6.15) is

$$\varphi\,(\mu, \lambda) = \underset{\mathbf{x}}{\text{minimum}} \left\{ \sum_{i=1}^{q} f_i\,(x_i) + \mu \sum_{i=1}^{q} g_i\,(x_i\,) + \lambda \sum_{i=1}^{q} h_i\,(x_i\,) \right\}$$

This minimization problem decomposes into the q separate *single variable* minimization problems

$$\underset{x_i}{\text{minimum}} \;\; f_i\,(x_i) + \mu\, g_i\,(x_i) + \lambda\, h_i\,(x_i),$$

which can be relatively easily solved. Thus, the dual function can be efficiently created and then maximized with respect to $\mu \ge 0$ and λ.

The use of duality on separable problems has been exploited in optimum design of structures. We present a truss in the following, example which is self-contained in that no prior knowledge on the part of the reader is assumed. Optimization of structures is discussed in more detail in Chapter 12.

Example 6.6

Consider the statically determinate truss shown in Fig. E6.6. The problem is to minimize the volume of the truss treating the cross-sectional area of each element as a design variable. The only constraint is a limit on the tip deflection. A lower bound is specified on each area. In the following, x_i = cross-sectional area of element i, L_i = length of element i, E = Young's modulus.

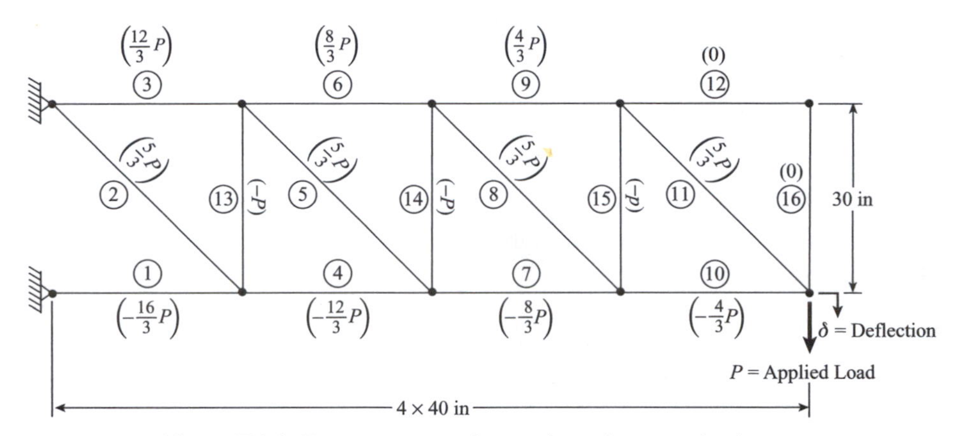

Figure E6.6. Truss structure; forces shows in parenthesis.

The problem is:

$$\text{minimize} \quad \sum_{i=1}^{16} L_i x_i \tag{a}$$

$$\text{subject to} \quad \delta(\mathbf{x}) \leq \delta_U \tag{b}$$

$$x_i > 0 \quad i = 1, \ldots, 16 \tag{c}$$

The bound constraint $x_i \geq 0$ is enforced as $x_i \geq x_i^L$, where $x_i^L = 10^{-6}$. The problem in (a)–(c) can also be stated as minimizing the weight of the truss structure subject to a upper limit on compliance (or lower limit on the stiffness).

The expression for the tip deflection δ is given by

$$\delta = \sum_{i=1}^{16} \frac{F_i f_i L_i}{E x_i}$$

where F_i and f_i are the forces in the element i due to the applied load P and due to a unit load at the point where δ is measured, respectively. Here, the tip deflection is evaluated at the point (and direction) where P is applied.

Thus, $F_i = Pf_i$. Moreover, f_i are independent of $\mathbf{x}$ because the structure is statically determinate. Thus, this problem is a *separable* problem. The values of f_i, obtained using force equilibrium at each node in the structure, are given in Fig. E6.6. In fact, if we denote

$$c_i = \frac{F_i f_i L_i}{E\delta_U}$$

then the convex constraint in (b) can be written as

$$\sum_{i=1}^{16} \left(\frac{c_i}{x_i} - \alpha \right) \quad \leq \quad 0 \tag{d}$$

where $\alpha = 1/16$. We choose: $P = 25{,}000$ lb, $E = 30 \times 10^6$ psi, $x^L = 10^{-6}$ in.^{2}, and $\delta_U = 1$ in. We define the dual function as

$$\varphi(\mu) = \underset{\mathbf{x} \geq \mathbf{x}^L}{\text{minimize}} \sum_{i=1}^{16} \left\{ L_i x_i + \mu \left(\frac{c_i}{x_i} - \alpha \right) \right\} \tag{e}$$

This leads to the individual problems

$$\underset{x_i \geq x_i^L}{\text{minimize}} \quad \psi \equiv L_i x_i + \mu \left(\frac{c_i}{x_i} - \alpha \right) \tag{f}$$

The optimality condition is evidently

$$x_i^* = \sqrt{\frac{\mu c_i}{L_i}} \quad \textit{if } c_i > 0 \tag{g}$$

$$x_i^* = x_i^L \quad \textit{if } c_i = 0 \tag{h}$$

Thus

$$\varphi(\mu) = 2 \sum_{i=1}^{16} \sqrt{c_i L_i} \sqrt{\mu} - \mu + \text{ constant} \tag{i}$$

The maximization of $\varphi(\mu)$ requires $d\varphi/d\mu = 0$, which yields

$$\mu = \left(\sum_i \sqrt{c_i L_i} \right)^2 \tag{j}$$

The closed-form solution is thus

$$\mu^* = 1358.2, \qquad f^* = 1358.2 \text{ in.}^3$$

$$\mathbf{x} = \begin{array}{l} [5.67 \quad 1.77 \quad 4.25 \quad 4.25 \quad 1.77 \quad 2.84 \quad 2.84 \quad 1.77 \\ \;\; 1.42 \quad 1.42 \quad 1.77 \quad 10^{-6} \quad 1.06 \quad 1.06 \quad 1.06 \quad 10^{-6}], \text{ in.}^2 \end{array}$$

Program TRUSS – OC in Chapter 12 may also be used to solve this problem.

Advantages of the dual approach on this problem over primal techniques in Chapter 5 is indeed noticeable for separable problems. Efficiency of dual methods can reduce significantly with increase in number of active constraints. The reader may experiment on the computer with both primal and dual methods to study these issues. Duality as presented involved differentiable dual functions and continuous variables. Extensions of the theory and application to problems with discrete (0 or 1) variables may be found in Lasdon [1970]. For discrete separable problems, the Lagrangian is minimized not over a (single) continuous variable $x_i \neq R^1$ but over a discrete set $[x_i^1, \ldots, x_i^p]$.

6.5 The Augmented Lagrangian Method

We have seen that in penalty function methods, convergence is forced by letting $r \rightarrow \infty$. This results in steeply curved functions (Fig. 6.1) whose accurate minimizations are hard to obtain. Theoretically, the Hessian of the function T becomes increasingly illconditioned as r increases, which leads to a poorer rate of convergence. On the other hand, duality is valid only when we make the strong assumption that the Lagrangian is locally convex, or $\nabla^2 L$ is positive definite at $\mathbf{x}^*$ – this assumption is more likely than not to be invalid in applications. Three researchers, namely. Powell, Hestenes, and Rockafellar developed the "augmented Lagrangian" method, which is based on combining duality with (exterior) penalty functions. See also the chapter by R. Fletcher in Gill and Murray [1974]. There are several variations of the method as given in Pierre and Lowe [1975].

The augmented Lagrangian method has now become very popular in finite element analysis applied to mechanics. See Simo and Laurren [1992] and references in that paper for applications in finite element analysis.

First, we will present the method for equality constraints for the readers convenience. Thus, we are considering problem (6.1) without the g. We define the augmented Lagrangian as

$$T(\mathbf{x}) = f(\mathbf{x}) + \sum_{j=1}^{\ell} \lambda_j h_j(\mathbf{x}) + \frac{1}{2} r \sum_{j=1}^{\ell} [h_j(\mathbf{x})]^2 \tag{6.16}$$

The first two terms form the Lagrangian L and the last term is the exterior penalty function. Hence, the name "augmented Lagrangian." The λ_j are Lagrange multipliers and r is a penalty parameter. Note that the gradient of T is

$$\nabla T(\mathbf{x}) = \nabla f(\mathbf{x}) + \sum_{j=1}^{\ell} \lambda_j \nabla h_j(\mathbf{x}) + r \sum_{j=1}^{\ell} h_j(\mathbf{x}) \nabla h_j$$

and the Hessian of T is

$$\nabla^2 T(\mathbf{x}) = \nabla^2 f(\mathbf{x}) + \sum_{j=1}^{\ell} \lambda_j \nabla^2 h_j(\mathbf{x}) + r \sum_{j=1}^{\ell} h_j(\mathbf{x}) \nabla^2 h_j + r \sum_{j=1}^{\ell} \nabla h_j^T \nabla h_j$$

Paralleling the development of the dual function $\phi(\boldsymbol{\mu}, \boldsymbol{\lambda})$ in Section 6.4, we can define a dual function $\psi(\boldsymbol{\lambda})$ as follows. If r is chosen large enough, then it can be shown that the Hessian $\nabla^2 T$ will be positive definite at $\mathbf{x}^*$. Thus, while we assume that the sufficient conditions for a minimum to the original problem are satisfied, we are not requiring the stronger condition that the Hessian of the Lagrangian $\nabla^2 L(\mathbf{x}^*)$ be positive definite as in pure duality theory. Then, for a given r and $\boldsymbol{\lambda}$, we can define $\mathbf{x}(\boldsymbol{\lambda})$ and hence $\psi(\boldsymbol{\lambda})$ from the minimization problem:

$$\psi(\boldsymbol{\lambda}) = \underset{\mathbf{x}}{\text{minimum}}\ T(\mathbf{x}) = f(\mathbf{x}) + \sum_{j=1}^{\ell} \lambda_j h_j(\mathbf{x}) + \frac{1}{2} r \sum_{j=1}^{\ell} [h_j(\mathbf{x})]^2 \tag{6.17}$$

We have

$$d\psi / d\lambda_i = \partial \psi / \partial \lambda_i + \nabla \psi^{\mathrm{T}} d\mathbf{x} / d\lambda_i$$

where ∇ represents derivatives with respect to $\mathbf{x}$ as usual. Since $\nabla \psi = \mathbf{0}$ from (6.17), we obtain the following expression for the gradient of the augmented Lagrangian in dual space:

$$\frac{d\psi}{d\lambda_i} = h_i(\mathbf{x}(\lambda)), \quad i = 1, \ldots, \ell \tag{6.18}$$

An expression for the Hessian of ψ can also be derived. We have

$$\frac{d^2\psi}{d\lambda_i\, d\lambda_j} = \frac{d}{d\lambda_j} (h_i) = \nabla h_i \left(\frac{d\mathbf{x}}{d\lambda_j} \right)$$

since $\frac{\partial h_i}{\partial \lambda_j} = 0$. To obtain an expression for $\frac{d\mathbf{x}}{d\lambda_j}$, we differentiate $\nabla \psi = \mathbf{0}$, which gives

$$[\nabla^2 T] \left\{ \frac{d\mathbf{x}}{d\lambda_j} \right\} = -\nabla h_j^{\mathrm{T}}$$

Combining the preceding two expressions, we get an expression for the dual Hessian

$$\frac{d^2\psi}{d\lambda_i d\lambda_j} = -\nabla h_i [\nabla^2 T]^{-1} \nabla h_j^{\mathrm{T}} \tag{6.19}$$

At a regular point $\mathbf{x}^*$, the dual Hessian is negative definite. Thus, a point that satisfies $\frac{d\psi}{d\lambda_i} = h_i\,(\mathbf{x}(\boldsymbol{\lambda})) = 0, i = 1, \ldots, \ell$ can be obtained by *maximizing* $\psi(\boldsymbol{\lambda})$. Moreover, the construction of $\psi(\boldsymbol{\lambda})$ requires that (6.17) be satisfied. This means that the point so obtained satisfies the optimality conditions for the primal problem.

The augmented Lagrangian method can now be stated. We start with a guess for $\boldsymbol{\lambda}$, solve the unconstrained minimization in (6.17) to define $\mathbf{x}(\boldsymbol{\lambda})$, and then solve

$$\text{maximize} \quad \psi(\boldsymbol{\lambda}) \tag{6.20}$$

using a suitable gradient method. Expressions for the gradient and Hessian are made use of. For instance, the Newton update will be

$$\lambda^{k+1} = \lambda^k - \left[\frac{d^2\psi}{d\lambda_i\, d\lambda_j}\right]^{-1} \mathbf{h} \tag{6.21}$$

For large $\mathbf{r}$, Powell has shown that the inverse of the dual Hessian is approximately given by r. This leads to the simpler update for the Lagrange multipliers

$$\lambda_i^{k+1} = \lambda_i^k + r_i h_i, \quad i = 1, \ldots, \ell \tag{6.22}$$

While r is not increased at every iteration as in penalty function methods, Powell suggests that in the beginning iterations it helps in obtaining convergence. The scheme suggested is that if $\|\ \mathbf{h}\ \|$ is not reduced by a factor of 4 then to increase r by a factor of (say) 10. Note that the update in (6.22) is also evident directly by rewriting modified Lagrangian gradient $\nabla T(\mathbf{x}) = \nabla f(\mathbf{x}) + \sum_{j=1}^{\ell} \lambda_j \nabla h_j(\mathbf{x}) + r\sum_{j=1}^{\ell} h_j(\mathbf{x})\nabla h_j$ as $\nabla T(\mathbf{x}) = \nabla f(\mathbf{x}) + \sum_{j=1}^{\ell} \{\lambda_j + r h_j(\mathbf{x})\}\nabla h_j$ and recognizing that the expression in curly brackets is expected to be a better estimate of the correct λ_j.

Example 6.7

Consider the design of a cylindrical water tank that is open at the top. The problem of maximizing the volume for a given amount of material (surface area) leads to the simple problem

$$\begin{aligned}
&\text{minimize} && V \equiv -\pi x_1^2 x_2 \\
&\text{subject to} && h \equiv 2x_1x_2 + x_1^2 - A_0/\pi = 0
\end{aligned}$$

where x_1 = radius and x_2 = height. Ignoring the constant π in the objective f and letting $A_0/\pi = 1$, we obtain the problem

$$\begin{aligned}
&\text{minimize} && f \equiv -x_1^2 x_2 \\
&\text{subject to} && h \equiv 2x_1x_2 + x_1^2 - 1 = 0
\end{aligned}$$

The KKT conditions require $\nabla L = \mathbf{0}$ or

$$-2x_1x_2 + 2\lambda x_2 + 2\lambda x_1 = 0 \tag{a}$$

$$x_1^2 + 2\lambda x_1 = 0 \tag{b}$$

From these, we obtain

$$x_1(\lambda) = x_2(\lambda) = 2\lambda$$

Substitution into the constraint $h = 0$ yields the KKT point

$$x_1^* = x_2^* = 2\lambda = \frac{1}{\sqrt{3}} = 0.577$$

with $-f^* = 0.192$.

We also have the Hessian of the Lagrangian

$$\nabla^2 L(\mathbf{x}^*) = \begin{bmatrix} -2x_2 + 2\lambda & -2x_1 + 2\lambda \\ -2x_1 + 2\lambda & 0 \end{bmatrix}_{\mathbf{x}^*} = \frac{-1}{\sqrt{3}} \begin{bmatrix} 1 & 1 \\ 1 & 0 \end{bmatrix}$$

To check the sufficiency conditions, we need to check if $\nabla^2 L$ is positive definite on the tangent subspace, or that $\mathbf{y}^{\mathrm{T}}\nabla^2 L\mathbf{y} > 0$ for every nonzero $\mathbf{y}$ that satisfies $\nabla h(\mathbf{x}^*)^{\mathrm{T}}\mathbf{y} = 0$. Since $\nabla h(\mathbf{x}^*) = 2/\sqrt{3}\ [2, 1]^{\mathrm{T}}$. Thus, our trial $\mathbf{y} = [1, -2]$. This gives $\mathbf{y}^{\mathrm{T}}\ \nabla^2 L\mathbf{y} = +\,3/\sqrt{3} > 0$. Hence, $\mathbf{x}^*$ is a strict local minimum.

However, we cannot apply duality theory since $\nabla^2 L$ is not positive definite (though it is nonsingular at $\mathbf{x}^*$). In fact, the dual function is given by $\varphi(\lambda) = f(\mathbf{x}^*(\lambda)) + \lambda\ h(\mathbf{x}^*(\lambda)) = 12\ \lambda^3 - 4\ \lambda^2 - \lambda$. A graph of the function is shown in Fig. E6.7a. Evidently, $\varphi\ (\lambda)$ is not a concave function.

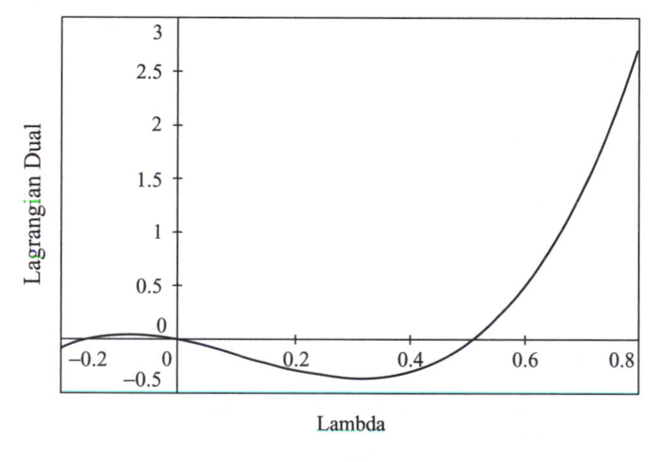

(a)

Figure E6.7A. The Lagrangian dual φ versus λ.

With the augmented Lagrangian, we have

$$T = -x_1^2 x_2 + \lambda\left(2x_1x_2 + x_1^2 - 1\right) + \tfrac{1}{2}r\left(2x_1x_2 + x_1^2 - 1\right)^2$$

We can show that $\nabla^2 T$ is positive definite for all r sufficiently large. In fact, by executing Program AUGLAG, we can experiment with different values of r: we see that $r = 5$ and above leads to success while $r = 0.1$ leads to failure. In other words, the augmented Lagrangian dual function $\psi(\lambda)$ is concave for sufficiently large r. Using the program, the plot of $\psi(\lambda)$ versus λ for two different values of

r are shown in Fig. E6.7b. At the optimum, the solution $\lambda = 1/(2\sqrt{3}) = 0.2887$, is the same regardless of the value of r as long as it is above the threshold.

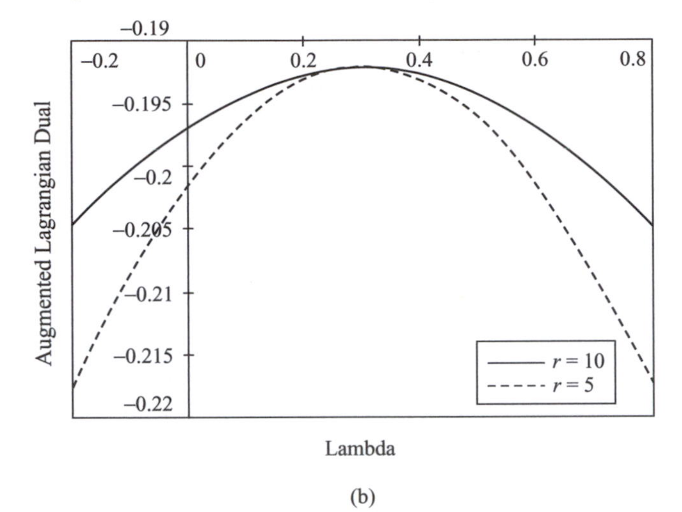

(b)

Figure E6.7B. The augmented dual function ψ versus λ for two different values of r.

Inequality Constraints

We consider the approach presented in Pierre and Lowe [1975] in the generalization of the aforementioned to handle inequality constraints. Extension of duality concepts to this case are given in the reference. This form of the augmented Lagrangian is found by the authors to be less sensitive to the accuracy of the T-function minimization owing to a certain "lock-on" feature associated with inequality constraints as discussed presently. We consider the general problem as in (6.1). We denote the T – function with only equality constraints by

$$T_e(\mathbf{x}) = f(\mathbf{x}) + \sum_{j=1}^{\ell} \lambda_j h_j(\mathbf{x}) + \frac{1}{2} r \sum_{j=1}^{\ell} [h_j(\mathbf{x})]^2 \tag{6.23}$$

and define the sets

$$I_1(\mu)) = \{i : \mu_i > 0, \quad i = 1, \ldots, m\}, \quad I_2(\mu) = \{i : \mu_i = 0, \quad i = 1, \ldots, m\}$$

The augmented Lagrangian for the general problem is now defined as

$$T(\mathbf{x}) = T_e(\mathbf{x}) + \sum_{i \in I_1} \left[\mu_i g_i(\mathbf{x}) + \frac{r_1}{2} [g_i(\mathbf{x})]^2\right] + \frac{r_2}{2} \sum_{i \in I_2} [\max(0, g_i(\mathbf{x})]^2 \tag{6.24}$$

Let $\boldsymbol{\lambda}_k, \boldsymbol{\mu}_k, r_1^k, r_2^k$ be the values of the multipliers and penalty parameters at the current iteration. We then minimize $T(\mathbf{x})$ in (6.24) with respect to $\mathbf{x}$ to obtain $\mathbf{x}(\boldsymbol{\lambda}_k, \boldsymbol{\mu}_k)$.

It is important to note that during this minimization, the inequality constraints g_i corresponding to the set $I_1(\mu_k)$ are "locked-on." That is, these are treated just as equality constraints regardless of their being < 0 or > 0 during the minimization. The last term in (6.24) accounts for constraints that are not locked on but which get violated during minimization. In Program AUGLAG, we use the method of Fletcher–Reeves to minimize T. We then update the Lagrange multipliers by the update formulas:

$$\lambda_i^{k+1} = \lambda_i^k + r_1^k h_i(\mathbf{x}(\lambda_k, \mu_k)), \qquad i = 1, \ldots, \ell$$

$$i \in I_1 : \mu_i^{k+1} = \mu_i^k + r_2^k g_i(\mathbf{x}(\boldsymbol{\lambda}_k, \boldsymbol{\mu}_k))$$

$$\text{IF } \mu_i^{k+1} < 0 \text{ THEN } \mu_i^{k+1} = 0$$

$$i \in I_2 : \mu_i^{k+1} = 0 \qquad \text{If } g_i(\mathbf{x}(\boldsymbol{\lambda}_k, \boldsymbol{\mu}_k)) \leq 0$$

$$\mu_i^{k+1} = r_2^k g_i(\mathbf{x}(\lambda_k, \mu_k)) \qquad \text{If } g_i(\mathbf{x}(\boldsymbol{\lambda}_k, \boldsymbol{\mu}_k)) > 0 \tag{6.25}$$

The update formula for locked-on constraints is the same as that for equality constraints except that we do not permit them to become negative. Also, if a constraint in I_2 becomes violated after a minimization of T, we assign it a positive μ_i for the subsequent minimization.

The initial values of the penalty parameters are chosen so that $\nabla^2 T(\mathbf{x}^0)$ is positive definite. In practice, any choice that avoids failure of the algorithm during the first minimization of T will work. Thereafter, we increase r_i as $r_i^{k+1} = \alpha r_i^k$ where $\alpha > 1$ (say, $\alpha = 2$). This is done until r_i reaches a predetermined upper limit r^{U}. In Program AUGLAG, lower and upper bounds on design variables are handled as general inequality constraints. These can be handled during minimization of the T-function but the modification is not straightforward.

6.6 Geometric Programming

Duality is not necessarily of the Lagrangian type discussed previously. The inequality between the arithmetic and geometric means also provides a beautiful duality. In the method of Geometric Programming, this duality is exploited for functions possessing a certain structure. There are some excellent books on Geometric Programming including [Duffin et al. 1967], [Zener 1971], [Papalambros and Wilde 1988]. Here, we present only the elementary concepts behind the method.

The inequality between the arithmetic and geometric means of n positive quantities ($u_i > 0$), denoted by $u_1, u_2, \ldots, u_n$, is stated as

$$M \geq G \tag{6.26}$$

where M is the arithmetic mean given by

$$M = \frac{u_1 + u_2 + \cdots + u_n}{n}$$

and G is the geometric mean given by the positive nth root

$$G = \sqrt[n]{u_1 u_2 \cdots u_n}$$

where $M = G$ only if all the u_i are equal. A simple illustration of (6.26) is the following problem: Among all rectangles with a prescribed perimeter, find one with the largest area. If x and y are the dimensions of a rectangle, then the perimeter is given by $2\,(x + y)$. Thus, (6.26) gives

$$\frac{x + y}{2} \geq \sqrt{xy}$$

Thus, for a given value of the left-hand side, the maximum value of the right-hand side will occur when $x = y$. That is, the rectangle with the largest area is a square, a well known result. Another instance of an inequality in geometry is: Given a closed curve of length L, the area A enclosed by the curve satisfies $A \leq L^2/4\pi$, with the equality sign holding only for a circle. This inequality is derived from a different criteria [Courant and Robbins 1981].

The inequality in (6.26) can be generalized using weighted means as:

$$\sum_{i=1}^{n} w_i x_i \geq \prod_{i=1}^{n} x_i^{w_i} \tag{6.27}$$

where $w_i > 0$ are positive weights summing to unity, that is, $\sum_{i=1}^{n} w_i = 1$. We now consider the application of the preceding inequality to unconstrained minimization of functions having a special form, namely, for *posynomials*:

$$f = C_1 x_1^{a_{11}} x_2^{a_{12}} \cdots x_n^{a_{1n}} + \ldots + C_t x_1^{a_{t1}} x_2^{a_{t2}} \ldots x_n^{a_{tn}} = \sum_{i=1}^{t} C_i \prod_{j=1}^{n} x_j^{a_{ij}} \tag{6.28}$$

where $C_i > 0$, a_{ij} = coefficient of the ith term belonging to the jth variable, n = number of variables, t = number of terms, and $x_i > 0$, and the coefficients a_{ij} are any real numbers (positive, negative, or fractional). Geometric programming problems (unconstrained or constrained) are convex problems. Thus, the necessary conditions of optimality are also sufficient, and a local minimum is also a global minimum. Further, duality theory becomes powerful. For instance, the dual function provides a global lower bound on the primal cost function. We introduce the concepts behind Geometric Programming with the following example.

Example 6.8 [Zener 1971]

The problem involves transporting a large quantity of iron ore by ship from one place to another. The company plans to use a single ship, with multiple trips. The total cost consists of three terms: renting the ship, hiring the crew, and fuel. Design variables are: ship tonnage T and ship velocity v. Here, the tonnage T determines the size and type of ship, total number of trips $\propto T^{-1}$, and time duration of each trip $\propto v^{-1}$. After some study, the following cost function has been developed to reflect total cost:

$$f = C_1 T^{0.2} v^{-1} + C_2 T^{-1} v^{-1} + C_3 T^{-1/3} v^2 \tag{a}$$

where $C_i > 0$, $T > 0$, and $v > 0$. Our task is to minimize f. Let the fraction of each cost term as part of the total cost be denoted by positive variables δ_i, $i = 1, 2, 3$. Now, f can be written as

$$f = \delta_1 \left(\frac{C_1 T^{0.2} v^{-1}}{\delta_1} \right) + \delta_2 \left(\frac{C_2 T^{-1} v^{-1}}{\delta_2} \right) + \delta_3 \left(\frac{C_3 T^{-1/3} v^2}{\delta_3} \right) \tag{b}$$

We can consider δ_i as "weights" noting that the "normality condition" $\sum \delta_i = 1$ is satisfied. From (6.26) we obtain

$$f \geq \left(\frac{C_1}{\delta_1} \right)^{\delta_1} \left(\frac{C_2}{\delta_2} \right)^{\delta_2} \left(\frac{C_3}{\delta_3} \right)^{\delta_3} T^{(0.2\delta_1 - \delta_2 - \delta_3/3)} v^{(-\delta_1 - \delta_2 + 2\delta_3)} \tag{c}$$

The right-hand side of the preceding inequality is a global lower-bounding function on the primal cost. However, its value is limited because it depends on T and v. Now consider the optimality conditions for f to have a minimum. In Eq. (a), we set $\partial f/\partial T = 0$ and $\partial f/\partial v = 0$. After differentiation and dividing through by f, we immediately arrive at the following equations:

$$\begin{aligned} 0.2\delta_1 - \delta_2 - \delta_3/3 &= 0 \\ -\delta_1 - \delta_2 + 2\delta_3 &= 0 \end{aligned} \tag{d}$$

Interestingly, these "orthogonality conditions" result in the lower-bounding function in (c) to become independent of T and v! We arrive at the useful result

$$f(\mathbf{x}) \geq V(\boldsymbol{\delta}) \tag{e}$$

where $\mathbf{x} = [T, v]$ and $\boldsymbol{\delta} = [\delta_1, \delta_2, \delta_3]^{\mathrm{T}}$, and

$$V(\boldsymbol{\delta}) = \left(\frac{C_1}{\delta_1} \right)^{\delta_1} \left(\frac{C_2}{\delta_2} \right)^{\delta_2} \left(\frac{C_3}{\delta_3} \right)^{\delta_3} \tag{f}$$

In this example, the orthogonality conditions in (*d*) and the normality condition $\sum \delta_i = 1$ result in our being able to solve for δ_i uniquely as

$$\boldsymbol{\delta} = [7/10.8, 0.2/10.8, 1/3]^{\mathrm{T}} = [0.648, 0.019, 0.333]^{\mathrm{T}} \tag{g}$$

The substitution of $\boldsymbol{\delta}$ from (*g*) into (*f*) leads to the maximum value of V that equals the minimum value of f, denoted by f^*. Furthermore, from (*g*), we see that about 65% of the optimum cost stems from ship rental, 0.2% from hiring the crew and 33% from fuel. These facts are independent of C_i! For instance, the hourly rate the crew is paid, reflected in C_2, will not change the fact that crew costs represent only 0.2% of the total optimized cost. In primal methods, we would have to know the values of C_i and solve (*a*) before we arrive at any conclusions. Thus, geometric programming offers certain insights not obtainable through conventional numerical procedures.

To recover the optimum values of the primal variables T and v, note that the three terms in parenthesis in (*b*) are equal to each other – a condition for the arithmetic-geometric means to be equal. In fact, in view of the definition of δ_i, each term is equal to f^*. Taking natural logarithms, we obtain

$$\begin{aligned} 0.2\log T - \log v &= \log(\delta_1 f^*/C_1) \\ -\log T - \log v &= \log(\delta_2 f^*/C_2) \\ -1/3\log T + 2\log v &= \log(\delta_3 f^*/C_3) \end{aligned}$$

We can evaluate f^* and use any two of the preceding equations. Or else, we can combine the first two equations to obtain

$$1.2\log T = \log(\delta_1/\delta_2) + \log(C_2/C_1),$$

which gives T. Combining the second and third equations, we obtain

$$-2/3\log T - 3\log v = \log(\delta_2/\delta_3) + \log(C_3/C_1),$$

which then gives v.

We note that if there are 4 terms in (*a*) and only 2 variables, then we will have 4 δ_i but only 3 equations for δ_i (2 orthogonality conditions and 1 normality condition). Thus, in this case, there will be one "free" variable δ_i and V has to be maximized with respect to it. But for any choice of this free variable, $V(\boldsymbol{\delta})$ will provide a global lower bound on f^*, which can be very useful.

We can now generalize the preceding example and state the Geometric Programming approach for unconstrained minimization as follows. Consider the unconstrained minimization of the posynomial f in (6.28). Then,

$$f(\mathbf{x}) \geq V(\delta) = \prod_{i=1}^{t} \left(\frac{C_i}{\delta_i}\right)^{\delta_i} \tag{6.29a}$$

where $\mathbf{x} = [x_1, x_2, \ldots, x_n]^{\mathrm{T}}$, and δ_i satisfy

$$\sum_{i=1}^{t} \delta_i = 1 \tag{6.29b}$$

$$\sum_{i=1}^{t} a_{ij}\delta_i = 0 \qquad j = 1, \ldots, n \tag{6.29c}$$

The maximum value of the dual function equals the minimum value of the primal cost function. The *degree of difficulty* of the problem is defined equal to

$$\text{degree} = t - (n + 1) \tag{6.30}$$

where t = number of terms in the posynomial and n = number of primal variables. This value was equal to zero in Example 6.8 resulting in the closed form solution of δ from the constraint conditions alone. If the degree of difficulty is large, then we need to use one of the previously discussed numerical procedures to solve (6.29). It should be emphasized that any δ that satisfies (6.29b–c) will result in $V(\boldsymbol{\delta})$ being a global lower bound to f^*. Further, especially in problems with zero degree of difficulty, geometric programming offers the engineer with unique insights.

Constrained Problems

Extension of the preceding theory to constrained problems is now presented. The problem must be posed as

$$\begin{aligned} &\text{minimize} \quad g_0(\mathbf{x}) \equiv \sum_{i=1}^{t_0} C_{0i} \prod_{j=1}^{n} x_j^{a_{0ij}} \\ &\text{subject to} \quad g_k(\mathbf{x}) \equiv \sum_{i=1}^{t_k} C_{ki} \prod_{j=1}^{n} x_j^{a_{kij}} \le 1 \qquad k = 1, \ldots, m \\ &\qquad \mathbf{x} > \mathbf{0} \end{aligned} \tag{6.31}$$

Preceding, f and $\{g_i\}$ are posynomials with all coefficients $C_{ki} > 0$. The notation on subscripts is as follows:

C_{ki} = coefficient of the kth function, ith term
a_{kij} = exponent corresponding to kth function, ith term, jth variable
n = number of primal variables
m = number of constraints
t_k = number of terms in the kth function
$k = 0$ corresponds to the cost function g_0

Note that constraints not in the form in (6.31) can often be changed to fit the form. For example

- $x_1 x_2 \geq 5 \quad \Rightarrow \quad 5x_1^{-1}x_2^{-1} \leq 1$
- $x_1 - x_2 \geq 10 \Rightarrow x_1 \geq 10 + x_2 \Rightarrow 10x_1^{-1} + x_1^{-1}x_2 \leq 1$

As in the unconstrained case we have

$$g_0(\mathbf{x}) \geq \prod_{i=1}^{t_0} \left(\frac{C_{0i}}{\delta_{0i}}\right)^{\delta_{0i}} \prod_{j=1}^{n} (x_j)^{\sum_{i=1}^{t_0} a_{0ij}\delta_{0i}} \tag{6.32}$$

Similarly, we have for each constraint function,

$$1 \geq g_k(\mathbf{x}) \geq \prod_{i=1}^{t_k} \left(\frac{C_{ki}}{\delta_{ki}}\right)^{\delta_{ki}} \prod_{j=1}^{n} (x_j)^{\sum_{i=1}^{t_k} a_{kij}\delta_{ki}} \quad k = 1, \ldots, m \tag{6.33}$$

with all δ being positive and satisfying the normality condition

$$\sum_{i=1}^{t_0} \delta_{ki} = 1 \qquad k = 0, \ldots, m \tag{6.34}$$

We now introduce Lagrange multipliers associated with each constraint function as $\mu_k, i = 1, \ldots, m$. Since $\mu_k \geq 0$ and $g_k \leq 1$, $(g_k)^{\mu_k} \leq 1$. Using this, (6.33) yields

$$1 \geq \prod_{i=1}^{t_k} \left(\frac{C_{ki}}{\delta_{ki}}\right)^{\mu_k\delta_{ki}} \prod_{j=1}^{n} (x_j)^{\sum_{i=1}^{t_k} a_{kij}\mu_k\delta_{ki}} \qquad k = 1, \ldots, m \tag{6.35}$$

Multiplying the extreme sides of inequality (6.32) with all the m inequalities in (6.35), and introducing the new quantities (for notational simplicity) Δ_{ki} as

$$\Delta_{ki} \equiv \mu_k \delta_{ki} \quad k = 0, 1, \ldots, m \text{ and } i = 1, \ldots, t_k \tag{6.36}$$

and in Eq. (6.36), defining

$$\mu_0 = 1 \tag{6.37}$$

we get

$$g_0(\mathbf{x}) \geq \prod_{k=0}^{m} \mu_k^{\mu_k} \prod_{i=1}^{t_k} \left(\frac{C_{ki}}{\Delta_{ki}}\right)^{\Delta_{ki}} \equiv V(\boldsymbol{\mu}, \boldsymbol{\Delta}) \tag{6.38}$$

where the dependence of the dual function V on $\mathbf{x}$ has been removed through the orthogonality conditions

$$\sum_{k=0}^{m} \sum_{i=1}^{t_k} a_{kij}\Delta_{ki} = 0 \quad j = 1, \ldots, n \tag{6.39}$$

In view of (6.36), the normality conditions in (6.34) take the form

$$\sum_{i=1}^{t_k} \Delta_{ki} = \mu_k, \quad k = 0, \ldots, m \tag{6.40}$$

The dual maximization problem involves maximization of the function $V(\boldsymbol{\mu}, \boldsymbol{\Delta})$ in (6.38) subject to (6.37), (6.39), (6.40) with dual variables $\boldsymbol{\mu} \geq \mathbf{0}$, $\boldsymbol{\Delta} \geq \mathbf{0}$. Any $\boldsymbol{\mu}$, $\boldsymbol{\Delta}$ satisfying the orthogonality and normality conditions will result in $V(\boldsymbol{\mu}, \boldsymbol{\Delta})$ being a global lower bound on the primal objective function. The degree of difficulty is given by (6.30) with t representing the total number of terms in all the posynomials. Illustration of the preceding dual and recovery of the primal solution are illustrated through the following example.

Example 6.9

We consider the design of the cylindrical water tank given in Example 6.7. The problem can be written as

$$\begin{aligned} &\text{minimize} \quad x_1^{-2} x_2 \\ &\text{subjectto} \quad 2x_1x_2 + x_1^2 \leq 1 \end{aligned}$$

where x_1 and x_2 represent the radius and height of the water tank, respectively. The degree of difficulty for this problem $= 3 - 2 - 1 = 0$. Thus, the solution can be obtained just from solving linear equations. The dual problem reduces to:

$$V(\boldsymbol{\delta}) = \mu^{\mu} \left(\frac{1}{\Delta_{01}}\right)^{\Delta_{01}} \left(\frac{2}{\Delta_{11}}\right)^{\Delta_{11}} \left(\frac{1}{\Delta_{12}}\right)^{\Delta_{12}} \tag{a}$$

Eq. (6.39) yields

$$\begin{aligned} -2\Delta_{01} + \Delta_{11} + 2\Delta_{12} &= 0 \quad (j = 1) \\ -\Delta_{01} + \Delta_{11} &= 0 \quad (j = 2) \end{aligned}$$

and (6.40) yields

$$\begin{aligned} \Delta_{01} &= 1 \quad (k = 0) \\ \Delta_{11} + \Delta_{12} &= \mu \quad (k = 1) \end{aligned}$$

The solution is $\Delta_{01} = 1$, $\Delta_{11} = 1$, $\Delta_{12} = 0.5$, $\mu = 1.5$. The dual objective is $V = 3\sqrt{3}$, which equals f^*. The primal variables may be recovered by noting that

$$\delta_{ki} = \frac{\text{term}_i}{g_k^*} \quad \text{or} \quad \text{term}_i = \frac{\Delta_{ki}}{\mu_k} g_k^* \quad k = 0, \ldots, m \tag{b}$$

We know that $g_1^* = 1$ since the Lagrange multiplier $\mu > 0$, which means the constraint is active. Thus, (b) yields

$$\begin{aligned} 2x_1x_2 &= 1/1.5 \\ x_1^2 &= 0.5/1.5, \end{aligned}$$

which gives $x_1^* = x_2^* = 1/\sqrt{3}$.

Of course, with several constraints, some μ may be zero. This means that the corresponding constraints are inactive and that the associated δ are zero.

COMPUTER PROGRAMS

SUMT, BARRIER, AUGLAG

PROBLEMS

P6.1. Consider

$$\begin{aligned} \text{minimize} \quad & x_1 + x_2 \\ \text{subject to} \quad & x_1^2 - x_2 \le 0 \\ & x_1 \ge 0 \end{aligned}$$

Plot the constraint and cost function contours in $x_1 - x_2$ space. Solve the problem using (i) quadratic exterior penalty function and (ii) interior penalty function based on (6.6). Use two different starting points. In each case, show the *trajectory of unconstrained minima* obtained for increasing values of r.

P6.2. Solve P6.1 using the log barrier interior penalty function in (6.7).

P6.3. Consider

$$\begin{aligned} \text{minimize} \quad & (x_1 - 2)^2 + (x_1 - 1)^2 \\ \text{subject to} \quad & x_1^2/4 + x_2 \le 1 \\ & -x_1 + 2x_2 \ge 1 \end{aligned}$$

Plot the constraint and cost function contours in $x_1 - x_2$ space. Solve the problem using (i) quadratic exterior penalty function and (ii) interior penalty function based on (6.6). Use (2.0, 2.0) as your starting point. In each case, show the *trajectory of unconstrained minima* obtained for increasing values of r.

P6.4. Consider the minimization of energy in a spring system with certain boundary conditions:

$$\begin{aligned}
\text{minimize} \quad & \pi = \tfrac{1}{2}100(Q_2 - Q_1)^2 + \tfrac{1}{2}100(Q_3 - Q_2)^2 + \tfrac{1}{2}300(Q_4 - Q_3)^2 - 25Q_2 \\
\text{subject to} \quad & Q_1 = 0 \\
& Q_3 = \alpha Q_4, \quad \text{where a is a known constant.}
\end{aligned}$$

As in Example 6.2, choose a large penalty parameter C and use the quadratic penalty function technique to write down the equilibrium equations (linear simultaneous equations).

P6.5. Determine the dual function $\varphi(\mu)$ for the problem

$$\begin{aligned}
\text{minimize} \quad & \sum_{i=1}^{n} i x_i \\
\text{subject to} \quad & \left(\sum_{i=1}^{n} \frac{1}{x_i} \right) \leq 1
\end{aligned}$$

P6.6. Derive the dual problem corresponding to the primal QP:

$$\begin{aligned}
\text{minimize} \quad & \tfrac{1}{2}\mathbf{x}^{\mathrm{T}}\mathbf{A}\mathbf{x} + \mathbf{c}^{\mathrm{T}}\mathbf{x} \\
\text{subject to} \quad & \mathbf{A}\mathbf{x} \geq \mathbf{b}
\end{aligned}$$

where **A** is symmetric and positive definite.

P6.7. Construct the dual problem corresponding to the primal:

$$\begin{aligned}
\underset{\mathbf{x}}{\text{minimize}} \quad & 5x_1 + 7x_2 - 4x_3 - \sum_{j=1}^{3} \ln(x_j) \\
\text{subject to:} \quad & x_1 + 3x_2 + 12x_3 = 37 \\
& \mathbf{x} > 0
\end{aligned}$$

P6.8. Generalize Example 6.6 to handle discrete set of choices for the cross-sectional areas. *Hint*: choose a discrete set for the areas, and minimize the Lagrangian for each value of μ by simple enumeration. Solve a suitable problem for verification.

P6.9. Consider the statically determinate truss shown in Fig. P6.9. The problem is to minimize the compliance of the truss treating the cross-sectional area of each element as a design variable. The only constraint is a limit on the total volume of the truss. A lower bound is specified on each area. In the following, x_i = cross-sectional area of element i, L_i = length of element i, V_0 = upper limit on truss volume, NE = number of elements, P = load applied at a node, δ = displacement where load P is applied. The problem is:

$$\begin{aligned} &\text{minimize} \quad P\delta \\ &\text{subject to} \quad \sum_{i=1}^{NE} L_i x_i \le V_0 \\ &x_i \le x_i^{U} \quad i = 1, \ldots, NE \end{aligned}$$

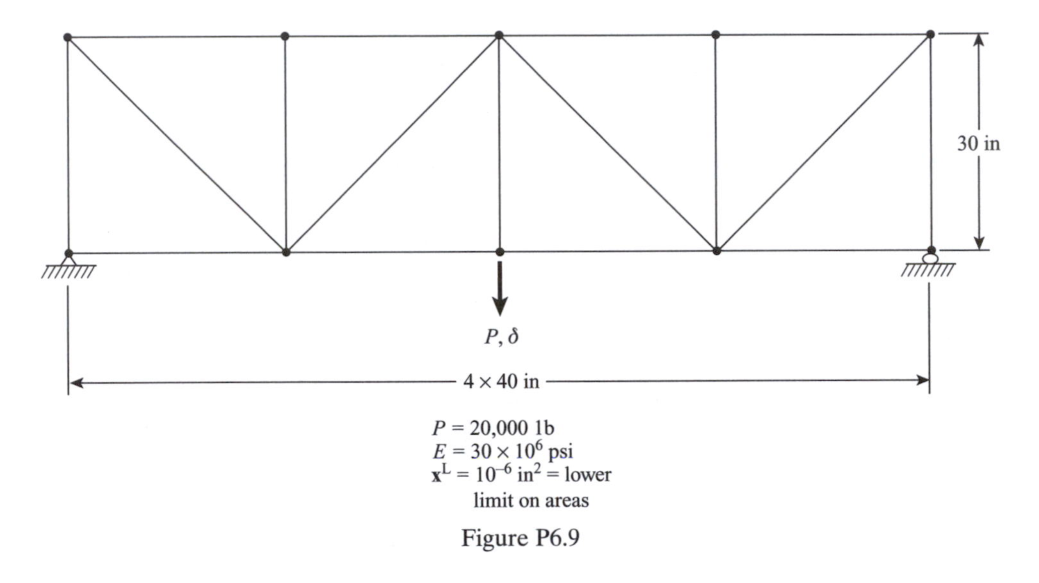

Figure P6.9

As in Example 6.6, apply the dual approach on this separable problem and determine $\mathbf{x}^*$ and f^*. Solve for $V_0 = 1000$ in.^{3} and for $V_0 = 1500$ in.^{3}

P6.10. Perform one iteration by hand calculations using the Augmented Lagrangian method. Use (0, 1) as the starting point. The problem is:

$$\begin{aligned} &\text{minimize} \quad -2x_1 - x_2 \\ &\text{subject to} \quad x_2 - 1 \le 0 \\ &4x_1 + 6x_2 - 7 \le 0 \\ &10x_1 + 12x_2 - 15 \le 0 \\ &x_1 + 0.25x_2^2 - 1 \le 0 \\ &-x_1 \le 0, \quad -x_2 \le 0 \end{aligned}$$

P6.11. Do Example 6.2 using the Augmented Lagrangian method instead of a penalty function.

P6.12. A thin-walled tubular column with radius x_1 and tube thickness x_2 is subjected to a compressive load. The column is to be designed for minimum weight subject to a direct stress constraint and a buckling constraint. The problem, with certain data, takes the form

$$\begin{aligned}
\text{minimize} \quad & f = x_1 x_2 \\
\text{subject to} \quad & g_1 = \frac{1}{x_1 x_2} - 156 \leq 0 \\
& g_2 = 1 - 6418 x_1^3 x_2 \leq 0 \\
& x_1 > 0, \quad x_2 > 0
\end{aligned}$$

(i) Plot the cost function contours and the constraint function in design space and obtain a graphical solution (*note*: there are infinite solutions along a curve).

(ii) Obtain a solution using Program AUGLAG with two starting points: (0.1, 0.2) and (0.1, 5.0). Discuss your solution.

(iii) Repeat (ii) but with the following additional constraint that the radius be at least 10 times the tube thickness: $10 - x_1/x_2 \leq 0$. Discuss.

P6.13. Consider Example E5.14 in Chapter 5. Do one or two iterations from the same starting point using a quadratic exterior penalty function.

P6.14. Consider Example E5.14 in Chapter 5. Do one or two iterations from the same starting point using a barrier function.

P6.15. Consider Example E5.14 in Chapter 5. Do two iterations from the same starting point using the augmented Lagrangian method.

P6.16. Compare the codes MFD, GRG, SQP, SUMT, and AUGLAG on selected examples from Chapter 5.

P6.17. Is the heat exchanger problem P4.18 in Chapter 4 a geometric programming problem? If so, proceed to obtain the solution analytically.

P6.18. There are four cost components in installing a power line, given by

$$f = C_1 x_1^{0.2} x_2^{-1} + C_2 x_1^{-1} x_2^{-1} + C_3 x_1^{-0.33} x_2^2 + C_4 x_1^{0.5} x_2^{-0.33}$$

Let f^* be the total minimum cost. Assuming that the first two components represent 30% each of f^* and that the last two components represent 20% each of f^*, derive a lower bound estimate of f^*.

P6.19. A closed cylindrical storage tank is required to hold a minimum of 3,000 m^3 of oil. Cost of material is \$1/m^2. The optimization problem for minimum cost is

$$\begin{aligned} &\text{minimize} \quad 2\pi x_2^2 + 2\pi x_1 x_2 \\ &\text{subject to} \quad \pi x_2^2 x_1 \geq 3000 \end{aligned}$$

Formulate and solve this using geometric programming.

P6.20. [Courtesy: Dr. J. Rajgopal]. A batch process consists of a reactor, feed tank, dryer, heat exchanger, centrifuge, and three pumps. Variables are V = product volume per batch in ft^3 and t_i = time required to pump batch through pump i, $i = 1, 2, 3$. Based on fitting a model to data, the following cost function has been developed:

$$\begin{aligned} f = &\ 590\, V^{0.6} + 580\, V^{0.4} + 1200\, V^{0.5} + 210\, V^{0.6} t_2^{-0.6} + 370\, V^{0.2} t_1^{-0.3} \\ &+ 250\, V^{0.4} t_2^{-0.4} + 250\, V^{0.4} t_3^{-0.4} + 200\, V^{0.9} t_3^{-0.8} \end{aligned}$$

The required average throughput is 50 ft^3/hr. The combined residence time for reactor, feed tank, and dryer is 10 hrs, or

$$500\, V^{-1} + 50\, V^{-1} t_1 + 50\, V^{-1} t_2 + 50\, V^{-1} t_3 \leq 1$$

with V and $t_i > 0$. What is the degree of difficulty? Write down the normality and orthogonality conditions.

REFERENCES

Bracken, J. and McCormick, G.P., *Selected Applications of Nonlinear Programming*, Wiley, New York, 1968.

Chandrupatla, T.R. and Belegundu, A.D., *Introduction to Finite Elements in Engineering*, 2nd Edition, Prentice-Hall, Englewood Cliffs, NJ, 1997.

Courant, R. and Robbins, H., *What Is Mathematics?* 4th Printing, Oxford University Press, Oxford, 1981.

Duffin, R.J., Peterson, E.L., and Zener, C., *Geometric Programming – Theory and Applications*, Wiley, New York, 1967.

Fiacco, A.V. and McCormick, G.P., *Nonlinear Programming: Sequential Unconstrained Minimization Techniques*, SIAM, Philadelphia, PA, 1990.

Gill, P.E. and Murray, W. (Ed.), *Numerical Methods for Constrained Optimization*, Academic Press, New York, 1974.

Haftka, R.T. and Gurdal, Z., *Elements of Structural Optimization*, 3rd Edition, Kluwer Academic Publishers, Dordrecht, 1992.

Lasdon, L.S., *Optimization Theory for Large Systems*, Macmillan, New York, 1970.

Luenberger, D.G., *Introduction to Linear and Nonlinear Programming*, Addison-Wesley, Reading, MA, 1973.

Mangasarian, O.L., *Nonlinear Programming*, McGraw-Hill, New York, 1969.

Papalambros, P.Y. and Wilde, D.J., *Principles of Optimal Design*, Cambridge University Press, Cambridge, 1988.

Pierre, D.A. and Lowe, M.J., *Mathematical Programming vs Augmented Lagrangians*, Addison-Wesley, Reading, MA, 1975.

Simo, J.C. and Laursen, T.A. An augmented Lagrangian treatment of contact conditions involving friction, *Computers & Structures*, **42** (1), 97–116, 1992.

Wolfe, P., A duality theorem for nonlinear programming, *Quarterly of Applied Mathematics*, **19**, 239–244, 1961.

Zener, C., *Engineering Design by Geometric Programming*, Wiley-Interscience, New York, 1971.

7

Direct Search Methods for Nonlinear Optimization

7.1 Introduction

Multivariable minimization can be approached using gradient and Hessian information, or using the function evaluations only. We have discussed the gradient- or derivative-based methods in earlier chapters. We present here several algorithms that do not involve derivatives. We refer to these methods as *direct methods*. These methods are referred to in the literature as *zero order methods* or *minimization methods without derivatives*. Direct methods are generally robust. A degree of randomness can be introduced in order to achieve global optimization. Direct methods lend themselves valuable when gradient information is not readily available or when the evaluation of the gradient is cumbersome and prone to errors. We present here the algorithms of cyclic coordinates, method of Hooke and Jeeves [1961], method of Rosenbrock [1960], simplex method of Nelder and Mead [1965], Powell's [1964] method of conjugate directions. The concepts of simulated annealing, genetic, and differential evolution algorithms are also discussed. Box's complex method for constrained problems is also included. All these algorithms are implemented in complete computer programs.

7.2 Cyclic Coordinate Search

In this method, the search is conducted along each of the coordinate directions for finding the minimum. If $\mathbf{e}_i$ is the unit vector along the coordinate direction i, we determine the value α_i minimizing $f(\alpha) = f(\mathbf{x} + \alpha\mathbf{e}_i)$, where α_i is a real number. A move is made to the new point $\mathbf{x} + \alpha_i\mathbf{e}_i$ at the end of the search along the direction i. In an n-dimensional problem, we define the search along all the directions as one stage. The function value at the end of the stage is compared to the value at the beginning of the stage in establishing convergence. The length of the move at

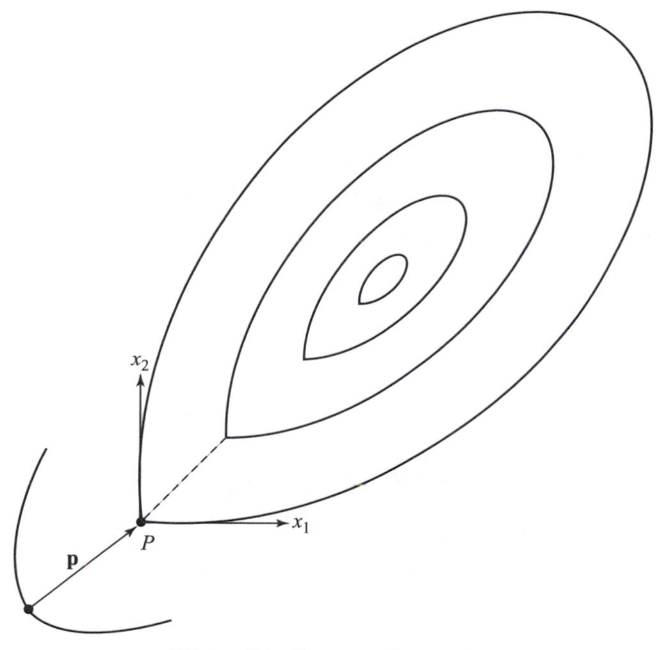

Figure 7.1. Zero gradient point.

each stage is another parameter for convergence consideration. The search for the minimum in each direction follows the steps of the establishment of a three-point pattern and the search for the minimum in the interval established. If the function contour is as shown in Fig. 7.1, the gradient appears to be zero at point P. The cyclic coordinate algorithm prematurely terminates here. We can safeguard this by introducing an *acceleration step* or pattern-search step consisting of one additional step along a direction developed by moves along the coordinate directions.

Cyclic Coordinate Algorithm

Let $\mathbf{x}^1$ be the starting point for the cycle. Set $i = 1, \mathbf{x} = \mathbf{x}^1, f = f_1$

Step 1. Determine α_i such that $f(\alpha) = (\mathbf{x} + \alpha \mathbf{e}_i)$ is a minimum
Note: α is permitted to take positive or negative values
Step 2. Move to the new point by setting $\mathbf{x} = \mathbf{x} + \alpha_i \mathbf{e}_i, f = f(\mathbf{x})$
Step 3. If $i = n$, go to step 4 else set $i = i + 1$ and go to step 1
Step 4. Acceleration Step: Denote direction $\mathbf{d} = \mathbf{x} - \mathbf{x}^1$
Find α_d such that $f(\mathbf{x} + \alpha \mathbf{d})$ is a minimum
Move to the new point by setting $\mathbf{x} = \mathbf{x} + \alpha_d \mathbf{d}, f = f(\mathbf{x})$
Step 5. If $\|\mathbf{x} - \mathbf{x}^1\| < \varepsilon_x$, or $|f - f_1| < \varepsilon_f$ go to step 6

Step 6. Set $\mathbf{x}^1 = \mathbf{x}$, $f_1 = f$, go to step 1
Step 7. Converged

The algorithm is implemented in the computer program CYCLIC, which is listed at the end of the chapter.

Example 7.1
Given $f = x_1^2 + 2x_2^2 + 2x_1x_2$, a point $\mathbf{x}^1 = (0.5, 1)^T$, with $f_1 \equiv f(\mathbf{x}^1) = 3.25$, apply the cyclic coordinate search algorithm. First, a search is made along x_1. Thus, $\mathbf{x}(\alpha) = [0.5 + \alpha, 1]^T$ and $f(\alpha) = (0.5 + \alpha)^2 + 2\,(.5 + \alpha) + 2$. Minimizing f over $\alpha \in R^1$ gives $\alpha = -1.5$ and a corresponding new point $\mathbf{x} = [-1, 1]^T, f = 1$. Next, searching along x_2; we have $\mathbf{x}(\alpha) = [-1, 1 + \alpha\,]^T$ and $f(\alpha) = 1 + 2(1 + \alpha)^2$ $-2(1 + \alpha)$ whose minimum yields $\alpha = -0.5$, a new point $\mathbf{x} = [-1, 0.5]^T, f = 0.5$. This completes one stage. As per the algorithm, an acceleration step is chosen in the direction $\mathbf{d} = [-1-0.5, 0.5-1]^T = [-1.5, -0.5]^T$. Thus, $\mathbf{x}(\alpha) = [-1 - 1.5\alpha,$ $0.5 - 0.5\alpha]^T$. Minimizing $f(\alpha)$ yields $\alpha = -0.1765$, $f = 0.3676$ and a new point $\mathbf{x}^1 = [-0.7352, 0.5882]^T$. The second stage now commences. The generated points in the example are shown in Fig. E7.1.

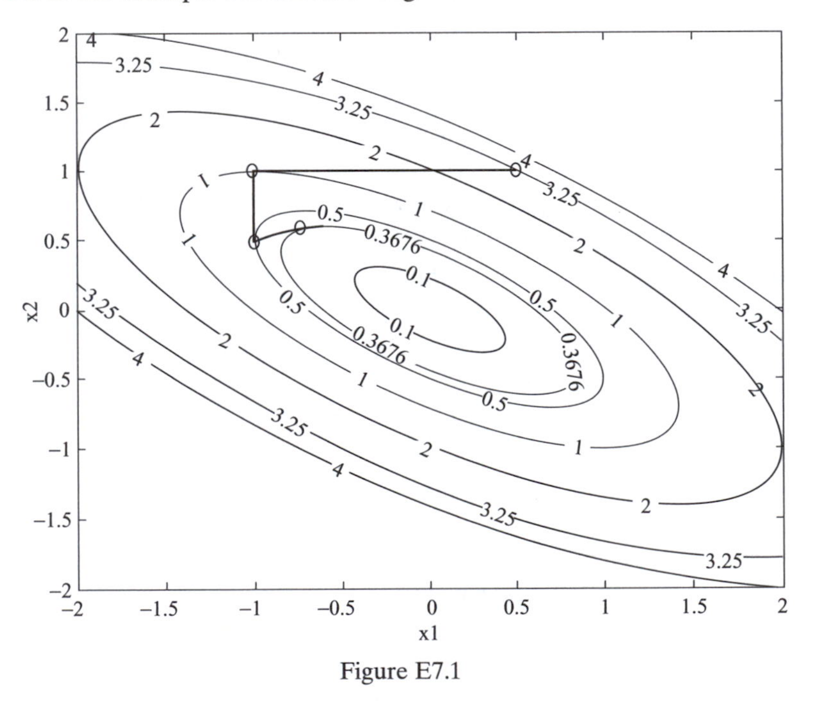

Figure E7.1

Example 7.2
Consider the problem of hanging weights from a cable, the ends of which are attached to fixed points in space. A specific configuration is shown in Fig. E7.2a.

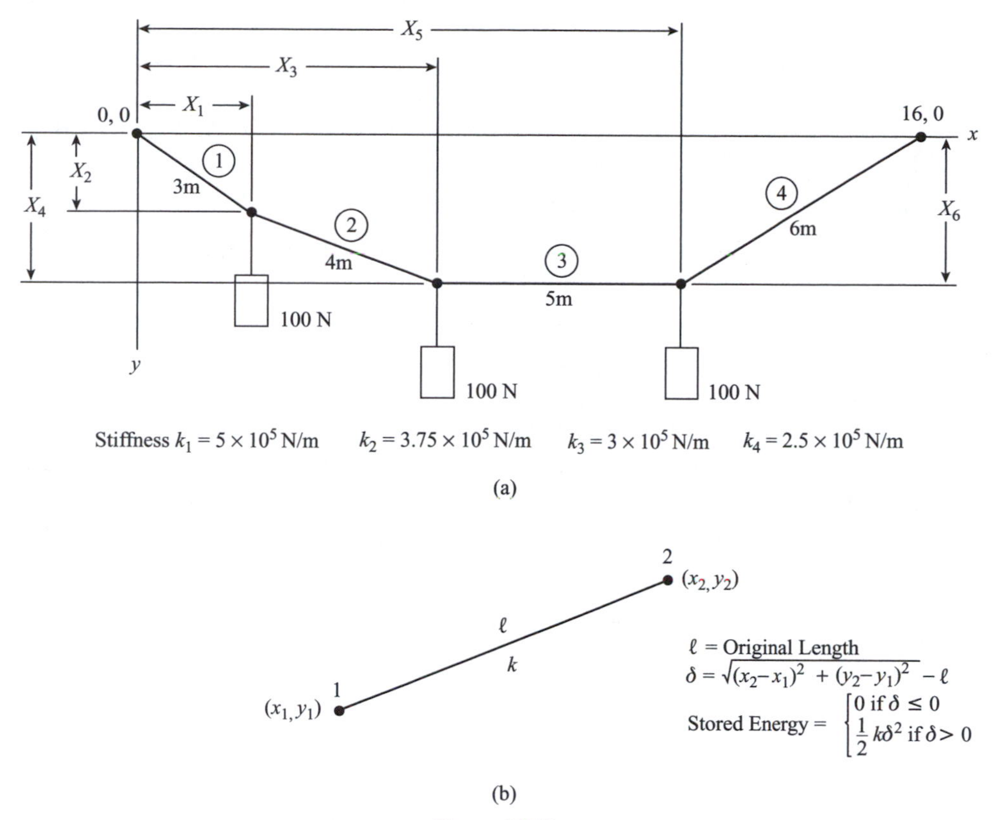

Figure E7.2

The stiffness k of a cable of length ℓ, area of cross section A and modulus of elasticity E is given by EA/ℓ. For the cable segment of original length ℓ and stiffness k shown in Fig. E7.2b, the change in length δ is given by

$$\delta = \sqrt{(x_1 - x_2)^2 + (y_1 - y_2)^2} - \ell \tag{a}$$

The stored energy (s.e.) is given by

$$\text{s.e.} = \{0 \quad \text{if} \quad \delta \leq 0; \quad \tfrac{1}{2}\text{k}\delta^2 \quad \text{if} \quad \delta > 0\} \tag{b}$$

The potential energy is given by

$$f = (\text{s.e.})_1 + (\text{s.e.})_2 + (\text{s.e.})_3 + (\text{s.e.})_4 - w_1 X_2 - w_2 X_4 - w_3 X_6 \tag{c}$$

If the equilibrium configuration corresponds to the one that minimizes f, determine the minimum value of f and the corresponding $\mathbf{X}$.

Solution

In the implementation in the computer program the segment loop goes from $i = 1$ to 4 and if $i = 1$, then $x_1 = 0$, $y_1 = 0$, otherwise $x_1 = X_{2(i-1)-1}$, $y_1 = X_{2(i-1)}$ and if $i = 4$ then $x_2 = 16$, $y_2 = 0$, otherwise $x_2 = X_{2i-1}$, $y_2 = X_{2i}$. Eq. (c) is then easily introduced in the function subroutine of the program. The results from the program are given in the following.

Function value = −870.292
$\mathbf{X} = (2.314, 1.910, 5.906, 3.672, 10.876, 3.124)$
Number of cycles = 200
Number of function evaluations = 17112

Initial step size for each cycle has been taken as 0.5 for the preceding run.

One characteristic of the cyclic coordinate method is that the function improves moderately in the beginning but the improvement is slow as we approach the minimum.

7.3 Hooke and Jeeves Pattern Search Method

In the Hooke and Jeeves method, an initial step size s is chosen and the search is initiated from a given starting point. The method involves steps of exploration and pattern search. We provide an explanation of the exploration step about an arbitrary point $\mathbf{y}$. The step size s may be chosen in the range 0.05 to 1. Values beyond 1 can be tried.

Exploration About a Point $\mathbf{y}$

Let $\mathbf{x} = \mathbf{y}$. If $\mathbf{e}_i$ is the unit vector along the coordinate direction x_i, then the function is evaluated at $\mathbf{x} + s\mathbf{e}_i$. If the function reduces, then $\mathbf{x}$ is updated to be $\mathbf{x} + s\mathbf{e}_i$. If the function does not reduce, then the function is evaluated at $\mathbf{x} - s\mathbf{e}_i$. If the function reduces, then $\mathbf{x}$ is updated to be $\mathbf{x} - s\mathbf{e}_i$. If both fail, the original $\mathbf{x}$ is retained. The searches are performed with i ranging from 1 to n. At this stage, the initial point is at $\mathbf{y}$ and the new point is at $\mathbf{x}$. Exploration is said to be successful if the function value at $\mathbf{x}$ is lower than the value at $\mathbf{y}$ by a predetermined amount (1e−8 is used in the computer program). We now explain the Hooke and Jeeves algorithm.

Pattern Search

The search is initiated from a given point B with coordinates $\mathbf{x}_B$. s is the step size and r (<1) is a step reduction parameter. Exploration is made about point B. If the exploration is not successful about the base point, the step is reduced ($s = rs$). If the exploration is successful as shown in Fig. 7.2, the pattern direction BP given by

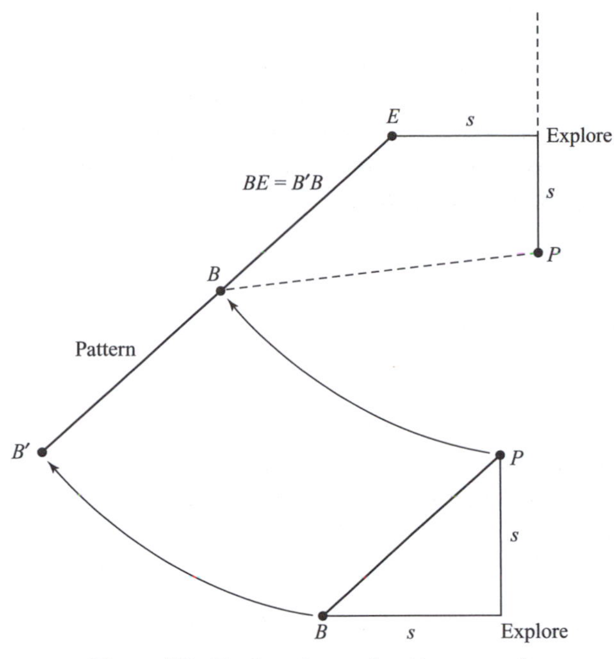

Figure 7.2. Exploration and pattern search.

$(\mathbf{x}_P - \mathbf{x}_B)$ is established. P is now made the new base point by renaming it as B, and the original point B is renamed as B' as shown in the figure. B' B is now extended along the pattern direction to point E as $\mathbf{x}_E = 2\mathbf{x}_B - \mathbf{x}_{B'}$. Exploration is now performed about point E. If the exploration is successful, that is, $f_P < f_B$, a new pattern direction BP is established as shown. We rename $P \to B$ and $B \to B'$. If $f_P > f_B$, however, we perform an exploration from point B itself. The steps keep reducing for each unsuccessful exploration. If ε_x (say 1e − 6) is the convergence parameter, we stop the search when the step size from the base point is less than ε_x.

The extension step in the algorithm can be made more general by setting $\mathbf{x}_E = \mathbf{x}_B + \alpha(\mathbf{x}_B - \mathbf{x}_{B'})$ with α ranging between 0.5 and 2. $\alpha = 1$ has been chosen in the preceding description.

The discrete version of the Hooke and Jeeves algorithm, as originally proposed, can be made more rigorous by implementing line search for exploration and search along the pattern direction.

Hooke–Jeeves Algorithm

Let $\mathbf{x}^1$ be the starting point for the cycle. Choose a relatively small termination parameter $\varepsilon > 0$, an initial step size s, a reduction parameter $r < 1$, and an acceleration parameter α.

Initial Step

set $\mathbf{x}_B = \mathbf{x}^1$. Do an exploration about $\mathbf{x}_B$. If successful, that is, function value reduces, define the new point generated as a new base point $\mathbf{x}_B$ and the old base point $\mathbf{x}_B$ as $\mathbf{x}_{B'}$. If the exploration is not successful reduce the step as $s = rs$. Repeat until successful and proceed to the typical step (performed within the iterative loop).

Typical Step

$\mathbf{x}_B$ and $\mathbf{x}_{B'}$ are the new and old base points, respectively.

(a) Generate a point $\mathbf{x}_E = \mathbf{x}_B + \alpha\,(\mathbf{x}_B - \mathbf{x}_{B'})$
(b) Perform an exploration about $\mathbf{x}_E$.
- If successful, define the new point generated as a new base point $\mathbf{x}_B$ and the old base point $\mathbf{x}_B$ as $\mathbf{x}_{B'}$. Go to step (a).
- If unsuccessful, perform an exploration about $\mathbf{x}_B$. Reduce s if needed until success or convergence. Name the new point generated as a new base point $\mathbf{x}_B$ and the old base point $\mathbf{x}_B$ as $\mathbf{x}_{B'}$ and go to step (a).

The algorithm is implemented in the computer program HOOKJEEV, which is listed at the end of the chapter.

Example 7.3
Given $f = x_1^2 + 2x_2^2 + 2x_1x_2$, a point $\mathbf{x}^1 = (0.5, 1)^T$, with $f_1 \equiv f(\mathbf{x}^1) = 3.25$, apply the Hooke and Jeeves algorithm. Assume step $s = 1$, $r = 0.25$, $\varepsilon = 0.001$, $\alpha = 1$.

Exploration about the starting point yields: $(-0.5, 1)$, $f = 1.25$ and then $(-0.5, 0)$, $f = 0.25$. Thus, $\mathbf{x}_{B'} = (0.5, 1)^T$, $\mathbf{x}_B = (-0.5, 0)^T$. The acceleration step $\mathbf{x}_E = \mathbf{x}_B + \alpha\,(\mathbf{x}_B - \mathbf{x}_{B'})$ yields a point $(-1.5, -1)^T$ with $f = 7.25$. Exploration about this point is unsuccessful; that is, points whose function value greater than 0.25 are generated. Thus, exploration about $\mathbf{x}_B$ itself is now done. $s = 1$ is not successful, so s is reduced to 0.5. This yields a new point $(0, 0)^T$, $f = 0$. Pattern point $\mathbf{x}_E = 2(0, 0)^T - (-0.5, 0)^T = (0.5, 0)^T$, $f = 0.25$, which is unsuccessful. Exploration about $(0, 0)^T$ remains unsuccessful for any value of s and, hence, the optimum point is $(0, 0)^T$ with $f = 0$.

Example 7.4
Solution to Example 7.2 using Hooke and Jeeves Discrete Algorithm:

Function value = −873.342
$\mathbf{X} = (2.202, 2.038, 5.891, 3.586, 10.869, 3.112)$
Number of contraction steps = 6
Number of base changes = 7

Number of pattern moves = 179
Number of function evaluations = 2339

7.4 Rosenbrock's Method

In the Hooke and Jeeves method, the pattern direction is established with a search in the coordinate directions. Once a pattern direction is established, we have some new information about the function. A new set of orthogonal directions can be developed using this information. Rosenbrock's method exploits this with some additional strategies for step-size decisions.

In Rosenbrock's method, the search is carried out in n orthogonal directions at any stage. New orthogonal directions are established at the next stage. The orthogonal setting makes this method robust and efficient. The main steps of initialization and performing one stage of calculations is explained in the following.

Initialization

The directions $\mathbf{v}_1, \mathbf{v}_2, \ldots, \mathbf{v}_n$, are set as the coordinate directions $\mathbf{e}_1, \mathbf{e}_2, \ldots, \mathbf{e}_n$. Step lengths s_i, $i = 1$ to n and step expansion parameter α (> 1) and a reduction parameter β (< 1) are selected. $\alpha = 3$ and $\beta = 0.5$ and $s_i = 0.55$ ($i = 1$ to n) are used in the computer program ROSENBRK. The search starts from a given point say $\mathbf{x}^0$. We set $\mathbf{x} = \mathbf{x}^0$, and $\mathbf{y} = \mathbf{x}^0$. Function is evaluated at $\mathbf{x}$, say its value is f_x. This defines the stage $k = 1$.

Stage k

The search during a stage is made cyclically as follows.

$i = 0$

Step 1. $i = i + 1$
If $i > n$, then set $i = 1$ (Note that this makes the evaluation cyclic).
A step of s_i is taken along the direction $\mathbf{v}_i$, say, $\mathbf{y} = \mathbf{x} + s_i\mathbf{v}_i$.
The function is evaluated at $\mathbf{y}$, say, its value is f_y.
If $f_y < f_x$, then the step is considered a *success*. In this case, the step size is reset as $s_i = \alpha\, s_i$ and $\mathbf{x}$ is replaced by $\mathbf{y}$, that is, $\mathbf{x} = \mathbf{y}$, $f_x = f_y$. Go to Step 2.
If $f_y \geq f_x$, then the step is considered a *failure*. The next trial, if needed, will be with a step equal to $-\beta\, s$.

Step 2. If there is at least one success and one failure in each of the n-directions, go to Step 3 else go to Step 1.

Step 3. Form n linearly independent directions $\mathbf{u}_1, \mathbf{u}_2, \ldots, \mathbf{u}_n$.

$\mathbf{u}_j = \sum_{k=j}^{n} s_k \mathbf{v}_k$, $j = 1$ to n. The length s of $\mathbf{u}_1$ ($s = \sqrt{\mathbf{u}_1^{\mathrm{T}} \mathbf{u}_1}$) is calculated. If this length is smaller than a small convergence parameter ε (= 1e − 6), the process is stopped. If convergence criterion is not satisfied, new orthogonal directions are constructed using Gram–Schmidt orthogonalization procedure.

$$\mathbf{v}_1 = \frac{\mathbf{u}_1}{s}$$

$$\mathbf{u}'_j = \mathbf{u}_j - \sum_{k=1}^{j-1} \left(\mathbf{u}_j^{\mathrm{T}} \mathbf{v}_k\right) \mathbf{v}_k \quad \text{for } j = 2 \text{to } n$$

$$\mathbf{v}_j = \frac{\mathbf{u}'_j}{\sqrt{\mathbf{u}_j^{\prime\mathrm{T}} \mathbf{u}'_j}} \tag{7.1}$$

The set of vectors $\mathbf{v}_1, \mathbf{v}_2, \ldots, \mathbf{v}_n$ form an orthonormal set ($\mathbf{v}_i^{\mathrm{T}} \mathbf{v}_j = 0$ for $i \neq j$, and $\mathbf{v}_i^{\mathrm{T}} \mathbf{v}_i = 1$). We set $k = k + 1$ and perform the next *stage* of calculations. Loss of orthogonality can be recognized by the size of the norm of vectors $\mathbf{u}'_j$ (if it is small, say, < 1e − 10). At this point the new stage is started with n coordinate directions as starting vectors.

In the algorithm described, the steps are taken in a cyclic manner and repeated till there is at least one success and one failure in each direction. Other strategies can be used in the discrete version. A more rigorous version of Rosenbrock's method is one in which exact line search is performed in each direction. In this case, if there is no step change in a certain direction, say, i, the original direction $\mathbf{v}_i$ is retained in $\mathbf{u}_i$.

Example 7.5

Given $f = x_1^2 + 2x_2^2 + 2\,x_1\,x_2$, a point $\mathbf{x}^1 = (0.5, 1)^{\mathrm{T}}$, with $f_1 = 3.25$, apply Rosenbrock's method with $\alpha = 3$, $\beta = 0.5$ and $s_i = 0.55$. In the initial stage, or stage 1, the orthogonal directions are the coordinate directions $(1, 0)^{\mathrm{T}}$ and $(0, 1)^{\mathrm{T}}$, respectively. A step to the point $(0.5 + s, 1)^{\mathrm{T}}$ results in $f = 5.2025$, which is a failure. Then, a step $(0.5, 1 + s)^{\mathrm{T}}$ gives $f = 6.605$, which is also a failure. The current point remains at $\mathbf{x}^1$. Then, a step of $-\beta\ s$ is made along directions where a previous failure occurred. We have $(0.5 - \beta\ s, 1) = (0.225, 1)^{\mathrm{T}}$, $f = 2.5006$, which is a success. Thus, the updated point is $\mathbf{x} = (0.225,1)^{\mathrm{T}}$. Next, a step to $(0.225, 1 - \beta\ s)^{\mathrm{T}}$ yields $f = 1.4281$, which is a success. The updated point is $\mathbf{x}^2 = (0.225, 0.725)^{\mathrm{T}}$, $f_2 = 1.4281$. Since one failure and one success has occurred in each of the search directions, this concludes the "first stage."

In stage 2, the first search direction is $\mathbf{d} = \mathbf{x}^2 - \mathbf{x}^1 = \sum_{k=1}^{2} s_k \mathbf{v}_k = -\beta s\binom{1}{0} - \beta s\binom{0}{1} = \binom{-0.275}{-0.275}$, which is normalized to $\mathbf{v}_1 = (-0.7071, -0.7071)^{\mathrm{T}}$. The second search direction is obtained by first defining $\mathbf{u}_2 = \sum_{k=2}^{2} s_k \mathbf{v}_k = -\beta s\binom{0}{1} = \binom{0}{-0.275}$ and then using Eqs. (7.1) to obtain $\mathbf{v}_2 = (0.7071, -0.7071)^{\mathrm{T}}$. The step size s is set equal to length of the first vector as $s = \|\mathbf{x}^2 - \mathbf{x}^1\| = \sqrt{\mathbf{d}^{\mathrm{T}}\mathbf{d}} = 0.3889$.

The first trial point is $\mathbf{x} = \mathbf{x}^2 + s\,\mathbf{v}_1 = (-0.05, 0.45)^{\mathrm{T}}, f = 0.3625 < 1.4281$; hence a success, and the current point is updated. Next trial is at $(-0.05, 0.45)^{\mathrm{T}} + s\mathbf{v}_2 = (0.225, 0.175)^{\mathrm{T}}, f = 0.1906 < 0.3625$, hence a success. With $\alpha = 3$, the next trial is at $(0.225, 0.175)^{\mathrm{T}} + 3\,s\,\mathbf{v}_1 = (-0.6, -0.65)^{\mathrm{T}}, f = 1.985$, a failure. Next point is at $(0.225, 0.175)^{\mathrm{T}} + 3\,s\,\mathbf{v}_2 = (1.05, -0.65)^{\mathrm{T}}, f = 0.5825$, a failure. Since a success and a failure are obtained along each search direction, the stage concludes with $\mathbf{x}^3 = (0.225, 0.175)^{\mathrm{T}}, f_3 = 0.1906$. The iterates are shown in Fig. E7.5. Iterations 3 – 6 are obtained from the computer program ROSENBRK.

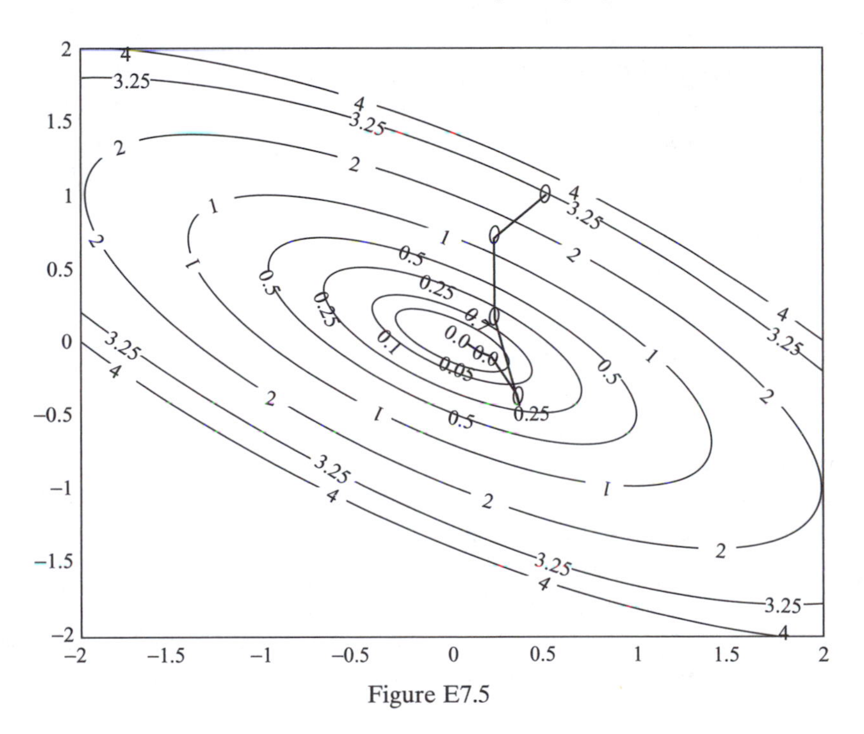

Figure E7.5

Example 7.6

Solution to Example 7.2 using Rosenbrock's method

Function value = – 872.3

$\mathbf{X} = (2.135, 2.108, 5.882, 3.510, 10.867, 3.109)$

Number of stages = 42

Number of function evaluations = 1924

7.5 Powell's Method of Conjugate Directions

Powell developed a method using the idea of conjugate directions defined with respect to the quadratic form. Among the zero order methods, Powell's conjugate direction method has similar characteristics as Fletcher–Reeves method in the gradient based methods. Convergence in n-iterations is easy to prove for the underlying quadratic form (see Chapter 3).

Consider the quadratic form

$$\text{minimize } q(\mathbf{x}) = \frac{1}{2}\mathbf{x}^{\mathrm{T}}\mathbf{A}\mathbf{x} + \mathbf{c}^{\mathrm{T}}\mathbf{x} \tag{7.2}$$

As discussed in Chapter 3, directions $\mathbf{d}_i$ and $\mathbf{d}_j$ are conjugate with respect to $\mathbf{A}$ if

$$\mathbf{d}_i^{\mathrm{T}}\mathbf{A}\mathbf{d}_j = 0 \quad \text{if} \quad i \neq j \tag{7.3}$$

If minimization is carried out along successive directions, which are conjugate with respect to all the previous directions, the convergence can be achieved in n-steps. Powell developed a neat idea of constructing the conjugate directions without using derivatives. We discuss the main ideas of this development. The gradient of the quadratic function q is given by

$$\mathbf{g} = \mathbf{A}\mathbf{x} + \mathbf{c} \tag{7.4}$$

If a function is minimized along the direction $\mathbf{d}$ as shown in Fig. 7.3, and if the minimum is achieved at point $\mathbf{x}^1$ where the gradient is $\mathbf{g}^1$, then

$$\mathbf{d}^{\mathrm{T}}\mathbf{g}^1 = 0 \tag{7.5}$$

Given a direction $\mathbf{d}$ and two points $\mathbf{x}_{\mathrm{A}}$ and $\mathbf{x}_{\mathrm{B}}$, let $\mathbf{x}^1$ and $\mathbf{x}^2$ be the points of minima along $\mathbf{d}$ from A and B, respectively, as shown in Fig. 7.4. Denoting $\mathbf{s} = \mathbf{x}^2 - \mathbf{x}^1$, we note that $\mathbf{d}^{\mathrm{T}}\mathbf{g}^1 = 0$, $\mathbf{d}^{\mathrm{T}}\mathbf{g}^2 = 0$ using the property in Eq. 7.5. Thus,

$$0 = \mathbf{d}^{\mathrm{T}}(\mathbf{g}^2 - \mathbf{g}^1) = \mathbf{d}^{\mathrm{T}}\mathbf{A}(\mathbf{x}^2 - \mathbf{x}^1) = \mathbf{d}^{\mathrm{T}}\mathbf{A}\mathbf{s} \tag{7.6}$$

Direction $\mathbf{s}$ is $\mathbf{A}$-conjugate with respect to the direction $\mathbf{d}$. This defines the *parallel subspace property*, which states that if two points of minima are obtained

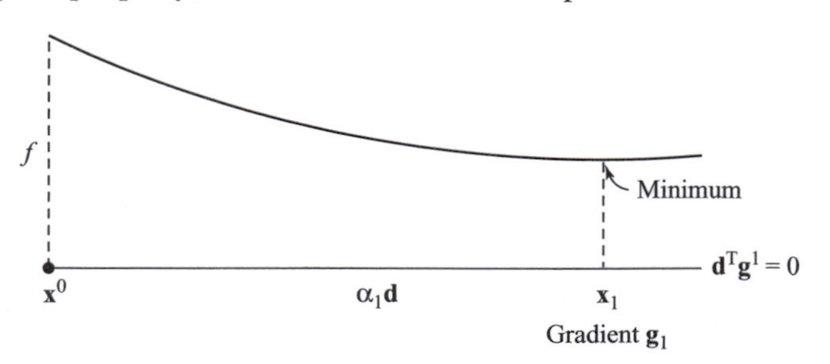

Figure 7.3. Minimum along $\mathbf{d}$.

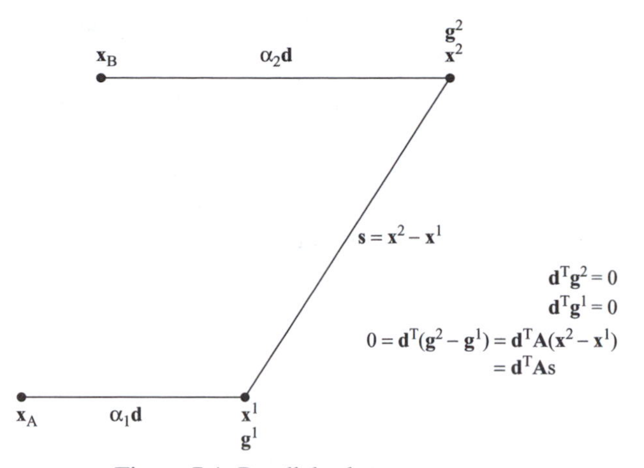

Figure 7.4. Parallel subspace property.

along parallel directions, then the direction given by the line joining the minima is **A**-conjugate with respect to the parallel direction.

In order to view the extended parallel subspace property, we refer to Fig. 7.5. $\mathbf{e}_1, \mathbf{e}_2, \mathbf{e}_3, \mathbf{e}_4$ are the unit vectors along the coordinate directions.

We first set $\mathbf{s}_1 = \mathbf{e}_1$, and then make line searches for minima along $\mathbf{s}_1$, $\mathbf{e}_2$, $\mathbf{e}_3$, $\mathbf{e}_4$ and $\mathbf{s}_1$ to reach point 6. Then the direction $\mathbf{s}_2 = (\mathbf{x}^6 - \mathbf{x}^2)$ is **A**-conjugate with $\mathbf{s}_1$ since these are the minima from points 1 and 5 along $\mathbf{s}_1$. That is, $\mathbf{s}_1^T\mathbf{A}\mathbf{s}_2 = 0$. In the next iteration, search starts from point 6 along $\mathbf{s}_2$, $\mathbf{e}_3$, $\mathbf{e}_4$, $\mathbf{s}_1$ and $\mathbf{s}_2$ to reach point 11. Conjugacy of $\mathbf{s}_2$ and $\mathbf{s}_3$ can be easily established. To establish the conjugacy of $\mathbf{s}_1$ and $\mathbf{s}_3$, we consider the minima point 6 from 5 and point 10 from 9, which are points of minima along $\mathbf{s}_1$. Thus,

$$0 = \mathbf{s}_1^T\mathbf{A}(\mathbf{x}^{10} - \mathbf{x}^6) = \mathbf{s}_1^T\mathbf{A}(\alpha_6\mathbf{s}_2 + \mathbf{s}_3 - \alpha_{10}\mathbf{s}_2) = \mathbf{s}_1^T\mathbf{A}\mathbf{s}_3 \tag{7.7}$$

$\mathbf{s}_1$ and $\mathbf{s}_3$ are, thus, conjugate. In the next iteration, searches are conducted along $\mathbf{s}_3$, $\mathbf{e}_4$, $\mathbf{s}_1$, $\mathbf{s}_2$ and $\mathbf{s}_3$ to obtain $\mathbf{s}_4$, and $\mathbf{s}_4$ is **A**-conjugate with $\mathbf{s}_1$, $\mathbf{s}_2$, and $\mathbf{s}_3$. A search now along $\mathbf{s}_4$ will complete the search along all the four conjugate directions.

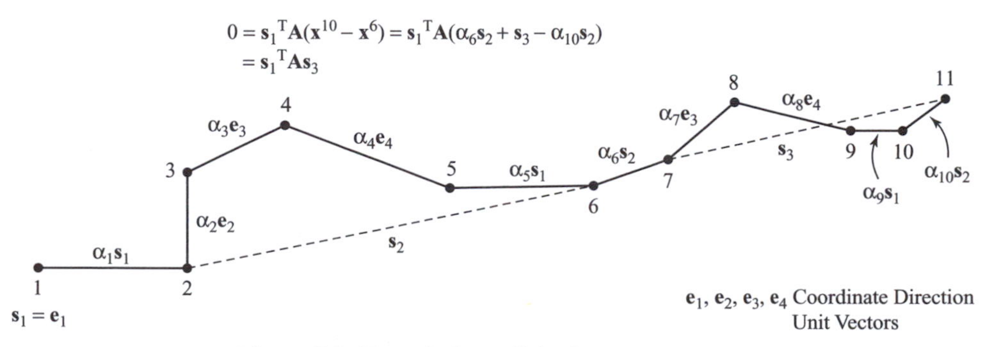

Figure 7.5. Extended parallel subspace property.

After searching along all the conjugate directions, a *spacer* step is introduced where a search is made from the current point along the *coordinate* directions. This is to ensure convergence. In the computer program, we repeat the preceding process – specifically, we denote n-iterations to represent one stage of the conjugate direction method, and a new stage is started from n coordinate directions. If a stage starts from point $\mathbf{x}^1$ and ends at point $\mathbf{x}^2$, then the convergence is said to be achieved when the norm $\|\mathbf{x}^2 - \mathbf{x}^1\|$ is less than a small parameter ε or by checking the function improvement from $\mathbf{x}^1$ to $\mathbf{x}^2$.

Powell's method of conjugate directions is implemented in the computer program POWELLCG.

Example 7.7

Given $f = x_1^2 + 2x_2^2 + 2\,x_1\,x_2$, a point $\mathbf{x}^1 = (0.5, 1)^T$, with $f_1 = 3.25$, apply Powells's Method.

The initial searches are along coordinate directions. Line search along x_1 yields $\mathbf{x} = (-1,\ 1)^{\mathrm{T}}$, $f = 1$, and line search along x_2 yields $\mathbf{x} = (-1,\ 0.5)^{\mathrm{T}}$, $f = 0.5$. Searching along x_1 again yields $\mathbf{x} = (-0.5, 0.5)^{\mathrm{T}}$, $f = 0.25$. Thus, $\mathbf{s}_2 = (-0.5 - (-1), 0.5 - 1)^{\mathrm{T}} = (0.5, -0.5)^{\mathrm{T}}$. Search along $\mathbf{s}_2$ from $(-0.5, 0.5)^{\mathrm{T}}$ yields $\mathbf{x} = (0, 0)^{\mathrm{T}}$, $f = 0$. Since this is a quadratic function in two variables, searches along $\mathbf{s}_1$ and $\mathbf{s}_2$ has produced the optimum. Note that $\mathbf{A}$–conjugacy is evident since $\mathbf{s}_1^T \mathbf{A} \mathbf{s}_2 = (1 \quad 0)\begin{bmatrix} 2 & 2 \\ 2 & 4 \end{bmatrix}\begin{pmatrix} 0.5 \\ -0.5 \end{pmatrix} = 0$. The iterates are plotted in Fig. E7.7.

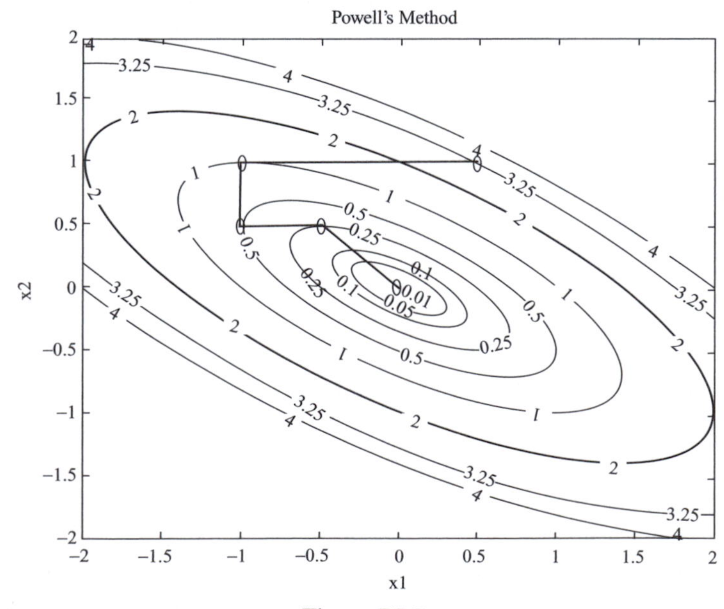

Figure E7.7

Example 7.8
Solution to Example 7.2 using Powell's method of conjugate directions:

Function value = −872.48
X = (2.267, 1.966, 5.926, 3.584, 10.910, 3.178)
Number of stages = 8
Number of function evaluations = 5004

The number of function evaluations is larger than Rosenbrock's method for this problem. Other trials show that this method is quite robust.

7.6 Nelder and Mead Simplex Method

The simplex method of Nelder and Mead, which is a further development of the method proposed by Spendley et al. [1962], uses geometric properties of the n-dimensional space. In an n-dimensional space, $n + 1$ points form a *simplex*. A triangle (three points) is an example of a simplex in two dimensions. In three dimensions, a tetrahedron (4 points) forms a simplex. An initial simplex in n-dimensions is easily created by choosing the origin as one corner and n points, each marked at a distance of c from the origin along the coordinate axes as shown in Fig. 7.6. For this simplex, the smaller sides along the coordinate directions are of length c and the n longer edges are each of length $c\sqrt{2}$.

A simplex with edges of equal length can be constructed by choosing one point at the origin and a point i at $(b, b, \ldots, b, a, b, \ldots b)$, where coordinate a is at location i, $i = 1$ to n (see Fig. 7.7). Denoting c as the length of the edge, a and b can be determined as

$$
\begin{aligned}
b &= \frac{c}{n\sqrt{2}}(\sqrt{n+1} - 1) \\
a &= b + \frac{c}{\sqrt{2}}
\end{aligned}
\tag{7.8}
$$

A regular simplex is easily created in space by adding the coordinates of a starting point to each of the $n + 1$ points (shifting the origin). In the simplex algorithm for minimization, we first evaluate the function at the $n + 1$ corner points of the simplex. Various strategies are then used to replace the point with the highest function value by a newly generated point. The newly generated point is obtained by the steps of *reflection*, *expansion*, or *contraction*. Let f_h, f_s, and f_ℓ be the highest, second highest, and the least function values among the $n + 1$ corners of the simplex respectively. Let $\mathbf{x}_h$, $\mathbf{x}_s$, and $\mathbf{x}_\ell$ be the corresponding coordinates. We calculate $\mathbf{x}_B$,

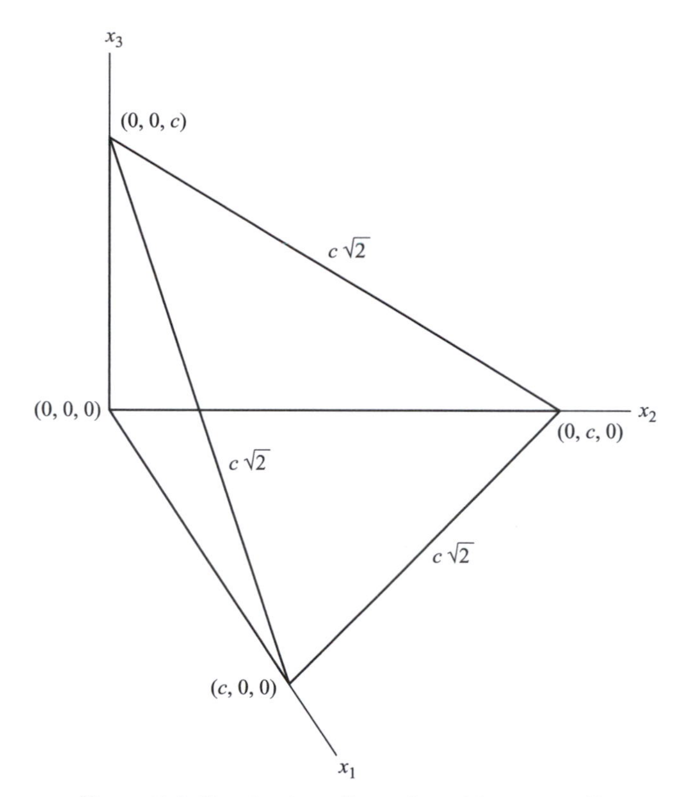

Figure 7.6. Simplex in n dimensions (shown $n = 3$).

the average of n points excluding the highest point $\mathbf{x}_h$. We may write $\mathbf{x}_B$ as

$$\mathbf{x}_B = \frac{1}{n}\sum_{\substack{i=1 \\ i \neq h}}^{n+1} \mathbf{x}_i$$

In computation, the following procedure for determining $\mathbf{x}_B$ does not allow the numbers to grow in magnitude. If $\mathbf{x}_B$ is the mean of i points and we introduce the $(i + 1)$th point $\mathbf{x}_{i+1}$, the updated mean of the $i + 1$ points is given by the relation

$$\mathbf{x}_B = \mathbf{x}_B + \frac{\mathbf{x}_{i+1} - \mathbf{x}_B}{i + 1} \tag{7.9}$$

While updating, the point with the highest value is skipped. If a point m is dropped from the previously calculated set, the update formula is

$$\mathbf{x}_B = \mathbf{x}_B - \frac{\mathbf{x}_m - \mathbf{x}_B}{i - 1} \tag{7.10}$$

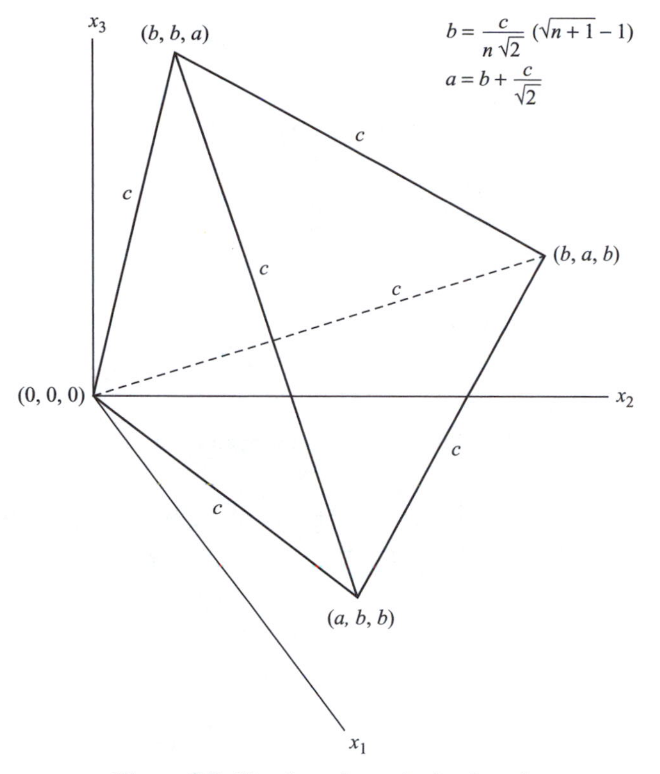

Figure 7.7. Simplex of equal edge lengths.

Reflection

After computing $\mathbf{x}_B$, we know that the line from $\mathbf{x}_h$ to $\mathbf{x}_B$ is a downhill or descent direction. A new point on this line is found by reflection. A reflection parameter $r > 0$ is initially chosen. The suggested value for r is 1. The reflected point is at $\mathbf{x}_r$ as shown in Fig. 7.8 and is given by

$$\mathbf{x}_r = \mathbf{x}_B + r\,(\mathbf{x}_B - \mathbf{x}_h) \tag{7.11}$$

If $f_\ell \le f_r \le f_h$, then we accept it wherein we define a new simplex by replacing $\mathbf{x}_h$ by $\mathbf{x}_r$.

Expansion

Let f_r be the function value at $\mathbf{x}_r$. If $f_r < f_\ell$, then this means that the reflection has been especially successful in reducing the function value, and we become "greedy"

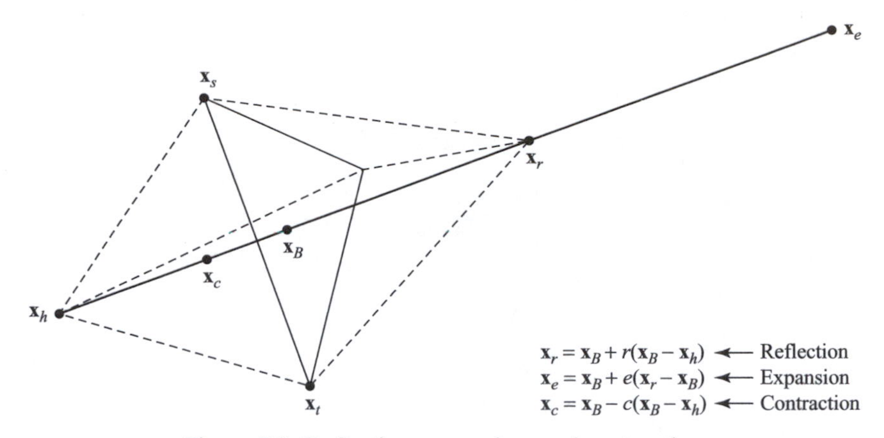

Figure 7.8. Reflection, expansion, and contraction.

and try an expansion. The expansion parameter $e > 1$ (with a suggested value of 1) defines a point farther from the reflection as shown in Fig. 7.8.

$$\mathbf{x}_e = \mathbf{x}_r + e\,(\mathbf{x}_r - \mathbf{x}_\mathrm{B}) \tag{7.12}$$

We accept $\mathbf{x}_e$ if $f_e < f_\ell$. Otherwise, we simply accept the previous reflected point.

Contraction – Case I

If $f_r > f_h$, then we know that a better point lies between point $\mathbf{x}_h$ and $\mathbf{x}_\mathrm{B}$, and we perform a contraction with a parameter $0 < c < 1$ (suggested value $c = 0.5$), yielding a point inside the current simplex.

$$\mathbf{x}_c = \mathbf{x}_\mathrm{B} + c\,(\mathbf{x}_h - \mathbf{x}_\mathrm{B}) \tag{7.13a}$$

Contraction – Case II

If $f_i < f_r \le f_h$ for all $i \ne h$, or equivalently $f_s < f_r \le f_h$ where f_s is the second highest value (this condition means that replacing $\mathbf{x}_h$ with $\mathbf{x}_r$ will make $\mathbf{x}_r$ the worst point in the new simplex), then the contraction is made as per

$$\mathbf{x}_c = \mathbf{x}_\mathrm{B} + c\,(\mathbf{x}_r - \mathbf{x}_\mathrm{B}) \tag{7.13b}$$

yielding a point outside the current simplex but between $\mathbf{x}_\mathrm{B}$ and $\mathbf{x}_r$.

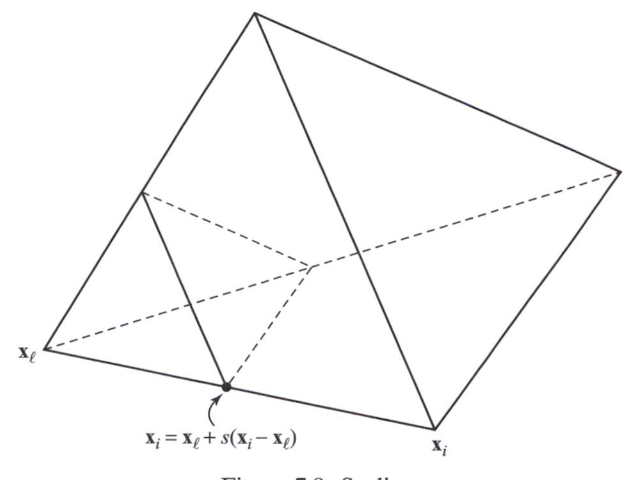

Figure 7.9. Scaling.

Scaling

We accept $\mathbf{x}_c$ unless (7.13a) yields a point with $f_c > f_h$ or (7.13b) yields a point with $f_c > f_r$, in which case a *scaling* operation is performed as follows.

The scaling operation retains the point with the least value and shrinks the simplex as shown in Fig. 7.9 with a scale parameter $0 < s < 1$ (0.5 is suggested). Point $\mathbf{x}_i$ is updated using

$$\mathbf{x}_i = \mathbf{x}_\ell + s\left(\mathbf{x}_i - \mathbf{x}_\ell\right) \tag{7.14}$$

We now present the algorithm, which is implemented in the program NELD-MEAD.

Nelder and Mead Simplex Algorithm

Step 1. Start with initial point $\mathbf{x}_1$. Function value f_1. Set $iter = 0$.
Step 2. Create a simplex with edge length of C by generating n new corner points. Evaluate function values at the n points. Set $iter = iter + 1$.
Step 3. Identify the highest $\mathbf{x}_h$, second highest $\mathbf{x}_s$, and the least $\mathbf{x}_l$ value points with function values f_h, f_s, and f_l, respectively.
Step 4. If $f_h - f_l < 1\text{e} - 6 + 1\text{e} - 4 * \text{abs}(f_l)$ go to step 20.
Step 5. Evaluate $\mathbf{x}_\text{B}$, the average of n points excluding the highest point $\mathbf{x}_h$.
Step 6. Perform *reflection* using Eq. (7.11), and evaluate the function value f_r.
Step 7. $f_r \geq f_l$ then go to step 12.

Step 8. Perform *expansion* using Eq. (7.12), and evaluate the function value f_e.
Step 9. If $f_e \geq f_l$, then go to step 11.
Step 10. Replace $\mathbf{x}_h$ by $\mathbf{x}_e$, f_h by f_e (accept expansion $\mathbf{x}_e$) go to step 3.
Step 11. Replace $\mathbf{x}_h$ by $\mathbf{x}_r$, f_h by f_r (accept reflection $\mathbf{x}_r$) go to step 3.
Step 12. If $f_r > f_h$ then go to step 14.
Step 13. Replace $\mathbf{x}_h$ by $\mathbf{x}_r$, f_h by f_r (accept $\mathbf{x}_r$ before checking if contraction is needed).
Step 14. If $f_r \leq f_s$ go to step 3.
Step 15. Perform *contraction* using Eq. (7.13), and evaluate the function value f_c.
Step 16. If $f_c > f_h$ go to step 18.
Step 17. Replace $\mathbf{x}_h$ by $\mathbf{x}_c$, f_h by f_c (accept contraction $\mathbf{x}_c$) go to step 3.
Step 18. Perform *scaling* using Eq. (7.14).
Step 19. Evaluate function values at the n new corners of the simplex and go to step 3.
Step 20. If *iter* > 1 and abs($f_{min} - f_l$) < 1e − 6 + 1e − 4*abs(f_l) go to step 24.
Step 21. $f_{min} = f_l$.
Step 22. Replace $\mathbf{x}_1$ by $\mathbf{x}_l$, f_1 by f_l.
Step 23. Reduce step reduction parameter and go to step 2.
Step 24. Convergence achieved. Stop.

At step 4, the convergence may be checked by calculating the value of the standard deviation σ in place of $f_h - f_l$. In this case, σ is calculated using

$$\sigma = \sqrt{\frac{\sum_{i=1}^{n+1} (f_i - \bar{f})^2}{n+1}} \tag{7.15}$$

where $\bar{f}$ is the mean of $n + 1$ function values. In view of the checks at Steps 4 and 20 in the algorithm implementation, the standard deviation approach did not show any advantages.

Nelder and Mead simplex method is implemented in the computer program NELDMEAD.

Example 7.9

Given $f = x_1^2 + 2x_2^2 + 2\,x_1\,x_2$, an initial simplex of three points $\mathbf{x}_1 = (0, 1)^{\mathrm{T}}$, $\mathbf{x}_2 = (0.5, 1)^{\mathrm{T}}$, $\mathbf{x}_3 = (0.5, 0.5)^{\mathrm{T}}$ apply Nelder and Meade's Method. Take $r = 1$, $e = 1$, $c = 0.5$, $s = 0.5$. In what follows, the superscript T is omitted on the coordinates.

We evaluate $f_1 = 2$, $f_2 = 3.25$, $f_3 = 1.25$. Thus, we define $\mathbf{x}_h = (0.5, 1)$, $\mathbf{x}_l = (0.5, 0.5)$, $f_h = 3.25$, $f_l = 1.25$. The centroid $\mathbf{x}_{\mathrm{B}} = 0.5[(0, 1) + (0.5, 0.5)] = (0.25,$

0.75), $f_B = 1.5625$. Reflection using (7.11) gives $\mathbf{x}_r = (0,\ 0.5)$, $f_r = 0.5$. Since $f_r < f_l$, an expansion is done as per (7.12) giving $\mathbf{x}_e = (-0.25,\ 0.25)$, $f_e = 0.0625 < f_l$. Thus, $\mathbf{x}_e$ is accepted and replaces $\mathbf{x}_h$. The new simplex is $(0, 1), (0.5, 0.5), (-0.25, 0.25)$.

We identify $\mathbf{x}_h = (0, 1)^T$, $x_l = (-0.25, 0.25)$, $f_h = 2$, $f_l = 0.0625$. $\mathbf{x}_B = (0.125, 0.375)$, $f_B = 0.3906$. $\mathbf{x}_r = (0.25, -0.25)$, $f_r = 0.0625$, which is accepted. Thus, the new simplex is $(0.5, 0.5), (-0.25, 0.25), (0.25, -0.25)$.

We identify $\mathbf{x}_h = (0.5,\ 0.5)$, $x_l = (-0.25, 0.25)$, $f_h = 1.25$, $f_l = 0.0625$ and the second highest value $f_s = 0.0625$. $\mathbf{x}_B = (0, 0)$, $f_B = 0$. $\mathbf{x}_r = (-0.5, -0.5)$, $f_r = 1.25$. Since $f_r > f_s$, contraction is performed based on (7.17) to yield $\mathbf{x}_c = (-0.25, -0.25)$, $f_c = 0.3125$. The simplex is: $(0.25, -0.25), (-0.25, 0.25), (-0.25, -0.25)$.

$\mathbf{x}_h = (-0.25,\ -0.25)$ with $f_h = 0.3125$, $f_l = 0.0625$. We get $\mathbf{x}_B = (0,\ 0)$, $\mathbf{x}_r = (0.25,\ 0.25)$, $f_r = 0.3125$. Contraction in (7.13) gives $\mathbf{x}_c = (-0.125,\ -0.125)$, $f_c = 0.078125$. The next simplex is: $(0.25,\ -0.25)$, $(-0.25,\ 0.25)$, $(0.125,\ 0.125)$. Decrease in objective function can be seen by computing the function values at the centroid of the simplex's generated: $\bar{f} = 2.0556, 0.7847, 0.1389, 0.0347, 0.0087$. The five simplex's generated are shown in Fig. E7.9 in the aforementioned. The reader may complete a few more iterations.

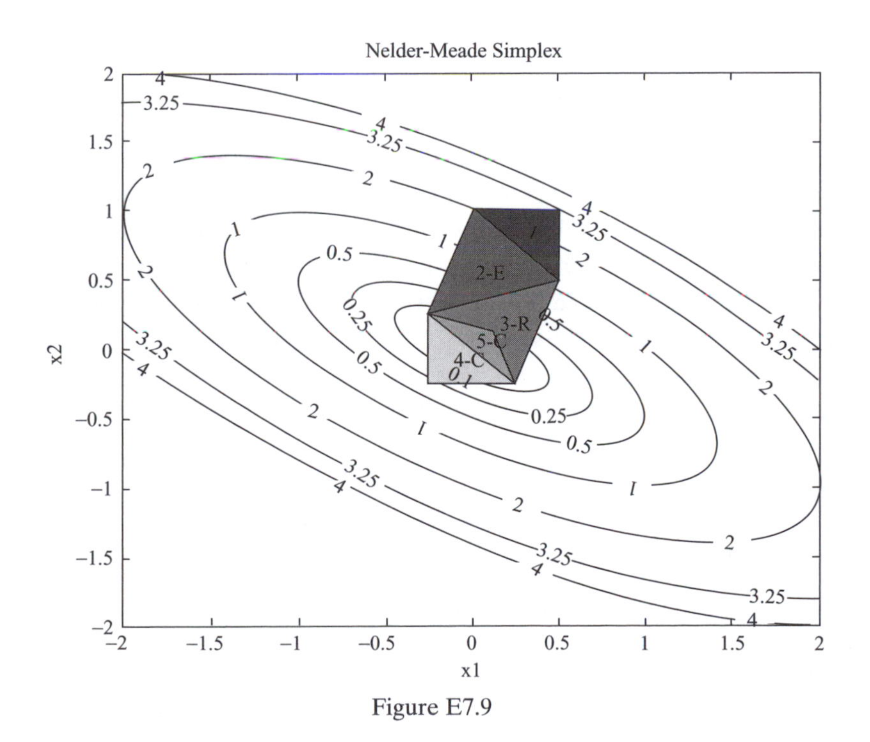

Figure E7.9

Example 7.10

Solution to Example 7.2 using Nelder and Mead simplex method:

Function value = −872.7
X = (2.250, 1.986, 5.929, 3.559, 10.915, 3.186)
Number of reflections = 1382
Number of expansions = 188
Number of contractions = 676
Number of scaling operations = 0
Number of function evaluations = 2307

Nelder and Mead simplex method is found to be quite efficient. Several variants of this method are available in the literature. One modification is to extend the method to solve optimization problems with constraints. By introducing some randomness, global optimization issues may be answered. A simple randomness idea is to generate the initial simplex in a random orthant. Let λ denote a vector of signs +1 or −1. The sign λ_i for direction i is decided based on a random number r $(0 \leq r \leq 1)$ generated. If $r < 0.5$ we set $\lambda_i = -1$ and $\lambda_i = +1$ otherwise. A regular simplex in a random orthant is obtained by multiplying the a or b defining direction i by λ_i. In two dimensions, we have four possible quadrants. There are 8 in three dimensions. The possibilities are 2^n in n-dimensions.

7.7 Simulated Annealing (SA)

In the conventional methods of minimization, we seek to update a point when the function has a lower value. This strategy generally leads to a local minimum. A robust method for seeking a global minimum must adopt a strategy where a higher value of a function is acceptable under some conditions. Simulated annealing provides such a strategy. Annealing is a process where stresses are relieved from a previously hardened body. Metal parts and parts made out of glass are subject to annealing process to relieve stresses. Parts with residual stresses are brittle and are prone to early failure. Metals, such as steel, have phase transformations when temperature changes. If a metal is heated to a high level of temperature, the atoms are in a constant state of motion. Controlled slow cooling allows the atoms to adjust to a stable equilibrium state of least energy. The probability $p(\Delta E)$ of change in energy ΔE is given by Boltzmann's probability distribution function given by

$$p(\Delta E) = e^{-\frac{\Delta E}{kT}} \tag{7.16}$$

where T is the temperature of the body and k is the Boltzmann's constant.

Metropolis observed that the probability of higher energy is larger at higher temperatures and that there is some chance of a high energy as the temperature drops. Energy in the annealing process sometimes increases even while the trend is a net decrease. This property applied to optimization problems is referred to as Metropolis algorithm. In optimization applications, we start from an initial state of temperature T, which is set at a high level. Boltzmann's constant may be chosen as equal to 1. The change Δf in the function value, is accepted whenever it represents a decrease. When it is an increase, we accept it with a probability of $p(\Delta f) = e^{-\frac{\Delta f}{T}}$. This is accomplished by generating a random number r in the range 0 to 1 and accepting the new value when $r \leq p$.

It is important for the reader to note that SA is not in itself an algorithm. Rather, it consists of the following:

i. A randomized search strategy
ii. Acceptance or rejection of points based on the Metropolis criterion
iii. A decreasing temperature schedule
iv. Storage of the best design point in one's "pocket" during the iterative process

Metropolis algorithm has been extended, coded, and further explored by other researchers. We present here the algorithm developed by Corana et al. [1987]. This algorithm is implemented in the program SIMANAL. The problem solved by the algorithm is of the form

$$\begin{aligned} &\text{minimize} \quad f(\mathbf{x}) \\ &\text{subject to} \quad l_i \leq x_i \leq u_i, \quad i = 1 \text{ to } n \end{aligned} \tag{7.17}$$

Each variable has lower and upper bounds. The search for the minimum is initiated at a feasible starting point $\mathbf{x}$. Function value f is evaluated at $\mathbf{x}$. We set $\mathbf{x}_{\min} = \mathbf{x}$, and $f_{\min} = f$. A vector $\mathbf{s}$ of step sizes with step-size s_i along the coordinate direction $\mathbf{e}_i$ is chosen. Initially, each s_i may be set equal to a step size s_T (=1) and a step reduction parameter r_s. A vector $\mathbf{a}$ of acceptance ratios with each element equal to 1 is defined. A starting temperature T and a temperature reduction factor r_T are chosen. r_s and r_T have been chosen as 0.9 and 0.5 in the program. Other values may be tried. The main loop is the temperature loop where the temperature is set as $r_T\, T$ and the step size set as $r_s s_T$ at the end of a *temperature step*. At each temperature, N_T (= 5 in the program) iterations are performed. Each iteration consists of N_C cycles. A cycle involves taking a random step in each of the n-directions, successively. A step in a direction i is taken in the following manner. A random number r in the range -1 to 1 is generated and a new point $\mathbf{x}_S$ is evaluated using

$$\mathbf{x}_S = \mathbf{x} + r\, s_i \mathbf{e}_i \tag{7.18}$$

If this point is outside the bounds, the i th component of $\mathbf{x}_S$ is adjusted to be a random point in the interval l_i to u_i. The function value f_s is then evaluated. If $f_s \leq f$, then the point is accepted by setting $\mathbf{x} = \mathbf{x}_S$. If $f_s \leq f_{\min}$ then $f_{\min}$ and $\mathbf{x}_{\min}$ are updated. If $f_s > f$ then it is accepted with a probability of

$$p = e^{\frac{f - f_s}{T}} \tag{7.19}$$

A random number r is generated and f_s is accepted if $r < p$. This is referred to as the *Metropolis criterion*. Whenever a rejection takes place, the acceptance ratio a_i, which is the ratio of the number of acceptances to the total number of trials for direction i, is updated as

$$a_i = a_i - \frac{1}{N_c} \tag{7.20}$$

At the end of N_C cycles, the value of the acceptance ratio a_i is used to update the step size for the direction. A low value implies that there are more rejections suggesting that the step size be reduced. A high rate indicates more acceptances, which may be due to small step size. In this case, step size is to be increased. If $a_i = 0.5$, the current step size is adequate with the number of acceptances at the same level as that of rejections. Once again drawing from the work of Metropolis on Monte Carlo simulations of fluids, our idea is to adjust the steps to achieve the ratio of acceptances to rejections equal to 1. A step multiplication factor, $g(a_i)$, is introduced as

$$s_i^{\text{new}} = g(a_i) s_i^{\text{old}} \tag{7.21}$$

We want $g(0.5) = 1$, $g(1) > 1$ (say, c), and $g(0) < 1$ (say, $1/c$). Corana et al. define g in such a way that $g = 1$ for $0.4 < a_i < 0.6$. See Fig. 7.10 and the following equation:

$$\begin{aligned} g\left(a_i\right) &= 1 + c\frac{a_i - 0.6}{0.4} && \text{for } a_i > 0.6 \\ &= \frac{1}{1 + c\dfrac{0.4 - a_i}{0.4}} && \text{for } a_i < 0.4 \\ &= 1 && \text{otherwise} \end{aligned} \tag{7.22}$$

where $c = 2$

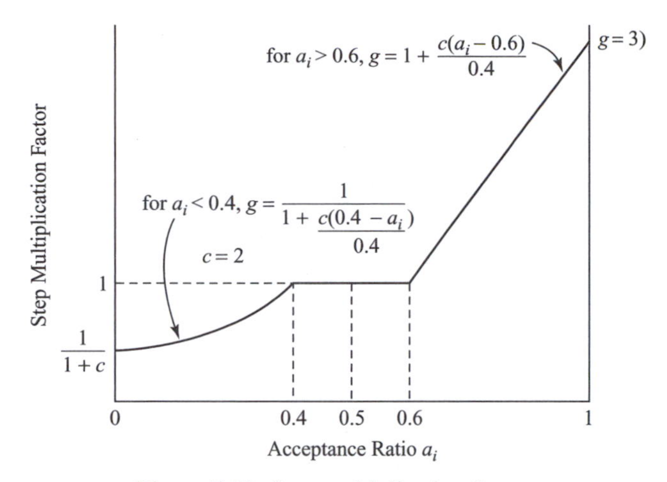

Figure 7.10. Step multiplication factor.

Tests show that this scheme works well. However, the step size may take low values eventually and may not permit a step into the higher energy rate even at moderate temperatures. A resetting scheme is introduced in the program developed. This scheme is simple. At the end of each temperature step, each step size is updated to be equal to s_T.

Example 7.11

A test function defining the form of a deflected corrugated spring in n-dimensions is given by

$$f = -\cos(kR) + 0.1R^2$$

where R is the radius. $R = \sqrt{(x_1 - c_1)^2 + (x_2 - c_2)^2 + \cdots + (x_n - c_n)^2}$, $\mathbf{c}$ is the minimum point, and k defines the periodicity nature of the corrugations. For the case $n = 2$, $c_1 = 5$, $c_2 = 5$, $k = 5$, use SA algorithm to find the minimum of f.

Solution

The periodic nature of the function can be seen by plotting the one variable function in R. The period is about $2\pi/5$ for $k = 5$. In n-dimensions, the function has a constant value at $R =$ constant. The global minimum is at a single point $R = 0$, with $f = -1$. The center location for the n-dimensional problem is at $x_1 = c_1, x_2 = c_2, \ldots, x_n = c_n$. We report here the results from 10 consecutive trials using the program SIMANAL. The range for each variable has been set at -10 to 10.

Function value	x_1	x_2	Number of function evaluations
−1	5.00010	4.99995	5401
−1	5.00001	4.99997	6801
−1	4.99996	5.00033	6001
−1	5.00034	5.00004	5201
−0.8433	6.05447	4.33497	3601
−1	4.99999	4.99999	5601
−0.9624	4.98573	4.94706	2001
−1	4.99984	4.99986	6001
−1	5.00008	5.00004	6001
−0.8433	5.94073	4.18200	4601

Two out of ten trials converged to an adjacent local minimum. Seven trials converged to the global minimum. The random points in one of the trials did not move toward the global minimum even though it captured the valley in one of the early iterations. The global minimum point is in a valley of radius 0.622 that amounts to 0.304% of the region. In Monte-Carlo simulations, the chance of a trial point hitting this region is thus limited. The degree of difficulty increases with increase in the number of dimensions of the problem.

Determination of Optimum Locations Using the Method of Simulated Annealing

In the preceding text, a new point was generated as using $\mathbf{x}_S = \mathbf{x} + r\, s_i \mathbf{e}_i$, where r is a random number between –1 and 1 and s_i is the step size along the ith coordinate direction. This update formula may be interpreted more generally for tackling the problem of finding optimal placements of objects (such as masses or sensors and actuators on a structure). An approach used by Constans [1998] was to interpret the step size s_i as the radius of a sphere centered at a nodal location and the "new point" as a randomly selected node or location within this sphere. The radius of the sphere is adjusted as per Eq. (7.23). Use of SA is attractive in vibration optimization problems where the objective function (such as amplitude or sound power) has sharp peaks owing to resonances.

7.8 Genetic Algorithm (GA)

Genetic algorithm is another optimization technique that draws its analogy from nature. The process of natural evolution intrigued John Holland of University of

Michigan in the mid-1960s. He developed computational techniques that simulated the evolution process and applied to mathematical programming. The genetic algorithms revolve around the genetic reproduction processes and "*survival of the fittest*" strategies.

Genetic algorithms follow the natural order of attaining the maximum. The problem for GA will be posed in the form

$$\begin{aligned} &\text{maximize} \quad f(\mathbf{x}) \\ &\text{subject to} \quad l_i \leq x_i \leq u_i, \quad i = 1 \text{ to } n \end{aligned} \tag{7.23}$$

We now attempt to present the Genetic algorithm by defining the various steps.

Coding and Decoding of Variables

In the most commonly used genetic algorithm, each variable is represented as a binary number, say, of m bits. This is conveniently carried out by dividing the feasible interval of variable x_i into $2^m - 1$ intervals. For $m = 6$, the number of intervals will be $N = 2^m - 1 = 63$. Then each variable x_i can be represented by any of the discrete representations

$$000000, 000001, 000010, 000011, \ldots, 111111 \tag{7.24}$$

For example, the binary number 110110 can be decoded as

$$x_i = l_i + s_i(1.2^5 + 1.2^4 + 0.2^3 + 1.2^2 + 1.2^1 + 0.2^0) = l_i + 54s_i \tag{7.25}$$

where s_i is the interval step for variable x_i, given by

$$s_i = \frac{u_i - l_i}{63} \tag{7.26}$$

The procedure discussed here defines the coding and decoding process.

Step 1. Creation of Initial Population

The first step in the development of GA is the creation of *initial population*. Each member of the population is a string of size $n*m$ bits. The variables in the binary form are attached end for end as follows:

$$\begin{aligned} &\textit{member } 1 \quad \underbrace{101101}_{x_1}\,\underbrace{101001}_{x_2}\,\underbrace{001010}_{x_3}\ldots\underbrace{101111}_{x_n} \\ &\textit{member } 2 \quad \underbrace{100101}_{x_1}\,\underbrace{101000}_{x_2}\,\underbrace{001010}_{x_3}\ldots\underbrace{101101}_{x_n} \\ &\qquad\qquad\qquad \ldots \\ &\textit{member } z \quad \underbrace{001101}_{x_1}\,\underbrace{100001}_{x_2}\,\underbrace{101010}_{x_3}\ldots\underbrace{111110}_{x_n} \end{aligned} \tag{7.27}$$

where z is the size of the population that defines the number of members. The size of population z is a parameter to be experimented with. The initial population of size z is created on the computer using a random number generator. We go successively from position 1 to position $6n$ in the creation of each member. A random number r in the range 0 to 1 is generated. If the generated number is ≤ 0.5, we put a zero in the position and 1 if it is > 0.5.

Step 2. Evaluation

In the evaluation phase, the values of variables for each member are extracted. Sets of six binary digits are read and decoding is done using Eq. 7.25. The function values $f_1, f_2, \ldots, f_z$ are evaluated. The function values are referred to as *fitness values* in GA language. The average fitness $\bar{f}$ is calculated.

The *reproduction* takes the steps of creation of a mating pool, crossover operation, and mutation.

Step 3. Creation of a Mating Pool

The reproduction phase involves the creation of a mating pool. Here, the weaker members are replaced by stronger ones based on the fitness values. We make use of a simple chance selection process by simulating a roulette wheel. The first step in this process is to have all the function values as positive entities. Some convenient scaling scheme may be used for this purpose.[1] The scaling scheme implemented in the computer program GENETIC is as follows. The highest value f_h and the lowest value f_l of the function are evaluated. The function values are converted to positive values by adding the quantity $C = 0.1\, f_h - 1.1\, f_l$ to each of the function values. Thus, the new highest value will be $1.1(f_h - f_l)$ and the lowest value $0.1(f_h - f_l)$. Each of the new values is then divided by $D = \max\,(1, f_h + C)$.

$$
\begin{aligned}
f_i' &= \frac{f_i + C}{D} \\
\text{where} \quad C &= 0.1\, f_h - 1.1\, f_l \\
D &= \max(1,\ f_h + C)
\end{aligned}
\tag{7.28}
$$

After performing the scaling operation, the sum of the scaled fitness values S is calculated using

$$
S = \sum_{i=1}^{z} f_i' \tag{7.29}
$$

[1] As an alternative to the preceding scaling procedure, the population may be "ranked" and fitness values can then be assigned based on the rank.

Roulette wheel selection is now used to make copies of the fittest members for reproduction. We now create a mating pool of z members as follows. The roulette wheel is turned z times (equal to the number of members in the population). A random number r in the range 0 to 1 is generated at each turn. Let j be the index such that

$$f_1' + f_2' + \cdots + f_{j-1}' \leq rS \leq f_1' + f_2' + \cdots + f_j' \tag{7.30}$$

The j th member is copied into the mating pool. By this simulated roulette wheel selection, the chance of a member being selected for the mating pool is proportional to its scaled fitness value.

Step 4. Crossover Operation

The crossover operation is now performed on the mating parent pool to produce offspring. A simple crossover operation involves choosing two members and a random integer k in the range 1 to $nm - 1$ ($6n - 1$ for six bit choice). The first k positions of the parents are exchanged to produce the two children. An example of this one-point crossover for 6 bits of three variables is:

$$\begin{array}{rl} \textit{parent } 1 & \underbrace{10110110}\,1001001010 \\ \textit{parent } 2 & \underbrace{11010001}\,0110111000 \\ \textit{child } 1 & 11010001\ 1001001010 \\ \textit{child } 2 & 10110110\ 0110111000 \end{array} \tag{7.31}$$

The crossover operation is performed with a probability c_p, which may be chosen as equal to 1.

Step 5. Mutation

We note that there is a chance that the parent selection and crossover operations may result in close to identical individuals who may not be the fittest. Mutation is a random operation performed to change the expected fitness. Every bit of a member of the population is visited. A bit permutation probability b_p of 0.005 to 0.1 is used. A random number r is generated in the range 0 to 1. If $r < b_p$, a bit is changed to 0 if it is 1 and 1 if it is 0.

Step 6. Evaluation

The population is evaluated using the ideas developed in Step 2. The highest fitness value and the corresponding variable values are stored in $f_{\max}$ and $\mathbf{x}_{\max}$. A *generation* is now completed. If the preset number of generations is complete, the process is stopped. If not, we go to Step 3.

In the computer program, a strategy of reducing the bounds of the variables is introduced. After each generation of calculations, the bounds are brought closer to the currently attained maximum point.

GA is a powerful tool in the optimization tool box. It provides robust solutions to continuous as well as discrete problems. We have presented a one point crossover operation. Various crossover techniques abound in the literature. Other crossover operations may be tried.

Example 7.12

We report here the results for Example 7.2 from 10 consecutive trials using the program GENETIC. Each trial is set for 4 region reductions, 40 generations of population 10. Thus, the number of function evaluations is fixed at 1600. 10 bit representation of each variable has been used in the run. The initial range for each variable is −10 to 10.

Table E7.12

Function value	x_1	x_2
0.9783	5.0415	5.0024
0.9871	5.0244	4.9792
0.9997	5.0049	5.0000
0.9917	4.9988	4.9743
0.9820	4.9646	5.0134
0.9644	4.9682	5.0428
0.9986	5.0061	5.0086
0.9783	5.0415	5.0012
0.9953	5.0110	4.9841
0.9289	5.0733	5.0183

Each trial is near the global minimum for the test problem.

Traveling salesman problem (TSP): Consider the visit of 8 cities named A, B, C, D, E, F, G, H. The salesman must start his travel at one city and return back without touching a city twice. The distances between cities are provided. The problem is to find the tour resulting in the least distance traveled. In this problem, a string BCAGDHEF defines the order of visiting cities, starting at B and ending with the final travel from F to B. A population of such strings can be created. Evaluation involves the calculation of the total distance for the tour. A crossover operation is more involved than in the case of binary strings. If a one-point crossover is performed, we must also see that the letters do not repeat in the string. As a simple example, consider a five city tour of cities A, B, C, D, E. Let CEABD and ABECD

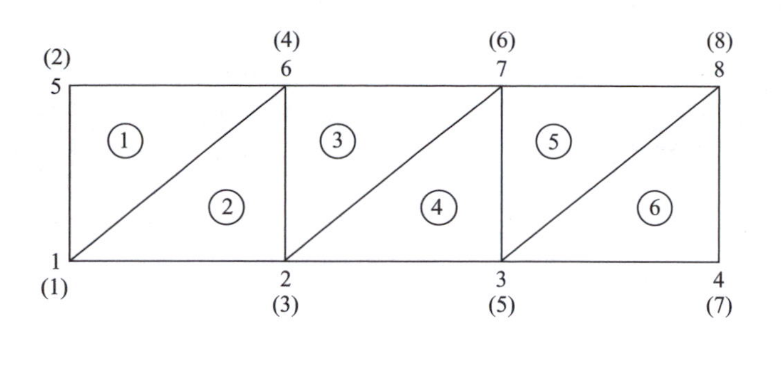

CONNECTIVITY

Element Number	3 Nodes
1	1 5 6
2	1 2 6
3	2 6 7
4	2 3 7
5	3 7 8
6	3 4 8

Bandwidth = 6 (= 6 − 1 + 1)
For node numbers shown in parenthesis
Bandwidth = 4 (= 4 − 1 + 1)

Figure 7.11. Bandwidth problem.

be the two members on which a single point crossover operation is performed at the end of first two characters. Then a possible crossover operation is to call the two offspring to be xxECD and xxABD. Noting that E replaces A in the first one, the second letter E is changed to A. C replaces B making the original C to B. The first offspring is BAECD. A similar logic makes the second offspring ECABD. GA may be used on reactor core refueling optimization as described in Chapter 1.

Bandwidth reduction is another interesting problem in finite element formulations in engineering. A region is defined as a collection of elements, which are connected at nodes. If the node numbers range from 1 to N, and the element connectivity is defined, the bandwidth of the numbering scheme is defined by the element with the largest difference among its node numbers. The problem now is to minimize the bandwidth. A six element 8 node configuration is shown in Fig. 7.11. We may handle this problem by defining an array of node numbers filled with 1 to N. The numbers in the array are then rearranged until a minimum is reached.

Real – Coded GAs

If the original design variables are continuous variables, it is possible to work directly with these rather than converting them to binary 0 −1 variables. Let $x_i^{(j)}$ and $x_i^{(k)}$ be the values of the ith variable in strings j and k, respectively. The crossover and mutation operators may be combined as

$$x_i^{\text{new}} = (1 - \lambda)x_i^{(j)} + \lambda x_i^{(k)} \tag{7.32}$$

where $0 \leq \lambda \leq 1$ is a random number. Other strategies are given in Deb [1998]. Various optimization algorithms based on SA and GA have been applied in computational molecular biology [Clote and Backofen 2001].

7.9 Differential Evolution (DE)

GAs described in Section 7.8 may be considered as belonging to the category of *evolutionary algorithms*. We describe another algorithm in this category, namely, DE [Storn and Price 1997]. Consider

$$\begin{aligned} &\text{minimize} \quad f(\mathbf{x}) \\ &\text{subject to} \quad x_j^L \le x_j \le x_j^U, \quad j = 1 \text{ to } n \end{aligned} \tag{7.33}$$

Like real – coded GAs, assume that an initial population of *npop* designs has been randomly generated, $npop \geq 4$. "Design" refers to the values of n design variables. This initial population is generation $G = 1$. In a typical generation, G during the evolutionary process, the population of designs may be depicted as

$$x_{i,G} = [x_{1,i,G}, x_{2,i,G}, \ldots, x_{n,i,G}], \quad i = 1, 2, \ldots, npop$$

Thus, a population of designs can be stacked up as

$$\begin{aligned} &[x_{1,1,G}, x_{2,1,G}, \ldots, x_{n,1,G}] \\ &[x_{1,2,G}, x_{2,2,G}, \ldots, x_{n,2,G}] \\ &\ldots\ldots\ldots\ldots\ldots\ldots\ldots\ldots \\ &[x_{1,npop,G}, x_{2,npop,G}, \ldots, x_{n,npop,G}] \end{aligned}$$

$[x_j^L, x_j^U]$ represent the lower and upper bounds on the jth variable. Crossover and mutation operators in GAs are replaced by mutation and recombination operators in DE.

for $i = 1, \ldots, npop$:

Mutation

Consider the ith design vector, or "target vector", $x_{i,G}$. Corresponding to this, randomly select three design vectors from the population $x_{r1,G}, x_{r2,G}, x_{r3,G}$ such that the indices i, $r1$, $r2$, $r3$ are distinct. Define a mutant vector v by adding the weighted difference of two vectors to the third vector as

$$\begin{aligned} &v_{i,G+1} = x_{r1,G} + F(x_{r2,G} - x_{r3,G}) \\ &\text{where } 0 \le F \le 2. \end{aligned} \tag{7.34}$$

Recombination

Define a trial vector as

$$
\begin{aligned}
t_{j,i,G+1} &= V_{j,i,G+1} \quad \text{if rand} \leq \text{CR or} \quad j = I_{\text{rand}} \\
&= x_{j,i,G} \quad \text{if rand} > \text{CR and} \quad j \neq I_{\text{rand}} \\
j &= 1, 2, \ldots, n
\end{aligned}
$$

where rand is a random number between 0 and 1, I_{rand} is a random integer from $\{1, 2, \ldots, n\}$ and ensures that the trial vector is not the same as the target vector.

Selection

The target vector is compared to the trial vector. The one with the smaller objective is chosen for the next generation as

$$
\begin{aligned}
x_{i,G+1} &= t_{j,i,G+1} \quad \text{if } f(t_i) \leq f(x_{i,G}) \\
&= x_{i,G} \quad \text{otherwise}
\end{aligned}
$$

7.10 Box's Complex Method for Constrained Problems

Box's complex method is a zero order method for solving problems with bounds on variables and inequality constraints. The problem is of the form

$$
\begin{aligned}
&\text{minimize} \quad f(\mathbf{x}) \\
&\text{subject to} \quad g_j(\mathbf{x}) \leq 0 \\
&\qquad\qquad l_i \leq x_i \leq u_i, \quad i = 1 \text{ to } n
\end{aligned}
\tag{7.35}
$$

If there are no bounds on variables, initial bounds may be set with a wide margin. An equality constraint can be considered if it can be used to express one of the variables explicitly in terms of the others.

> *The main assumption in the Box method is that the inequality constraints define a convex region.*

Convexity implies that given any two feasible points, every point along the line joining the two points is also feasible. The user must provide an initial feasible point.

Generation of a Complex of k (= $2n$) Points

Let the initial feasible point be $\mathbf{x}^1$. We now proceed to generate a *complex* of $k(=2n)$ corner points. We know that $n + 1$ points define a *simplex* in an n-dimensional space. The convex hull of $n + 2$ or more points is conveniently referred to as a *complex*. Let us consider the stage when $i - 1$ feasible points have been generated and that their mean called *center* is $\mathbf{x}^C$. The next point $\mathbf{x}^i$ in the set is generated as follows. We generate a random number r in the range 0 to 1 by visiting each of the directions $j = 1$ to n. The jth component is defined using the relationship

$$x_j^i = l_j + r(u_j - l_j) \tag{7.36}$$

Having generated the n coordinates, the feasibility is checked. If the point is not feasible, $\mathbf{x}^i$ is updated using the bisection step

$$\mathbf{x}^i = 0.5(\mathbf{x}^i + \mathbf{x}^C) \tag{7.37}$$

until $\mathbf{x}^i$ is feasible. Convexity assumption of the feasible region guarantees that we get a feasible point. Then the center $\mathbf{x}^C$ is updated to add the new point to the accepted set using

$$\mathbf{x}^C = \mathbf{x}^C + \frac{(\mathbf{x}^i - \mathbf{x}^C)}{i} \tag{7.38}$$

The process is continued till all the k points of the complex are generated.

We then determine the function values at the k points of the complex. The next steps of the algorithm are:

Step 1. The point $\mathbf{x}^H$ with the highest function value f_H is identified. The centroid $\mathbf{x}^0$ of the other $k - 1$ points is calculated using

$$\mathbf{x}^0 = \mathbf{x}^C + \frac{(\mathbf{x}^C - \mathbf{x}^H)}{k - 1} \tag{7.39}$$

Step 2. The *reflection* $\mathbf{x}^R$ of the highest point is calculated with respect to the centroid point $\mathbf{x}^0$ using the reflection parameter $\alpha(> 1)$. Box suggests a value of 1.3.

$$\mathbf{x}^R = \mathbf{x}^0 + a(\mathbf{x}^0 - \mathbf{x}^H) \tag{7.40}$$

Step 3. A check for the feasibility of $\mathbf{x}^R$ is conducted. If l_j is violated, we set $x_j^R = l_j + 10^{-6}$, and if u_j is violated, we set $x_j^R = u_j - 10^{-6}$. If $\mathbf{g}$ is violated, the bisection step $\mathbf{x}^R = 0.5\ (\mathbf{x}^R + \mathbf{x}^0)$ is repeated until a feasible point is obtained. The function value f_R is now evaluated.

Step 4. If $f_R > f_H$, then the bisection step $\mathbf{x}^R = 0.5\,(\mathbf{x}^R + \mathbf{x}^0)$ is repeated until $f_R < f_H$. $\mathbf{x}^H$ is replaced by $\mathbf{x}^R$.

Step 5. Check for convergence is made using the standard deviation σ or the variance σ^2 of the function value over the complex or the simple entity $f_H - f_{min}$. If this is less than a small preselected value ε, convergence is achieved and stop the process, if not go to Step 1.

Box's method is a convenient method for a quick solution of problems with inequality constraints. However, ensuring the convexity of the feasible region is the key. Following example problem is due to Rosenbrock.

Example 7.13.

If x_1, x_2, and x_3 are the lengths of edges of a rectangular box, the post office packing problem is stated as

$$\begin{aligned} &\text{maximize} && x_1 x_2 x_3 \\ &\text{subject to} && x_1 + 2x_2 + 2x_3 \leq 72 \\ & && 0 \leq x_1 \leq 20, \quad 0 \leq x_2 \leq 11, \quad 0 \leq x_3 \leq 42 \end{aligned}$$

Solution

The problem is input into the computer program BOX. Function f to be minimized is given as $-x_1x_2x_3$. The main constraint is defined as $g_1 = x_1 + 2x_2 + 2x_3 - 72$ (≤ 0). The bounds are defined in the computer program. The starting point is given as (1, 1, 1). The solution is obtained as

Minimum function value = 3298.2
Number of function evaluations = 121
Number of iterations = 94
Minimum point (19.9975, 10.999, 15.011)
Constraint value G(1) = −4.458e−4

If the upper limits on each of the variables is set as 50, the solution obtained is $f = 3456$ and the variable values are (24.033, 12.017, 11.966).

Transformation of Variables To Remove Some Bounds and Constraints

Direct methods can be used on problems with bounds on variables and some constraints. We will consider some useful transformations here. A simplified version of Riemann's stereographic projection for a unit circle is shown in Fig. 7.12. Since triangles ACQ and PQA are similar to triangle POB, we have $QC = ux$, $PC = u/x$. In

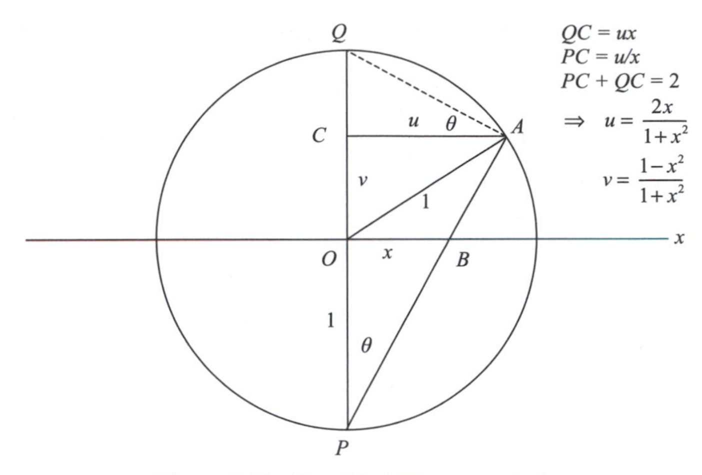

Figure 7.12. Simplified Riemann circle.

view of $PC + QC = 2$, we get

$$u = \frac{2x}{1+x^2}$$

Since $v + ux = 1$, we also have

$$v = \frac{(1-x^2)}{(1+x^2)}$$

u and v also satisfy the equality

$$u^2 + v^2 = 1$$

While x varies from $-\infty$ to $+\infty$, we have $-1 \leq u \leq +1$ and $-1 < v \leq +1$. While both u and v are in the interval -1 to $+1$, v never reaches -1, which is a characteristic of this projection.

This suggests that bounds on a variable x_i given by

$$l_i \leq x_i \leq u_i \tag{7.41}$$

can be treated using the new unconstrained variable y_i

$$x_i = \frac{u_i + l_i}{2} + \frac{u_i - l_i}{2}\left(\frac{2y_i}{1 + y_i^2}\right) \tag{7.42}$$

The transformation works well. Sisser [1981] used the transformation even with gradient-based methods.

Constraints of the type

$$x_1^2 + x_2^2 + \cdots + x_n^2 = 1 \tag{7.43}$$

often occur where the vector the point is restricted to lie on a unit hypersphere. An extension of the Riemann construction depicted in Fig. 7.12 is given by

$$x_1 = \frac{2y_1}{1+p}, \quad x_2 = \frac{2y_2}{1+p}, \quad x_{n-1} = \frac{2y_{n-1}}{1+p}, \quad x_n = \frac{1-p}{1+p}$$
$$\text{where } p = y_1^2 + y_2^2 + \cdots + y_{n-1}^2 \tag{7.44}$$

In the two-variable case, this precisely corresponds to the variables u and v defined earlier.

Example 7.14

The equilibrium state of n electrons positioned on a conducting sphere is obtained by minimizing

$$f(x, y, z) = \sum_{i=1}^{n-1} \sum_{j=i+1}^{n} ((x_i - x_j)^2 + (y_i - y_j)^2 + (z_i - z_j)^2)^{-\frac{1}{2}}$$
$$\text{subject to} \quad x_i^2 + y_i^2 + z_i^2 = 1, \qquad i = 1, 2, \ldots, n$$

Solution

The problem can be solved by the $3n$ variables (x_i, y_i, z_i), $i = 1$ to n, using $2n$ unrestricted variables (u_i, v_i), $i = 1$ to n.

$$x_i = \frac{2u_i}{1+u_i^2+v_i^2} \quad y_i = \frac{2v_i}{1+u_i^2+v_i^2} \quad z_i = \frac{1-u_i^2-v_i^2}{1+u_i^2+v_i^2}, \quad i = 1 \text{ to } n$$

COMPUTER PROGRAMS

CYCLIC, HOOKJEEV, ROSENBRK, POWELLCG, NELDMEAD, SIMANL, GENETIC, DE, BOX

PROBLEMS

P7.1. Discuss the convergence aspects of the cyclic coordinate search on a separable function of the form $f = \sum_{i=1}^{n} g_i(x_i)$. Justify your remarks by considering the example: Min. $f = x_1^2 + 5x_2^2$, starting point $\mathbf{x}^0 = (3, 1)^T$.

P7.2. Do a second iteration of the cyclic coordinate search algorithm in Example 7.1, consisting of a 2nd stage and an acceleration step.

P7.3. Consider $f = (x_1 - 2)^4 + (x_1 - 2x_2)^2$, starting point $(0, 3)^T$. Do one iteration of cyclic coordinate search by hand–calculations.

P7.4. Consider $f = (x_1 - 2)^4 + (x_1 - 2x_2)^2$, starting point $(0, 3)^T$. Do one iteration of Hooke and Jeeves algorithm by hand–calculations.

P7.5. Explore the number of function evaluations in Example 7.4 (Hooke–Jeeves) with different values of s and r.

P7.6. Consider $f = (x_1 - 2)^4 + (x_1 - 2x_2)^2$, starting point $(0, 3)^T$. Do two iterations of Rosenbrock's search by hand–calculations.

P7.7. Consider $f = (x_1 - 2)^4 + (x_1 - 2x_2)^2$, starting point $(0, 3)^T$. Do two iterations of Powell's search by hand calculations.

P7.8. Find the minimum value and its location for the following test functions:

(a) Rosenbrock's function

$$100\left(x_1^2 - x_2\right)^2 + (1 - x_1)^2 \text{ starting point } (-1.2, 1)$$

(b) Powell's function

$$(x_1 + 10x_2)^2 + 5(x_3 - x_4)^2 + (x_2 - 2x_3)^4 + (10x_1 - x_4)^4$$

starting point (3, –1, 0, 1)

P7.9. Solve problem 3.14 of chapter 3 using a direct method.

P7.10. Solve problem 3.17 of chapter 3 using a direct method.

P7.11. Solve problem 3.22 of chapter 3 using a direct method.

P7.12. In the evaluation of the geometric form error of circularity of an object, the coordinates x_i, y_i of points $i = 1$ to n are measured. Then if (a, b) represents the location of the center of a circle, we define $r_i = \sqrt{(x_i - a)^2 + (y_i - b)^2}$. With a, b being the variables for the problem, circularity is defined as

$$\min\left[\max_{i=1 \text{ to } n}(r_i) - \min_{i=1 \text{ to } n}(r_i)\right]$$

Determine the circularity for the following data

x	2	5	7	8	5	3	1	0
y	1	1	2	5	8	9	7	4

P7.13. In reliability studies, failure data are fitted to a two-parameter Weibull curve given by

$$R(t) = \exp\left[-\left(\frac{t}{\alpha}\right)^{\beta}\right]$$

where t is the time, R is the reliability, and scale parameter α and the shape parameter β are the two Weibull parameters. The failure data for 100 items ($N = 100$) is given in the following table.

Cumulative no. of failures n	5	15	32	52	73	84
Time in hours t	8	16	24	32	40	48

If $R(t)$ is estimated as $\frac{N-n}{N}$, determine the Weibull parameters corresponding to the least squared error in logarithmic plot. (*Note*: Taking logorithm twice we get $\ln(-\ln(R(t)) = \beta\,[\ln(t) - \ln(\alpha)]$. Use this form in the formulation)

P7.14. Three springs of stiffness values 100N/mm, 200N/mm, and 300N/mm are suspended from the points A, B, and C, respectively. The coordinates of A, B, and C are (500, 0, 0) mm, (0, 500, 0) mm, (–200, –200, 0) mm, respectively. The other ends of the springs are tied together at point D. This point D is at (0, 0, –600) mm and the springs are under no tension or compression in this configuration. If a load of 50kN is applied in the downward direction ($P_z = -50$kN), determine the displacements of D. (*Note*: If l_i' is the stretched length of member i, and l_i is its original length, and q_1, q_2, and q_3, are the x, y, z displacements of D, the potential energy is $\pi = \sum_i \frac{1}{2}k_i\,(l_i' - l_i)^2 - P_z q_3$. The equilibrium condition corresponds to the minimum potential energy)

P7.15. Find the solution of the following problem, due to Colville, using a direct method:

$$\begin{aligned}\min f = {} & 100\left(x_2 - x_1^2\right)^2 + (1 - x_1)^2 + 90\left(x_4 - x_3^2\right)^2 + (1 - x_3)^2 \\ & + 10.1\,(x_2 - 1)^2 + 10.1\,(x_4 - 1)^2 + 19.8\,(x_2 - 1)\,(x_4 - 1)\end{aligned}$$

P7.16. Solve the following constrained problem, using Box's method.

$$\begin{aligned}\text{Minimize}\quad & (x_1 - 2)^2 + (x_2 - 1)^2 \\ \text{Subject to}\quad & x_1^2/4 + x_2 \le 1 \\ & -x_1 + 2x_2 \ge 2\end{aligned}$$

P7.17. Minimum weight tubular column problem with radius x_1 and thickness x_2 subjected to stress and bucking constraints reduces to the following form:

$$\begin{aligned}\text{minimize}\quad & x_1 x_2 \\ \text{subject to}\quad & 156 x_1 x_2 \ge 1 \\ & x_2 \ge 1 \\ & x_1 > 0,\ x_2 > 0\end{aligned}$$

Solve the problem using a direct approach, using penalty functions.

P7.18. Consider the design of a two-bar truss as shown in Fig. P7.18. The load P applied at node A causes member AB to be in tension and member AC to be in compression. A simple minimum weight design problem here would be to minimize the weight of the truss subject to: (i) ensuring that each member does not yield, (ii) that member AC, which is in compression does not buckle. As design variables, we will choose $\mathbf{x} = [A_1, A_2, H]^T$. Thus, the problem may be stated as:

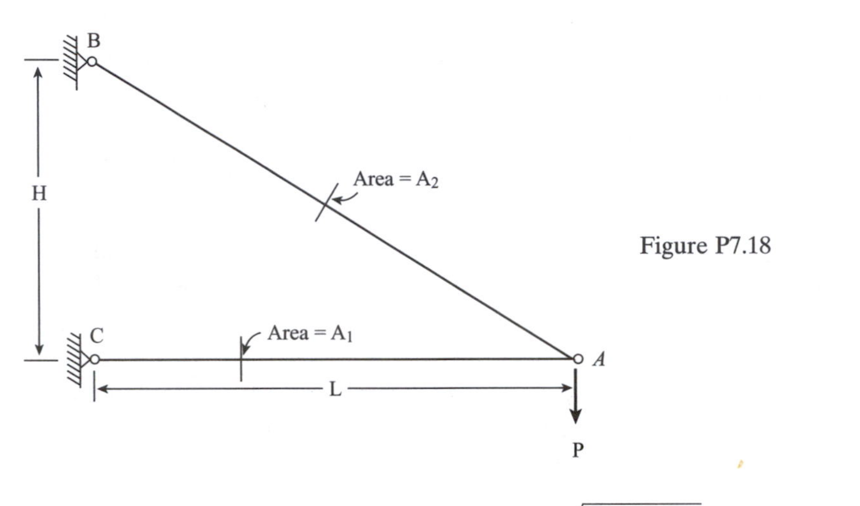

Figure P7.18

$$\text{minimize } \gamma A_1 L + \gamma A_2 \sqrt{(L^2 + H^2)} \tag{a}$$

$$\text{subject to} \quad P/(A_2 \sin\theta) \le S_y/F_s \tag{b}$$

$$P/(A_1 \tan\theta) \le S_y/F_s \tag{c}$$

$$P/\tan\theta \le \pi^2\, EI_1/(L^2 F_s) \tag{d}$$

$$A_1, A_2, H \ge 0 \tag{e}$$

where γ = weight density, S_y = yield strength, F_s = factor of safety, E = Young's modulus and I_1 = moment of inertia. For a circular cross-section, we have $I_1 = A_1^2/4\pi$. Substituting for θ in terms of L and H we have

$$\text{minimize} \quad f = \gamma A_1 L + \gamma A_2 \sqrt{(L^2 + H^2)} \tag{a}$$

$$\text{subject to} \quad g_1 = PF_s[\sqrt{(L^2 + H^2)}]/(A_2 H S_y) - 1 \le 0 \tag{b}$$

$$g_2 = PF_s L/(A_1 H S_y) - 1 \le 0 \tag{c}$$

$$g_3 = 4PF_s L^3/(\pi H E A_1^2) - 1 \le 0 \qquad (d)$$

$$A_1, A_2, H \ge 0 \qquad (e)$$

The solution to this problem will depend on the data. For steel- 0.2%C HR, we have $E = 30 \times 10^6$ psi, $\gamma = 0.2836$ lb/in.^{3}, $S_y = 36{,}260$ psi, $F_s = 1.5$, $P = 25000$lb. Take $L = 5$ in. Determine the optimum solution with various starting points. Use the method of simulated annealing (SA).

P7.19. A cantilever beam is welded onto a column as shown in Fig. P7.19. The weld has two segments, each of length ℓ and size b. The beam of length L is subjected to a load F. The problem of minimizing weld volume subject to a maximum shear stress limit in the weld reduces to

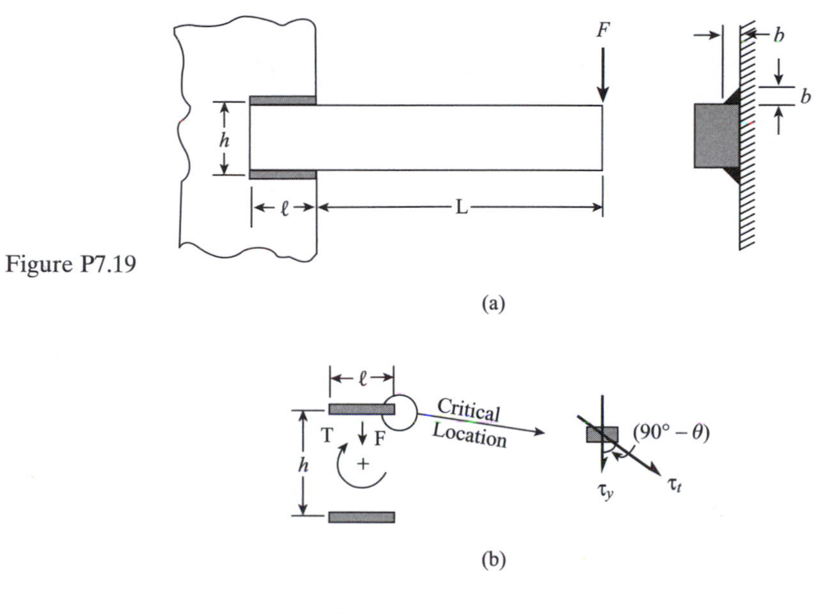

Figure P7.19

$$\begin{array}{ll} \text{minimize} & f = 2\,b^2\ell \\ \text{subject to} & \tau \le \tau^{\mathrm{U}} \\ \text{and} & b^{\mathrm{L}} \le b \le b^{\mathrm{U}}, \quad \ell^{\mathrm{L}} \le \ell \le \ell^{\mathrm{U}} \end{array}$$

Take $\tau^{\mathrm{U}} = 30{,}000.$ psi, $F = 6{,}000.$ lb, $h = 3$ in, $L = 14$ in. The expression for τ is given as:

$$\tau_y = F/(b\ell\sqrt{2}),$$

$$\tau_t = [6F(L + 0.5\ell)\sqrt{(h^2 + \ell^2)}]/[\sqrt{2}b\,\ell(\ell^2 + 3h^2)]$$

$$\tau = \sqrt{\tau_y^2 + 2\tau_y\tau_t\cos\theta + \tau_t^2}, \quad \cos\theta = \ell/(h^2 + \ell^2)$$

Carefully choose reasonable values for b^L, b^U, ℓ^L, and ℓ^U (you may refer to a handbook). Obtain the solution graphically and also using SA and/or GA.

P7.20. Implement Hooke and Jeeves algorithm with line searches (use quadfit subroutine from Chapter 2)

P7.21. Implement Rosenbrock's algorithm with line searches (use quadfit subroutine from Chapter 2)

P7.22. Modify the GA code and implement a real–coded GA based on Eq. (7.32).

P7.23. A finite element mesh consisting of two-noded truss elements is shown in Fig. P7.23. The problem is to renumber the nodes to minimize the bandwidth nbw, which is given by the formula $nbw = 2\max_{e} |\, j - i + 1 \,|$, where i, j are the two nodes associated with element e. That is, $nbw =$ twice the (maximum difference between node numbers connecting an element $+1$). Use Genetic algorithm and Simulated Annealing codes used for Traveling Salesman problems.

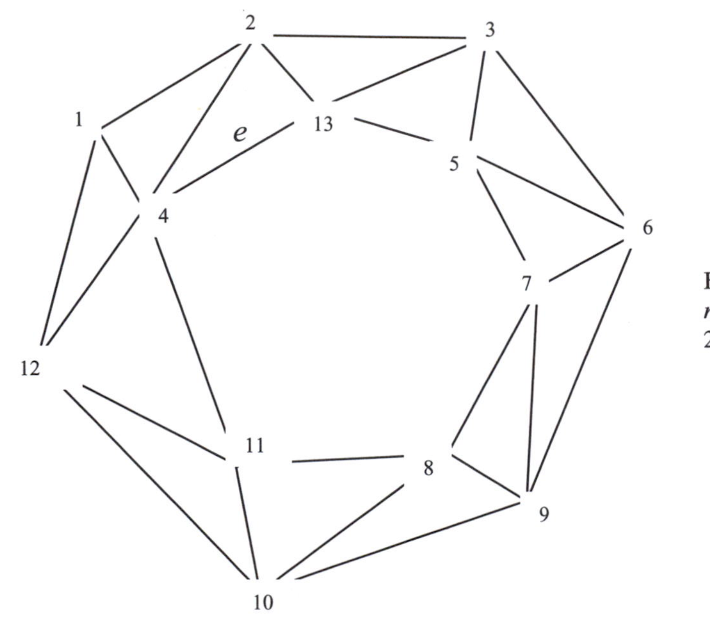

Figure P7.23. (current $nbw = 2(13 - 2 + 1) = 24$).

P7.24. In GAs, after crossover, a new "child" is always accepted regardless of its fitness. Instead, develop a hybrid GA-SA algorithm. That is, implement the Metropolis criterion in selecting a new individual into the population after crossover in the GA. Compare with the pure GA and pure SA algorithms in the text on a set of test problems. Refer to: Ron Unger and John Moult. Genetic Algorithms for Protein Folding Simulations. *Journal of Molecular Biology*, 231: 75–81, 1992.

P7.25. The four bar mechanism in Fig. P7.25 is to be designed to be a function generator. Specifically, if ψ is the output (rocker) angle and ϕ is the input (crank) angle, let $\psi = G(\phi, \mathbf{a})$ be the generated function corresponding to the design variable vector $\mathbf{a} = [a_1, a_2, a_3, a_4]^T$. The objective is to optimize $\mathbf{a}$ so that G matches a prescribed function $F(\phi)$ over a range for $\phi \in [\phi_0, \phi_m]$ where ϕ_0 corresponds to the rocker being at the right extreme position, and $\phi_m = \phi_0 + 90°$. The difference between G and F, which is to be minimized, may be measured by a mean square error norm as

$$f = \sqrt{\frac{I}{(\varphi_m - \varphi_0)}}, \quad \text{where } I = \int_{\varphi_0}^{\varphi_m} [F(\varphi) - G(\varphi, \mathbf{a})]^2 d\varphi$$

or by a maximal error norm as

$$f = \max_{\varphi_0 \le \varphi \le \varphi_m} |F(\varphi) - G(\varphi, \mathbf{a})|$$

Constraints are:

(i) Prescribed function: $F(\varphi) = \psi_0 + \frac{2}{3\pi}(\varphi - \varphi_0)^2$ where ψ_0 corresponds to φ_0
(ii) The crank should rotate 360°
(iii) The transmission angle must satisfy $45° \le \mu \le 135°$
(iv) Link lengths must be positive

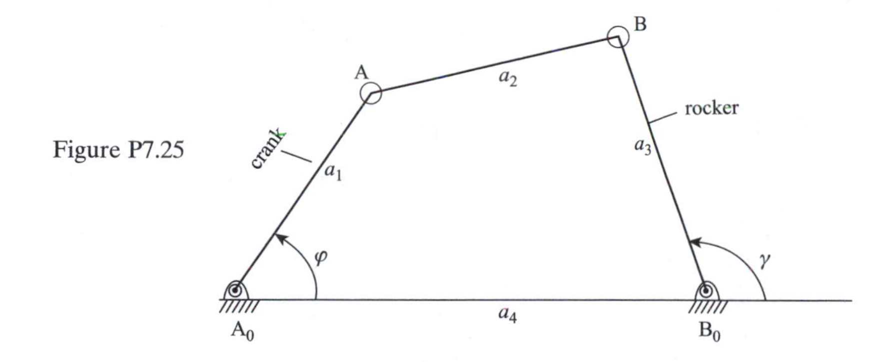

Figure P7.25

Note: First, assume $a_1 = 1$, $a_4 = 5$ and verify your nonlinear programming (NLP) solution graphically. Then allow all four $\{a_i\}$ to be design variables.

P7.26. A high speed robot arm is being rotated at constant angular velocity ω, (Fig. P7.26). An end-effector with payload at E weighs m_E. This is to be considered as a flexible mechanism. The (steel) link OE has a length L, has a tubular cross-section, and is modeled using, say, two beam finite elements. The outer diameter of the cross-section D is to be considered as a design variable, with thicknesses $t = 0.1D$. Objective and constraint functions include the mass of the link, the resultant deflection at the end effector E during its motion, and bending stress in each beam element, $|\sigma| \leq \sigma_{yield}$. Formulate and solve the problem.

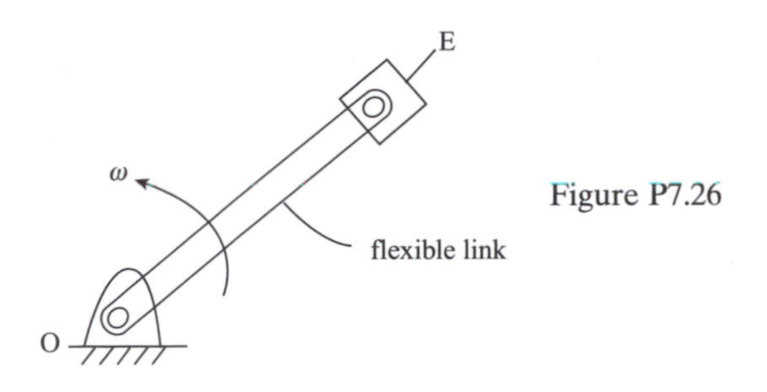

Figure P7.26

REFERENCES

Box, M.J., A New Method of Constrained Optimization and a Comparison with Other Methods, *The Computer Journal*, **8**, 42–52, 1965.

Clote, P. and R. Backofen., *Computational Molecular Biology – An Introduction*, John Wiley & Sons, Ltd., 2001.

Constans, E.W., Minimizing radiated sound power from vibrating shell structures: theory and experiment, Ph.D. Thesis, The Pennsylvania State University, University Park, PA, May 1998.

Corana, A., Marchesi, M., Martini, C., and Ridella, S., Minimizing multimodal functions of continuous variables with simulated annealing algorithm, *ACM Transactions on Mathematical Software*, **13**, 262–280, 1987.

Davis, L., *Handbook of Genetic Algorithms*, Van Nostrand Reinhold, New York, 1991.

Goldberg, D.E., *Genetic Algorithms in Search, Optimization, and Machine Learning*, Addison-Wesley, New York, 1989.

Deb, K., *Optimization for Engineering Design*, Prentice-Hall of India, New Delhi, 1998.

Holland, J.H., *Adaptation in Natural and Artificial Systems*, University of Michigan Press, Ann Arbor, 1975.

Dolan, E.D., More, J.J., and Munson, T.S., Benchmatking optimization software with COPS 3.0, Argonne National Laboratory, Technical Report ANL/MCS-TM-273, 2002.

Hooke, R. and Jeeves, T.A., Direct search solution of numerical and statistical problems, *Journal of the ACM*, **8**, 212–229, 1961.

Metropolis, N., Rosenbluth, A., Rosenbluth, M., Teller, A., and Teller, E., Equation of state calculations by fast computing machines, *Journal of Chemical Physics*, **21**, 1087–1090, 1953.

Nelder, J.A. and Mead, R., A simplex method for function minimization, *The Computer Journal*, **7**, 308–313, 1965.

Powell, M.J.D., An efficient method for finding the minimum of a function of several variables without calculating derivatives, *The Computer Journal*, **7**, 303–307, 1964.

Rosenbrock, H.H., An automatic method for finding the greatest or least value of a function, *The Computer Journal*, **3**, 175–184, 1960.

Sisser, F.S., Elimination of bounds in optimization problems by transforming variables, *Mathematical Programming*, **20**, 110–121, 1981.

Spendley, W., Hext, G.R., and Himsworth, F.R., Sequential application of simplex designs in optimization and evolutionary operation, *Technometrics*, **4**, 441–461, 1962.

Storn, R. and Price, K. Differential evolution – a simple and efficient heuristic for global optimization over continuous spaces, *Journal of Global Optimization*, **11**, 341–359, 1997.

8

Multiobjective Optimization

8.1 Introduction

The engineer is often confronted with "simultaneously" minimizing (or maximizing) different criteria. The structural engineer would like to minimize weight and also maximize stiffness; in manufacturing, one would like to maximize production output or quality while minimizing cost or production time. In generating a finite element mesh, we may wish to optimize the nodal locations in the mesh based on distortion parameter, aspect ratio, or several other mesh quality criteria simultaneously. Mathematically, the problem with multiple objectives (or "criteria" or "attributes") may be stated as

$$\text{minimize} \quad \mathbf{f} = [f_1(\mathbf{x}), f_2(\mathbf{x}), \ldots, f_m(\mathbf{x})] \tag{8.1a}$$

$$\text{subject to} \quad \mathbf{x} \in \Omega \tag{8.1b}$$

Sometimes, the criteria do not conflict with one another, in which case a single optimum $\mathbf{x}^*$ minimizes all the objectives simultaneously. More often, there is a conflict between the different criteria. We usually tackle the situation by weighting the criteria as

$$\begin{aligned} &\text{minimize} \quad f = w_1 f_1(\mathbf{x}) + w_2 f_2(\mathbf{x}) + \cdots \\ &\text{subject to} \quad \mathbf{x} \in \Omega \end{aligned} \tag{8.2}$$

where the weights w_i are ≥ 0, $\sum w_i = 1$. The weights are chosen from experience. The other popular approach to handle multiple criteria is to designate one of the objectives as the primary objective and constrain the magnitudes of the others as

$$\begin{aligned} &\text{minimize} \quad f_1(\mathbf{x}) \\ &\text{subject to} \quad f_2(\mathbf{x}) \leq c_2 \\ &\qquad \ldots \end{aligned}$$

$$f_m(\mathbf{x}) \le c_m$$
$$\mathbf{x} \in \Omega \tag{8.3}$$

However, there is always an uncomfortable feeling when we choose a specific set of weights w_i or limits c_i and obtain a single optimum point $\mathbf{x}^*$, namely, what if our selection of the weights were different? The sensitivity results in Chapter 5 only relate small changes in the problem parameters to changes in the optimum solution and cannot satisfy us. More importantly, what are the implications in a weighted or constraint approach with regard to our preferences? This chapter is important in engineering, as it deals with the definition of the objective function itself. All too often, a problem is posed as in (8.3) when it should be posed as in (8.1). That is, constraints should represent "hard" criteria, such as dimensions, a dollar limit, etc. while constraints whose right-hand side limits are uncertain should be considered as objectives.

A central concept in multiobjective optimization involves the concept of Pareto optimality, which is first presented in the following (Section 8.2). Subsequently, methods of generating the Pareto frontier (Section 8.3) and methods to identify a single best compromise solution (Section 8.4) are discussed.

8.2 Concept of Pareto Optimality

Consider the following example in production:

> Each unit of product Y requires 2 hours of machining in the first cell and 1 hour in the second cell. Each unit of product Z requires 3 hours of machining in the first cell and 4 hours in the second cell. Available machining hours in each cell = 12 hours. Each unit of Y yields a profit of $0.80, and each unit of Z yields $2. It is desired to determine the number of units of Y and Z to be manufactured to maximize:
>
> (i) total profit
> (ii) consumer satisfaction, by producing as many units of the superior quality product Y.

If x_1 and x_2 denote the number of units of Y and Z, respectively, then the problem is:

$$\begin{aligned} \text{maximize} \quad & f_1 = 0.8\,x_1 + 2\,x_2 \quad \text{and} \quad f_2 = x_1 \\ \text{subject to} \quad & 2x_1 + 3\,x_2 \le 12 \\ & x_1 + 4\,x_2 \le 12 \\ & x_1, x_2 \ge 0 \end{aligned}$$

The preceding problem is shown graphically in Fig. 8.1. Point A is the solution if only f_1 is to be maximized, while point B is the solution if only f_2 is to be maximized. For every point in $x_1 - x_2$ space, there is a point $(f((x_1), f(x_2))$ in $f_1 - f_2$ or *criterion* (or attribute) space. The image of the feasible region Ω and points A, B are shown

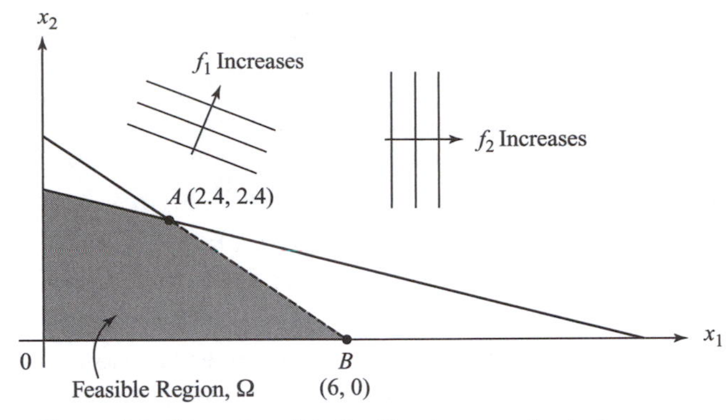

Figure 8.1. Line $A - B$ is the Pareto curve in design space.

in the criterion space as Ω', A' and B' (Fig. 8.2). Referring to Fig. 8.2, we observe an interesting aspect of points lying on the line $A' - B'$: no point on the line is "better" than any other point on the line with respect to *both* objectives. A point closer to A' will have a higher value of f_1 than a point closer to B' but at the cost of having a lower value of f_2. In other words, no point on the line "dominates" the other. Furthermore, a point P in the interior is dominated by all points within the triangle as shown in Fig. 8.2. That is, given a point in the interior, there are points on the line $A' - B'$, which have a higher f_1 and/or f_2 without either function having a lower value. Thus, the line segment $A' - B'$ represents the set of "nondominated" points or *Pareto* points in Ω'. We refer to the line $A' - B'$ as the Pareto curve in criterion space. This curve is also referred to as the Pareto efficient frontier or the nondominated frontier. The general definition for Pareto optimality now follows.

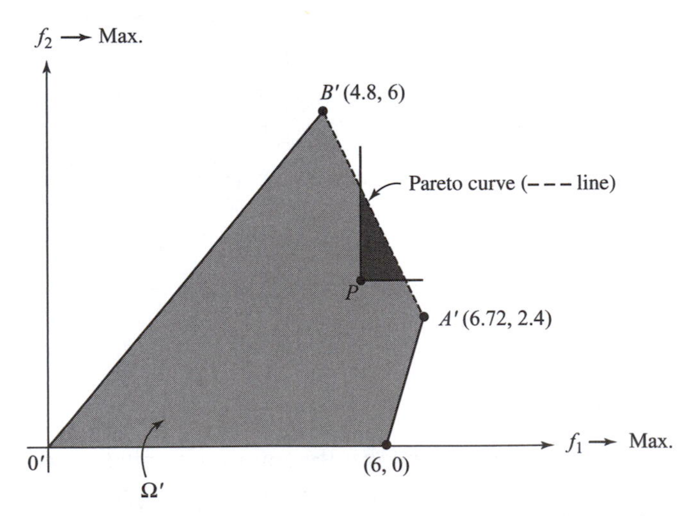

Figure 8.2. Line $A' - B'$ is the Pareto curve in criterion space: $f_1^{\max} = 6.72$, $f_2^{\max} = 6.0$.

Definition of Pareto Optimality

A design variable vector $\mathbf{x}^* \in \Omega$ is Pareto optimal for the problem (8.1) if and only if there is no vector $\mathbf{x} \in \Omega$ with the characteristics

$$\begin{aligned} & f_i(\mathbf{x}) \le f_i(\mathbf{x}^*) && \text{for all } i,\ i = 1, 2, \ldots, m \\ \text{and} \quad & f_j(\mathbf{x}) < f_j(\mathbf{x}^*) && \text{for at least one } j,\ 1 \le j \le m \end{aligned}$$

Example 8.1

Referring to Fig. E8.1a, consider the feasible region Ω' and the problem of maximizing f_1 and f_2. The (disjointed) Pareto curve is identified and shown as dotted lines in the figure. If f_1 and f_2 were to be minimized, then the Pareto curve is as shown in Fig. E8.1b.

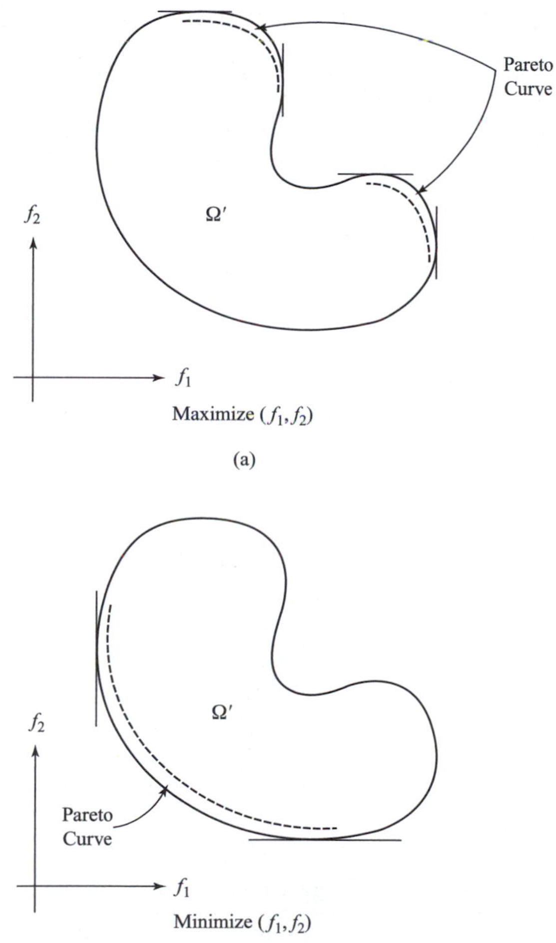

Figure E8.1. Pareto curves.

Example 8.2
A shoe manufacturer has determined a "comfort" index and a "wear resistant" index for a set of six different walking shoes. Higher the indices, better is the shoe. The measurements are:

Comfort index	Wear resistance index
10	2
8	2.25
5	2.5
5.5	2.5
4	4.5
3.5	6.5
3	8

Are there any shoes that are dominated and which can be discontinued? The answer is "Yes," since the shoe with indices (5, 2.5) is dominated by (5.5, 2.5) and can be discontinued. This is also readily apparent by visualizing the nondominated front for the given data (Fig. E8.2).

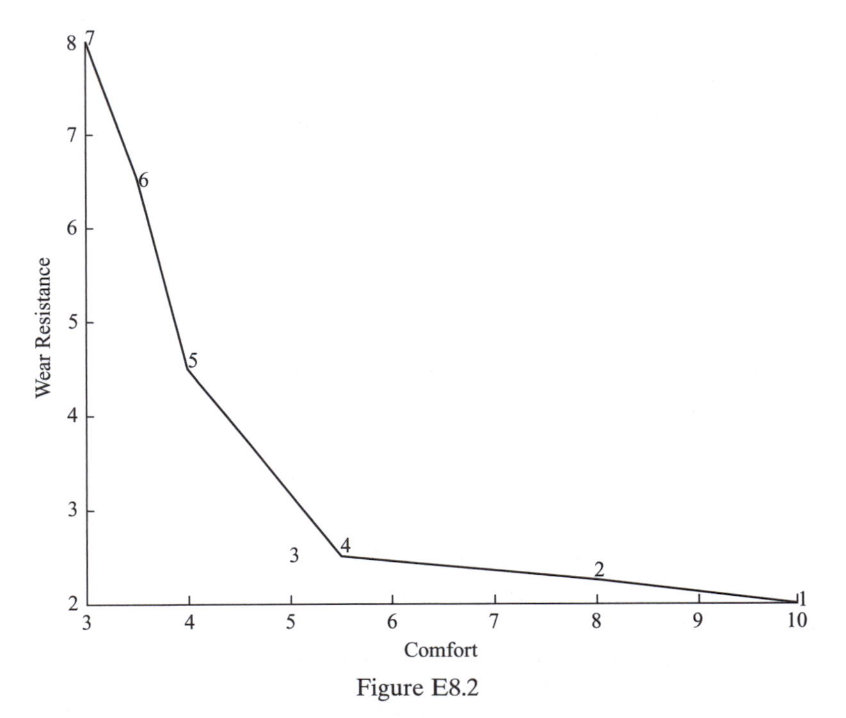

Figure E8.2

Pareto curves often give an insight into the optimization problem that simply cannot be obtained by looking at a single point in the design or criterion space. The

Figure 8.3. An illustrative Pareto curve with a "knee."

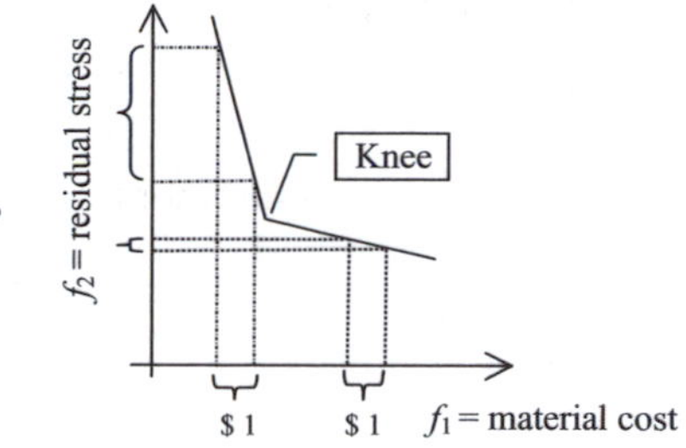

Pareto curve is, in fact, a "trade-off curve." In a paper by Belegundu and Salagame [1995], a Pareto curve with a "knee" was obtained as schematically shown in Fig. 8.3, where f_1 represented residual stress after curing a laminated composite and f_2 represented the material (fiber) cost. The curve has an obvious "knee," which is an attractive Pareto point to choose, since very little addtional reduction in stress is obtainable even with increased volume of fiber (i.e., cost). In other words, a bigger return on the dollar is obtainable before the "knee" as compared to after (Fig. 8.3). The knee signifies a point of diminishing marginal utility.

Pareto curves have been used in designing product families Simpson et al. [1999] and Chen et al. [2000]. The Pareto curve in design space represents different product designs. One approach is to use a filter to reduce the Pareto frontier to a minimal set of distinct designs. The filters perform pairwise comparisons of designs in a Pareto set. When the comparison concludes that the points are overly similar in criterion space, one is removed from the set. The idea is to remove any point within a region of insignificant tradeoff, and repeat the filter operation.

8.3 Generation of the Entire Pareto Curve

In Section 8.1, we presented a "weighting" approach in (8.2) and a "constraint" approach in (8.3) as natural and popular approaches to handling multiple criteria. We raised the question as to what the effect of different weights would have on the solution. The answer is: for each choice of weights w_i or for each set of limits c_i, the solution of (8.2) or (8.3), respectively, is a Pareto point. In fact, as illustrated in Fig. 8.4a–b, we can generate the entire Pareto curve by varying the weights or constraint limits. However, the weighting approach will not generate the entire curve for nonconvex problems (Fig. 8.5a) [Koski 1985]. Also, for certain choices of c_i, no feasible solutions may exist.

It is also possible to use genetic algorithms for generating the Pareto curve or, more precisely, a set of nondominated points. The concept is simple: given a set of points (the population), the set of nondominated points is determined and given a rank 1. These points are excluded from the population and the remaining

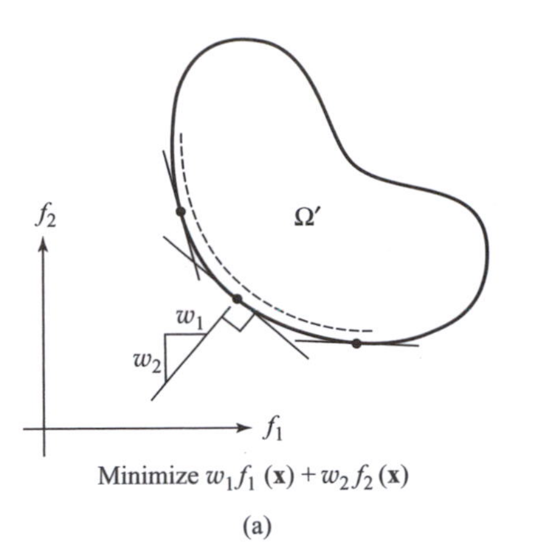

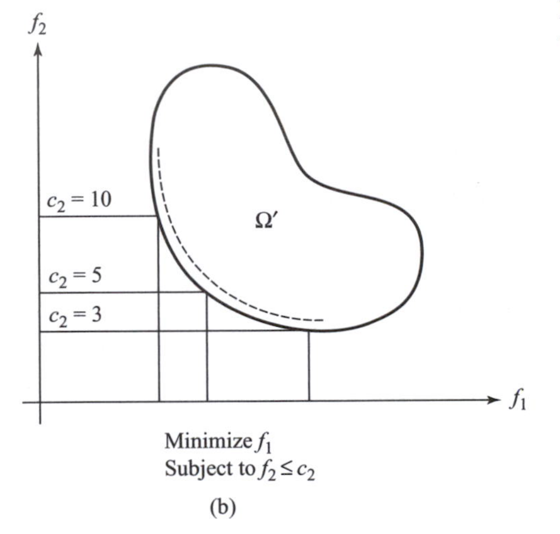

Figure 8.4. Generation of a Pareto curve by varying w_i or c_i.

points are screened for the nondominated points, which are then given a rank of 2. We proceed in this manner till the entire population is ranked. A fitness value is assigned to each point based on its rank. The key concept is that nondominated individuals are mated (crossover) with other nondominated individuals to give rise to new nondominated points. After repeating this process of ranking, crossover and mutation a few times (generations), the final rank 1 individuals will be the Pareto optimal front. Two aspects are important in implementing this scheme: (1) At each generation, a pool or "bank" of nondominated points is maintained. Thus, a good design that is nondominated, which is not picked up for the next generation will nevertheless be in the pool. (2) Individuals that are too close to one-another (either in criterion space or in design variable space) must be given a smaller fitness value to avoid the nondominated points from forming clusters. If this happens, then the final Pareto curve will not represent the entire curve. Niche and

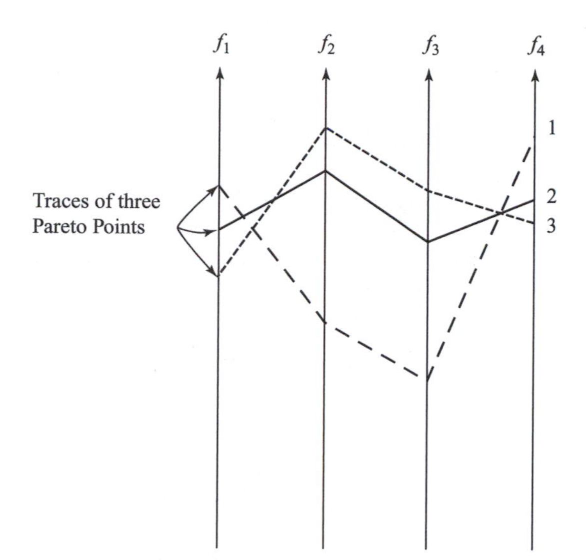

Figure 8.5. Graphical approach with several objectives.

Speciation are two techniques that may be used in this context. See Deb [2001], Zalzala and Fleming [1997], and Goldberg [1989] for additional discussions and references.

While generation of the entire Pareto curve gives us a valuable insight, its visualization for problems with more than two objectives is impractical. With $m = 3$, we have a Pareto surface in 3-D. Beyond that, alternative techniques have to be employed. For instance, we can select a sample of the Pareto optimal or nondominated points and plot traces of each point as illustrated in Fig. 8.5.

8.4 Methods to Identify a Single Best Compromise Solution

In this section, approaches to identify a single Pareto point, or more generally, a single best compromise solution, are discussed.

Min-Max Pareto Solution

It is natural to ask if there is a "best" Pareto point from among the entire Pareto set. Ideally, the entire Pareto set is determined and engineering intuition or discussions among the engineers and managers provides an obvious choice. However, there are quantitative criteria whereby one can select a single best compromise Pareto design. The most popular criterion is the min-max method. Consider a Pareto point Q with (f_1, f_2) as its coordinates in criterion space (Fig. 8.6). Let $f_1^{\min}$ be the minimum of $f_1(\mathbf{x})$, $\mathbf{x} \in \Omega$ and $f_2^{\min} = \min_{\mathbf{x} \in \Omega} f_2(\mathbf{x})$. Then, we can define the deviations from the

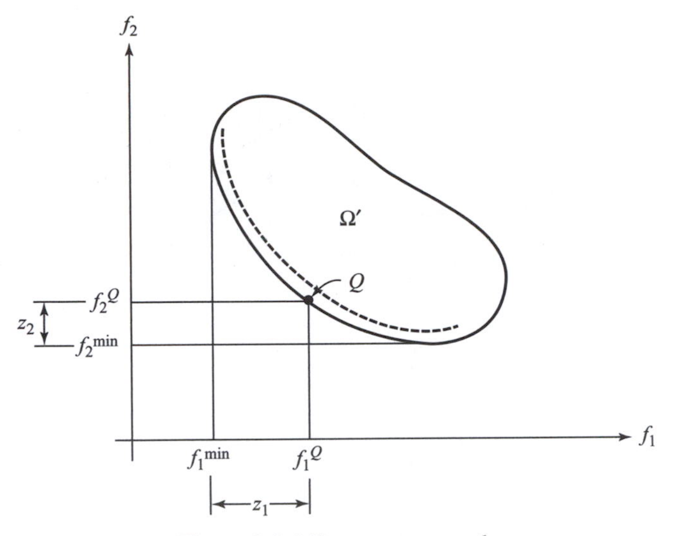

Figure 8.6. Min-max approach.

best values as $z_1 = | f_1 - f_1^{min}|$, and $z_2 = | f_2 - f_2^{min}|$. In the min-max approach, we seek to find the location of Q such that this maximum deviation is minimized. That is, we seek to find a single Pareto point $\mathbf{x}^*$ from min [max $\{z_1, z_2\}$]. With m objectives, the problem is to determine $\mathbf{x}^*$ from

$$\min\,[\max\{ z_1,\ z_2,\ \ldots,\ z_m\}] \tag{8.4}$$

subject to $\mathbf{x} \in \Omega$, where $z_i = f_i(\mathbf{x}) - f_i^{min}(\mathbf{x})$. It is usual to normalize the deviations z_i as

$$z_i = \frac{|f_i(\mathbf{x}) - f_i^{min}|}{f_i^{min}} \quad \text{or} \quad z_i = \frac{|f_i(\mathbf{x}) - f_i^{min}|}{f_i^{max} - f_i^{min}} \tag{8.5}$$

Example 8.3

In the production example considered in Section 8.2 (Fig. 8.1), the min-max approach results in finding $\mathbf{x}^*$ from

$$\min\max\left\{ \frac{|0.8x_1 + 2x_2 - 6.72|}{6.72},\ \ \frac{|x_1 - 6|}{6} \right\}$$

$$\begin{aligned} \text{subject to}\quad & 2\,x_1 + 3\,x_2 \le 12 \\ & x_1 + 4\,x_2 \le 12 \\ & x_1,\ x_2 \ge 0 \end{aligned}$$

This gives $\mathbf{x}^* = (4.84, 0.77)$ with $\mathbf{f}^* = (5.412, 4.84)$.

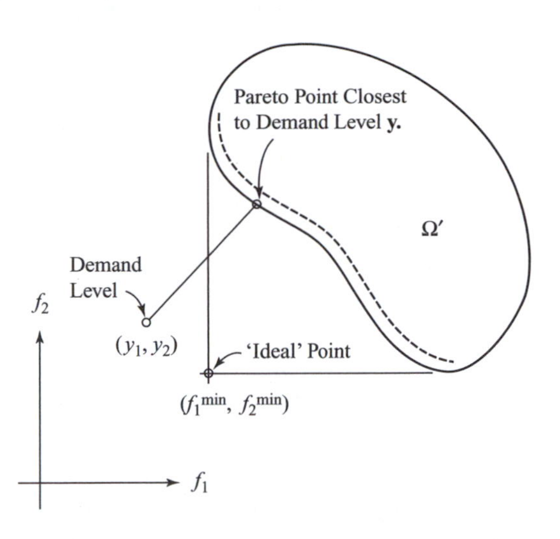

Figure 8.7. General concept of "best" Pareto point.

It is important to note that we can replace the so-called "ideal" point ($f_1^{\min}$, $f_2^{\min}$) with a reasonable demand vector $\mathbf{y} = (y_1, y_2)$ or "goal" in the preceding analysis (Fig. 8.7). This makes it unnecessary to determine $f_i^{\min}$, which involves the solution of separate optimization problems whose solution can be difficult, especially for nonconvex problems. There are other methods for finding the best compromise Pareto point, which may all be graphically interpreted as different distance measures between a point on the Pareto curve and a demand level **y**.

Weighted-sum Approach

As in (8.2), the multiobjective problem is formulated as

$$\begin{aligned} &\text{minimize} \quad f = \sum_{i=1}^{m} w_i \, f_i(\mathbf{x}) \\ &\text{subject to} \quad \mathbf{x} \in \Omega \end{aligned}$$

where weights $w_i \geq 0$, $\sum w_i = 1$. For a given choice of weights, a single solution $\mathbf{x}^*$ is obtained. However, there are implications that the designer (decision-maker) must be aware of as the resulting solution may not be satisfactory, as will now be explained [Grissom et al. 2006]. Referring to Fig. 8.8, for $m = 2$, the optimum point in attribute space is characterized by a Pareto point at which the f-contour is tangent to the feasible region. From $df = 0$, we get

$$\left.\frac{df_i}{df_j}\right|_{\text{optimum}} = -\frac{w_j}{w_i} \tag{8.6}$$

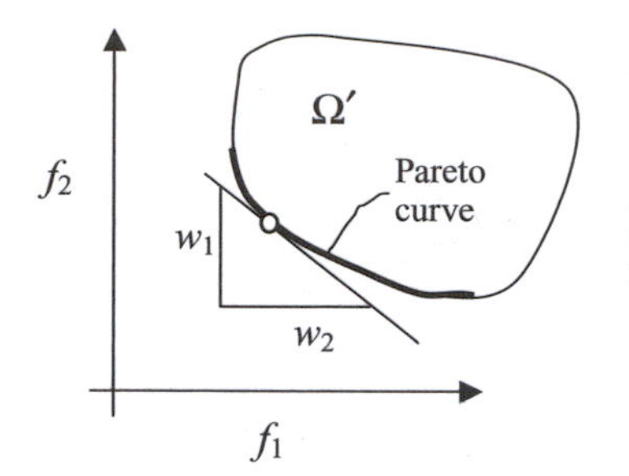

Figure 8.8. Optimal solution using the weighted-sum method.

Thus, a tradeoff between any pair f_i and f_j is independent of the actual values of f_i and f_j, implying a *constant tradeoff* where diminishing marginal utility is ignored; the designer does not include any preference on the level of each attribute in the design. The only preference is on the tradeoff. For example, in the context of noise reduction, if we are prepared to pay \$100 for each dB reduction in sound pressure, then this will remain the amount we will pay regardless of the actual dB levels in the machine or surroundings. Selection of weights that satisfy designer's preferences is possible using pairwise comparisons and other techniques discussed in the literature on decision theory.

Constraint Approach

As in (8.3), Fig. 8.4, we formulate the problem as

$$\begin{aligned} &\text{minimize} \quad f_1(\mathbf{x}) \\ &\text{subject to} \quad f_j(\mathbf{x}) \leq c_j, \quad j = 2, \ldots, m \end{aligned}$$

For a given choice of **c**, a single solution $\mathbf{x}^*$ is obtained. From the optimality conditions in attribute space, we have at optimum

$$\left. \frac{df_1}{df_j} \right|_{\text{optimum}} = -\lambda_j, \quad j = 2, \ldots, m \tag{8.7}$$

where λ_j is a Lagrange multiplier. Thus, the constraint approach arrives at the tradeoff condition for the solution in an arbitrary manner. The preference is dictated by a set of maximum attribute levels. For example, with $m = 2$, $f_1 \equiv$ dB reduction and $f_2 \equiv$ cost, with $c_2 = \$100$, then we can expect a 10-dB reduction with \$100, but a design where 9-dB reduction is possible with only \$50 will be overlooked. Again, diminishing marginal utility is not incorporated into the approach.

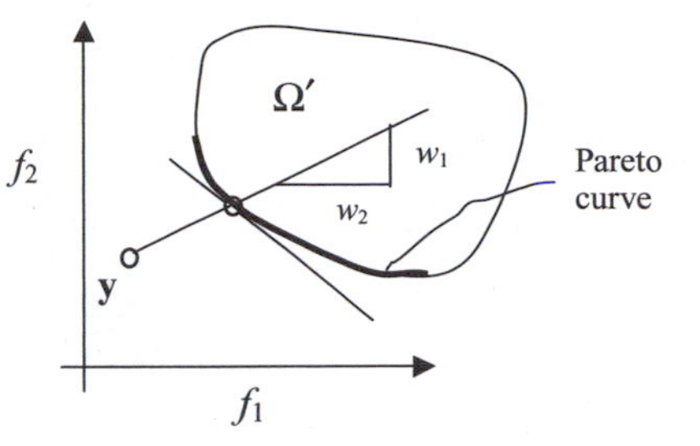

Figure 8.9. Optimal solution using goal programming ($r = \infty$).

Goal Programming Approach

Referring to Fig. 8.7, denoting the deviations from the goal as $d_i = f_i(\mathbf{x}) - y_i$, and assuming that the weights are the same above and below the goals, the problem takes the form

$$\begin{aligned} &\underset{\mathbf{x}}{\text{minimize}} && \| w_1 d_1, w_2 d_2, \ldots, w_m d_m \|_r \\ &\text{subject to} && d_i = f_i(\mathbf{x}) - y_i \\ &\text{and} && \mathbf{x} \in \Omega \end{aligned} \tag{8.8}$$

where r defines the type of norm or distance measure. Specifically, if $\mathbf{z}$ is a vector with m components, then $\|\mathbf{z}\|_1 = \sum_i |z_i|$, $\|\mathbf{z}\|_2 = \sqrt{\sum_i z_i^2}$, $\|\mathbf{z}\|_\infty = \max_i |z_i|$. The case $r = 1$ is identical to the weighted-sum method. For $r = \infty$, an optimum point exists where the weighted distances are equally far from the goal point. It can be shown that at optimum, for each pair of attributes

$$\frac{d_i}{d_j} = \frac{w_j}{w_i} \tag{8.9}$$

This equation defines a line passing through the goal point and having a slope equal to the ratio of the weights placed on each attribute, as shown in Fig. 8.9.

In Eq. (8.9), we see that attribute levels do enter into determining the optimum. However, diminishing marginal utility over the range of the attribute level is not clearly incorporated. The success of goal programming hinges critically on the choice of the ideal point and the weights.

Interactive Approaches

The basic idea behind an interactive technique is presented here by considering the Constraint Approach with $m = 2$ [Sadagopan and Ravindran 1982]. The problem takes the form

$$\begin{aligned} &\text{minimize} && f_1(\mathbf{x}) \\ &\text{subject to} && f_2(\mathbf{x}) \le c \\ &\text{and} && \mathbf{x} \in \Omega \end{aligned} \tag{8.10}$$

where $c_{\min} \le c \le c_{\max}$. The "best" multiobjective solution is a single variable problem for the best value of c. Determining $c_{\min}$ and $c_{\max}$ involves solving two additional NLPs (Nonlinear Programming Problems). Referring to Fig. 8.4, $c_{\min} = f_2(\mathbf{x}^{**})$ where $\mathbf{x}^{**}$ is the solution of

$$\begin{aligned} &\text{minimize} && f_2(\mathbf{x}) \\ &\text{and} && \mathbf{x} \in \Omega \end{aligned} \tag{8.11}$$

and $c_{\max} = f_2(\mathbf{x}^{***})$ where $\mathbf{x}^{***}$ is the solution of

$$\begin{aligned} &\text{minimize} && f_1(\mathbf{x}) \\ &\text{and} && \mathbf{x} \in \Omega \end{aligned} \tag{8.12}$$

Golden Section search can be initiated as detailed in Chapter 2. For every pair of values of c within the current interval, the designer *interactively* chooses the better value. We assume that the implicit utility function is unimodal. At the start, two points c_1 and c_2 are introduced, with $c_{\min} \le c_1 \le c_2 \le c_{\max}$. NLP in (8.10) is solved for c_1 and c_2 with the corresponding solutions $\mathbf{x}^1$ and $\mathbf{x}^2$, respectively. The decision-maker examines $[\mathbf{x}^1, f_1(\mathbf{x}^1), f_2(\mathbf{x}^1)]$ and $[\mathbf{x}^2, f_1(\mathbf{x}^2), f_2(\mathbf{x}^2)]$ and chooses either $\mathbf{x}^1$ or $\mathbf{x}^2$ as the more preferred design. If $\mathbf{x}^1$ is preferred, then the new interval is $[c_{\min}, c_2]$. Now, with c_1 already in the right location, one new point is introduced and the pairwise comparisons proceed until convergence. With $m = 3$ objectives, the interactive problem is a two-variable problem.

Preference Functions Approach

The aim is to obtain a realistic value or preference function

$$V(\mathbf{f}) = V(f_1,\ f_2, \ldots,\ f_m) \tag{8.13}$$

The fundamental property of V is that $\mathbf{f}^a$ is preferred to $\mathbf{f}^b$ if and only if $V(\mathbf{f}^a) > V(\mathbf{f}^b)$. The associated NLP problem is

$$\begin{aligned} &\text{maximize} && V(\mathbf{f}(\mathbf{x})) \\ &\text{subject to} && \mathbf{x} \in \Omega \end{aligned} \tag{8.14}$$

Depending on the value functions, the optimum may lie in the interior or on the boundary of Ω. In the latter case, the optimum lies on the value contour that touches

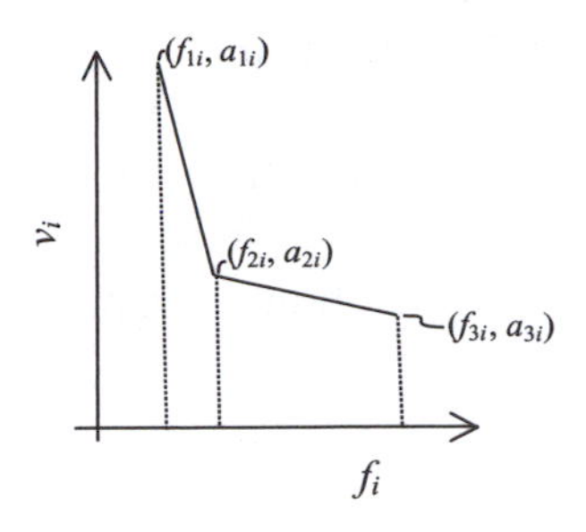

Figure 8.10. Piecewise linear representation of v_i in terms of part-worths.

Ω. A contour of V is referred to as an indifference curve or surface. A special case is the additive model when certain assumptions are met:

$$V = \sum_{i=1}^{m} v_i(f_i) \tag{8.15}$$

where v_i is a consistently scaled-value function associated with attribute i only. A well-known assumption that ensures the existence of a separable form as in Eq. (8.15) is that the attributes are mutually preferentially independent. This assumption implies that the individual preferences on attributes satisfy the "tradeoff condition," which is that the tradeoff between any pair of attributes f_i and f_j, keeping the levels of other attributes fixed, does not depend on the fixed levels. It should be realized that, in Eq. (8.15), a highly valued option in one attribute can compensate for an unattractive option on another attribute. The individual value functions may be determined in several ways. One technique is to use conjoint analysis as explained in the following.

Conjoint Analysis

This is a method based on the additive model, Eq. (8.15). The idea is to characterize a product by a set of attributes (e.g., reduction in sound pressure levels in different frequency bands), each attribute having a set of levels (e.g., small, medium, and large reduction). The aggregated value function V is developed by direct interaction with the customer/designer; the designer is asked to rate, rank order, or choose a set of product bundles. The bundles may be real objects, a computer representation, or simply a tabular representation, and must be reasonably small in number. The procedure is as follows [Lilien and Rangaswamy 1997]. Assume f_i has three levels f_{1i}, f_{2i}, and f_{3i}. Let $a_{ji}, j = 1, 2, 3$, be the "attribute part-worths" such that the value function v_i is represented as a piecewise linear function (Fig. 8.10).

Thus, if we were to plot V in **f**-coordinates, it will have a faceted shape. Now, based on the ratings of the product bundles, the attribute part-worths are determined using regression. For example, among a total of K bundles, let a specific bundle B_k with $f_1 = f_{21}$, $f_2 = f_{22}$, and $f_3 = f_{23}$ have a rating of, say, 75. Then the associated model error is $e_k = a_{21} + a_{22} + a_{23} - 75$. Regression based on minimizing $\sum_{k=1}^{K} e_k^2$ will then give us $[a_{ij}]$, which defines the total aggregate function V. Here, note that V is defined only within the bounds on f_i, which were used in the conjoint study. Optimization may be pursued based on

$$\begin{aligned} &\text{maximize} && V(\mathbf{f}(\mathbf{x})) \\ &\text{subject to} && \mathbf{f}^{\mathrm{L}} \leq \mathbf{f}(\mathbf{x}) \leq \mathbf{f}^{\mathrm{U}} \\ &\text{and} && \mathbf{x} \in \Omega \end{aligned} \tag{8.16}$$

The set Ω may represents bounds on variable as $\mathbf{x}^{\mathrm{L}} \leq \mathbf{x} \leq \mathbf{x}^{\mathrm{U}}$. The preceding process was carried out in the design of a mug in Example 1.4 in Chapter 1.

Example 8.4

In a portfolio optimization problem, it is desired to construct an objective function based on conjoint analysis. The objectives are to maximize profit, P, and to minimize risk, R. Levels for P are chosen to be [0.1, 1.0, 2.0], while levels for risk are chosen to be [0.1, 0.5, 1.0]. Ratings for a set of product bundles representing an orthogonal array are given in the following. The ratings are scaled to be within 0–100.

Table E8.4

Sample	P	R	Rating	Scaled rating
1	0.10	0.10	40.00	41.18
2	0.10	0.50	10.00	5.88
3	0.10	1.00	5.00	0.00
4	1.00	0.10	78.00	85.88
5	1.00	0.50	60.00	64.71
6	1.00	1.00	30.00	29.41
7	2.00	0.10	90.00	100.00
8	2.00	0.50	68.00	74.12
9	2.00	1.00	40.00	41.18

Adopting the procedure discussed above, regression for the part-worths yields the component value functions in Figs. E8.4a–b. Contours of total value V are in Fig. E8.4c. It is evident that these are not straight lines implying that the tradeoffs are not constant. In fact, the contours curving upward show that the decision-maker is risk-averse. In this example, the focus has been on formulating the objective function.

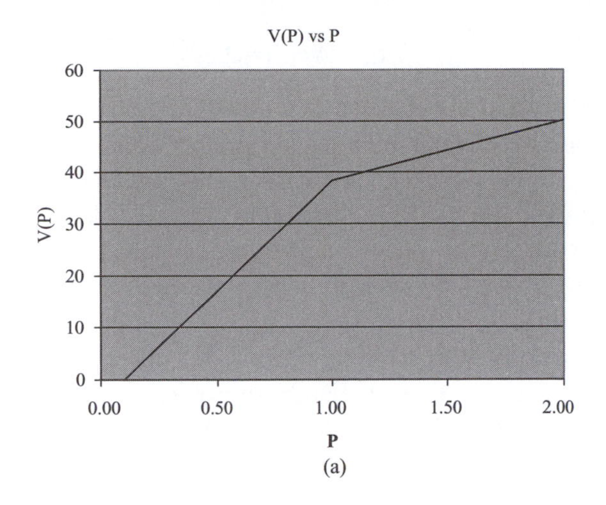

(a)

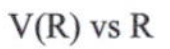

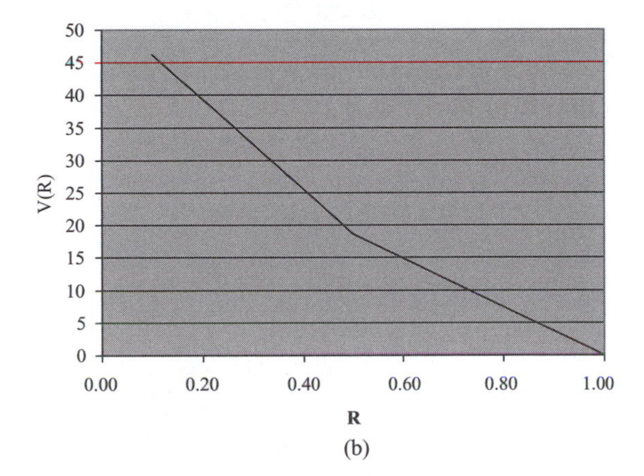

(b)

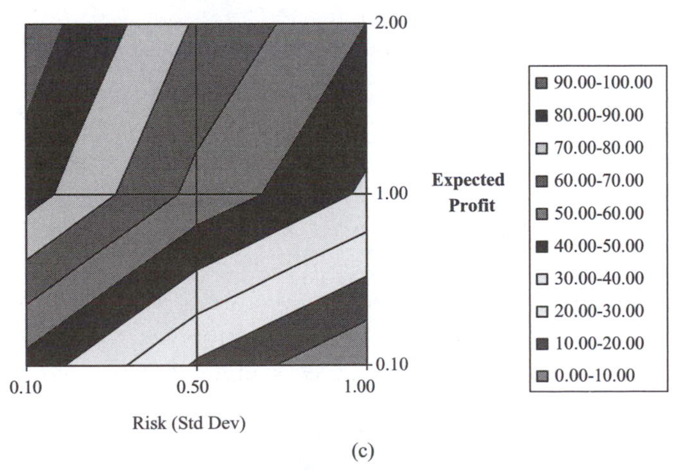

(c)

Figure E8.4

Other Approaches

Multiattribute utility theory and evaluation driven design analysis use lottery questions and probabilities to elicit a designer's preferences [Locascio and Thurston 1998]. In these methods, diminishing marginal utility is accounted for and, hence, represent an improvement over weighted-sum, constraint, and goal programming. Further, risk is accounted for. The main drawback is that the lottery questions may be a little confusing to pose to engineers, especially on problems that are somewhat deterministic.

Preference functions approach is powerful as it lends itself readily to quantification of diverse issues, such as ease of manufacturability (for example, to limit variety of component items). For this reason, it is popular in management and economics and has potential in engineering.

PROBLEMS

P8.1. Describe the Pareto optimal set for a bicriteria problem where the two objectives attain their optimum at the same $\mathbf{x}^*$.

P8.2. In Fig. E8.1, sketch the Pareto curves for the problem:

$$\{\text{minimize } f_1, \text{maximize } f_2\}$$

P8.3. Consider the problem: maximize $\{f_1(x), f_2(x)\}$, $0 \le x \le 1$, f_1 and f_2 as plotted in the following:

(i) Plot the feasible region Ω' in attribute space, by plotting the points $[f_1, f_2]$ for each value of x.

(ii) Identify the Pareto frontier on your plot from (i).

(iii) Determine the min-max Pareto point, without normalization.

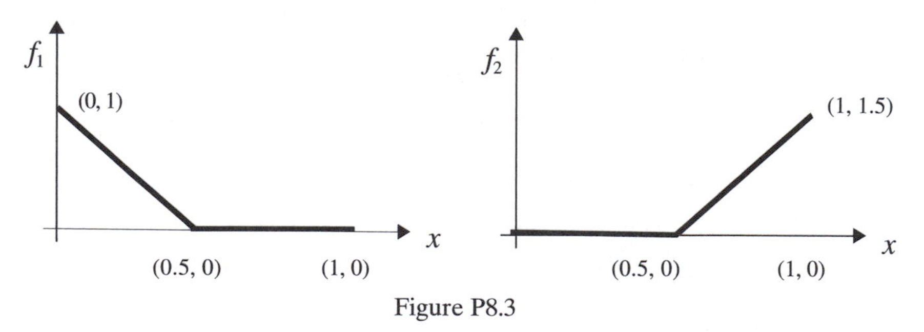

Figure P8.3

P8.4. Consider

$$\text{maximize } \{-(x-0.3)^2, -(x-0.7)^2\}$$
$$0 \le x \le 1$$

With plots of f_1 versus f_2, and f_1 and f_2 versus x, identify the Pareto set in both design space and in criterion space and show these in the plots.

P8.5. Plot the Pareto curve for:

$$\begin{aligned} &\max && 3x_1 + x_2 + 1 \\ &\max && -x_1 + 2x_2 \\ &\text{subject to} && 0 \le x_1 \le 3, \quad 0 \le x_2 \le 3 \end{aligned}$$

P8.6. Four candidates were interviewed for a teaching position in a middle school by a panel. Selection was to be based on three criteria: Attitude towards children, technical competence, and ability as a team player. The scores were on a scale of 1–10 with 10 = highest or best. The scores are:

Candidate	Care toward children	Technical competence	Team player
A	6.0	9.0	5.0
B	8.5	6.0	6.5
C	8.5	3.0	8.0
D	4.0	8.0	4.0

Based on the min-max approach, determine the best candidate.

P8.7. The architect has presented Mrs. Smith with a floor plan for a portion of the house. The portion includes a living room, a kitchen and a dining room as shown in Fig. P8.7. The variables that have to be determined by Mrs. Smith are the living room width x_1 and length of kitchen x_2. The criteria or objectives are to minimize cost and to maximize kitchen area. The unit prices for the different rooms are: \$60/ft^2 for the kitchen, \$45/ft^2 for the living room and \$30/ft^2 for the dining. Thus, the bicriterion problem becomes

$$\begin{aligned} &\text{minimize} && f_1 = 750\,x_1 + 60\,(25 - x_1)\,x_2 + 45\,(25 - x_1)\,(25 - x_2) \\ &\text{maximize} && f_2 = (25 - x_1)\,x_2 \\ &\text{subject to} && 10 \le x_1 \le 15\,\text{ft} \\ &&& 10 \le x_2 \le 18\,\text{ft} \end{aligned}$$

Show a graph of f_1 versus $-f_2$ and indicate the Pareto curve. Obtain values of x_1, x_2 based on a min-max criterion.

[*Reference*: Mitchell W.J., et al. Synthesis and optimization of small rectangular floor plans, *Environment and Planning B*, **3**, 37–70, 1976].

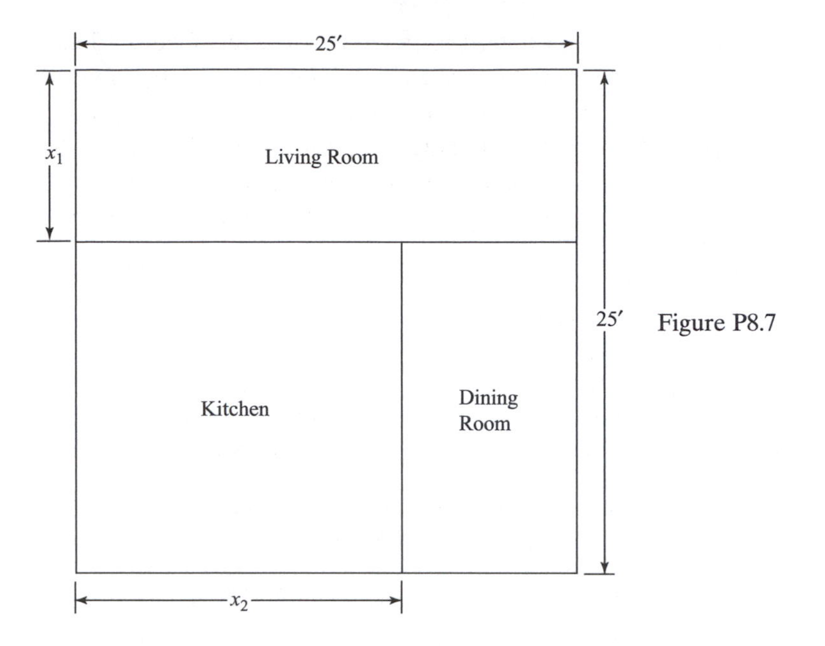

Figure P8.7

P8.8. Consider the truss shown in Fig. P8.8. Obtain a plot of the Pareto curve, showing the feasible region, in criterion space. The problem is:

$$\begin{aligned} &\text{minimize} && \{V,\ \Delta\} \\ &\text{subject to} && 10\,\text{mm}^2 \le A_i \le 200\,\text{mm}^2 \qquad i = 1, 2, 3 \\ & && -100\,\text{MPa} \le \sigma_i \le 100\,\text{MPa} \quad i = 1, 2, 3 \end{aligned}$$

where the volume $V = \sum A_i\, L_i$, Δ = deflection of node 1 along the direction of the applied load, σ_i are stresses. Take Young's modulus $E = 200{,}000$ MPa, $F = 15{,}000$ N.

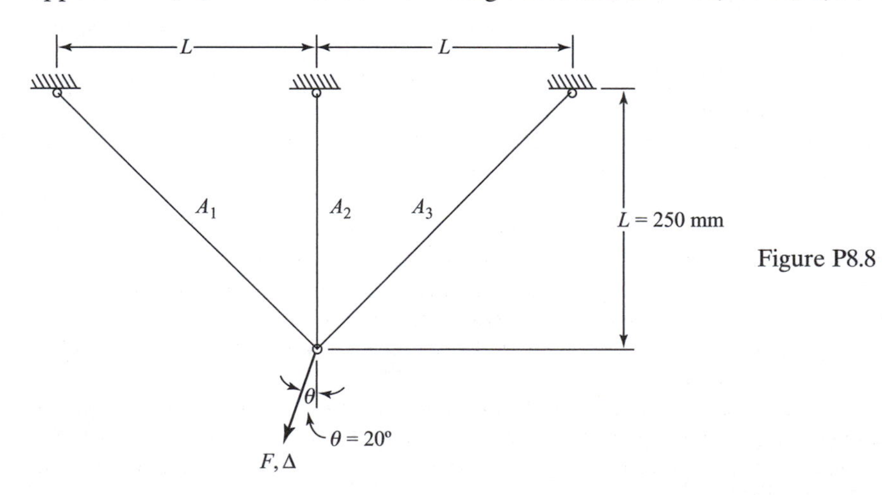

Figure P8.8

P8.9. Carry out the calculations in the preference functions based mug design – Example 1.4, Chapter 1.

P8.10. Assume you in a dilemna with regard to purchase of a car. Choose 5 conflicting attributes, such as price, ride/handling, noise, reliability, and quality of headlights. Each attribute has a low (L) and a high (H) value. Similar to the procedure in Example 8.4, develop a set of attribute bundles (consisting of a fraction of the $2^5 = 32$ possible combinations), rate the bundles and develop a preference function. Provide plots $v(f_i)$ versus f_i. Compare your preferences with that of your friends.

REFERENCES

Belegundu, A.D. and Salagame, R.R., Optimization of laminated ceramic composites for minimum residual stress and cost, *Microcomputers in Civil Engineering*, **10**, 301–306, 1995.

Chen, W., Sahai, A., Messac, A., and Sundararaj, G.J., Exploration of the effectiveness of physical programming in robust design, *ASME Journal of Mechanical Design*, **122**(2), 155–163, 2000.

Cohon, L.L., *Multiobjective Programming and Planning*, Academic Press, New York, 1978.

Deb, K., *Multi-Objective Optimization Using Evolutionary Algorithms*, Wiley, New York, 2001.

Eschenauer, H., Koski, J., and Osyczka, A. (Eds.), *Multicriteria Design Optimization*, Springer, Berlin, 1990.

Geoffrion, A.M., Dyer, J.S., and Feinberg, A., An interactive approach for multicriterion optimization, with an application to the operation of an academic department, *Management Science*, **19**(4), 357–368, 1972.

Goicoechea, A., Hansen, D.R., and Duckstein, L., *Multiobjective Decision Analysis with Engineering and Business Applications*, Wiley, New York, 1982.

Goldberg, D.E., *Genetic Algorithms in Search, Optimisation and Machine Learning*, Addison-Wesley, Reading, MA, 1989.

Grissom, M.D., Belegundu, A.D., Rangaswamy, A., Koopmann, G.H., Conjoint analysis based multiattribute optimization: application in *acoustical design, Structural as Multi-disciplinary Optimization* (31), pp. 8–16, 2006.

Ignizio, J.P., Multiobjective mathematical programming via the MULTIPLEX model and algorithm, *European Journal of Operational Research*, **22**, 338–346, 1985.

Keeney, R.L. and Raiffa, H., *Decisions with Multiple Objectives: Preference and Value Trade-offs*, Wiley, New York, 1976.

Koski, J., Defectiveness of weighting method in multicriterion optimization of structures, *Communications in Applied Nunerical Methods*, **1**, 333–337, 1985.

Lilien, G.L., Rangaswamy, A., *Marketing Engineering*, Addison-Wesley, Reading, MA, 1997.

Locascio, A. and Thurston, D.L., Transforming the house of quality to a multiobjective optimization formulation, *Structural Optimization*, **16**(2–3), 136–146, 1998.

Marler, R.T., and Arora, J.S., Survey of multi-objective optimization methods for engineering, *Structural and Multidisciplinary Optimization*, **26**, 369–395 (2004).

Messac, A., Physical programming: effective optimization for computational design, *AIAA Journal*, **34**(1), 149–158, 1996.

Mistree, F., Hughes, O.F., and Bras, B., "Compromise decision support problem and the adaptive linear programming algorithm," M.P. Kamat (Ed.), *Structural Optimization: Status and Promise, Progress in Astronautics and Aeronautics*, Vol. 150, AIAA, Washington, DC, pp. 247–289, 1993.

Pareto, V., *Manual of Political Economy*, Augustus M. Kelley Publishers, New York, 1971. (Translated from French edition of 1927.)

Rardin, R.L., *Optimization in Operations Research*, Prentice-Hall, Englewood Cliffs, NJ, 1998.

Sadagopan, S. and Ravindran, A., Interactive solution of bicriteria mathematical programs, *Naval Research Logistics Quarterly*, **29**(3), 443–459, 1982.

Simpson, T., Maier, J., and Mistree, F., A product platform concept exploration method for product family design, *Proceedings of ASME Design Engineering Technical Conferences*, Las Vegas, NV, DTM-8761, 1999.

Statnikov, R.B. and Matusov, J.B., *Multicriteria Optimization and Engineering*, Chapman & Hall, New York, 1995.

Thurston, D.L., A formal method for subjective design evaluation with multiple attributes, *Research in Engineering Design*, **3**, 105–122, 1991.

Zalzala, A.M.S. and Fleming, P.J. (Eds.), *Genetic Algorithms in Engineering Systems*, *Control Engineering Series 55*, The Institution of Electrical Engineers, London, 1997, Chapter 3.

Zeleny, M., *Multiple Criteria Decision Making*, McGraw-Hill, New York, 1982.

9

Integer and Discrete Programming

9.1 Introduction

There are many problems where the variables are not divisible as fractions. Some examples are the number of operators that can be assigned to jobs, the number of airplanes that can be purchased, the number of plants operating, and so on. These problems form an important class called *integer programming* problems. Further, several decision problems involve binary variables that take the values 0 or 1. If some variables must be integers and others allowed to take fractional values, the problem is of the *mixed integer* type. *Discrete programming* problems are those problems where the variables are to be chosen from a discrete set of available sizes, such as available shaft sizes and beam sections, engine capacities, pump capacities.

In integer programming problems, treating the variable as continuous and rounding off the optimum solution to the nearest integer is easily justified if the involved quantities are large. Consider the example of the number of barrels of crude processed where the solution results in 234566.4 barrels. This can be rounded to 234566 barrels or even 234570 without significantly altering the characteristics of the solution. When the variable values are small, such as the number of aircraft in a fleet, rounding is no longer intuitive and may not even yield a feasible solution. In problems with binary variables, rounding makes no sense as the choice between 0 or 1 is a choice between two entirely different decisions. The following two variable example provides some characteristics of integer requirements.

Example 9.1

A furniture manufacturer produces lawn chairs and tables. The profit from the sale of each chair is \$2 and the profit from the sale of a table is \$3. Each chair weighs 4 pounds and each table weighs 10 pounds. A supply of 45 pounds of material is available at hand. The labor per chair is 4 hours and the labor per

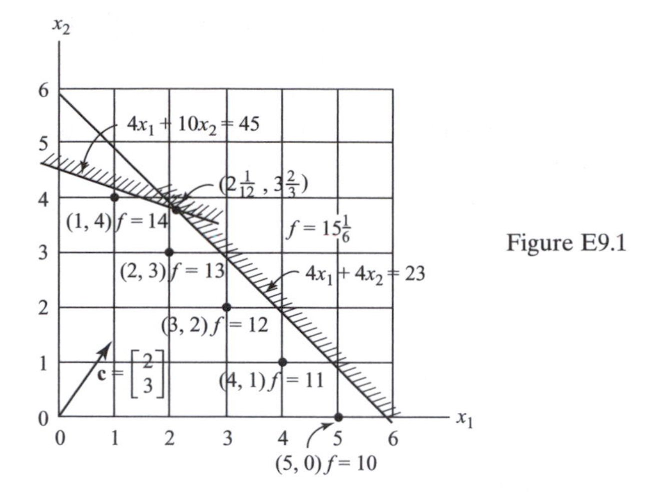

Figure E9.1

table is 4 hours. Only 23 hours of labor is available. Determine the number of chairs and tables for realizing maximum profit.

Solution

The problem may be put in the form

$$\begin{aligned} \text{maximize} \quad & 2x_1 + 3x_2 \\ \text{subject to} \quad & 4x_1 + 10x_2 \leq 45 \\ & 4x_1 + 4x_2 \leq 23 \\ & x_1, x_2 \geq 0 \text{ and integer} \end{aligned}$$

The graphical representation of the problem is shown in Fig. E9.1.

The solution without the integer requirement follows the steps of Chapter 4. The simplex solution is $2^1/_{12}$, $3^2/_3$, when the integer requirement is relaxed. The corresponding profit is $15^1/_{16}$. Feasible integer solutions are tabulated as follows.

Number of chairs	Number of tables	Profit
1	4	14
2	3	13
3	2	12
4	1	11
5	0	10

The intuitive round off results in the solution 2, 3, which gives a profit of 13. The integer solution 1, 4 results in the maximum profit of 14.

The simplest integer programming problems are those where the variables take values of zero and one. We start our discussion with these problems. Our focus here is on linear problems. Subsequently, the branch and bound method for mixed integer problems is discussed. The branch and bound method discussed herein can be generalized to handle nonlinear problems, as also simulated annealing and genetic algorithms of Chapter 7 (see also Section 9.6). For nonlinear problems involving monotonic functions, see Section 9.5.

9.2 Zero–One Programming

Several integer programming problems can be brought into a form where each the variable takes the values 0 or 1. Some modeling examples are given as follows.

Satisfaction of a Set of Alternative Constraints

Given a set of inequality constraints $g_1(\mathbf{x}) \leq 0$, $g_2(\mathbf{x}) \leq 0$, $g_3(\mathbf{x}) \leq 0$, $\mathbf{x} \in R^n$ it is required that at least one of them be satisfied. This can be formulated in a form suitable for numerical optimization as

$$
\begin{aligned}
g_1(\mathbf{x}) \leq My_1, \quad g_2(\mathbf{x}) \leq My_2, \quad & g_3(\mathbf{x}) \leq My_3, \\
& y_1 + y_2 + y_3 \leq 2 \\
& y_i = 0 \quad \text{or} \quad 1, \quad i = 1, 2, 3
\end{aligned}
$$

where M is a large number. Thus, there are a total of $n + 3$ variables, n continuous and 3 binary.

Logical Conditions

Let X_i represent a logical proposition and x_i its associated binary variable. For instance, these are involved when representing shipping or mailing costs. A few instances are given as follows.

(a) If X_1 or X_2 is true, then X_3 is true:

$$x_1 \leq x_3, \quad x_2 \leq x_3$$

where $x_1 = 1$ if X_1 is true and $= 0$ if false.

(b) If X_3 is true, then X_1 or X_2 is true:

$$x_1 + x_2 \geq \mathrm{x}_3$$

(c) If both X_1 and X_2 are true, then X_3 is true:

$$1 + x_3 \geq x_1 + x_2$$

Integer Variables Using 0–1 Approach

Regular integer problems can be solved using 0–1 programming. If a nonnegative integer variable Y is bounded above by 11 ($Y \leq 11$) then, we first determine a K such that $2^{K+1} \geq 11 + 1$. This gives $K = 3$. Then, we can represent Y as

$$Y = 2^0 x_1 + 2^1 x_2 + 2^2 x_3 + 2^3 x_4 \quad \text{with} \quad x_i = 0 \quad \text{or} \quad 1 \quad \text{for} \quad i = 1 \text{ to } 4$$

The constraint then becomes

$$x_1 + 2x_2 + 4x_3 + 8x_4 \leq 11$$

This method, however, increases the number of variables.

Polynomial 0–1 Programming

Any 0–1 variable x_i raised to a positive integer p has the property $x_i^p = x_i$. Thus, general polynomials can be reduced to the product form $x_1 x_2 \ldots x_k$.

$$x_1^{p_1} x_2^{p_2} \ldots x_k^{p_k} = x_1 x_2 \ldots x_k$$

A product can be substituted by y with two constraints that ensure that y is zero if any of x_i is zero, and it is 1 only when all x_i are 1.

$$\begin{aligned}
&\text{replace } x_1 x_2 \ldots x_k \text{ by } y \\
&x_1 + x_2 + \cdots + x_k - y \leq k - 1 \\
&x_1 + x_2 + \cdots + x_k \geq ky \\
&y = 0 \quad \text{or} \quad 1
\end{aligned}$$

Polynomial expressions can be simplified using this idea and solved using the algorithm developed previously. As an example, consider the problem

$$\begin{aligned}
\text{Maximize} \quad & 3x_1^2 x_2 x_3^3 \\
\text{subject to} \quad & 4x_1 + 7x_2^2 x_3 \leq 12 \\
& x_1, x_2, x_3 \text{ are 0–1 variables}
\end{aligned}$$

Using the polynomial property, the problem can be put in the form

$$\begin{aligned} &\text{Maximize} && 3x_1x_2x_3 \\ &\text{subject to} && 4x_1 + 7x_2x_3 \leq 12 \\ & && x_1, x_2, x_3 \text{ are 0–1 variables} \end{aligned}$$

Now substituting $y_1 = x_1\, x_2\, x_3$, and $y_2 = x_2\, x_3$, and making use of the preceding relations, we get the equivalent problem in the form

$$\begin{aligned} &\text{Maximize} && 3y_1 \\ &\text{subject to} && 4x_1 + 7y_2 \leq 12 \\ & && x_1 + x_2 + x_3 - y_1 \leq 2 \\ & && x_1 + x_2 + x_3 \geq 3y_1 \\ & && x_2 + x_3 - y_2 \leq 1 \\ & && x_2 + x_3 \geq 2y_2 \end{aligned}$$

This problem is linear in the variables.

Implicit enumeration is a popular method for solving these problems. We consider these problems in the standard form

$$\begin{aligned} &\text{minimize} && \mathbf{c}^{\mathrm{T}}\mathbf{x} \\ &\text{subject to} && \mathbf{Ax} - \mathbf{b} \geq \mathbf{0} \\ & && x_i = 0 \quad \text{or } 1 \quad (i = 1, 2, \ldots, n) \end{aligned} \tag{9.1}$$

In the standard form, each of the coefficients c_i is nonnegative ($c_i \geq 0$). The search for the solution starts from $\mathbf{x} = \mathbf{0}$, which is optimal. If this optimal solution is also feasible, then we are done. If it is not feasible, then some strategies to adjust the variable values are developed so that feasibility is attained. Optimal solution is identified among these. In order to illustrate the main ideas, let us consider an illustrative problem and its explicit enumeration.

Consider the problem

$$\begin{aligned} &\text{minimize} && f = 4x_1 + 5x_2 + 3x_3 \\ &\text{subject to} && g_1 = x_1 - 3x_2 - 6x_3 + 6 \geq 0 \\ & && g_2 = 2x_1 + 3x_2 + 3x_3 - 2 \geq 0 \\ & && g_3 = x_1 + x_3 - 1 \geq 0 \\ & && x_i = 0 \quad \text{or } 1 \quad (i = 1, 2, 3) \end{aligned} \tag{9.2}$$

Implicit Enumeration for 0–1 Programming (Non-LP Base Algorithm)

Before discussing implicit enumeration let us take look at explicit enumeration where all the possible values are tried. The values are shown in the table.

x_1	x_2	x_3	g_1	g_2	g_3	f
0	0	0	6	−2	−1	0
0	0	1	0	1	0	3
0	1	0	3	1	−1	5
0	1	1	−3	4	0	8
1	0	0	7	0	0	4
1	0	1	1	3	1	7
1	1	0	4	3	0	9
1	1	1	−2	6	1	12

The solution from explicit enumeration is rather obvious. Among all possible feasible solutions (not shaded in the table), the point 0, 0, 1 gives a minimum function value of 3.

We now discuss implicit enumeration. The algorithm does not require an LP solver, in contrast with the algorithm in Section 9.3. We start at 0, 0, 0, which gives the lowest function value. This is represented by 0 in Fig. 9.1.

If this is feasible, we are done; otherwise we perform *fathoming*, a systematic procedure to determine whether improvements could be obtained by changing the levels of the variables in succession. Let us decide that the variable x_1 be raised to 1. A heuristic basis for this choice will be given later. This gives a feasible solution with function value 4 at node 1 in Fig. 9.1. Raising an additional variable will only increase the function. So the branch with x_1 alone raised to 1 can be terminated. The branch is said to be fathomed. The minimum value and the corresponding $\mathbf{x}$ are now stored. The next step is *backtracking* where x_1 is brought to level 0, and the node number is changed to -1 to denote that $x_1 = 0$. With x_1 held at zero, we note from the positive coefficients of x_2 and x_3 that there is some chance of achieving feasibility and function improvement. If there is no chance, the branch would be fathomed

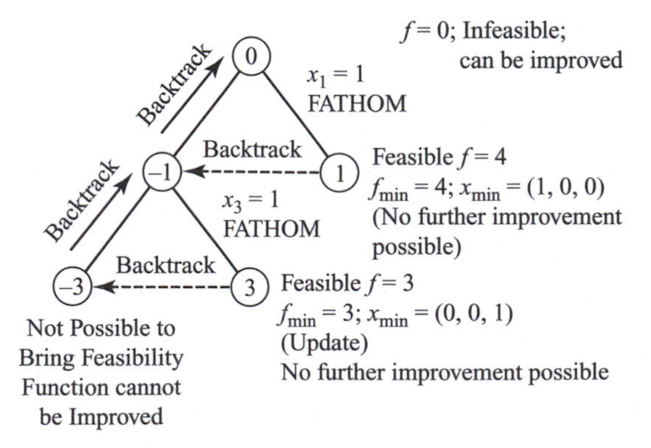

Figure 9.1. Illustrative problem.

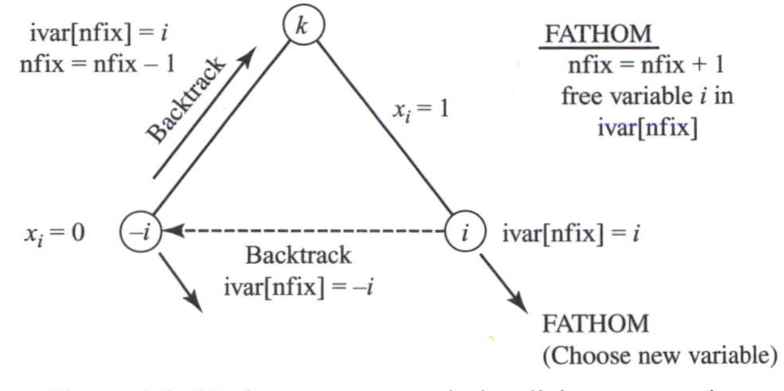

Figure 9.2. Node management in implicit enumeration.

and we would then backtrack. Raising x_3 brings feasibility here. We decide to raise it to 1 and continue. The function value is lower than before and, therefore, it is updated. This brings us to node 3 ($x_3 = 1$). The branch (0)↔(−1) ↔ (3) indicates that x_1 is fixed at 0 and x_3 is fixed at 1. Raising any other variable increases the function value. The branch ending at node 3 is, thus, fathomed. Backtracking takes us to −3 with x_3 now fixed to a value of zero. With both x_1 and x_3 fixed at 0, it is not possible to achieve feasibility by raising x_2. Also the function value cannot be improved. The search is complete and there are no more free nodes. The enumeration process is illustrated in Fig. 9.1. We have thus observed the main elements of implicit enumeration.

We explain here the main steps of the implicit enumeration process. Let us say that from node k, we chose variable x_i to be raised to 1 and we are at node i as shown in Fig. 9.2.

The node management aspects are given in the figure. Note that the process starts at node 0 where all variables are free. The following possibilities exist:

(1) The point is feasible.
(2) Feasibility is impossible from this branch.
(3) The function cannot be improved from a previous feasible value.
(4) Possibilities for feasibility and improvement exist.

If the node is in one of the first three states, then the branch is fathomed and we must attempt to backtrack. If the point is feasible, that is, condition of case 1 is satisfied, the function value and the point are recorded the first time. Later feasible solutions are compared to the previous best value and the function value and the point are recorded whenever it is better. Further search down the branch is not necessary since the function value will only increase. Cases 2 and 3 also forbid further search. *Backtracking* is to be performed at this stage. The test for Case 2 is easily performed. We check to see if raising the free variables corresponding to positive coefficients in the violated constraints maintains feasibility. If not, the impossibility is established. The test for case 3 needs a look at the objective function coefficients. If the smallest coefficient of a free variable (i.e., one that has not been fixed at 0 or 1) raises the function to a value more than previous low value, there is no chance for improvement.

Case 4 is the default when 1, 2, and 3 are not satisfied. In this case, further search is necessary. A new variable to be raised is to be chosen from the free variables.

Backtracking is a simple step. If we are at node i, which is the end of the right leg $x_i = 1$, we take the backtrack step by fixing $x_i = 0$ and naming the node $-i$. At this node the variable levels are same as at the previous node, say, k. The feasibility check 1 need not be performed. Checks for 2 and 3 are performed. Backtracking from a negative node goes up to k as shown in Fig. 9.1. If the node number is negative, we keep moving up. If backtracking takes us to the point where the number of fixed variables is zero (node 0), then we are done. The node numbers are simply the variable numbers stored in an array. Only an array of size n is necessary. The rule "last in first out (LIFO)" is imbedded in the backtracking step described previously.

Fathoming is the step where a new variable to be brought in is chosen. A heuristic commonly used is to choose the free variable that is closest to feasibility. Consider a free variable x_i and determine the value of each constraint when x_i is raised to 1. From among the constraint values that are zero and negative, the least absolute value is called the distance from feasibility. The free variable closest to feasibility is chosen and raised and fixed at a value of 1. The tests for backtracking or fathoming are now performed.

Bringing a Problem into Standard Form

The steps to bring a general problem into the standard form in Eq. (9.1) are rather simple. An LE ($\leq$) constraint is brought to the standard form by multiplying throughout by -1 and bringing the right-hand side to the left. If there are equality constraints, the equality is changed to $\geq$ and one additional constraint (sum of all the equality constraints) is put in the "$\leq$" form as discussed in Chapter 4 (Eq. 4.56) and transformed to $\geq$. If there are negative coefficients in the objective function, say, a coefficient corresponding to variable x_i, then the variable is replaced by $x_i' = (1 - x_i)$. This adds a constant to the objective function. This transformation of the variable is also applied to each of the constraints.

The zero–one algorithm described previously is implemented in the program INT01. We try an example problem using this program.

Example 9.2

(Tolerance Synthesis) We are to select tolerances for the shaft shown in Fig. E9.2a. Each dimensional tolerance can be obtained using two process alternatives (Fig. E9.2b). In Fig. E9.2b, t_1 and t_2 denote the three-sigma standard deviations of processes A and B used to generate dimension 1, respectively, while t_3 and t_4 are the deviations of the two processes to generate dimension 2, respectively. The 0–1 programming problem reduces to:

$$\begin{aligned} \text{minimize} \quad & f = 65x_1 + 57x_2 + 42x_3 + 20x_4 \\ \text{subject to} \quad & x_1 + x_2 = 1 \\ & x_3 + x_4 = 1 \\ & x_i = 0, 1 \end{aligned}$$

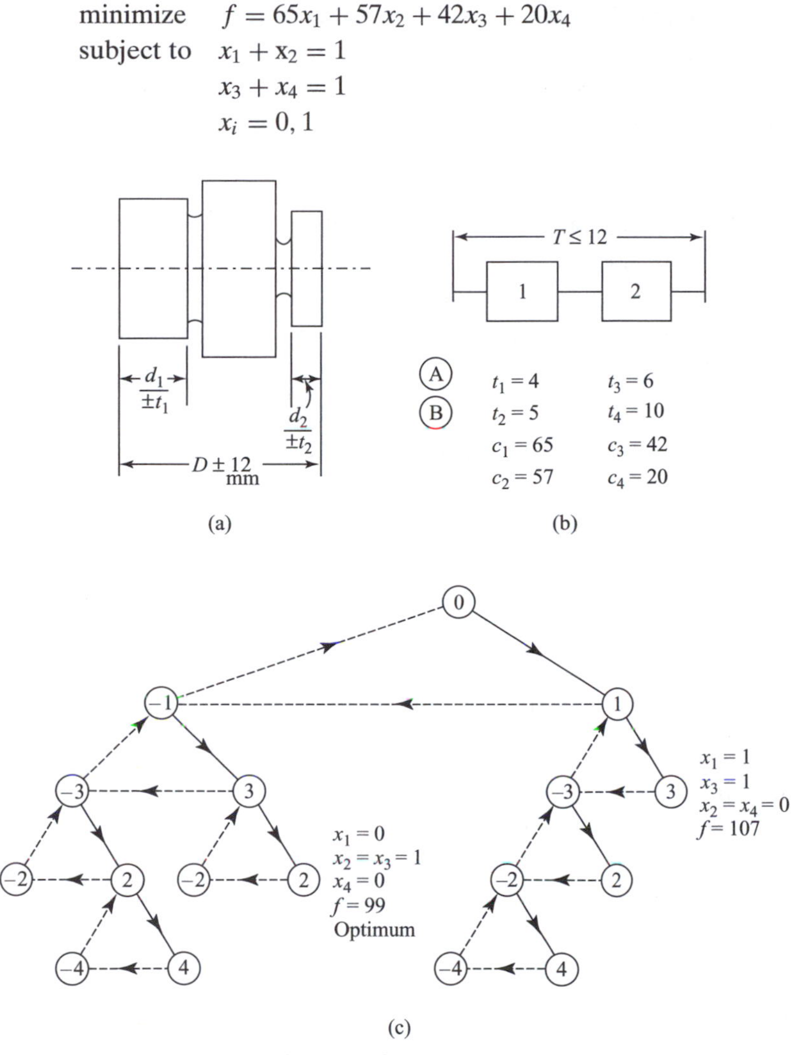

Figure E9.2

where $x_3 = 1$ if process A is selected to generate dimension 2 and $= 0$ if not, etc. The optimum solution is $f_{\min} = 99$, $\mathbf{x} = [0, 1, 1, 0]$. The complete branch and bound logic is shown in Fig. E9.2c. The reader may verify that this is the path taken by the program INT01 by applying a breakpoint at the start of each iteration and printing out the first "NFIX" values of the "IVAR" array.

Example 9.3

(Student Class Schedule Problem) A student must choose four courses Physics, Chemistry, Calculus, and Geography for his fall semester. Three sections of Physics, and two sections each of Chemistry, Calculus, and Geography are available. Following combinations of sections are not possible because of overlap.

Section 1 of Physics, section 1 of Chemistry, and section 1 of Calculus
Section 2 of Physics and section 2 of Calculus
Section 1 of Chemistry and section 2 of Geography

The student preference weightage values for the various sections are shown in parenthesis. Physics – section 1 (5), section 2 (3), section 3 (1), Chemistry – section 1 (3), section 2 (5), Calculus – section 1 (2), section 2 (5), Geography – section 1 (5), section 2 (2). Find the course schedule that is most satisfactory to the student.

Solution

We designate the variables for the various sections as follows. Physics – section 1 (x_1), section 2 (x_2), section 3 (x_3); Chemistry – section 1 (x_4), section 2 (x_5); Calculus – section 1 (x_6), section 2 (x_7); Geography – section 1 (x_8), section 2 (x_9). The requirements of taking one course from each group and the overlap conditions lead to the following constrained 0–1 integer programming problem.

$$\begin{aligned} \text{maximize} \quad & 5x_1 + 3x_2 + x_3 + 3x_4 + 5x_5 + 2x_6 + 5x_7 + 5x_8 + 2x_9 \\ \text{subject to} \quad & x_1 + x_4 + x_6 \le 1 \\ & x_2 + x_7 \le 1 \\ & x_4 + x_9 \le 1 \\ & x_1 + x_2 + x_3 = 1 \\ & x_4 + x_5 = 1 \\ & x_8 + x_9 = 1 \end{aligned}$$

$$x_i = 0 \quad \text{or} \quad 1 \qquad (i = 1, 2, \ldots, 9) \tag{9.3}$$

The problem is solved using the program INT01. The solution is $f_{\max} = 20$, and the variable values are $x_1 = 1, x_2 = 0, x_3 = 0, x_4 = 0, x_5 = 1, x_6 = 0, x_7 = 1, x_8 = 1, x_9 = 0$. The solution satisfies all the constraints.

9.3 Branch and Bound Algorithm for Mixed Integers (LP-Based)

In the branch and bound algorithm for pure or mixed integer problems, a relaxed linear programming problem is solved at every stage. Additional constraints are

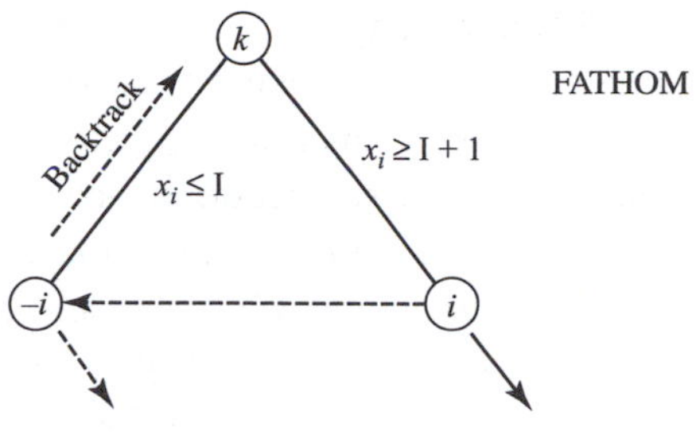

Figure 9.3. Node management in branch and bound method.

then imposed. The problem is first solved as an LP problem without imposing that any of the variables be integers. From among the variables required to be integers, we choose a variable x_i, which has a fractional value in the current solution. Let x_i be represented by

$$x_i = I + \alpha_i \tag{9.4}$$

where $I = \lfloor x_i \rfloor$ is the largest integer bounded aforementioned by x_i and α_i is the fractional part ($0 < \alpha_i < 1$). If $x_i = 2.3$, then $I = \lfloor x_i \rfloor = 2$ and $\alpha_i = 0.3$, and if $x_i = -2.3$, then $I = \lfloor x_i \rfloor = -3$ and $\alpha_i = 0.7$. Two constraints are now introduced as bounds on the variable x_i.

$$x_i \geq I + 1 \tag{9.5}$$

$$x_i \leq I \tag{9.6}$$

Each condition added to the current set of constraints defines a new subproblem. Two linear programming problems are thus created. This dichotomy is the key to the branch and bound algorithm. The conditions (9.5) and (9.6) exclude a region in which there are no integer solutions. Consider the node i in Fig. 9.3, which has been reached from node k by adding the constraint Eq. (9.5) to the set of constraints used in the solution at node k.

The linear programming problem is solved at node k. The attempt to solve the problem at node i leads to the following possibilities:

(1) The problem has a solution and the required variables are integers.
(2) No feasible solution exists.
(3) The function cannot be improved from a previous feasible integer solution.
(4) Some required variables have fractional values and the function value may be improved.

Note that the situation is almost identical to the one seen in implicit enumeration in Section 9.2. The first occurrence of case 1 provides us an integer feasible solution that we record. This will be used as a bound for future comparisons. Later integer solutions are compared to the previous best value and the bound and the point are updated if the solution is an improvement. Further trials are not necessary from this branch. Cases 2 and 3 also suggest that fathoming from this branch is not necessary. For cases 1, 2, and 3 fathoming is completed and backtracking must be performed. Case 4 suggests fathoming. Fathoming involves identifying a new integer variable that has a fractional value and the two conditions Eqs. 9.5 and 9.6 are imposed on this variable. Backtracking involves switching the condition from 9.5 to 9.6 and labeling the new node as $-i$. Fathoming is carried out at this node. Back tracking from a node with a negative number takes us to the parent node that is node k in Fig. 9.3. When the backtracking takes us back to a point where the constrained variable set is empty, the search is complete. We illustrate the complete algorithm using the problem presented in Example 9.1.

Example 9.4

(Solution of Example 9.1 using the LP-based branch and bound approach). The branch and bound solution to the problem is illustrated in Fig. E9.4.

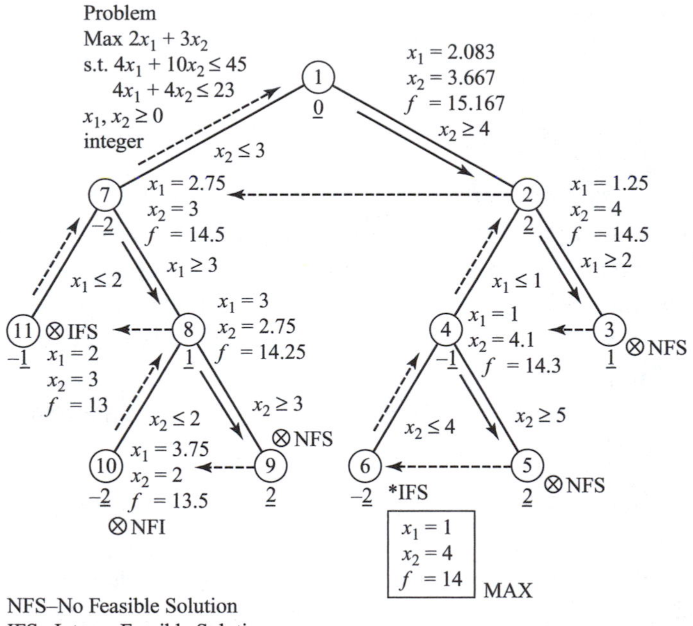

Figure E9.4

In the development of the branch and bound tree, the revised simplex program is used at each stage. Phase I of the solution is used in identifying if a feasible solution exists. The relaxed problem is solved at node 1 at which stage zero variables are imposed to be integers. Among the integer variables with fractional values, we choose the one that has a fractional value close to 0.5. This is a heuristic. Other heuristics may be tried. Variable x_2 is chosen. We impose the condition $x_2 \geq 4$ to move to node 2 where we solve the problem. An identifier of 2 for variable x_2 is used at node 2. From node 2, we move to node 3 by adding the constraint $x_1 \geq 2$ (identifier 1). There is no feasible solution at node 3. We backtrack to node 4 by dropping the constraint $x_1 \geq 2$ and adding $x_1 \leq 1$ (identifier -1). The problem is solved at node 4 and adding the GE constraint on variable 2, we move to node 5 (identifier 2). There is no feasible solution at node 5. We backtrack to node 6 (identifier -2). At node 6 there is an integer feasible solution. The solution $x_1 = 1, x_2 = 4, f = 14$ provides us the bound. We record this and note that further fathoming is not necessary. We backtrack to node 4, which is the end of the left leg from its previous node (identifier -1). Backtracking continues till a node with a positive identifier is found at node 2. We jump to node 7 (identifier -2) and then to 8, 9, 10, 11, 7, and 1. At node 10, it is not an integer solution but the function value 13.5 is worse than the previous integer solution value of 14. This suggests backtracking. At node 11 there is an integer feasible solution where the function value is inferior to the previous best one at node 6. Once we reach, node 1 with identifier 0, the solution process is complete. This algorithm shows that we can stop the process if the solution at node 1 is an integer solution. The maximum solution is $x_1 = 1, x_2 = 4, f = 14$.

The branch and bound algorithm is implemented in MINTBNB, which can be used for solving mixed integer problems in linear programming. We notice that the current solution at any node is infeasible with respect to the two subproblems defined by constraints 9.5 and 9.6. Introduction of each constraint needs the addition of a row and two columns: one for the surplus or slack variable and one for an artificial variable. The phase I of the simplex method is used. If a feasible solution exists, the artificial variables are removed before applying phase II. This approach is used in the program. Alternatively, dual simplex method may be used without introducing the artificial variables.

Example 9.5

(Generalized Knapsack Problem) The value of carrying certain grocery items in a sack is to be maximized, subject to the constraint on the total weight. Variables $x_1, x_2, x_3,$ and x_4 are to be integers and x_5, and x_6 can have fractional values. The problem is put in the mathematical form as

$$\begin{array}{lll} \text{maximize} & 14x_1 + 12x_2 + 20x_3 + 8x_4 + 11x_5 + 7x_6 & \text{(Value)} \\ \text{subject to} & 1.1x_1 + 0.95x_2 + 1.2x_3 + 2x_4 + x_5 + x_6 \leq 25 & \text{(Weight)} \\ & x_1, x_2, \ldots, x_6 \geq 0, \quad x_1, x_2, x_3, x_4 & \text{Integers} \end{array}$$

Solution

The problem is solved using MINTBNB. The branch and bound algorithm visits 10 nodes in the branch tree in arriving at the solution $x_1 = 1, x_3 = 20, x_5 = 0.05$, and other variables are each equal to zero. The maximum value of the function is 412.55.

Branch and bound algorithm is of a combinatorial nature and is robust. The cutting plane method of Gomory introduces one constraint at a time and keeps cutting the feasible region closer and closer to the integer hull of the region.

9.4 Gomory Cut Method

Gomory's method starts from a linear programming solution where the integer requirement is relaxed. The current updated constraint matrix is now available in the form

$$\mathbf{x} + \mathbf{A}\mathbf{y} = \mathbf{b} \tag{9.7}$$

where $\mathbf{x}$ is the set of basic variables and $\mathbf{y}$ is the set of nonbasic variables. The solution is $\mathbf{x} = \mathbf{b}$ with $\mathbf{y} = \mathbf{0}$. The idea of Gomory's method is to introduce a constraint which:

(i) cuts the polytope (i.e., makes Ω smaller) without excluding any integer solutions
(ii) makes the current optimum solution infeasible with respect to the newly added constraint

Figure 9.4 illustrates the concept of a cut. In the figure, the dashed line represents the objective function contour. Currently, point A is the optimum. The cutting plane moves the optimum from A to B, which is closer to an integer optimum solution. The idea behind cutting plane algorithms is to apply the cuts iteratively.

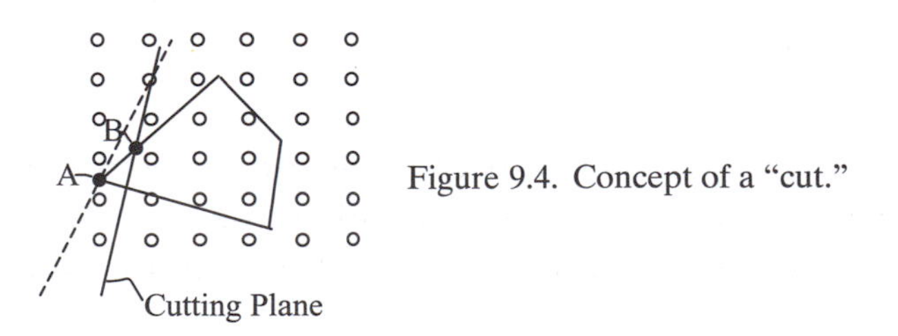

Figure 9.4. Concept of a "cut."

We first consider the all integer case and then develop a cut for mixed integer case.

Cut for an All Integer Problem

We present here the fractional cut approach. Consider a pure integer problem. Select from among the basic variables, the one having the largest fractional part β_i, $0 < \beta_i < 1$, such that $b_i = \lfloor b_i \rfloor + \beta_i$. The corresponding equation from Eq. 9.7 can be put in the form

$$x_i + \sum_j a_{ij} y_j = b_i, \quad \text{or}$$

$$x_i + \sum_j (\lfloor a_{ij} \rfloor + \alpha_{ij}) y_j = \lfloor b_i \rfloor + \beta_i \tag{9.8}$$

where α_{ij} is the fractional part of a_{ij} and $\lfloor a_{ij} \rfloor$ is the largest integer smaller than a_{ij}. For example, $\lfloor 2.5 \rfloor = 2$, $\lfloor -4.3 \rfloor = -5$. Since all variables are nonnegative in a feasible solution, any set of feasible values that satisfy (9.8) must also satisfy

$$x_i + \sum_j \lfloor a_{ij} \rfloor y_j \leq b_i \tag{9.9}$$

And since the left-hand side of (9.9) must be integer valued in a feasible solution, (9.9) can be strengthened to

$$x_i + \sum_j \lfloor a_{ij} \rfloor y_j \leq \lfloor b_i \rfloor \tag{9.10}$$

More conveniently, x_i can be eliminated from (9.10) using (9.8) to yield the fractional cut

$$\sum_j \alpha_{ij} y_j \geq \beta_i \tag{9.11}$$

The reason this is a "cut" can be understood from the following. First, note that the current (fractional) solution has all $y_j = 0$ so that the left-hand side of Eq. (9.11) equals zero, which clearly violates (9.11). Thus, adding this constraint cuts out the current solution to the LP relaxation. Moreover, it does not cut out any integer-feasible solution to the original problem. To see this, any integer solution [**x**, **y**] that satisfies Eq. (9.8) will satisfy (9.11) as well.

Example 9.6

Consider $x_i + 4\frac{1}{3}y_1 + 2y_2 - 3\frac{2}{3}y_3 = 5\frac{1}{3}$. The corresponding fractional cut is $\frac{1}{3}y_1 + \frac{1}{3}y_3 \geq \frac{1}{3}$, which can also be written using a slack variable s as $\frac{1}{3} - \frac{1}{3}y_1 - \frac{1}{3}y_3 + s = 0$, $s \geq 0$. Note that the current LP solution $x_i = 5\frac{1}{3}$, $y_1 = y_2 = y_3 = 0$ is infeasible with respect to the cut.

Cut for a Mixed-Integer Problem

Consider again (9.8) where x_i must still be integer valued but b_i is fractional valued. A more complex cut than (9.11) is required. Omitting a lengthy derivation, the cut is given as

$$\sum_j A_{ij} y_j \geq \beta_i \tag{9.12}$$

where

$$A_{ij} = \begin{cases} \alpha_{ij} & \text{for } \alpha_{ij} \leq \beta_i \text{ and } y_j \text{ integer} \\ \dfrac{\beta_i(1-\alpha_{ij})}{(1-\beta_i)} & \text{for } \alpha_{ij} > \beta_i \text{ and } y_j \text{ integer} \\ a_{ij} & \text{for } a_{ij} \geq 0 \text{ and } y_j \text{ continuous} \\ \dfrac{-\beta_i a_{ij}}{(1-\beta_i)} & \text{for } a_{ij} < 0 \text{ and } y_j \text{ continuous} \end{cases}$$

This cut is applied iteratively in obtaining the mixed integer solution. Dual simplex method can be used in order to solve the problem with the added constraint.

Example 9.7

Determine the first all integer Gomory cut for the two variable problem

$$\begin{aligned} \text{maximize} \quad & x_1 + x_2 \\ \text{subject to} \quad & 2x_1 - x_2 \leq 3 \\ & x_1 + 2x_2 \leq 7 \\ & x_1, x_2 \geq 0 \quad \text{and integer} \end{aligned}$$

Solution

The graphical representation of the problem is shown in Fig. E9.7. The initial tableau to obtain an LP solution without any integer constraints is

	x_1	x_2	y_1	y_2	b
f	−1	−1	0	0	0
y_1	2	−1	1	0	3
y_2	1	2	0	1	7

where y_1 and y_2 are slack variables. Using the simplex method in Chapter 4, x_1 is selected as the incoming variable (x_2 may also be chosen). The ratio test involves choosing the smaller of 3/2 and 7/1, resulting in y_1 as the outgoing variable. The pivot operation is achieved by premultiplying the tableau with a (3×3) matrix T (see Chapter 1 under Elementary Row Operations)

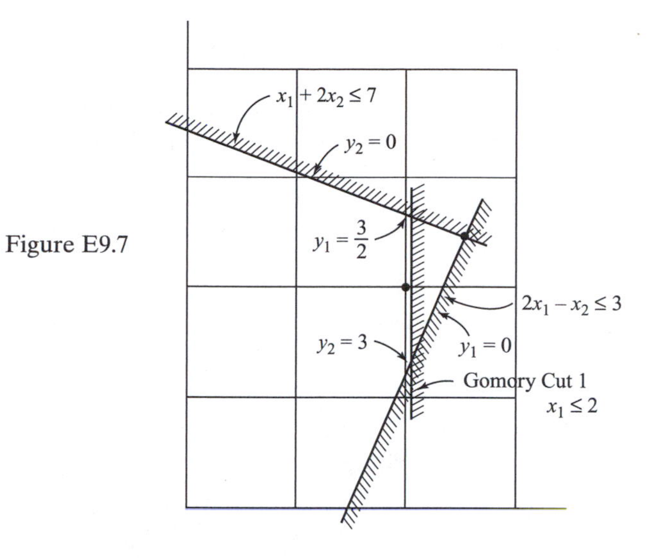

Figure E9.7

	x_1	x_2	y_1	y_2	b
f	0	−1.5	0.5	0	1.5
x_1	1	−0.5	0.5	0	1.5
y_2	0	2.5	−0.5	1	5.5

Next, x_2 is incoming and y_2 is outgoing, resulting in

	x_1	x_2	y_1	y_2	b
f	0	0	0.2	0.6	4.8
x_1	1	0	0.4	0.2	2.6
x_2	0	1	−0.2	0.4	2.2

Since all objective coefficients are nonnegative, this is optimal: $x_1 = 2.6$, $x_2 = 2.2$, $f = 4.8$. Evidently, this is not an integer solution. The largest fractional part is in b_1. The Gomory cut as per Eq. (9.11) is

$$0.4y_1 + 0.2y_2 \geq 0.6$$

or, using a nonnegative slack variable y_3,

$$-0.4y_1 - 0.2y_2 + y_3 = -0.6$$

The current optimum point is now infeasible. The tableau is

	x_1	x_2	y_1	y_2	y_3	b
f	0	0	0.2	0.6	0	4.8
x_1	1	0	0.4	0.2	0	2.6
x_2	0	1	−0.2	0.4	0	2.2
y_3	0	0	−0.4	−0.2	1	−0.6

Instead of a two-phase simplex technique, the dual-simplex method will be used. The latter technique is more efficient since the objective coefficients are nonnegative implying optimality while $b_3 < 0$ implying infeasibility. Most –ve right-hand side corresponds to y_3, which is the outgoing variable. Ratio test involves determining smaller of $\{0.2/(-0.4)$ and $0.6/(-0.2)\}$, which gives y_1 as the incoming variable. Elementary row operations or premultiplication by a (4×4) T matrix gives

	x_1	x_2	y_1	y_2	y_3	b
f	0	0	0	0.5	0.5	4.5
x_1	1	0	0	0	1	2
x_2	0	1	0	0.5	−0.5	2.5
y_1	0	0	1	0.5	−2.5	1.5

The reader may perform two additional tableau operations to obtain the final tableau

	x_1	x_2	y_1	y_2	y_3	y_4	b
f	0	0	0	0	0	0.5	4
x_1	1	0	0	0	1	0	2
x_2	0	1	0	0	−1	0.5	2
y_1	0	0	1	0	−3	0.5	1
y_2	0	0	0	1	1	−1	1

which corresponds to the integer optimum $x_1 = 2$, $x_2 = 2$, $f = 4$. This problem has another integer solution with the same objective value: $x_1 = 1$, $x_2 = 3$, $f = 4$. This is evident from Fig. E9.7 since the objective contour passes through two integer points.

The cutting plane algorithm is generally efficient. The main problem is in the round off errors while handling the fractional parts. The coefficients become smaller and smaller and the round off errors have a larger effect as more and more cuts are applied. Another problem is that each cut introduces a constraint. Noting that the original dimension n of the problem is not changed, we always have n-nonbasic variables. Each additional constraint adds one row and, thus, brings in an additional basic variable. The mixed integer algorithm using Gomory cut has been implemented in the program MINTGOM. Revised simplex routine is used for solving the original problem as well as the problems with Gomory constraints.

Example 9.8

Solve the knapsack problem in Example 9.4, using Gomory cut method.

Solution

The problem is solved using MINTGOM. The second cut itself results in the solution $x_1 = 1$, $x_3 = 20$, $x_5 = 0.05$, and other variables are each equal to zero. The maximum value of the function is 412.55.

9.5 Farkas' Method for Discrete Nonlinear Monotone Structural Problems

There are many structural design problems in which the objective function is monotone in terms of the design variables. The weight of a tubular shaft is monotone with respect to the tube thickness, the diameter, and the length. The weight of an I-beam increases with increase in flange thickness, web thickness, and length. The method of Farkas is useful in solving discrete nonlinear problems in structural design. The minimization problem is of the form

$$\begin{aligned} &\text{minimize} \quad f(\mathbf{x}) \\ &\text{subject to} \quad g_j(\mathbf{x}) \leq 0 \quad j = 1 \text{ to } m \\ &\qquad\qquad x_i \in (x_{i1}, x_{i2}, \ldots, x_{iq}) \quad i = 1 \text{ to } n \end{aligned} \tag{9.13}$$

where n is the number of variables, m is the number of constraints, q is the number of levels of each variable arranged in an ascending order such that $x_{i1} < x_{i2} < \cdots < x_{iq}$. The function increases with increase in each variable. For this presentation, each variable is taken to have the same number of steps. In the program implementation, the number of levels of each variable can be different. q is an integer. The steps in each variable need not be uniform. The program requires each level of the variable to be specified even if they are uniform. The solution starts with all variables set at the highest level and this combination must also be feasible. The order of arrangement of these variables also plays an important role.

First the feasibility is checked for the highest level of the variables, and the corresponding function value $f_{\min}$ is set as an upper bound for f and the corresponding $\mathbf{x}$ is stored in $\mathbf{x}_{\min}$. The variable levels are lowered one at a time successively $x_1, x_2, \ldots,$ x_{n-1}. The last variable is handled in a special way to be discussed later. Bisection approach is used in trying out the various levels of a variable. Let us first consider variable 1. We first evaluate the constraints at level $l_1 = q$, and level $l_2 = 1$. If feasibility is satisfied at the lowest level l_2, we set the first variable at that level and proceed to the second variable. With level l_1 feasible and if level l_2 is not feasible, then the next trial is made at the bisection point $l = \frac{l_1+l_2}{2}$, where l is evaluated as an integer. If the level l is feasible, l_1 is set equal to l and set equal to l_2 otherwise. This process is continued till l_1 and l_2 differ by 1. l_1 is the final feasible level for the variable x_1. We leave it at that level and turn our attention to variable x_2, which is at level q. The bisection process is applied to this variable to decide the lowest feasible level. This process of determination of successive lowest levels is *fathoming*. The fathoming process is continued till $n - 1$ variables are completed. With the first $n - 1$ variables set at the current levels, the last variable is calculated as that value, which gives us a function value equal to the bound f_{min}. Treating the objective function f as dependent on the variable x_n, we determine the zero of the function

$$f(x_n) - f_{\text{min}} = 0 \tag{9.14}$$

This calculated value x_n may fall into any of the following categories:

1. x_n is larger than the highest value x_{nq}.
2. x_n is lower than the smallest value x_{n1}.
3. x_n lies in the interval x_{n1} to x_{nq}.

For case 1, we already know that x_{nq} is feasible. The variable is decreased one step at a time till the lowest feasible level is identified. The function value f is then evaluated. If this value is less than the previous bound f_{min}, f_{min} and the corresponding $\mathbf{x}_{\text{min}}$ are updated. Fathoming is complete. We backtrack. In case 2, we note that the function value cannot be improved for this choice of levels. Fathoming is complete. We then go to the backtracking step. In case 3, the constraints are evaluated at the next higher step and, if feasible, we go on to calculate at the next lower value. If not, fathoming operation is stopped.

Backtracking is performed by raising the previous variable to the next higher level. When the last but one variable is raised, the next operation is the last variable check described previously. Whenever the level of a variable $n - 2$ or below is raised, the fathoming step involves the bisection steps for variables up to $n - 1$ and then the last variable calculation.

The algorithm has been implemented in the computer program BAKTRAK. The program requires the entry of the objective function to be minimized, the constraints, the variable level data, and the last variable equation.

Example 9.9
An I-beam of the type shown in Fig. E9.9 is to be chosen in a structural application.

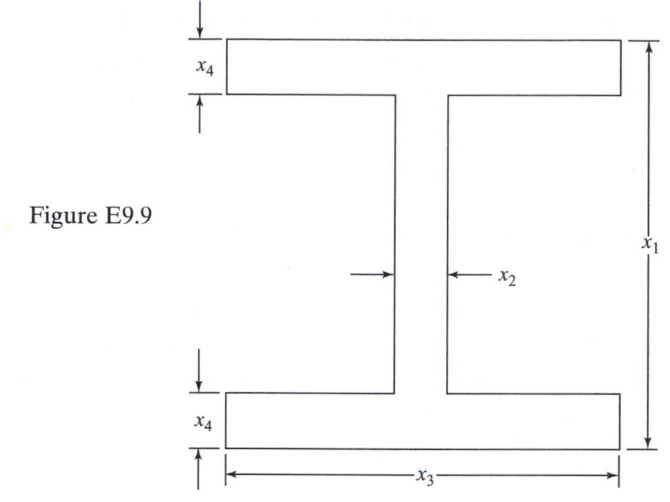

Figure E9.9

The data and the relevant relationships are given as follows:

Cross sectional area $A = x_1 x_2 + 2x_3 x_4 \text{ cm}^2$

Section modulus $S = x_1 (x_3 x_4 + x_1 x_2/6) \text{ cm}^3$

Bending Moment $M = 400 \text{ kN m}$, Axial Force $P = 150 \text{ kN}$

Bending stress $\sigma_B = \dfrac{1000M}{S}$ MPa

Axial stress $\sigma_P = \dfrac{10P}{A}$ MPa

Stress constraint $\sigma_B + \sigma_P - 200 \leq 0$ MPa

$$\text{Buckling constraint } \frac{x_1}{x_2} - 145 \left(\frac{\left(1 + \dfrac{\sigma_P}{\sigma_B}\right)^2}{1 + 173\left(\dfrac{\sigma_P}{\sigma_B}\right)^2} \right)^{\frac{1}{4}} \leq 0$$

The plate widths and thicknesses are available in following sizes:

x_1 60 to 78 in steps of 2
x_2 0.4 to 0.9 in steps of 0.1
x_3 16 to 52 in steps of 4
x_4 0.4 to 0.9 in steps of 0.1

Choose the beam of minimum cross section.

Solution

The objective function $f\ (= A)$ to be minimized, the constraints $g_j\ (\leq 0)$ and the input data, and the last variable equation are entered in the computer program. The last variable is calculated using the current bound $f_{\min}$ of the function

$$x_4 = 0.5\frac{f_{\min} - x_1 x_2}{x_3}$$

The output from the program gives the optimum selection $x_1 = 70$ cm, $x_2 = 0.6$ cm, $x_3 = 28$ cm, $x_4 = 0.9$ cm, and the corresponding minimum cross-sectional area $f_{\min} = 85.2\ \text{cm}^2$. The number of function evaluation for this problem is 1100. The total number of combinations for the exhaustive enumeration of the problem is 3600.

Farkas' approach provides a robust method for a class of discrete minimization problems in structural engineering.

9.6 Genetic Algorithm for Discrete Programming

The main idea of the genetic algorithm presented in Chapter 7 is discretization of the variables. We had used it for continuous problems by dividing each variable range into fine uniform steps. The algorithm can also be used for discrete variables, which are nonuniformly distributed. We store the available sizes of a variable in an array. As an example, the variable x_1 may be stored in an array $x_{11}, x_{12}, \ldots, x_{1q}$, where q is an integer expressible as 2^k. In the decoding operation, once the binary number is converted to the base 10 integer, the value of the variable can be accessed from the array. If the number of levels of a variable is not an integer power of 2, we allocate the closest larger integer power of 2 and fill the extra locations by repeating the closest variable values. Constraints may be taken care of using penalty approach.

A general development of integer programming for nonlinear problems is highly involved. Sequential integer linear programming techniques have been successfully tried. Integer penalty approach, augmented Lagrange multiplier methods, Lagrangian relaxation methods are some of the techniques in use for nonlinear integer programming.

COMPUTER PROGRAMS

INT01, MINTBNB, MINTGOM, BAKTRAK

PROBLEMS

P9.1. Given that x is a continuous variable in the interval $0 \leq x \leq 1$, formulate the requirement that: if $x > 0$ then $x \geq 0.02$.

P9.2. Represent the following cost function (e.g., shipping cost) using binary variable(s) in standard programming form as $f = f(x, y)$:

$$f = \begin{cases} 3 + x, & \text{if } x \leq 1 \\ 8 + 2x, & \text{if } x > 1 \end{cases}$$

P9.3. A candy manufacturer produces four types of candy bars in a batch using 2000 g of sugar and 4000 g of chocolate. The sugar and chocolate requirements for each type of bar and the profit margin are given in the table. Determine the number of bars of each type for maximum profit.

	Sugar g	Chocolate g	Profit $
Bar I	30	65	0.04
Bar II	40	75	0.08
Bar III	30	40	0.05
Bar IV	35	50	0.06

(The number of bars must be integers)

P9.4. A tourist bus company is equipping its fleet with a mix of three models – a 16 seat van costing $35,000, 30 seat minibus costing $60,000, and a 50 seat big bus at cost of $140,000. The total budget is $10 million. A total capacity of at least 2000 seats is required. At least half the vehicles must be the big buses. The availability of drivers and the garage space limit the total vehicles to a maximum of 110. If the profit dollars per seat per month values on the big bus, minibus, and the van are $4, $3, and $2, respectively, determine the number of vehicles of each type to be acquired for maximum profit.

P9.5. Cargo given in the following table is to be loaded into a ship. The total weight is not to exceed 200 tons, and the volume is not allowed to exceed 180ft^3. Assuming that each item is available in unlimited numbers, determine the loading for carrying maximum value.

Item i	Weight, tons	Volume, ft^3	Value thousands of $
1	8	7	6
2	6	9	8
3	5	6	5
4	9	7	8

P9.6. A manufacturer of gear boxes is planning the production of three models. The production capacities in number of operator hours for the departments of machining, assembly, painting, and packing departments are 2000, 3000, 1500, and 1800, respectively. The requirements for each model and the profit value are given in the table. Determine the product mix for maximum profit. Also determine the amount of production capacity that is not used.

	Operator Hours per Unit				
Gear box type	Machining	Assembly	Painting	Packing	Profit $
I	4	8	1	0.5	25
II	5	9	1	0.6	35
III	6	7	2	0.8	40

P9.7. Solve the following mixed integer programming problem

$$\begin{aligned}\text{maximize}\quad & 4x_1 + 5x_2 + x_3 + 6x_4\\ & 7x_1 + 3x_2 + 4x_3 + 2x_4 \leq 6\\ & 2x_1 + 5x_2 + x_3 + 5x_4 \leq 4\\ & x_1 + 3x_2 + 5x_3 + 2x_4 \leq 7\\ & x_1, x_2, x_3, x_4 \geq 0,\ x_1, x_2 \text{ integers}\end{aligned}$$

P9.8. Solve the example in Section 9.2 (see Eq. (9.2)) using the LP-based branch and bound procedure in Section 9.3. (Show detailed hand calculations as in Example 9.4)

P9.9. Solve the following pure integer programming problem using (LP-based) branch and bound procedure:

$$\begin{aligned}\text{minimize}\quad & 14x_1 + 12x_2\\ & 35x_1 + 24x_2 \geq 106\\ & 14x_1 + 10x_2 \geq 43\\ & x_1, x_2 \geq 0 \text{ and integer}\end{aligned}$$

P9.10. Solve the 0–1 problem in Example 9.2 of the text using Gomory's method

P9.11. Solve P9.5 using Gomory's method.

P9.12. For the art gallery layout shown in Fig. P9.12, an attendant placed at a doorway can supervise the two adjacent rooms. Determine the smallest number of attendants for supervising all the rooms.

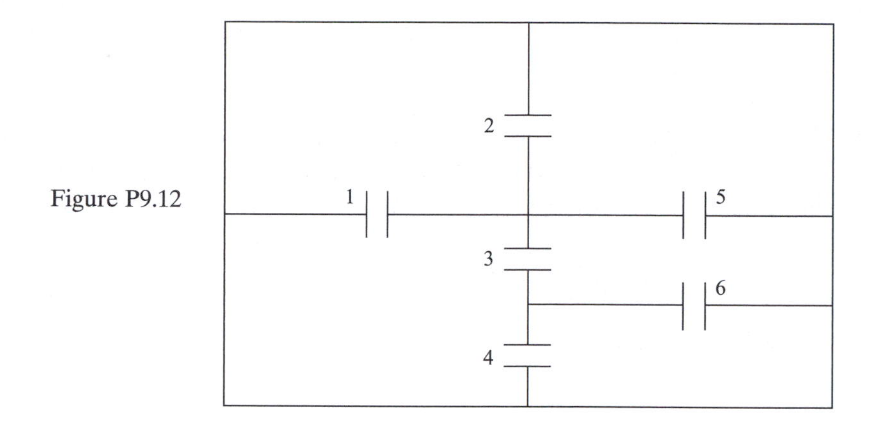

Figure P9.12

P9.13. Determine the optimal tableau for the linear programming problem

$$\begin{aligned} \text{maximize} \quad & 5x_1 + 6x_2 + 4x_3 \\ & 2x_1 + 4.6x_2 + x_3 \leq 12 \\ & x_1, x_2, x_3 \geq 0, \; x_1, x_3 \text{ integers} \end{aligned}$$

and then obtain the inequality defining the first fractional Gomory cut.

P9.14. A total amount of $18,000 is available for investment. The five investment options are available with the following breakup.

Investment	Investment	Net present value of yield
1	$5,000	$12,000
2	$6,000	$15,000
3	$4,000	$10,000
4	$8,000	$20,000
5	$3,000	$ 9,000

Determine the investment for maximum yield.

P9.15. A tourism company conducts boat tours. The company is considering adding two types of small boats to its fleet. The first type of boat costs $30000, has a seasonal capacity of 1400 and an estimated seasonal profit per boat is $5000. The second type of boat costs $90000, has a seasonal capacity of 2500 and an estimated seasonal profit per boat is $9000. The number of additional tourists projected to be served by the company is 8000. If the company plans to invest $250000, determine the additional number of boats of each type for maximum profit.

P9.16. Solve the following problem by the branch and bound algorithm.

$$\begin{aligned} \text{maximize} \quad & 16x_1 + 12x_2 + 8x_3 - 4x_4 + 5x_5 \\ \text{subject to} \quad & 12x_1 + 9x_2 + 6x_3 + 8x_4 + 5x_5 \leq 35 \\ & x_i = 0 \text{ or } 1, \quad i = 1, 2, \ldots, 5 \end{aligned}$$

P9.17. Solve Example 10.2, Chapter 10, using integer programming.

P9.18. A minimum cross-section hollow circular column fixed at one end and free at the other end is to be designed. An axial compressive force $P = 60{,}000$ N is applied at the free end. The area of the cross-section to be minimized is πDt. The axial stress $\sigma = \frac{P}{\pi Dt}$ should not exceed 200MPa (1 MPa $= 10^6$ N/m^2). The axial load P should not exceed 50% of the Euler buckling load P_{cr}, where $P_{cr} = \frac{\pi^2 EI}{4l^2}$, where E is the modulus of elasticity 210 GPa (1GPa $= 10^9$N/m^2), and the moment of inertia I for the thin walled tube is given by $I = \frac{\pi D^3 t}{8}$. If the diameter D is available in sizes 70 to 200 mm in steps of 10 mm and the thickness t is available in the range 3 mm to 15 mm in steps of 2 mm, find the optimum dimensions.

REFERENCES

Avriel, M. and Golany, B. (Eds.), *Mathematical Programming for Industrial Engineers*, Marcel Dekker, New York, 1996.

Balas, E., An additive algorithm for solving linear programs with zero-one variables, *Operations Research*, **13**, 517–546, 1965.

Farkas, J., *Optimum Design of Metal Structures*, Ellis Horwood, Chichester, 1984.

Gomory, R.E., An algorithm for the mixed integer problem, Rand Report, R.M. 25797, July 1960.

Land, A.H. and Doig, A., An automatic method of solving discrete programming problems, *Econometrica*, **28**, 497–520, 1960.

Taha, H.A., *Integer Programming*, Academic Press, New York, 1975.

10

Dynamic Programming

10.1 Introduction

Dynamic programming evolved out of the extensive work of Richard Bellman in the 1950s. The method is generally applicable to problems that break up into stages and exhibit the Markovian property. A process exhibits the *Markovian property* if the decisions for optimal return at a stage in the process depend only on the current state of the system and the subsequent decisions. A variety of problems in engineering, economics, agriculture and science exhibit this property. Dynamic programming is the method of choice in many computer science problems, such as the Longest Common Subsequence problem that is used frequently by biologists to determine the longest common subsequence in a pair of DNA sequences; it is also the method of choice to determine the difference between two files.

When applicable, advantages of dynamic programming over other methods lie in being able to handle discrete variables, constraints, and uncertainty at each subproblem level as opposed to considering all aspects simultaneously in an entire decision model, and in sensitivity analysis. However, computer implementation via dynamic programming does require some problem-dedicated code writing.

Dynamic programming relies on the *principle of optimality* enunciated by Bellman. The principle defines an optimal policy.

Bellman's Principle of Optimality

An *optimal policy* has the property that whatever the initial state and initial decisions are, the remaining decisions must constitute an optimal policy with respect to the state resulting from the first decision.

We start with a simple multistage example of the creation of a book tower defined by a single variable at each stage.

Example 10.1. (Farthest Reaching Book Tower Problem)
The task at hand is to stack ten identical books one on top of the other to create a tower so that the end of the top book reaches the farthest from the edge of the table. Each of the books is10 inches long.

From the principles of physics, we find that at any stage, the combined center of gravity of the books above must fall inside the current book. Solving the problem using conventional formulation looks formidable. This problem is easily solved using the physical principle turned into an optimal policy. The *optimal policy* at any stage is to let the combined center of gravity of the books above to fall at the outer edge of the current book (we are looking at the limit case without the consideration of instability). The problem is illustrated in Fig. E10.1.

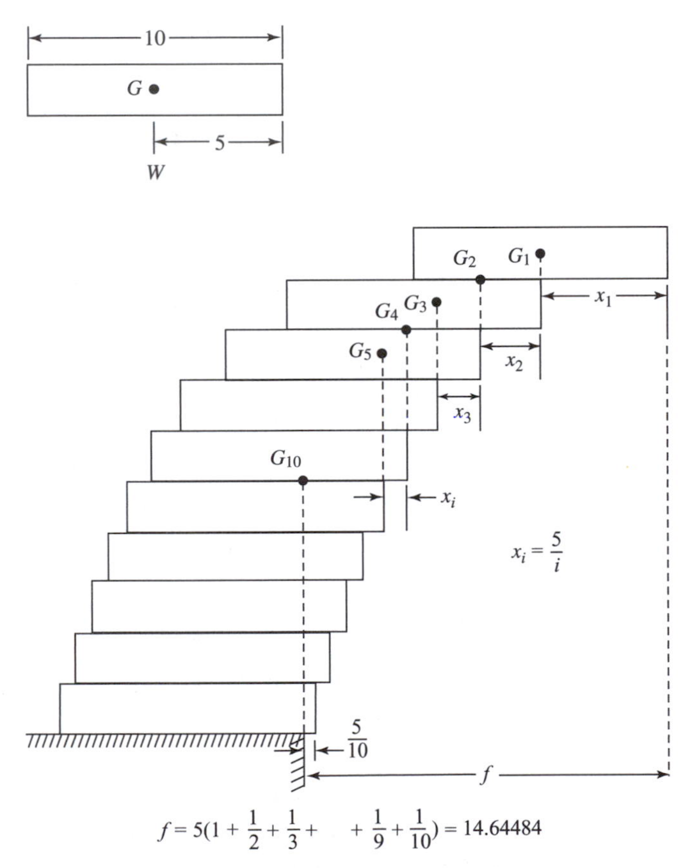

$$f = 5\left(1 + \frac{1}{2} + \frac{1}{3} + \quad + \frac{1}{9} + \frac{1}{10}\right) = 14.64484$$

Figure E10.1. Book tower problem.

Since the policy says that the decision is based on the future placements, it will be easy if we start from the last book and work backwards. We define the variable x_i as the distance of the combined center of gravity of the top i books when each is placed at the stability limit based on the optimal policy. The solutions are optimal for all intermediate system configurations. The objective function for the problem is

$$f = f_{10} = x_1 + x_2 + \cdots + x_9 + x_{10}$$

The distance of the center of gravity of top i members from the outer edge of member $i+1$ can be derived as $x_i = \frac{5}{i}$. The objective function can be defined using the recursive relation as follows. We start with $f_0 = 0$ and define

$$f_0 = 0$$
$$f_i = \frac{5}{i} + f_{i-1}, \quad i = 1 \text{ to } 10$$

The solution is obtained in a straightforward manner as

$$f_{\min} = 5\left(1 + \frac{1}{2} + \frac{1}{3} + \frac{1}{4} + \frac{1}{5} + \frac{1}{6} + \frac{1}{7} + \frac{1}{8} + \frac{1}{9} + \frac{1}{10}\right) = 14.64484$$

For the ten book problem, the overhang length is larger than the book. We note from mathematical analysis that the series $\sum_n \frac{1}{n}$ does not have a limit as n reaches infinity. Thus, if we have enough number of books and if stability can be held at the limit placement, we can reach very far into outer space.

The book stacking problem illustrates the basic characteristics of dynamic programming. There is a start point and an end point. Placement of each book is a stage in this problem. There are no iterative steps for the optimization. We directly shoot for the optimum at every stage. Each typical problem in dynamic programming represents a class of problems with wide range of applications. We proceed to discuss the general aspects of dynamic programming and go on to discuss several example classes.

10.2 The Dynamic Programming Problem and Approach

We have already seen that the problem for dynamic programming problem must be decomposable into stages. A general multistage system is shown in Fig. 10.1.

The stages may be sequence in space, such as in a shortest route problem or a sequence in time, such as in construction and project management. Looking at the stage i of the system, $\mathbf{s}_i$ is the set of *input state variables* to stage i, and $\mathbf{s}_{i+1}$ is the *output state variables* from stage i. The input $\mathbf{s}_1$ and $\mathbf{s}_{\text{out}}$ $(= \mathbf{s}_{n+1})$ are generally specified. $\mathbf{x}_i$ represents the set of *decision variables*, and r_i is the *return* from stage i. Here, "return" is understood to be an objective function that is minimized. Thus, if r represents cost, it is minimized, while if it represents profit, then it is maximized. Our aim is to select the decision variables $\mathbf{x}_i$ to optimize a function of f, which depends on

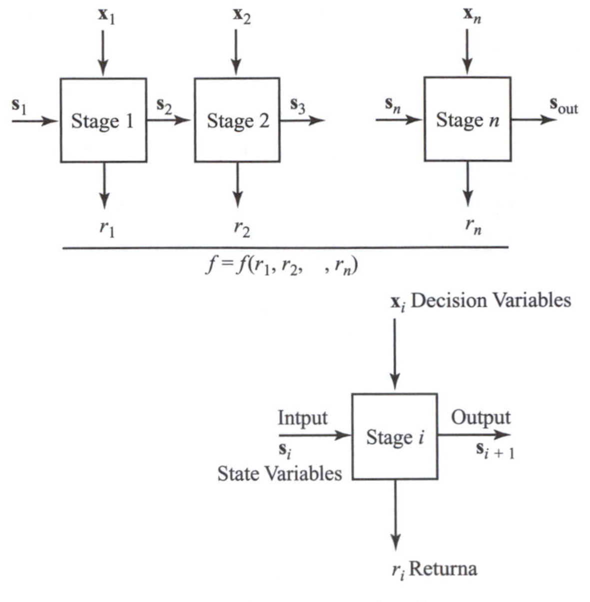

Figure 10.1. A multistage system.

the r_i. In most problems, the function is a summation of the returns. Let us consider the problem posed as a minimization problem.

$$f = \min \sum_{i=1}^{n} r_i \tag{10.1}$$

The optimization can be tackled for a single stage and then the optimum for the connected problem can be developed. In the backward recursion development, we proceed from the last stage where the output $\mathbf{s}_{out}$ is specified and the input $\mathbf{s}_n$ may be varied over some possible values. The optimum for the final stage can be obtained for each component of $\mathbf{s}_n$ and can be written in the form

$$f_n(\mathbf{s}_n) = \min_{\mathbf{x}_n} [r_n(\mathbf{x}_n, \mathbf{s}_n)] \tag{10.2}$$

Then the optimal f values at each stage can be obtained by the *recursion* relation

$$f_i(\mathbf{s}_i) = \min_{\mathbf{x}_i} [r_i(\mathbf{x}_i, \mathbf{s}_i) + f_{i+1}(\mathbf{s}_{i+1})], \quad i = n-1, \quad n-2, \ldots, 1 \tag{10.3}$$

Equation 10.3 is called a *functional equation* in the dynamic programming literature. In the preceding relation, $\mathbf{s}_{i+1}$ is a function of $\mathbf{x}_i$ and $\mathbf{s}_i$. We note that this evaluation is combinatorial in the sense that the second term changes with each $\mathbf{x}_i$. As we move to f_1, the function will have one value since $\mathbf{s}_1$ is specified. Example 10.2 will illustrate the preceding equation.

All problems for which the principle of optimality holds satisfy the *monotonicity condition*, which states that the increase in any of the f_{i+1} terms on the right-hand side of Eq. (10.3) cannot decrease f_i.

Forward recursive relations can be developed where applicable. In problems of reliability type, the objective function may be represented as a product of the returns. In this case

$$f = \max \prod_{i=1}^{n} r_i \tag{10.4}$$

For this function, the addition in the recursive relation changes into a product. Other types of functions, such as combinations of summation and product forms may be used.

Example 10.2

A duct system shown in Fig. E10.2 is designed for a pressure drop of 600 Pa from point 1 to point 5. Duct costs for various pressure drops in the sections is given in Table E10.2a. Determine the minimal cost and the corresponding pressure drops. Duct flow requirement states that there should be a pressure drop in each section.

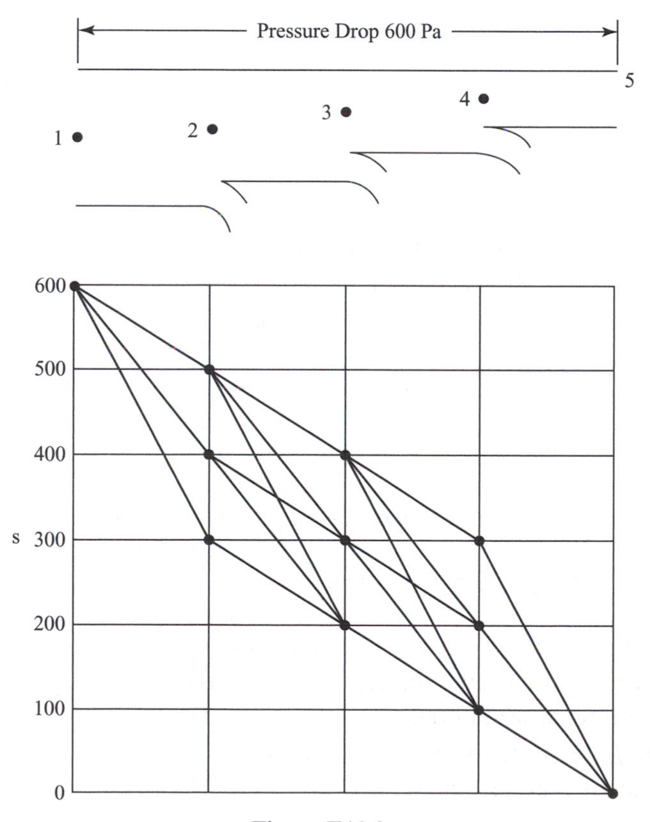

Figure E10.2

Table E10.2a

Section	Cost, $		
	Pressure drop 100 Pa	Pressure drop 200 Pa	Pressure drop 300 Pa
1–2 (stage 1)	235	215	201
2–3 (stage 2)	190	172	161
3–4 (stage 3)	141	132	118
4–5 (stage 4)	99	91	84

Solution

The total pressure drop needed is 600 Pa. There are 4 sections, thus, $n = 4$. The state variable for stage i is the operating pressure at the start of stage i, and we denote this as P. That is, $s \equiv P$. Without loss in generality, we set the pressure at 1 as $P_1 = 600$, and the output pressure ($P_{\text{out}} \equiv P_5 = 0$). Note that $0 \leq P_i \leq 600$ Pa. The design variable x_i is the discrete pressure drop (100 Pa or 200 Pa or 300 Pa) that may be chosen for stage i, where $i = 1, 2, 3, 4$. That is, there are 3 duct choices at each of the 4 stages. The cost function is represented by f. For instance, if $x_1 = x_2 = x_3 = 100$ and $x_4 = 300$, then $f = 235 + 190 + 141 + 84 = \$650$.

We start our dynamic programming approach at the final stage. The solution proceeds directly from the backward recursion relations in Eq. (10.3). In fact, Eq. (10.3) can be rewritten here as

$$f_i\,(P_i) = \min_{\mathbf{x}_i}\left[r_i\,(x_i) + f_{i+1}\,(P_i - x_i)\right],\ \ i = n,\ n-1,\ \ n-2, \ldots, 1 \text{ with } f_{n+1} = 0. \tag{a}$$

where r represents cost. The preceding equation follows from the fact that if the inlet pressure at stage i is P_i, and has a pressure drop x_i, then the inlet pressure for the following stage will be $P_i - x_i$. The meaning of $r_i(x_i)$ is clear if we look at Table E10.2a. For instance, $r_2(100) = \$190$, $r_2(300) = \$161$, and $r_3(200) = 132$.

From the data in Table E10.2a, at the last (4th) stage, we have

$$f_4(P) = \begin{cases} 99, & \text{if} \quad P = 100\,\text{Pa} \\ 91, & \text{if} \quad P = 200\,\text{Pa} \\ 84, & \text{if} \quad P = 300\,\text{Pa} \end{cases}$$

The design variables are pressure drops that yield f_4:

$$x_4(P) = \begin{cases} 100, & \text{if} \quad P = 100\,\text{Pa} \\ 200, & \text{if} \quad P = 200\,\text{Pa} \\ 300, & \text{if} \quad P = 300\,\text{Pa} \end{cases}$$

Now, we have from Eq. (a) at stage $i = 3$:

$$
f_3(P) = \begin{cases} 141 + f_4(100) & - & - & \text{if} \quad P = 200\,\text{Pa} \\ 141 + f_4(200) & 132 + f_4(100) & - & \text{if} \quad P = 300\,\text{Pa} \\ 141 + f_4(300) & 132 + f_4(200) & 118 + f_4(100) & \text{if} \quad P = 400\,\text{Pa} \\ - & 132 + f_4(300) & 118 + f_4(200) & \text{if} \quad P = 500\,\text{Pa} \\ - & - & 118 + f_4(300) & \text{if} \quad P = 600\,\text{Pa} \end{cases}
$$

Based on information *thus far*, an operating pressure $P_3 = 100$ Pa is not valid and does not appear above since otherwise there would not be any pressure drop in stage 4. Substituting for f_4 into the preceding equation yields

Table E10.2b. *Stage 3.*

$P_3 \downarrow x \rightarrow$	100	200	300	x_3	f_3, \$
200	240	–	–	100	240
300	232	231	–	200	231
400	225	223	217	300	217
500	–	216	209	300	209
600	–	–	202	300	202

The last two columns contains the minimum cost solution for stage 3. The reader is encouraged to verify that Eq. (10.3) with i set equal to 2 yields (in tabular form).

Table E10.2c. *Stage 2.*

$P_2 \downarrow x \rightarrow$	100	200	300	x_2	f_2
300	430	–	–	100	430
400	421	412	–	200	412
500	407	403	401	300	401
600	399	389	392	200	389

and setting $i = 1$ yields (while only $P_1 = 600$ Pa is of interest to get the solution, the complete table allows us to do sensitivity calculations as shown subsequently):

Table E10.2d. *Stage 1.*

$P_1 \downarrow x \rightarrow$	100	200	300	x_1	f_1
400	665	–	–	100	665
500	647	645	–	200	645
600	636	627	631	200	627

The last column gives the optimum cost $f_1 = \$627$ corresponding to input operating pressure $P = 600$ Pa. The optimum pressure drops are readily obtainable from the preceding tables, since $x_1 = 200$ implies that $P_2 = 600 - 200 = 400$. Table E10.2c gives $x_2 = 200$. This then implies $P_3 = 400 - 200 = 200$. Table E10.2b gives $x_3 = 100$. Lastly, $x_4 = 100$. Thus, $[x_1, x_2, x_3, x_4] = [200, 200, 100, 100]$ Pa.

Sensitivity Analysis

The *paths* from start to end for different input operating pressures are readily determined from above in the same manner as was done for $P_1 = 600$. Sensitivity of optimum solution to different starting operating pressures is given in Table E10.2e.

Table E10.2e

P_1	f_1	$\mathbf{x}$
600	627	[200, 200, 100, 100]
500	645	[200, 100, 100, 100]
400	665	[100, 100, 100, 100]

We note from the preceding example problem that feasibility is checked at every stage and we are aiming at the optimum. The tabular form of calculations can easily be implemented in a computer program to solve a class of problems of the type discussed earlier. Problems in chemical engineering where separation and filtration, and distillation columns are involved have multistage configurations. Multistage turbines and generators in power generation and transmission also fall in this category.

10.3 Problem Modeling and Computer Implementation

There are a number of problems that can be efficiently solved using the dynamic programming ideas presented above. We present here some general classes and the computational aspects of the solution

PM1: Generalized Resource Allocation Problem

In the generalized resource allocation problem, the maximum amount of resource w is specified. There are various activities in which the resource may be allocated in order to get a return. Activity i can be implemented at a level x_i, for which it uses a resource amount of $g(x_i)$, and gives a return of $r_i(x_i)$. If there are N stages in the system, the problem may be put in the form

$$\begin{aligned} &\text{maximize} \quad \sum_{i=1}^{N} r_i(x_i) \\ &\text{subject to} \quad \sum_{i=1}^{N} g_i(x_i) \le w \end{aligned} \tag{10.5}$$

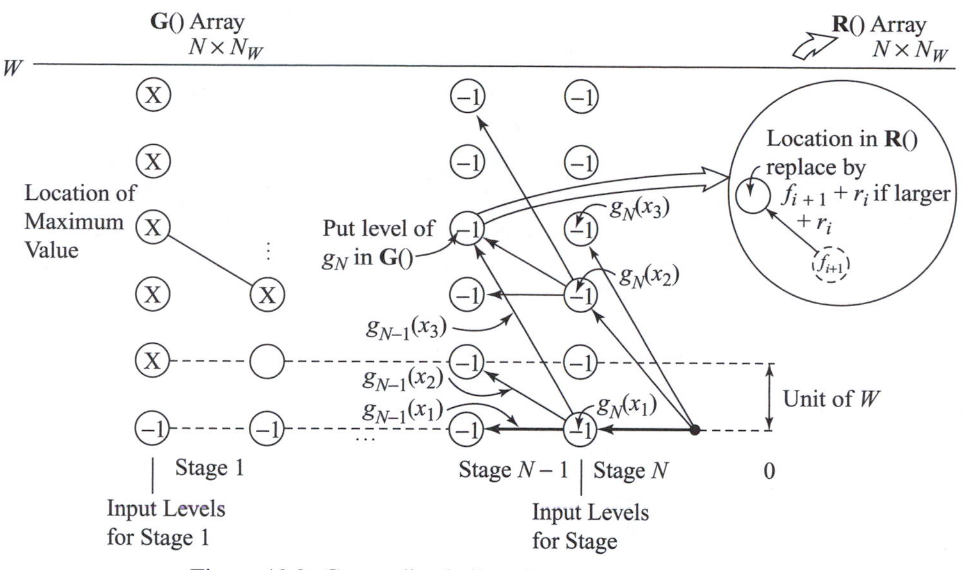

Figure 10.2. Generalized allocation array management.

This generalized problem provides us with the framework for investment, capital budgeting, generalized knapsack, cargo loading in ships, and other related problems. The problem can be solved by the forward recursion or backward recursion. We have implemented the backward recursion scheme.

Algorithm and Implementation

This problem is conveniently solved by using dynamic programming. The complete recurrence process and its implementation is shown in Fig. 10.2. We draw the input points to each stage on vertical lines on which each level of resource in the base units is represented by array locations. We create two rectangular arrays **G** and **R**. **G** relates resource information and **R** records return levels. To start with, each location is set as -1 in **G**. We start at zero level at the output of stage N. The g_N levels are used to access the appropriate levels in the column N of **G**. At the level of location accessed, the originating value of 0 is inserted. At the corresponding locations of the column N of **R**, the return values are added to obtain f_N. This gets the recurrence calculations started.

Consider the work at stage $N - 1$. We start from every location in column N where we have a nonnegative number and access the locations in column $N - 1$ for each of the $g_{N-1}(x_i)$ levels noting that we never exceed the level w. We add the return value r_{N-1} to the f_N value of column N and compare it with the value in the corresponding column in **R**. For the maximization problem, we replace the value if the new one is larger. If the value is entered for the first time, it is compared to -1, and if there is a value already in the location, the comparison and replacement take care of the updating. Simultaneously, the level of location of the originating point is put into the location in **G**. When this process is completed, we have the information about the final values in column 1 of **R**. We determine the location with the largest value, which is the optimum desired. Finding the path is a simple matter. The corresponding location in column 1 of **G** provides the level location at the output of the stage. We go to the corresponding location in column 2, which in turn gives the level of origin in column 3. This process is continued till we reach the zero level at the output of the last stage. The complete solution is now at hand.

This algorithm has been implemented in the computer program DPALLOC. The program can easily be modified for a minimization where the initial entries in **R** are set as large numbers and whenever a new value is compared, the smaller wins to occupy the location.

Example 10.3

An investor would like to invest 50 thousand dollars and he is considering three choices from which mix and match is possible.

Investment	levels:	1	2	3	4	5	6
1	g_1	0	10	18	24	28	30
2	g_2	0	8	16	24	30	34
3	g_3	0	12	24	34	42	46
Return	levels:	1	2	3	4	5	6
1	r_1	0	5	7	9	12	16
2	r_2	0	6	8	12	14	15
3	r_3	0	10	14	17	20	24

Find the investment level choices for maximizing the total return.

Solution

We input the data to the computer program DPALLOC and solve the problem. The allocation obtained is:

Maximum return = 32 thousand dollars
Type 1 Level of activity 30 and Investment number 6
Type 2 Level of activity 8 and Investment number 2
Type 3 Level of activity 12 and Investment number 2

In the preceding result, we see that the zero level is the investment number 1.

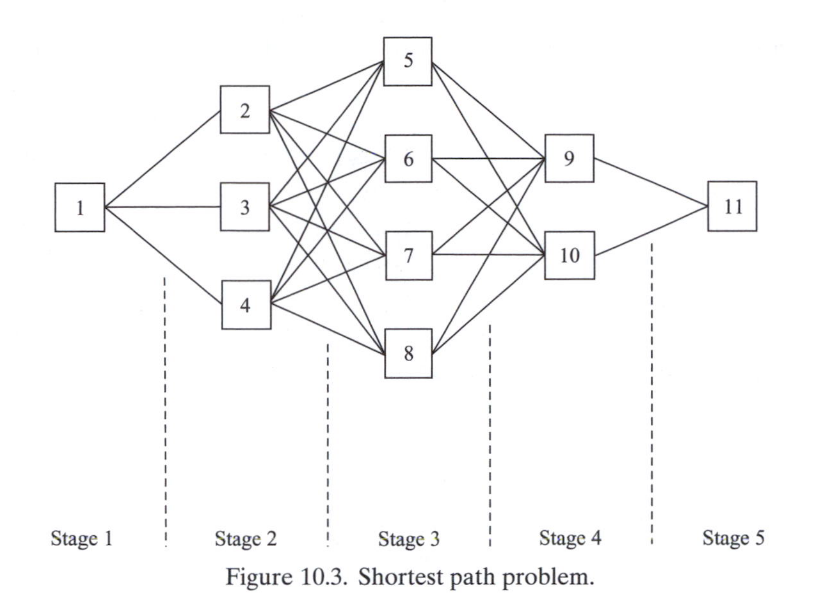

Figure 10.3. Shortest path problem.

PM2: Shortest Path Problem (Stagecoach Problem)

Shortest distance traveled by a stage coach from a starting city to a destination city, touching one city from each of the groups of cities is to be determined. Shortest wire for connecting two points in a circuit board, shortest pipeline layout type problems also fall in this category. This problem differs somewhat from the traveling salesman problem. Consider the problem in Fig. 10.3.

The problem can be worked, using forward or backward recursion. The optimal policy in the backward iteration scheme is to find the optimum for a given node from the final destination node. If the optimal distances for the Stage 3 are already computed as $f(5), f(6), f(7), f(8)$, then $f(2)$ can be found to be

$$f(2) = \min[d_{25} + f(5), d_{26} + f(6), d_{27} + f(7), d_{28} + f(8)] \tag{10.6}$$

We note here that this computation is straightforward. The above minimization is meaningful over the nodes that are connected. When City 1 is reached using this process, the shortest distance is obtained. We may create a path array in which the node number in Eq. 10.6 that gives the minimum is entered into the location 2. Thus, once we reach City 1, the path is obtained, using this array.

PM3: Reliability Problem

Consider the design of a machine that has N main components. Each component has a certain number of choices. The cost c_j and the reliability R_j for each choice of the component are known. The problem is to come up with a design of the highest reliability. The problem is stated as

$$\begin{aligned} &\text{maximize} \quad \prod_{j=1}^{N} R_j \\ &\text{subject to} \quad \sum_{j=1}^{N} c_j\,(x_j) \le C \end{aligned} \tag{10.7}$$

The problem requires that each of the reliability values be positive numbers. Reliability values are numbers in the range $0 < R < 1$. In view of this, we can take the logarithm of the objective function in Eq. (10.7). This turns into summation of logarithms as follows:

$$\begin{aligned} &\text{maximize} \quad \sum_{j=1}^{N} \ln\,(R_j) \\ &\text{subject to} \quad \sum_{j=1}^{N} c_j\,(x_j) \le C \end{aligned} \tag{10.8}$$

This problem is similar to the resource allocation problem PM1. A small modification in the program DPALLOC may be needed in order to handle the negative numbers. At the placement stage, we enter the number into the location if the previous entry is zero. If the previous entry is a negative number, comparison is made to select the larger one.

A large variety of problems can be solved by dynamic programming considerations. We have presented three important classes in our problem modeling. The computer program included can be modified to suit other problems. Generally speaking, dynamic programming is applicable if the optimal solution to the entire problem contains within it optimal solutions to subproblems, and when a recursive algorithm can be generated to solve the subproblems over and over again. The method directly aims at the optimum and feasibility is maintained. It is of a combinatorial nature. One of the main drawbacks of the method is that it is slow when the number of variables is large. On small problems it is efficient and direct.

COMPUTER PROGRAMS

DPALLOC

PROBLEMS

P10.1. A mass slides under the action of gravity along a chute (or a slide) as shown in Figure P10.1. Assume a smooth, frictionless ride for the mass. The problem is to determine the heights y_A and y_B so that the mass reaches the bottom in minimum time. The mass is released at the top of the chute with zero initial velocity. Is this problem a candidate for dynamic programming (say, using 3 stages)? Explain your answer.

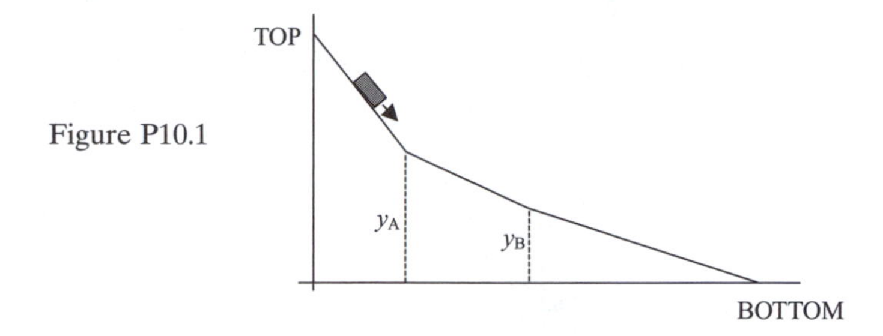

Figure P10.1

P10.2. Does the functional equation

$$f_i(\mathbf{s}_i) = \min_{\mathbf{x}_i} [r_i(\mathbf{x}_i, \mathbf{s}_i) - f_{i+1}(\mathbf{s}_{i+1})] \quad i = n-1, \quad n-2, \ldots, 1$$

violate the monotonocity condition?

P10.3. Extend the sensitivity analysis in Example 10.2, Table 10.2e to include starting operating pressures of $P_1 = 700$ Pa and $P_1 = 800$ Pa.

P10.4. Do Example 10.2 with the following change: A cost of $(0.25\ P_{i+1})$ is added to the cost values in Table E10.2a, where P_{i+1} is the operating pressure at the end of the stage under consideration. For instance, if the operating pressure in stage 1 is 500 Pa, and $x = 200$ Pa pressure drop, then the cost is $r = 215 + (0.25)(300) = \$290$. Other aspects remain the same. Specifically, $P_1 = 600$, $P_5 = 0$.

P10.5. A sales person wishes to pack maximum value of items into the suitcase for travel. The weight of the items should not exceed 50 lb. Determine the optimum selection assuming that more than one item of a type is permitted.

Item no.	Weight, lb	Value, $
1	3	100
2	5	200
3	2	60
4	4	150
5	6	250

P10.6. A television manufacturer has a contract to deliver television units during the next three months. The delivery must be 1st month 3000 units, 2nd month 4000 units, 3rd month 4000 units. For each television produced, there is a variable cost of \$25. The inventory cost is \$5 per television in stock at the end of the month. Set up cost each month is \$900. Assuming that production during each month must be a multiple of 500, and that the initial inventory is zero, determine an optimal production schedule. (*Note*: Televisions produced in the past can be delivered any time in the future.)

P10.7. For the stage coach problem shown in Fig. P10.7, the numbers on the arcs indicate the cost of travel. Determine the route from city 1 to city 11 for minimum cost, using dynamic programming.

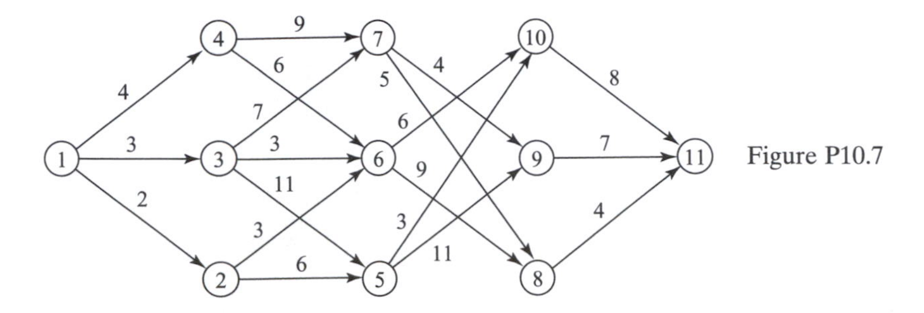

Figure P10.7

P10.8. Use forward recursion approach for solving the problem 10.7. Implement the forward recursion scheme in a computer program.

P10.9. In an investment problem, amount invested is in units of \$1000, and total \$6000 is available for investment. There are three possible investments with return amounts indicated.

Investment \$1000s	0	1	2	3	4	5	6
Return 1	0	10	25	35	55	75	100
Return 2	0	5	15	25	35	60	80
Return 3	0	3	10	15	25	45	65

Find the optimum investment distribution.

P10.10. Consider the matrix product $A_1\ A_2\ A_3\ A_4\ A_5\ A_6$. The matrices $A_1 \ldots A_6$ are square with dimensions [5, 10, 3, 12, 5, 50, 6], respectively. Develop a computer code based on dynamic programming and find an optimal parenthesization for matrix chain product.

[*Hint*: For example, consider the chain $A_1\ A_2\ A_3$ where A_1 is 10×10, A_2 is 10×5, A_3 is 5×1. Then, $(A_1\ A_2)A_3$ implies that A_1 and A_2 are multiplied first and the resulting matrix is premultiplied by A_3 giving rise to $500 + 50 = 550$ ops, where an op equals a multiplication plus an addition of two numbers. On the other hand, $A_1\ (A_2\ A_3)$ involves only 150 ops.]

REFERENCES

Bellman, R.E., *Dynamic Programming*, Princeton University Press, Princeton, NJ, 1957.

Bellman, R.E., On the theory of dynamic programming, *Proceedings of the National Academy of Science, USA*, **38**, 716–719, 1952.

Bellman, R.E. and Dreyfus, S.E., *Applied Dynamic Programming*, Princeton University Press, Princeton, NJ, 1962.

Cormen, T.H., Leiserson, C.E., and Rivest, R.L., *Introduction to Algorithms*, MIT, Cambridge, MA, 1990.

Wagner, M.W., *Principles of Operations Research*, 2nd Edition, Prentice-Hall, Englewood Cliffs, NJ, 1975.

11

Optimization Applications for Transportation, Assignment, and Network Problems

11.1 Introduction

Transportation, assignment, and network problems are of great significance in engineering industry, agriculture, global economy, and commerce. These problems can be solved by the conventional simplex method of linear programming. However, because of their simpler structures, they are more efficiently solved by special methods. In most cases, the calculations involved are simple additions and subtractions and the methods are effective and robust. Many problems in totally different situations may be closely related to these models and the methods may be used for their solution. We consider the transportation, assignment, and network flow problems with a view to develop the formulation of the problems, the solution algorithms, and their implementation in simple computer codes.

11.2 Transportation Problem

There are m supply points where quantities $s_1, s_2, \ldots, s_m$, respectively, are produced or stocked. There are n delivery points or destinations where respective quantities of $d_1, d_2, \ldots, d_n$, are in demand. Cost c_{ij} is incurred on an item shipped from origination point i to the destination j. The problem is to determine the amounts x_{ij} (≥ 0) originating at i and delivered at j so that the supply and demand constraints are satisfied and the total cost is a minimum. The problem is shown in Fig. 11.1.

The linear program may be put in the form

$$\text{minimize} \quad \sum_{i=1}^{m} \sum_{j=1}^{n} c_{ij} x_{ij}$$

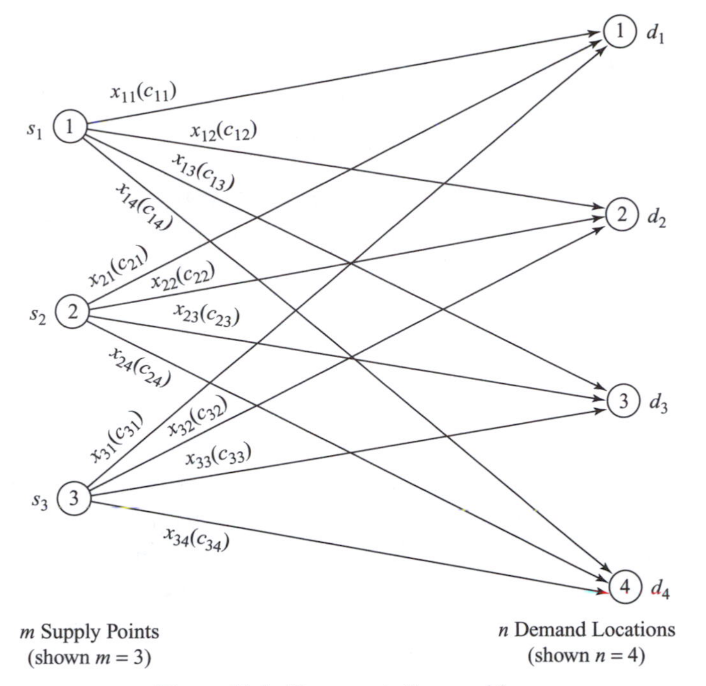

Figure 11.1. Transportation problem.

$$
\begin{aligned}
\text{subject to} \quad & \sum_{j=1}^{n} x_{ij} = s_i \quad i = 1, 2, \ldots, m \quad \text{Supply constraints} \\
& \sum_{i=1}^{m} x_{ij} = d_j \quad j = 1, 2, \ldots, n \quad \text{Demand constraints} \\
& x_{ij} \geq 0, \quad i = 1, 2, \ldots, m; \quad j = 1, 2, \ldots, n
\end{aligned}
\tag{11.1}
$$

The problem posed in the preceding is a *balanced transportation problem* where the total supply is equal to the total demand

$$
\sum_{i=1}^{m} s_i = \sum_{j=1}^{n} d_j \tag{11.2}
$$

If the supply and demand are not in balance, the case may be handled by adding a dummy demand point or a dummy supply station as needed. The costs associated with the dummy station may be taken as zero. If meaningful distribution is necessary, some penalty costs may be introduced for this case.

This problem can be solved using the conventional simplex method but the size of the constraint coefficient matrix becomes large $(m+n) \times mn$. The basic structure of the problem can be seen by writing the constraints in the conventional form as

$$
\begin{array}{llllll}
x_{11}+x_{12}+\cdots+x_{1n} & & & & = & s_1 \\
 & x_{21}+x_{22}+\cdots+x_{2n} & & & = & s_2 \\
 & & \ddots & & \vdots & \\
 & & & x_{m1}+x_{m2}+\cdots+x_{mn} & = & s_m \\
x_{11} & x_{21} & & x_{m1} & = & d_1 \\
\quad x_{12} & \quad x_{22} & \ddots & \quad x_{m2} & = & d_2 \\
\qquad \ddots & \qquad \ddots & & \qquad \ddots & \vdots & \\
\qquad\quad x_{1n} & \qquad\quad x_{2n} & & \qquad\quad x_{mn} & = & d_n
\end{array}
\tag{11.3}
$$

By adding the first m equations, then subtracting the next $n - 1$ equations and making use of the balance relation Eq. (11.2), we get the last equation. In fact, by manipulating any of the $m + n - 1$ equations, the remaining equation can be obtained. $m + n - 1$ of the equations are linearly independent. Thus, any basic solution to the transportation problem has $m + n - 1$ basic variables. An efficient optimal solution algorithm for the transportation algorithm switches between the dual and the primal formulations. We associate m dual variables $u_1, u_2, \ldots, u_m$, with the m source constraints and n dual variables $v_1, v_2, \ldots, v_n$, with the n demand constraints. The dual problem can be put in the form

$$
\begin{aligned}
&\text{maximize} \ \sum_{i=1}^{m} s_i u_i + \sum_{j=1}^{n} d_j v_j \\
&\text{subject to} \ \ u_i + v_j \le c_{ij}, \ \ i = 1 \ \text{to} \ m, \quad j = 1 \ \text{to} \ n
\end{aligned}
\tag{11.4}
$$

We note from Eq. (11.3) that the matrices are sparsely filled and the computer handling is inefficient if solution is attempted using the standard LP model. All the information of the primal and the dual problems can be handled by the compact form given in Fig. 11.2.

The first step in finding the optimum is to establish a basic feasible solution for the primal problem (11.3). The complementary slackness condition provides us with the link that establishes a neat algorithm for finding the optimum. The *complementary slackness condition* states that for every x_{ij} satisfying feasibility of (11.3) and u_i and v_j satisfying the feasibility of (11.4),

$$x_{ij}(u_i + v_j - c_{ij}) = 0 \tag{11.5}$$

In particular, if x_{ij} is a basic variable, then

$$u_i + v_j = c_{ij} \quad \text{for all } x_{ij} \text{ basic} \tag{11.6}$$

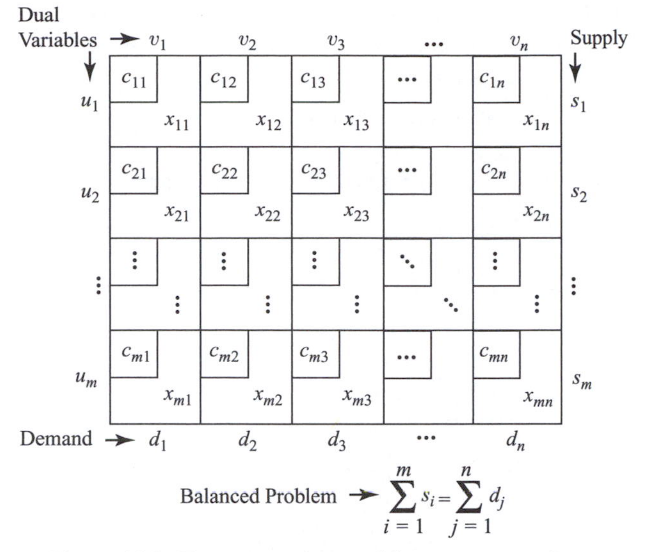

Figure 11.2. Transportation problem – compact form.

Eq. (11.6) can be used for finding u_i and v_j components. Even for the dual variables, only $m + n - 1$ variables are independent. Any one of the $m + n$ variables can be taken as zero. Without loss in generality, we may set $u_1 = 0$. We will comment on the solution of Eq. (11.6) more specifically as we solve a problem. For nonbasic variable x_{ij}, the *shadow costs* or the *reduced cost coefficients* are calculated as

$$r_{ij} = c_{ij} - u_i - v_j \quad \text{for all } x_{ij} \text{ nonbasic} \tag{11.7}$$

We have all the preparation needed to solve the problem.

Transportation Simplex Method

The transportation problem is solved, using the simplex method. The basic steps given below are similar to the ones used in the regular simplex method. However, the special structure of the problem leads to simpler calculations.

Step 1. Find a basic feasible solution
Step 2. Check for optimality. If not optimal, determine the variable entering the basis.
Step 3. Determine the variable leaving the basis.
Step 4. Find the new basic feasible solution and go to Step 2.

We illustrate each of the steps by means of an example.

Consider the transportation problem shown below. Goods manufactured at three stations are to be delivered at four locations. The supply and demand matrix and the corresponding cost-coefficient matrix are shown separately. We would like to obtain the optimal allocations for minimal cost.

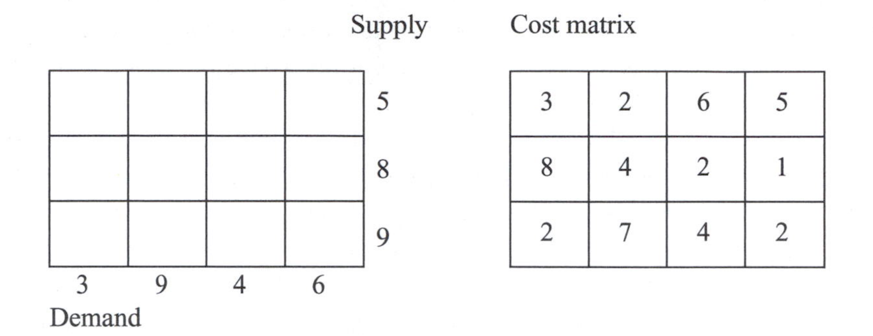

Supply:

				Supply
				5
				8
				9
3	9	4	6	
Demand				

Cost matrix:

3	2	6	5
8	4	2	1
2	7	4	2

Step 1: Obtain Basic Feasible Solution (bfs)

Northwest Corner Approach

In the first cell located at the northwest corner, we put the smaller of the supply (5) or the demand (3) values, which is 3. Then, since the supply is larger we move to the next column and put 2 (= 5 − 3). The row total is matched. We then go down one row and put in a value 7 (= 9 − 2). The second column total matches the demand for the station. The row total is short by 1. We put 1 (= 8 − 7) in the next column. We go down one row and put in the value 3 (= 4 − 1). The final value in the next column is 6, which satisfies both the row and column totals, simultaneously. The *bfs* is defined in six variables. The values are shown in the updated table as follows.

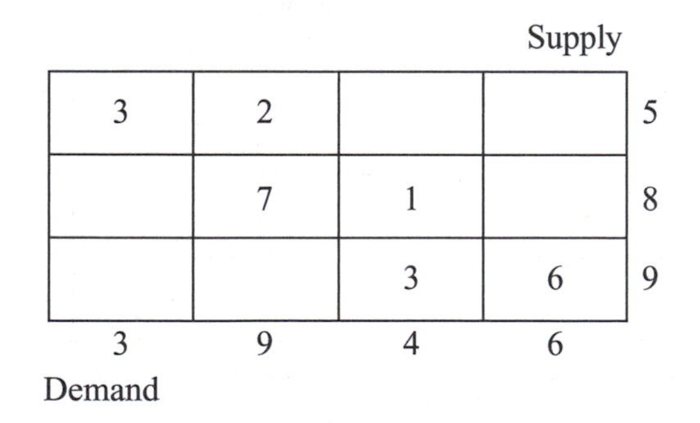

				Supply
3	2			5
	7	1		8
		3	6	9
3	9	4	6	
Demand				

We note that at each entry we move by a column or a row. Since there are m rows and n columns, we precisely need $m + n - 1$ independent entries, and then the last one is automatically determined. The generation of the basic feasible solution is similar to the back-substitution process in Gauss elimination. The basis matrix of the $m + n - 1$ equations in 11.3 being triangular, and each of the coefficients being 1, the back-substitution is one of addition and subtraction. Every basis of the transportation problem has this property, called the *triangular property*. This provides us with a starting basic feasible solution. Northwest corner approach is the simplest of the methods for generating a basic feasible solution. There are other methods that provide a bfs, which is closer to the optimum solution.

In some situations, when a number is entered, it may satisfy both the supply (row) and demand (column) values even before approaching the last entry. In this case, we proceed to the next column if the previous move was a row move or the next row if the previous move was a column move and make an entry of *zero* and then proceed further as discussed previously. If it is the starting location, the move can be in the row or column direction. This is a *degenerate basic feasible solution*. We need to distinguish this zero from the zeros in other nonbasic locations. In the computer implementation, we put in a number -2 in each of the nonbasic locations.

Step 2. Determine the Variable Entering the Basis

The determination of the variable entering the basis also includes the step to check for optimality. For the basic feasible solution, we will now calculate the dual variables u_1, u_2, u_3, and v_1, v_2, v_3, v_4. The solution is easily obtained by placing the cost coefficient in the array and then calculating the dual variables. We arbitrarily set $u_1 = 0$ and then determine the other dual variables in the order indicated in parenthesis.

$u_i \backslash v_i$	(2) $v_1 = 3$	(3) $v_2 = 2$	(5) $v_3 = 0$	(7) $v_4 = -2$
(1) $u_1 = 0$	3	2		
(4) $u_2 = 2$		4	2	
(6) $u_3 = 4$			4	2

Costs in basic variable locations $c_{ij} = u_i + v_j$

The dual variable Eqs. (11.6) also exhibit the *triangular property*; thus, the evaluation is a simple process of back-substitution with only one variable unknown at a time. We now evaluate the reduced cost coefficients in the nonbasic variable locations using Eq. (11.7).

$u_i \backslash v_i$	(2) $v_1 = 3$	(3) $v_2 = 2$	(5) $v_3 = 0$	(7) $v_4 = -2$
(1) $u_1 = 0$			$6 - 0 - 0 = 6$	$5 - 0 + 2 = 2$
(4) $u_2 = 2$	$8 - 2 - 3 = 3$			$1 - 2 + 2 = 1$
(6) $u_3 = 4$	$2 - 4 - 3 = -5$	$7 - 4 - 2 = 1$		

Costs in basic variable locations $r_{ij} = c_{ij} - u_i - v_j$

The most negative variable x_{31} in location 3–1 is the incoming variable. *If all reduced costs (shadow costs) are nonnegative, we are done.* Now, we need to raise the level of the incoming variable and see its effects.

Step 3. Determine the Variable Leaving the Basis

Consider the effect of raising the location 3–1 by a. If we then decrease the location 1–1 (in column 1) by a, increase the location 1–2 by a, decrease location 2–2 by a, increase location 2–3 by a, and decrease location 3–3 by a, we have completed the loop 3–1, 1–1, 1–2, 2–2, 2–3, 3–3 (to 3–1). This change does not affect the row and column sums. We also need to ensure that when variable x_{31} comes in, another variable must exit. Among the basic variable values reduced, we choose the least one (equal to 3) and set $a = 3$.

				Supply
$3 - a$	$2 + a$			5
	$7 - a$	$1 + a$		8
a		$3 - a$	6	9
3	9	4	6	
Demand				

In this example, two of the variables are becoming zero simultaneously, showing that it is degenerate. We decide that one of these, say, 1–1 as the variable exiting the basis. By the triangular property, for every basic variable there is at least one other variable in the row or column containing that variable. This property ensures the existence of a unique closed loop for each nonbasic variable. On the computer, we start from the location corresponding to the incoming variable and move up or down. We reach a base point when we hit a basic variable. At every base point, we have two choices of paths – in the vertical mode it is up or down, and in the horizontal mode it is right or left. If we first start in the vertical mode, the goal is to reach the starting row in the final vertical search, when the loop is complete. If we are in the vertical mode, the next path from a base point is horizontal and vice versa. If the next column or row falls outside the matrix, no progress is possible and we backtrack to the current base point and try first in the other direction. If this is also not successful, we backtrack and search for a new basic variable location in the previous direction. The whole process is like an interesting video game on the computer. Once the loop is established, the starting location is named 1. We then look for the location with the smallest value in the even points of the loop. This location is the leaving variable.

Step 4. Find the New Basic Feasible Solution

This value is then inserted at the starting location, subtracted from the locations at the even points and added to the locations at the other odd points. The location at the variable that exited is identified as nonbasic. We have now obtained a new improved basic solution and are ready for the next iteration.

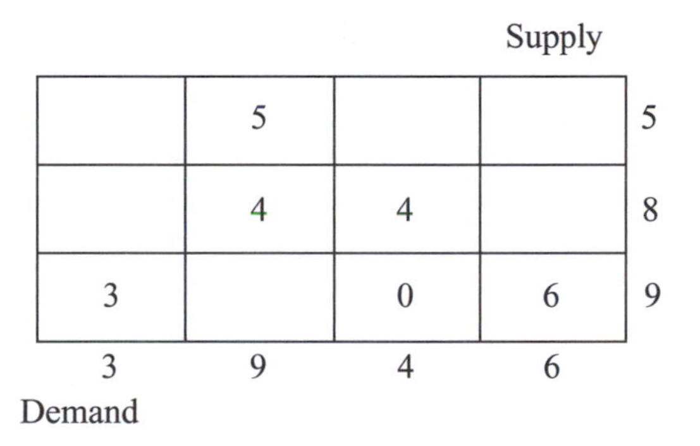

				Supply
	5			5
	4	4		8
3		0	6	9
3	9	4	6	

Demand

We now go to step 2. The next iteration tables are given as follows.

$u_i \backslash v_i$	(6) $v_1 = -2$	(2) $v_2 = 2$	(4) $v_3 = 0$	(7) $v_4 = -2$
(1) $u_1 = 0$		2		
(3) $u_2 = 2$		4	2	
(5) $u_3 = 4$	2		4	2

Costs in basic variable locations $c_{ij} = u_i + v_j$

$u_i \backslash v_i$	(6) $v_1 = -2$	(2) $v_2 = 2$	(4) $v_3 = 0$	(7) $v_4 = -2$
(1) $u_1 = 0$	$3 - 0 + 2 = 5$		$6 - 0 - 0 = 6$	$5 - 0 + 2 = 7$
(3) $u_2 = 2$	$8 - 2 + 2 = 6$			$1 - 2 + 2 = 1$
(5) $u_3 = 4$		$7 - 4 - 2 = 1$		

Costs in nonbasic variable locations $r_{ij} = c_{ij} - u_i + v_j$

Costs in nonbasic variable locations $r_{ij} = c_{ij} - u_i + v_j$

All the reduced cost coefficients are positive. The solution is, thus, a minimum. The optimum solution from the last basic solution is

Source number	Destination	Quantity
1	2	5
2	2	4
	3	4
3	1	3
	4	6

The minimum cost is 52.

The transportation algorithm has been implemented in the computer program TRANSPO. Dummy station introduction is automatically taken care of in the computer program.

Transshipment Problem

In a transshipment problem, there are points where the goods are shipped to, stored, and later delivered to final destinations. These intermediate points are called transshipment points. The transshipment problem can be handled using the regular transport algorithm as follows. The transshipment points are introduced both as source stations and supply stations. The supply and demand are each set at a sufficiently large value, usually taken as the total supply (= demand). We thus ensure that what goes in comes out at these points.

11.3 Assignment Problems

The assignment problems arise in manufacturing industries and businesses. The general class of problems are similar to the assignment of n jobs to n machines in a production operation. Other problems may be like assignment of teachers to courses, workers to jobs, pilots to flights, and so on. There is cost c_{ij} associated with job i assigned to machine j. $\mathbf{c}$ is referred to as the *effectiveness matrix*. The corresponding variable x_{ij} takes the value of 1. The general problem can be put in the form

$$\begin{aligned}
\text{minimize} \quad & \sum_{i=1}^{n} \sum_{j=1}^{n} c_{ij} x_{ij} \\
\text{subjectto} \quad & \sum_{j=1}^{n} x_{ij} = 1 \quad i = 1, 2, \ldots, n \\
& \sum_{i=1}^{n} x_{ij} = 1 \quad j = 1, 2, \ldots, n \\
& x_{ij} = 0 \ \text{or} \ 1 \quad i = 1, 2, \ldots, n; \quad j = 1, 2, \ldots, n
\end{aligned} \tag{11.8}$$

We may view the problem as a special case of the transportation problem. If the number of jobs is not equal to the number of machines, dummy jobs or machines may be added as needed to balance the problem. The associated costs may be left as zeros. The requirement that each variable be an integer equal to 0 or 1 is automatically satisfied by the triangular property of the basis. The problem can be solved using the transportation algorithm. However, the assignment problem has a special structure that enables us to solve by simpler manipulations. We present here the method originally developed by two Hungarian mathematicians.

The Hungarian Method

In developing the Hungarian method for the solution of the assignment problem, we develop two preliminary results.

Effect of Subtracting p_i from Row i and q_j from Column j

If we define the new effectiveness coefficient as $c'_{ij} = c_{ij} - p_j - q_j$, and consider the objective function with these coefficients, then

$$\begin{aligned}
\sum_{i=1}^{n} \sum_{j=1}^{n} c'_{ij} x_{ij} &= \sum_{i=1}^{n} \sum_{j=1}^{n} c_{ij} x_{ij} - \sum_{i=1}^{n} p_i \sum_{j=1}^{n} x_{ij} - \sum_{j=1}^{n} q_j \sum_{i=1}^{n} x_{ij} \\
&= \sum_{i=1}^{n} \sum_{j=1}^{n} c_{ij} x_{ij} - \sum_{i=1}^{n} p_i - \sum_{j=1}^{n} q_j
\end{aligned} \tag{11.9}$$

The objective function is reduced by the sum of the row and column subtractions. The minimum point does not change, but the value gets reduced by the sum of values subtracted. This sum must be added to the new optimum value to get the original optimum. The strategy is to subtract first the row minimum from each row, successively, and then the column minimum from each column, successively. Then each row and each column has at least one zero in it. If we can perform the assignments at the zero locations, the minimum of the new objective is zero. If assignment is not possible, additional manipulations are necessary. One other manipulation involves partitioned matrices of **c**.

Manipulation of a Partition of **c**

We develop an identity for the indices defined as follows.

$$\begin{aligned}
0 &= -p\sum_{i=1}^{m}\sum_{j=1}^{r} x_{ij} + p\sum_{i=1}^{m}\sum_{j=1}^{r} x_{ij} = -p\sum_{i=1}^{m}\sum_{j=1}^{r} x_{ij} + p\sum_{i=1}^{m}\left(1 - \sum_{j=r+1}^{n} x_{ij}\right) \\
&= -p\sum_{i=1}^{m}\sum_{j=1}^{r} x_{ij} + mp - p\sum_{i=1}^{m}\sum_{j=r+1}^{n} x_{ij} \\
&= mp - p\sum_{i=1}^{m}\sum_{j=1}^{r} x_{ij} - p\sum_{j=r+1}^{n}\left(1 - \sum_{i=m+1}^{n} x_{ij}\right) \\
&= p(m + r - n) - p\sum_{i=1}^{m}\sum_{j=1}^{r} x_{ij} + p\sum_{i=m+1}^{n}\sum_{j=r+1}^{n} x_{ij}
\end{aligned} \tag{11.10}$$

The result of the above evaluation is the identity

$$-p\sum_{i=1}^{m}\sum_{j=1}^{r} x_{ij} + p\sum_{i=m+1}^{n}\sum_{j=r+1}^{n} x_{ij} = -p(m + r - n) \tag{11.11}$$

Eq. (11.11) can be interpreted as follow. Subtracting p from every element from the partition c_{ij}, $i = 1$ to m, $j = 1$ to r, and adding p to each of the elements of the partition c_{ij}, $i = m + 1$ to n, $j = r + 1$ to n, reduces the objective function by the constant value $p(m + r - n)$. If such a manipulation is done, this constant quantity must be added to the objective function value to get the value of the original problem.

With the concepts of the basic manipulations fixed, we present the steps in the assignment calculations.

Consider the problem of assignment of five jobs to five machines with the following effectiveness matrix.

		1	2	3	4	5	<= Machine
	1	5	2	4	3	6	
	2	7	4	3	6	5	
Job =>	3	2	4	6	8	7	
	4	8	6	3	5	4	
	5	3	9	4	7	6	

Step 1. *Subtract row minimum from each row*

The row minima 2, 3, 2, 3, 3 are subtracted from the respective rows to obtain

3	0	2	1	4
4	1	0	3	2
0	2	4	6	5
5	3	0	2	1
0	6	1	4	3

The value of the function is set as $2 + 3 + 2 + 3 + 3 = 13$

Step 2. *Subtract column minimum from each row.*

The column minima 1,1 are subtracted from the columns 4 and 5. If the minimum is a zero, there is no change in the column.

3	0	2	0	3
4	1	0	2	1
0	2	4	5	4
5	3	0	1	0
0	6	1	3	2

The value of the function now becomes $13 + 1 + 1 = 15$

Step 3. *Cover the zeros by minimum number of lines drawn through columns and rows.*

The idea is to have maximum number of nonzero elements exposed. We follow a simple strategy, which is implemented on the computer. We first determine the row or column that has the minimum number of zeros. If it is a row that has the minimum number of zeros, lines are passed through the columns with the zeros and vice versa. After that a similar procedure is followed for the exposed values. In the problem at hand, the second row has a single zero (minimum) at column 3. We draw a vertical line through column 3. The last row has a single zero at column 1, so we draw a vertical line through column 1. In the uncovered matrix, the second column has a single zero at row 1. We draw a horizontal line through row 1. There is one zero left at row 4 of the original matrix at column 5. We draw a vertical line through column 5.

3	0	2	0	3
4	1	0	2	1
0	2	4	5	4
5	3	0	1	0
0	6	1	3	2

The minimum number of lines is 4. *In an n × n problem, assignment can be made if the minimum number of lines covering the zeros is n*. Since the number of lines is not 5, we perform the manipulation suggested in Eq. (11.11). If we subtract a number from the exposed matrix, we add the same number from the double-crossed elements. We choose the smallest number, which is 1 from the exposed matrix. Subtracting this from the eight exposed locations and adding to the three double-crossed locations, we have a modified effectiveness matrix with two additional zeros. The function value is set as $15 + 1^*(4 + 2 - 5) = 16$.

We repeat step 3. The strike-off procedure gives the following matrix.

4	0	3	0	4
4	0	0	1	1
0	1	4	4	4
5	2	0	0	0
0	5	1	2	2

The minimum value 1 of the exposed matrix is subtracted from the exposed locations and added to the three locations where the lines intersect. The function value is updated as $16 + 1^*(4 + 2 - 5) = 17$. After the first two strike-offs, across from single zero columns, we get.

5	0	3	0	4
5	0	0	1	1
0	0	3	3	3
6	2	0	0	0
0	4	0	1	1

Now the minimum number of zeros is two in every direction. We choose row 2 and draw two vertical lines through columns 2 and 3. The remaining two zeros require another vertical line.

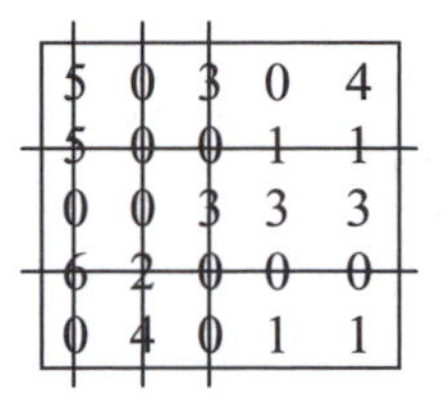

We have now required 5 lines to cover the zeros. We are now ready to assign.

Step 4. *Assign.*

Assignment is very similar to the previous strike-off except that we strike-off both row and column and look for a zero to be assigned in the remaining elements. There is only one zero in column 5. We assign job 4 to machine 5 and strike off row 4 and column 5.

$$\begin{array}{|ccccc|}
\hline
5 & 0 & 3 & (0) & 4 \\
5 & (0) & 0 & 1 & 1 \\
(0) & 0 & 3 & 3 & 3 \\
6 & 2 & 0 & 0 & (0) \\
0 & 4 & (0) & 1 & 1 \\
\hline
\end{array}$$

In the remaining matrix, we have a single zero at column 4 and we assign job 1 to machine 4. In the remaining matrix, we have two zeros in each row and column. In this situation, the assignment is not unique. We can assign job 2 to machine 2 or 3. We decide to assign job 2 to machine 2 and strike off row 2 and column 2. Then there is a single zero in row 3, which says that we assign job 3 to machine 1. We strike off row 3 and column 1. The remaining zero is at job 5 assigned to machine 3. Referring to the original matrix, the sum of the values at the assigned locations is $3 + 4 + 2 + 4 + 4 = 17$. This is the cost as expected.

The Hungarian method for assignment is implemented in the computer program HUNGARYN. If the number of jobs does not match with the number of machines, zeros may be added to make the effectiveness matrix a square matrix.

11.4 Network Problems

A *network* consists of a set of *nodes* connected by *arcs*. An *arc* is an ordered pair of nodes often indicating the direction of motion between the nodes. The network

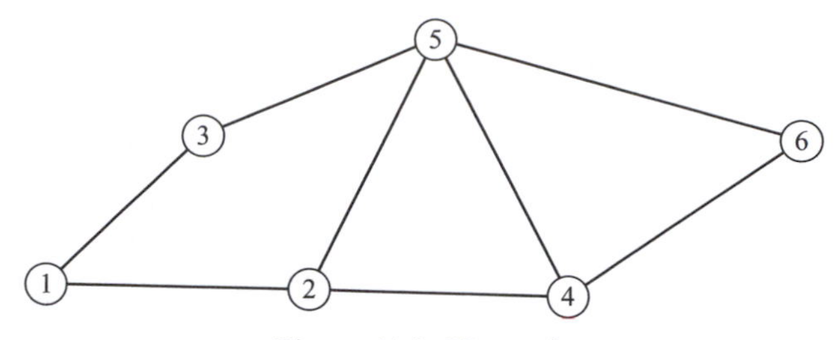

Figure 11.3. Network.

may be represented by a graph $G = (\mathbf{N}, \mathbf{A})$, where $\mathbf{N}$ is the set of nodes and $\mathbf{A}$ is the set of arcs. The network shown in Fig. 11.3 may be represented by

$$\begin{aligned} \mathbf{N} &= [1, 2, 3, 4, 5, 6] \\ \mathbf{A} &= [(1, 2), (1, 3), (2, 5), (2, 4), (3, 5), (5, 4), (4, 6), (5, 6)] \end{aligned} \tag{11.12}$$

A *chain* between two nodes is a sequence of arcs connecting them. A chain from 1 to 6 in Fig. 11.3 is given by (1, 2), (2, 4), (6, 4). A *path* is a chain, in which the terminal node of each node is the initial node of the next arc. (1, 2), (2, 4), (4, 6) is both a chain and a path. A node j of a network is *reachable* from node i if there is a path from node i to node j. A *cycle* or a *loop* is a chain starting from a node and ending at the same node. (2, 4), (4, 5), (5, 2) represents a cycle. A network is said to be *connected* if there is a chain between any two nodes. A network or a graph is a *tree* if it is connected and has no cycles. A *spanning tree* is a connected network that includes all the nodes in the network with no loops. A *directed* arc is one, in which there is a positive flow in one direction and a zero flow in the opposite direction. A *directed* arc is represented by a line with an arrow pointing in the direction of flow. A directed network is shown in Fig. 11.4. Two-way flow with different capacities for each direction is denoted by two arrows.

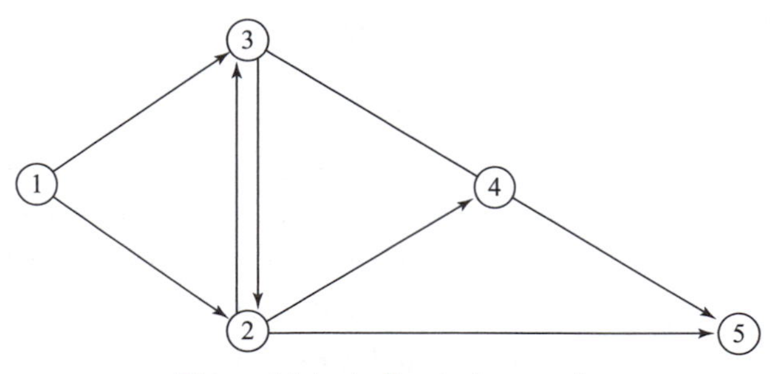

Figure 11.4. A directed network.

We now proceed to discuss some procedures and methods useful in solving network problems.

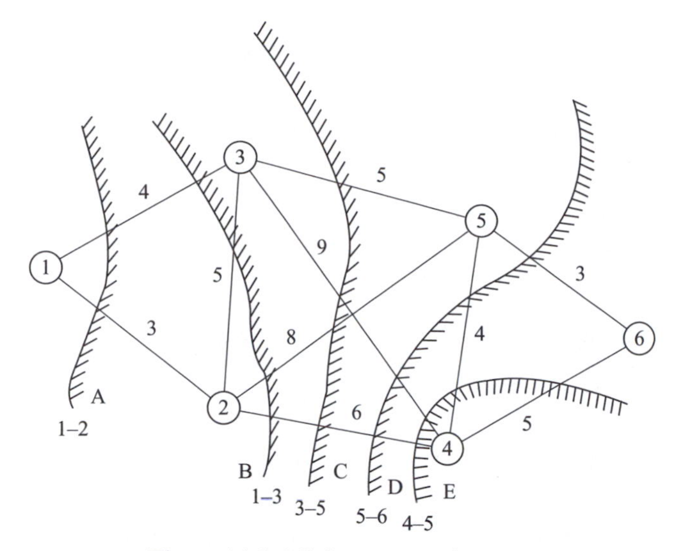

Figure 11.5. Minimum spanning tree.

Tree Procedure

Reachability is a key step in finding feasible solutions to flow problems in networks. The reachability of a node n from node 1 can be established by a tree procedure. We label the node 1 as, say, 1. All other nodes are unlabeled. Go to a labeled node, say, i and put the node number i as the label for all the connected unlabeled nodes. We then change the sign of the label of node i. If in the process of labeling, node n is labeled, we are done. If no unlabeled nodes can be labeled and node n has not been reached, then we conclude that there is no connecting path exists. The path can be easily traced in the reverse direction by proceeding from the last node and reading off the absolute values of the labels. The tree procedure is implemented in the program TREE.

Minimum Spanning Tree

Finding the minimum spanning tree is in itself an important problem. Building a network of roads to connect n cities so that the total length of the road built is a minimum is a minimum spanning tree problem. Similar problems include television and telephone cable laying, pipeline layout, and so on. We consider the problem of building roads connecting six towns shown in the Fig. 11.5.

The towns are situated at the nodes. The possible arc connections representing the roads are given. The distances are shown in the table as follows. We note that $i = j$ is not necessary in the table. We also use a convention that an entry of zero (0) in the location (i, j) indicates that towns i and j are not connected.

	Town 2	Town 3	Town 4	Town 5	Town 6
Town 1	3	4	0	0	0
Town 2		5	6	8	0
Town 3			9	5	0
Town 4				4	5
Town 5					3

At any stage in the algorithm for establishing the minimum spanning tree, we have two sets of nodes, one set labeled and the other unlabeled. At the start, one of the nodes, say, 1 is labeled. The idea is to find a node from the labeled set, say, i and a node from the unlabeled set, say, j such that $i - j$ is the arc of least length. $i - j$ is the next arc for the minimum spanning tree. Node j is labeled and moved into the labeled set. The minimum tree is obtained when all nodes are covered. In the problem shown in Fig. 11.5, we start at node 1, draw a wave front A, which separates labeled nodes from the unlabeled ones. The closest nodes are $1 - 2$ with a distance of 3 and, thus, the first link is established. The front now moves to B enclosing node 2, which is labeled. The closest nodes are $1 - 3$, which is the second link with a distance of 4. This procedure establishes the minimum spanning tree $3 - 5$, $5 - 6$, and $4 - 5$, connecting all the nodes. The total length of the road is $3 + 4 + 5 + 3 + 4 = 19$. The algorithm has been implemented in the program TREEMIN.

Shortest Path – Dijkstra Algorithm

Shortest path problems are some of the common network problems. In directed networks, cyclic paths may occur as shown in Fig. 11.6 in the path $2 - 4 - 5 - 2$.

Dijkstra's algorithm also referred to as *cyclic algorithm* solves the problems with cyclic loops. We present here the algorithm by referring to the problem in Fig. 11.6. The moving iteration front separates the permanently labeled nodes from the unlabeled and temporary labeled nodes. At the beginning, node 1 is permanently labeled. Labeling is done, using two parameters. The first parameter is the current shortest distance to the current node and the second parameter is the number of the node from which the current node has been reached. The node 1 is labeled (0, 0). At the first iteration, nodes 2 and 3, which can be reached from the current permanently labeled node are temporarily labeled (40, 1) and (90, 1), respectively. From among the current temporarily labeled nodes, the node with the shorter distance, which is node 2, is permanently labeled. At the second iteration, we look for the connections from the current permanently labeled node 2. We label node 4

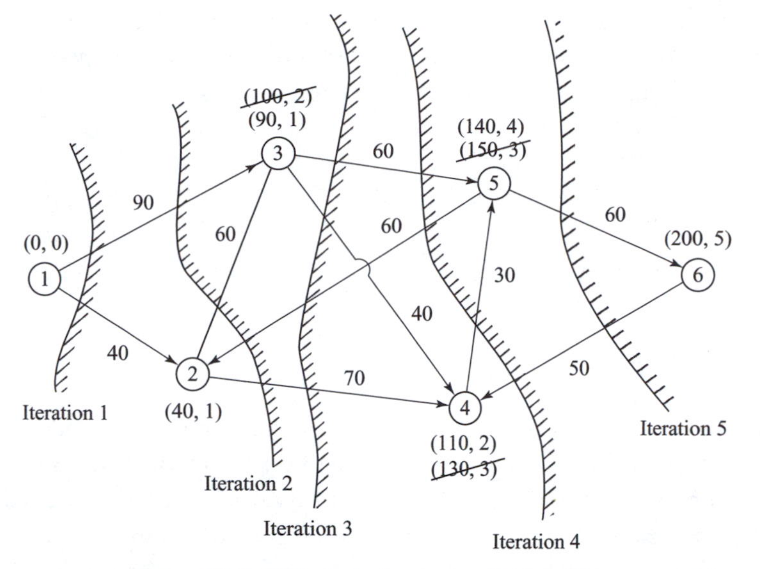

Figure 11.6. Dijkstra's shortest path algorithm.

as (110, 2) where 110 is 40 + 70. Label (90, 1) of node 3 is compared with (40 + 60, 2) and the smaller distance label is retained. Node 3 becomes the permanent node to initiate iteration 3. The new (130, 3) is compared to (110, 2) at node 4 and (110, 2) is retained. At node 5, the temporary label is (150, 3). Node 4 is now permanently labeled. The path from 4 to 5 gives (140, 4) which replaces the previous temporary label (150, 3). Node 5 is now permanently labeled. At iteration 5, node 6 is labeled (200, 5) and the process is complete. The shortest distance from node 1 to 6 is 200, and the path is easily obtained in the reverse direction by the second parameter. We start at node 6. The second label 5 for node 6 is the previous node of the path. We then go to node 5. The second label 4 of node 5 is the previous node of the path. This process is continued till we reach node 1. The path for the problem is 6 – 5 – 4 – 2 – 1. The algorithm has been implemented in the computer program DIJKSTRA. In summary, we also note that

1. At iteration *i*, we find the *i*th closest node to the node from which we start (node 1). Thus, Dijkstra's method gives the shortest path from node 1 to every other node.
2. Dijkstra's method works only for nonnegative costs (or distances).

Maximal Flow in Networks

In a capacitated network, the most common problem is to determine the maximal flow from a *source* node say 1 and a *sink* node, say, *m*. Suppose, there is a flow *f* into source node 1 and outflow f at the sink node. k_{ij} is the capacity of arc (i, j) and x_{ij} the flow in the arc. In the definition of the network flow problem, the two directions of arc with nodes i and j are represented by (i, j) and (j, i). The maximum flow problem can be stated as

$$\begin{aligned} &\text{maximize} && f \\ &\text{subject to} && \sum_{j=1}^{n} x_{ij} - \sum_{j=1}^{n} x_{ji} = b_i \quad i = 1 \text{ to } n \\ & && 0 \leq x_{ij} \leq k_{ij} \quad \text{all } i, j \text{ pairs defining arcs} \end{aligned} \tag{11.13}$$

where b_i is the source at node i. If there are no sources or sinks at nodes other than 1 and m, then $b_1 = f$, and $b_m = -f$. This must be kept in view while solving. This problem can be solved by the simplex method. We present here an efficient algorithm, which is based on the tree algorithm. The tree procedure is used to establish a path from node 1 to m. The maximal flow f_1 for this path is minimum of k_{ij} over all the valid arcs (i, j) for this path. Then the capacities k_{ij} and k_{ji} of arc (i, j) along the path are replaced by k'_{ij}, and k'_{ji} respectively, where $k'_{ij} = k_{ij} - f_1$, and $k'_{ji} = k_{ji} + f_1$. With these new capacities, the tree procedure is repeated and the new path gives a flow f_2. The process is repeated until no path can be established from 1 to m. The maximal flow is the sum of all the maximal flows $f_1 + f_2 + \cdots$ of all the iterations. The algorithm has been implemented in the computer program MAXFLO.

Example 11.1

Determine the maximal flow from node 1 to node 5 for the network configuration shown in Fig. E11.1.

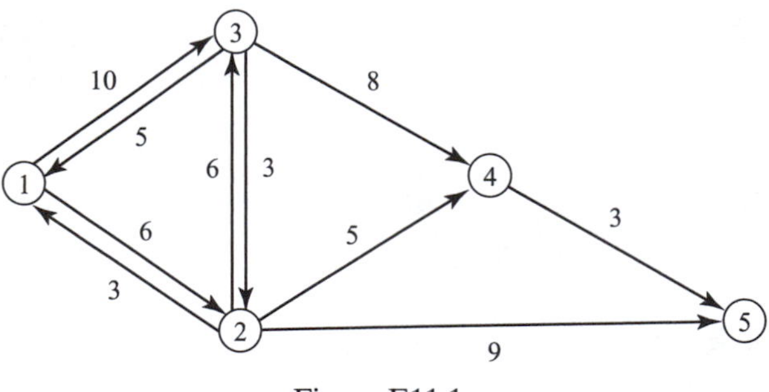

Figure E11.1

Solution

We edit the data of the problem in the computer program MAXFLO.

```
Number of nodes NN, Number of arcs NA, Source
  node NS, Sink node NE
DATA 5, 7, 1,  5
Arc node 1, node 2, Flow 1–2, Flow 2–1
DATA 1, 2, 6,  3
DATA 1, 3, 10, 5
DATA 2, 3, 6,  3
DATA 2, 4, 5,  0
DATA 3, 4, 8,  0
DATA 2, 5, 9,  0
DATA 4, 5, 3,  0
```

The solution is obtained in three iterations as

```
Maximum flow = 12
Arc        Flow
1–2        6
1–3        6
3–2        3
2–4        0
3–4        3
2–5        9
4–5        3
```

The computer output also shows the new arc capacities at each iteration.

Minimum Cost Network Flow Analysis

In the general form of network flow problem, a cost c_{ij} is associated with the flow in the arc (i, j). We present here the minimum cost network flow problem without going into the details of its solution.

$$\begin{aligned} &\text{minimize} \quad c_{ij}x_{ij} \\ &\text{subject to} \quad \sum_{j=1}^{n} x_{ij} - \sum_{j=1}^{n} x_{ji} = b_i \qquad \text{for each node } i \\ &\qquad\qquad l_{ij} \le x_{ij} \le u_{ij} \quad \text{all } i, j \quad \text{pairs defining arcs} \end{aligned} \tag{11.14}$$

b_i is positive for a source or supply point and negative for a sink or a demand point. Transportation, assignment, and transshipment points can be put in the form of Eq. (11.14). The problem (11.14) can be solved, using the conventional simplex method or using network simplex method, which uses primal dual relations. The network simplex has many steps similar to the transportation algorithm.

COMPUTER PROGRAMS

TRANSPO, HUNGARYN, TREE, TREEMIN, DIJKSTRA, MAXFLO

PROBLEMS

P11.1. Goods produced at three locations are to be supplied at four locations. The supplies and the requirements are given in the table as follows.

	Costs				Supply
	18	40	52	25	16
	55	32	48	40	6
	15	12	51	66	10
Demand	12	8	7	5	

Find the optimum solution for the problem.

P11.2. In *the minimum cost method* of finding a basic feasible solution for a transportation problem, the variable x_{ij} with the least cost is chosen and the largest possible value (the minimum of the supply-demand value) is assigned. The value is subtracted from the supply and demand values. Zero level row or column (only one if both reach zero simultaneously) is removed from further consideration. The next variable is chosen from the remaining cells. This process is continued till the last cell when the row and column values are equal. Implement this approach in the computer program for the basic feasible solution. Find a basic feasible solution for problem 11.1 using the minimum cost approach.

P11.3. *Vogel's* method for finding the basic feasible solution of a transportation takes the following steps.

Step 1. For each row, compute the row penalty given by the difference between the two smallest cost elements in the row. Also compute the column penalties.

Step 2. In the row or column that has the highest penalty, assign the largest possible units. This value is subtracted from the supply and demand values.

Step 3. Zero level row or column (only one if both reach zero, simultaneously) is removed from further consideration. If only one cell remains, assign value and stop. Otherwise go to Step 1.

Find a basic feasible solution to the transportation problem 11.1, using Vogel's method.

P11.4. Find the optimal solution to the transportation problem with 4 supply points and 5 demand stations given in the table.

	Costs					Supply
	12	20	32	22	20	5
	15	22	18	30	16	6
	10	14	12	40	14	2
	15	12	31	33	18	8
Demand	4	4	6	2	4	

(*Note*: The problem is not balanced)

P11.5. The relative times taken by four secretaries for handling four word processing jobs are given in the table.

	1	2	3	4 $\leq=$ Jobs
Secretary 1	8	24	16	11
Secretary 2	12	28	6	22
Secretary 3	28	18	17	14
Secretary 4	20	23	20	10

Assign the jobs for minimizing the total time.

P11.6. For the network shown in Fig. P11.6, find the shortest distance from node 1 to node 10.

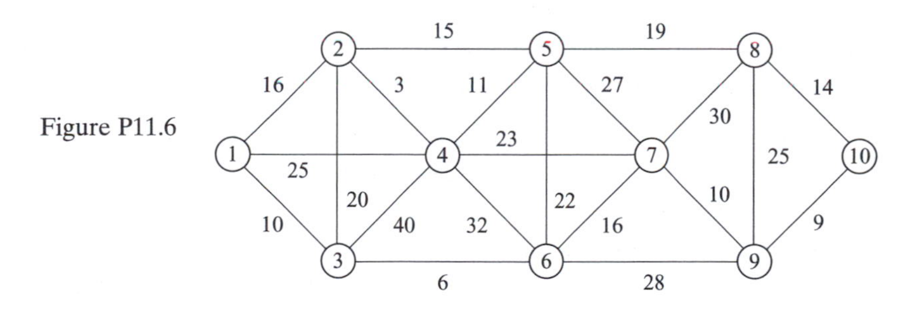

Figure P11.6

P11.7. Find the minimal spanning tree for the network given in problem 11.6.

P11.8. Find the maximal flow from node 1 to node 12 for the network shown in Fig. P11.8.

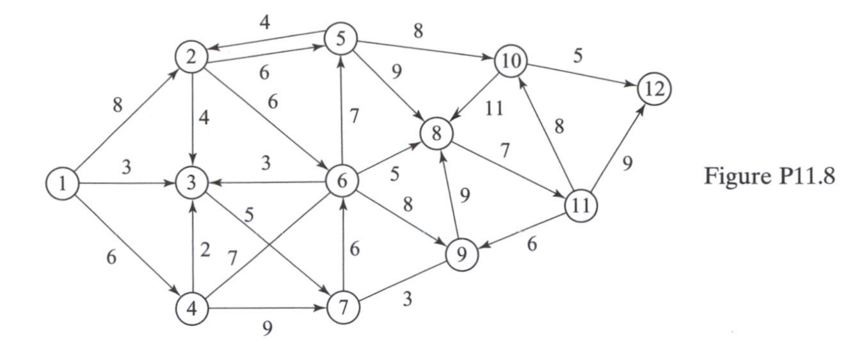

Figure P11.8

P11.9. Five workers are available to perform four jobs. Assign the jobs to minimize total time.

	1	2	3	4 ← Jobs
Worker 1	6	20	14	10
Worker 2	12	18	6	20
Worker 3	18	14	15	12
Worker 4	12	8	9	18
Worker 5	10	12	20	14

P11.10. construction firm must complete three projects during the next five months. Project 1 requires 10 labor months. Project 2 must be completed within five months and requires 12 labor months. Project 3 must be completed no later than three months and requires 14 labor months. 10 workers are available each month. Formulate a maximal flow problem to check if the projects can be completed on time. (*Hint*: Maximal flow must exceed the total labor months $10 + 12 + 14$)

P11.11. machine tools are available for producing four different components. Each component can be produced on any one of the machines and the machining times are given below:

	Component			
	1	2	3	4
Machine 1	14	12	10	11
Machine 2	10	16	13	15
Machine 3	9	15	12	16
Machine 4	8	17	14	12

Find the minimum total time and the corresponding assignments.

P11.12. A sudden strike in an airline company has left its fleet of 30 jumbo jets in three cities as follows: 8 in city A, 9 in city B, and 13 in city C. An early settlement of the dispute is expected and to start the operations afresh, the aircraft will be required at cities D, E, F, and G as follows: 3 in city D, 9 in city E, 8 in city F, and 10 in city G. The distances in hundreds of miles between the cities is given as follows:

	City D	City E	City F	City G
City A	4	20	15	12
City B	10	35	16	18
City C	8	19	9	14

Determine the allocation of planes so that the total miles is a minimum.

REFERENCES

Gillett, B., *Introduction to Operations Research: A Computer Oriented Algorithmic Approach*, McGraw-Hill, New York, 1976.

Luenberger, D.G., *Linear and Nonlinear Programming*, 2nd Edition, Addison-Wesley, Reading, MA, 1984.

Murty, K.G., *Network Programming*, Prentice-Hall, Englewood Cliffs, NJ, 1992.

12

Finite Element-Based Optimization

12.1 Introduction

Finite elements is a well known tool for analysis of structures. If we are given a structure, such as an airplane wing, a building frame, a machine component, etc., together with loads and boundary conditions, then finite elements can be used to determine the deformations and stresses in the structure. Finite elements can also be applied to analyze dynamic response, heat conduction, fluid flow, and other phenomena. Mathematically, it may be viewed as a numerical tool to analyze problems governed by partial differential equations describing the behavior of the system. Lucien Schmit in 1960 recognized the potential for combining optimization techniques in structural design [Schmit 1960]. Today, various commercial finite element packages have started to include some optimization capability in their codes. In the aircraft community, where weight and performance is a premium, special codes combining analysis and optimization have been developed. Some of the commercial codes are listed at the end of Chapter 1. The methodology discussed in this chapter assumes that a discretized finite element (or a boundary element) model exists. This is in contrast with classical approaches to optimization of structures that directly work with the differential equations governing equilibrium and aim for an analytical solution to the optimization problem.

Organization of the Chapter

The goal of this chapter, as with the rest of this text, is to give the reader theory along with computer codes to provide hands-on experience. Section 12.2 discusses derivative calculations (also called sensitivity analysis in the literature). Section 12.3 focuses on fast optimality criteria methods for singly constrained parameter problems. Handling of general constraints via nonlinear programming methods of

Chapter 5 is then discussed. Section 12.4 deals with topology optimization for maximum stiffness (or minimum compliance). The problem is shown to reduce to a finite dimensional parameter optimization problem that can be efficiently solved using the optimality criteria method presented in Section 12.3. Section 12.5 focuses on shape optimization with general constraints where again, it is shown how to reduce it to a finite dimensional parameter problem whose solution can be determined, using nonlinear programming methods.

For an introduction with finite element programs, see the text by [Chandrupatla and Belegundu 1997].

The General Problem with Implicit Dependence on Variables

The general problem being considered herein may be understood by considering the problem in stress analysis. Once this is understood, extension to other types of problems is relatively easy. We are given a body, also referred to as a continuum or structure, which occupies a region V in x, y, z space, and is bounded by a surface S. Loads, boundary conditions, and material distribution are specified. The analysis task is to determine the displacement field $\mathbf{u}(x, y, z)$ = the displacement vector at each point in V. From this, we can determine strains and stresses. The "body" referred to here typically assumes one of the following forms: a 3-D continuum (Fig. 12.1a), a 2-D plane strain approximation of a long body (Fig. 12.1b), a 2-D plane stress approximation of a thin plate (Fig. 12.1c), a shell structure (Fig. 12.1d), a frame structure (Fig. 12.1e), or a pin-jointed truss (Fig. 12.1f). Design variables $\mathbf{x} = [x_1, x_2, \ldots, x_n]$ can take on various forms as well. In the plane stress case (Fig. 12.1c) or in the shell (Fig. 12.1d), x_i can be the thicknesses of each finite element used to mesh the region. In a framed structure (Fig. 12.1e), x_i can be the cross-sectional area or moment of inertia of the element. In a pin-jointed truss (Fig. 12.1f), x_i can be the cross-sectional area. However, if our aim is to optimize the shape or topology of the body, then x_i are variables that control the shape or topology of the body.

Optimization problems involving finite elements can be generally expressed as

$$\begin{array}{lll} \text{minimize} & f(\mathbf{x}, \mathbf{U}) & \\ \text{subject to} & g_i(\mathbf{x}, \mathbf{U}) \leq 0 & i = 1, \ldots, m \\ \text{and} & h_j(\mathbf{x}, \mathbf{U}) = 0 & j = 1, \ldots, \ell \end{array} \tag{12.1}$$

where $\mathbf{U}$ is an $(ndof \times 1)$ nodal displacement vector from which the displacement field $\mathbf{u}(x, y, z)$ is readily determined. We refer to $ndof$ as the number of degrees of freedom in the structure. The relation between $\mathbf{u}$ and $\mathbf{x}$ is governed by partial

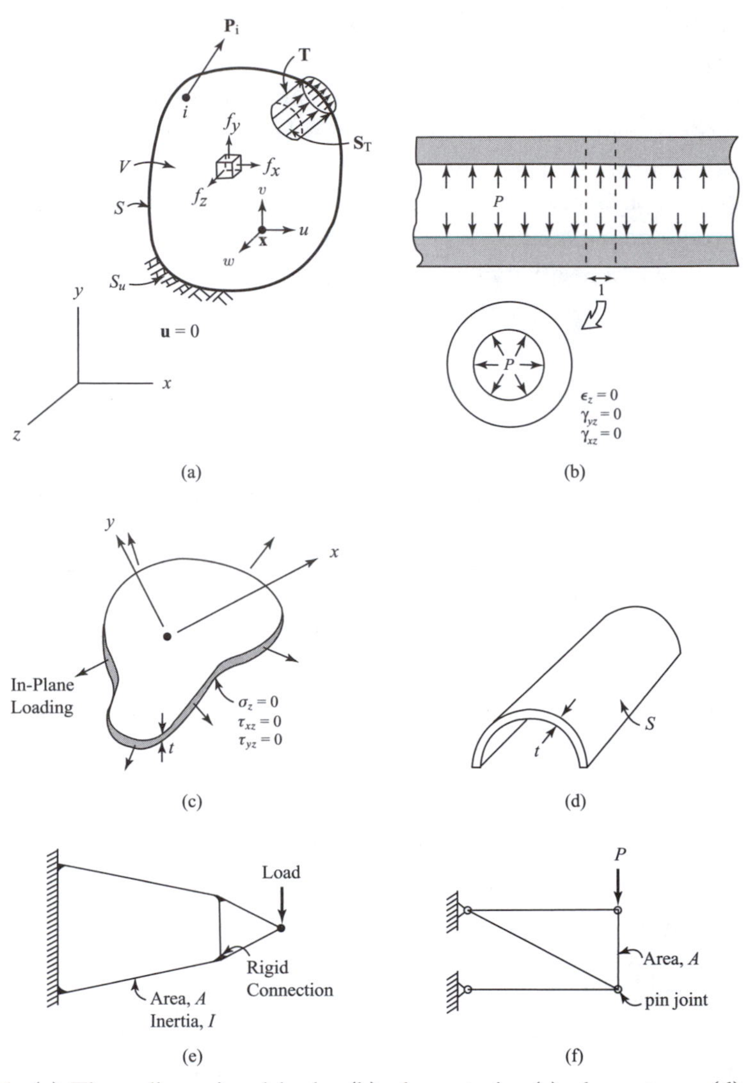

Figure 12.1. (a) Three-dimensional body, (b) plane strain, (c) plane stress, (d) thin shell, (e) frame, and (f) truss.

differential equations of equilibrium. Finite element discretization of these differential equations take the form of linear simultaneous equations for **U**:

$$\mathbf{K}(\mathbf{x})\mathbf{U} = \mathbf{F}(\mathbf{x}) \tag{12.2}$$

where **K** is a $(ndof \times ndof)$ square stiffness matrix and **F** is a $(ndof \times 1)$ load vector. It is essential to observe that the functions f, g_i, h_j are *implicit* functions of design variables **x**. They depend explicitly on **x**, and also implicitly through **U** as given in (12.2).

In (12.1), the "cost" function f is typically the weight of the structure, and g_i are constraints, reflecting limits on stress and displacement. It is important that these constraints be expressed in normalized form. Thus, a stress constraint $\sigma \leq 30000$ psi should be expressed as $g \equiv \frac{\sigma}{30000} - 1 \leq 0$. This will ensure that all constraints, when satisfied, will have values in the interval [0, −1]. The constraint thickness parameter used to determine "active sets" will also be meaningful. The normalization is particularly important if stress and displacement limits are imposed simultaneously in the optimization problem.

Often, there are multiple load cases. For instance, a structure must be designed for wind loading from different directions; a valve must be designed for different operating conditions – under pressure, under seismic loading, under no load; or an engine mount must be designed under idle shake and also due to different road inputs. We formulate such problems by increasing the number of constraints as:

$$
\begin{aligned}
\text{minimize} \quad & f(\mathbf{x}) \\
\text{subject to} \quad & g_i(\mathbf{x}) \leq 0 \quad i = 1, \ldots, m \quad \text{(load case 1)} \\
& g_i(\mathbf{x}) \leq 0 \quad i = m+1, \ldots, 2m \quad \text{(load case 2)} \\
& \ldots \\
& g_i(\mathbf{x}) \leq 0 \quad i = (NLC-1)^* \, m+1, \ldots, NLC^* \, m \\
& \qquad\qquad\quad \text{(load case } NLC\text{)}
\end{aligned}
$$

12.2 Derivative Calculations

In this section, expressions for the derivatives of implicit functions in Problem (12.1) will be derived. Design variable x_i is a parameter, such as the element thickness, area of cross-section, moment of inertia of a beam cross-section, or a parameter that controls the shape or topology. The parameter x_i affects the stiffness, mass, and load matrices. While certain dynamic response problems can be better tackled using nongradient methods, for the most part, the finite element-based optimization problem in (12.1) is best solved, using gradient methods especially since both number of degrees of freedom *ndof* and number of design variables n can be large. Further, the derivatives or sensitivity coefficients are important to design engineers in themselves, without optimization.

Efficient computation of gradients, ∇f and ∇g_i, is important owing to the size of the matrices. Specifically, consider an implicit function q, which can represent a cost or constraint function in (12.1), as

$$
q = q(\mathbf{x}, \mathbf{U}) \tag{12.3}
$$

where **U** is dependent on **x** through (12.2).

Let $\mathbf{x}^0$ be the current design. There are two principal methods of evaluating the gradient at $\mathbf{x}^0$ analytically:

$$\nabla q \equiv \frac{dq}{d\mathbf{x}} = \left[\frac{dq}{dx_1}, \frac{dq}{dx_2}, \ldots, \frac{dq}{dx_n}\right] \text{ evaluated at } \mathbf{x}^0$$

namely, the *direct method* and the *adjoint method.*

Direct Method: Differentiating (12.3) we get

$$\frac{dq}{dx_i} = \frac{\partial q}{\partial x_i} + \frac{\partial q}{\partial \mathbf{U}}\frac{d\mathbf{U}}{dx_i}, \quad i = 1, \ldots, n \tag{12.4}$$

Partial differentiation with respect to x_i is done while keeping $\mathbf{U}$ fixed and vice versa. In (12.4), the derivatives $\frac{\partial q}{\partial x_i}$ and $\frac{\partial q}{\partial \mathbf{U}}$ are readily available from the definition of q. For example, if $q = 3x_1\,U_3 + 2x_2^3 + 2.5\,U_5^2$, $n = 3$, $ndof = 6$, then

$$\frac{\partial q}{\partial \mathbf{x}} = \left[3\,U_3^0,\ 6\left(x_2^0\right)^2,\ 0\right]$$

$$\frac{\partial q}{\partial \mathbf{U}} = \left[0,\ 0,\ 3\,x_1^0,\ 0,\ 5\,U_5^0,\ 0\right]$$

where $\mathbf{U}^0$ is the displacement vector at the current point, obtained by solving $\mathbf{K}(\mathbf{x}^0)$ $\mathbf{U}^0 = \mathbf{F}(\mathbf{x}^0)$. The main hurdle in (12.4) is to evaluate the displacement derivatives $d\mathbf{U}/dx_i$. This is obtained by differentiating (12.2) with respect to x_i at $\mathbf{x}^0$:

$$\mathbf{K}(\mathbf{x}^0)\frac{d\mathbf{U}}{dx_i} + \frac{\partial \mathbf{K}}{\partial x_i}\mathbf{U}^0 = \frac{d\mathbf{F}}{dx_i}$$

which can be written as

$$\mathbf{K}(\mathbf{x}^0)\frac{d\mathbf{U}}{dx_i} = -\frac{\partial \mathbf{K}}{\partial x_i}\mathbf{U}^0 + \frac{d\mathbf{F}}{dx_i}, \quad i = 1, \ldots, n \tag{12.5}$$

The direct method consists of first solving for $d\mathbf{U}/dx_i$ from (12.5) and then substituting this into (12.4).

Regarding the computational effort needed to obtain ∇q, note that a finite element analysis has to be performed anyway, and hence (12.2) has already been solved. This means that the factorized $\mathbf{K}$ matrix is available. We can use this same factorized matrix in (12.5). Thus, the solution of (12.5) is analogous to solving a matrix system with n right-hand side "load" vectors. Thus, the operations count may be summarized as involving

$$\textit{direct method}: \quad n \text{ right-hand side vectors} \tag{12.6}$$

Each right-hand side vector involves $ndof^2$ operations (an operation being an addition together with a multiplication – the multiplication of a $1 \times ndof$ vector with a $ndof \times 1$ vector will thus involve $ndof$ operations) [Atkinson 1978]. Thus, the direct method involves $n\ ndof^2$ operations.

Adjoint Method: Here, we avoid calculation of $\mathrm{d}\mathbf{U}/\mathrm{d}x_i$ in obtaining ∇q. We multiply (12.2) by the transpose of an $(ndof \times 1)$ *adjoint vector* $\boldsymbol{\lambda}$:

$$\boldsymbol{\lambda}^{\mathrm{T}}\,\mathbf{K}(\mathbf{x})\,\mathbf{U} = \boldsymbol{\lambda}^{\mathrm{T}}\,\mathbf{F}(\mathbf{x}) \tag{12.7}$$

Differentiating (12.7) with respect to x_i at $\mathbf{x}^0$ and using the fact that we have equilibrium at $\mathbf{x}^0$ or that $\mathbf{K}(\mathbf{x}^0)\,\mathbf{U}^0 = \mathbf{F}(\mathbf{x}^0)$, we have

$$\boldsymbol{\lambda}^{\mathrm{T}}\,\mathbf{K}(\mathbf{x}^0)\,\frac{d\mathbf{U}}{dx_i} = -\boldsymbol{\lambda}^{\mathrm{T}}\,\frac{d\mathbf{K}}{dx_i}\,\mathbf{U}^0 + \boldsymbol{\lambda}^{\mathrm{T}}\,\frac{d\mathbf{F}}{dx_i} \tag{12.8}$$

From (12.8) and (12.4), we note that if we choose the as yet arbitrary vector $\boldsymbol{\lambda}$ such that

$$\boldsymbol{\lambda}^{\mathrm{T}}\,\mathbf{K} = \frac{\partial q}{\partial \mathbf{U}}$$

or,

$$\mathbf{K}(\mathbf{x}^0)\,\boldsymbol{\lambda} = \frac{\partial q}{\partial \mathbf{U}}^{\mathrm{T}} \tag{12.9}$$

then the gradient of q is given by

$$\frac{dq}{dx_i} = \frac{\partial q}{\partial x_i} - \boldsymbol{\lambda}^{\mathrm{T}}\,\frac{d\mathbf{K}}{dx_i}\,\mathbf{U}^0 + \boldsymbol{\lambda}^{\mathrm{T}}\,\frac{d\mathbf{F}}{dx_i} \qquad i = 1, \ldots, n \tag{12.10}$$

We have thus avoided computation of $\mathrm{d}\mathbf{U}/\mathrm{d}x_i$ in obtaining ∇q. In summary, we solve the adjoint equations in (12.9) for the adjoint vector $\boldsymbol{\lambda}$ and substitute for this into (12.10). Again, the stiffness matrix $\mathbf{K}(\mathbf{x}^0)$ has already been factorized while solving (12.2). Thus, only a single right-hand side vector is involved in (12.9). Of course, if there are several constraints functions $\mathbf{q} = [q_1, q_2, \ldots, q_p]$ whose derivatives are needed, then we have

$$\textit{adjoint method}: \quad p \text{ right-hand side vectors} \tag{12.11}$$

or $p\ ndof^2$ operations. Evidently, if $p < n$, then the adjoint method is more efficient than the direct method. This is often the case as the number of active constraints is less than the number of design variables, at least near the optimum.

Example 12.1

Consider the function $f = U_3$ where $\mathbf{U}$ is determined from $\mathbf{K}(\mathbf{x})\,\mathbf{U} = \mathbf{F}$ as:

$$\begin{bmatrix} 5x_1 & -5x_1 & 0 \\ -5x_1 & 5x_1 + 10x_2 + 5x_3 & -5x_3 \\ 0 & -5x_3 & 5x_3 + 5x_4 \end{bmatrix} \begin{Bmatrix} U_1 \\ U_2 \\ U_3 \end{Bmatrix} = \begin{Bmatrix} 20 \\ 0 \\ 15 \end{Bmatrix}$$

Given $\mathbf{x}^0 = [1.0, 1.0, 1.0, 1.0]^{\mathrm{T}}$, determine the gradient $df/d\mathbf{x}$ using: (i) the direct method, and (ii) the adjoint method.

In either method, the first step is analysis: we solve $\mathbf{K}(\mathbf{x}^0)\ \mathbf{U}^0 = \mathbf{F}(\mathbf{x}^0)$ to obtain $\mathbf{U}^0 = [6.2, 2.2, 2.6]^{\mathrm{T}}$. The derivatives of $\mathbf{K}$ with respect to x_i are:

$$\frac{\partial \mathbf{K}}{\partial x_1} = \begin{bmatrix} 5 & -5 & 0 \\ -5 & 5 & 0 \\ 0 & 0 & 0 \end{bmatrix}, \quad \frac{\partial \mathbf{K}}{\partial x_2} = \begin{bmatrix} 0 & 0 & 0 \\ 0 & 10 & 0 \\ 0 & 0 & 0 \end{bmatrix},$$

$$\frac{\partial \mathbf{K}}{\partial x_3} = \begin{bmatrix} 0 & 0 & 0 \\ 0 & 5 & -5 \\ 0 & -5 & 5 \end{bmatrix}, \quad \frac{\partial \mathbf{K}}{\partial x_4} = \begin{bmatrix} 0 & 0 & 0 \\ 0 & 0 & 0 \\ 0 & 0 & 5 \end{bmatrix}$$

Multiplying each of the preceding matrices with $\mathbf{U}^0$ gives

$$\frac{\partial(\mathbf{K}\mathbf{U}^0)}{\partial x_1} = \begin{pmatrix} 20 \\ -20 \\ 0 \end{pmatrix}, \quad \frac{\partial(\mathbf{K}\mathbf{U}^0)}{\partial x_2} = \begin{pmatrix} 0 \\ -22 \\ 0 \end{pmatrix},$$

$$\frac{\partial(\mathbf{K}\mathbf{U}^0)}{\partial x_3} = \begin{pmatrix} -2 \\ 2 \\ 0 \end{pmatrix}, \quad \frac{\partial(\mathbf{K}\mathbf{U}^0)}{\partial x_4} = \begin{pmatrix} 0 \\ 0 \\ 13 \end{pmatrix}$$

The preceding four vectors form the right-hand side of (12.5) for the direct method. Solving (12.5), we get

$$\frac{d\mathbf{U}}{dx_1} = \begin{pmatrix} 4 \\ 0 \\ 0 \end{pmatrix}, \quad \frac{d\mathbf{U}}{dx_2} = \begin{pmatrix} -1.76 \\ -1.76 \\ -0.88 \end{pmatrix}, \quad \frac{d\mathbf{U}}{dx_3} = \begin{pmatrix} -0.4 \\ 0 \\ 0 \end{pmatrix}, \quad \frac{d\mathbf{U}}{dx_4} = \begin{pmatrix} 0.52 \\ 0.52 \\ 1.56 \end{pmatrix}$$

Thus, since $f = U_3$, $\nabla f = (0.0, -0.88, 0.0, 1.56)$.

In the adjoint method, we would solve (12.9). Here, the right-hand side of (12.9) is $\partial f/\partial \mathbf{U} = (0, 0, 1)$. Solving, we obtain the adjoint vector $\boldsymbol{\lambda} = (0.04, 0.04, 0.12)^{\mathrm{T}}$. Equation (12.10) gives

$$\frac{df}{dx_i} = -\boldsymbol{\lambda}^{\mathrm{T}} \frac{d\mathbf{K}}{dx_i}\mathbf{U}^0 = -\boldsymbol{\lambda}^{\mathrm{T}} \frac{d(\mathbf{K}\mathbf{U}^0)}{dx_i}$$

which gives $\nabla f = (0.0, -0.88, 0.0, 1.56)$, which is the same result as that obtained using the direct method.

Example 12.2

Determine an expression for the adjoint vector corresponding to $q = \mathbf{F}^{\mathrm{T}}\mathbf{U} =$ "compliance." We have from (12.9), $\mathbf{K}(\mathbf{x}^0)\ \boldsymbol{\lambda} = \frac{\partial q}{\partial \mathbf{U}}^{\mathrm{T}} = \mathbf{F}$. However, this is identical to the equilibrium equations in (12.2) and, hence, $\boldsymbol{\lambda} = \mathbf{U}^0 =$ displacement vector at the current point. This result will be used in Sections 12.3 and 12.4.

Table 12.1. *Operations count for computing ∇q, q being a $p \times 1$ vector of functions.*

Direct method	Adjoint method	Forward difference
$n\, ndof^2$	$(p)\, ndof^2$	$\sim (n)\, ndof^3$

Divided Differences to Evaluate ∇q

Considering only ease of implementation in a computer program, the simplest way to obtain derivatives is to use the forward difference formula

$$\frac{dq}{dx_i} \approx \frac{q(x_1^0,\, x_2^0, \ldots,\, x_i^0 + \varepsilon, \ldots,\, x_n^0) - q(\mathbf{x}^0)}{\varepsilon}, \quad i = 1, 2, \ldots, n \tag{12.12}$$

where ε may be chosen equal to 1% of x_i ($\varepsilon = 0.01\, x_i$), but not less than a minimum value of, say, 0.001. However, (12.12) is only advocated for small-sized problems. This is because it entails solution of (12.2) with n different stiffness matrices, each matrix being of the form $\mathbf{K}(x_1^0,\, x_2^0, \ldots,\, x_i^0 + \varepsilon,\, \ldots, x_n^0)$. This is expensive since factorizing each $\mathbf{K}$ involves on the order of $ndof^3$ operations. For large $ndof$, the operations far exceed the analytical approaches presented earlier. This technique is useful for "debugging" code with analytical expressions. Table 12.1 summarizes the operations count for the three methods discussed so far.

Derivatives of Penalty Functions

The T-function that we minimize sequentially in penalty and augmented Lagrangian methods is an example of a composite function – a function that is made up from individual functions. Consider a function of the form

$$T = T(q_1(\mathbf{x}, \mathbf{U}), q_2(\mathbf{x}, \mathbf{U}), \ldots\, q_p(\mathbf{x}, \mathbf{U})) \tag{12.13}$$

where q_i are implicit functions. Now, the adjoint method can be used to efficiently compute ∇T without having to compute each individual ∇q_i [Belegundu and Arora 1981]. That is, it is not necessary to compute ∇T as $\nabla T = \sum_{i=1}^{p} \partial T/\partial q_i \bullet \nabla q_i$, with ∇q_i evaluated using (12.10). Instead, we may think of T as *a single function* and directly use the adjoint method on it. Thus, we find an adjoint vector from

$$\mathbf{K}(\mathbf{x}^0)\, \boldsymbol{\lambda} = \frac{\partial T}{\partial \mathbf{U}}^{\mathrm{T}} \tag{12.14}$$

and then substitute $\boldsymbol{\lambda}$ into (12.10), with T replacing q. This is computationally less expensive as we are solving only for one adjoint vector involving $ndof^2$ operations as opposed to $p\; ndof^2$ operations if we were to evaluate the gradient of T from individual ∇q_i.

Physical Significance of the Adjoint Vector

Consider a general function $q(\mathbf{U})$. The dependence of q on $\mathbf{x}$ is not pertinent to this discussion. As was shown by Belegundu [1986] and Belegundu and Arora [1981], the adjoint vector $\boldsymbol{\lambda}$ associated with the given $q(\mathbf{U})$ has an interesting meaning. To see this, we write $\mathbf{U} = \mathbf{K}^{-1}\mathbf{F}$ from (12.2). Thus, $\frac{dq}{d\mathbf{F}} = \frac{dq}{d\mathbf{U}}\frac{d\mathbf{U}}{d\mathbf{F}} = \frac{dq}{d\mathbf{U}}\,\mathbf{K}^{-1}$ which yields $\mathbf{K}\frac{dq}{d\mathbf{F}}^{\mathrm{T}} = \frac{\partial q}{\partial \mathbf{U}}^{\mathrm{T}}$. Comparing this to the adjoint equation in (12.9), we arrive at the result

$$\boldsymbol{\lambda} = \frac{dq}{d\mathbf{F}}^{\mathrm{T}} \tag{12.15}$$

Equation (12.15) tells us that the adjoint vector represents the derivative (or sensitivity) of q with respect to the forcing function. Furthermore, if q is a linear function of $\mathbf{F}$ and $q = 0$ when $\mathbf{F} = 0$ (e.g., when q = stress in an element or a reaction force at a support), then we obtain from (12.15) the result λ_i = value of q due to a unit load applied along degree of freedom i In view of (12.16), we say that λ_i is an *influence coefficient* associated with q.

Sensitivity Analysis and Derivatives

Derivative calculations or *sensitivity analysis* as discussed earlier can be directly used to make design improvements, without iterative optimization. The derivative $\partial f/\partial x_i$ is the approximate change in f due to a unit change in x_i and is valid in the local region where the slope was calculated. Similarly, $\frac{1}{f}\,\partial f/\partial x_i = \partial(\log f)/\partial x_i$ is the approximate percentage change in f due to a unit change in x_i. As an application, consider the sound pressure P at the driver's ear generated by a vibrating floor of a car. The floor is modeled using finite elements with thicknesses t_i. Then, a contour plot of the sensitivities $\partial P/\partial t_i$ will identify the critical zones in the floor panel where thickness adjustments can reduce sound pressure.

Engineers are often interested in changes due to substantial changes in the variables – for example, in the choice of different density foam materials or different types of composites (e.g., Kevlar vs Graphite–Spectra). Use of orthogonal arrays with statistical techniques involving analysis of means and analysis of variance (ANOM, ANOVA) can provide these more global sensitivity measures.

12.3 Sizing (i.e., Parameter) Optimization via Optimality Criteria and Nonlinear Programming Methods

In Chapter 6, Example 6.6, we showed that the minimum weight design of statically determinate trusses subject to a displacement limit can be transformed using duality into a single-variable problem that can be solved efficiently, regardless of the number of design variables in primal space. Here, we generalize this discussion to statically *indeterminate* trusses, and subsequently use this for topology optimization. The problem under consideration involves only the minimization of volume (or weight) subject to a limit on the compliance. More general constraints can be handled using NLP methods as discussed subsequently.

Optimality Criteria Methods for Singly Constrained Problems

The problem being considered is of the form

$$\begin{aligned} \text{minimize} \quad & f = \mathbf{F}^{\mathrm{T}} \mathbf{U} \\ \text{subject to} \quad & \sum L_i x_i \leq V^{\mathrm{U}} \\ & x_i \geq x^{\mathrm{L}} \end{aligned} \tag{12.16}$$

where V^{U} = upper limit on volume. In this method, $x_i{}^{\mathrm{L}}$ is a small lower bound, say 10^{-6}. The volume constraint will be expressed in normalized form as $g \equiv \frac{\sum L_i x_i}{V^{\mathrm{U}}} - 1 \leq 0$. Such problems occur frequently since we are often interested in finding the stiffest structure for a given volume of material. As discussed in Chapter 8, a given value V^{U} will generate a specific point on the Pareto curve. Note that $\mathbf{U}$ is obtained from (12.2). Employing KKT optimality conditions, we have to minimize the Lagrangian L (not to be confused with L_i, which represents element length)

$$\underset{\mathbf{x}^{\mathrm{L}} \leq \mathbf{x}}{\text{minimize}} \; L = \mathbf{F}^{\mathrm{T}} \mathbf{U} + \mu \left(\frac{\sum L_i x_i}{V^{\mathrm{U}}} - 1 \right) \tag{12.17}$$

where the summation is over all the finite elements in the model. Noting that the adjoint vector associated with the compliance $\mathbf{F}^{\mathrm{T}}\, \mathbf{U}$ is $\boldsymbol{\lambda} = \mathbf{U}$ (see Example 12.2), we have from (12.10) assuming that external loads are independent of $\mathbf{x}$,

$$\frac{d(\mathbf{F}^{\mathrm{T}} \mathbf{U})}{dx_i} = -\,\mathbf{U}^{\mathrm{T}} \frac{d\mathbf{K}}{dx_i} \mathbf{U} = -\,\frac{\mathbf{q}^{\mathrm{T}} \mathbf{k}\, \mathbf{q}}{x_i} = -\,\frac{\varepsilon_i}{x_i} \tag{12.18}$$

where use has been made of the fact that for trusses and certain types of structures, the stiffness matrix is linearly dependent on x_i and, hence, $\frac{d\mathbf{K}}{dx_i} = \frac{\hat{\mathbf{K}}}{x_i}$ where $\hat{\mathbf{K}} = \mathbf{k}$ = element stiffness matrix with zeroes in other locations. Above, ε_i = strain energy in the ith element (ignoring a factor of ½). A truss finite element code is used to compute ε_i. Thus, the gradient of the Lagrangian is given by

$$dL/dx_j = -\,\varepsilon_j / x_j + \mu L_j / V^{\mathrm{U}}$$

Setting this equal to zero yields $\frac{\varepsilon_i}{x_i\, L_i} = \frac{\mu}{V^{\mathrm{U}}}$ = constant. That is, at optimum, the strain energy per unit volume or strain energy density is the same for all elements whose areas are not at their lower bounds. This condition can be expressed as

$$\varphi_i \equiv \left(\frac{V^{\mathrm{U}} \varepsilon_i}{L_i x_i \mu} \right) = 1 \tag{12.19}$$

At this stage, it is assumed that the dependence of ε_i on $\mathbf{x}$ is weak and can be ignored for the current iteration. Then, Eq. (12.19) serves as a basis for a recurrence relation to improve the current estimate of the ith variable. This is done by multiplying both sides of the equation $\varphi_i = 1$ by x_i^p and then taking the pth root to obtain a "re-sizing" scheme

$$x_i^{(v+1)} = x_i^{(v)} \left[\varphi_i^{(v)} \right]^{1/p} \tag{12.20}$$

where v is a iteration index. We get

$$x_i^{(v+1)} = \left(\frac{V^{\mathrm{U}} \varepsilon_i x_i^{(p-1)}}{\mu L_i} \right)^{1/p} \tag{12.21}$$

At most one or two iterations are needed in the preceding equation. Moreover, ε_i are not updated and, thus, expensive structural finite element analysis is not involved in (12.21).

Equation (12.19) defines $\mathbf{x}(\mu)$. The Lagrange multiplier μ is updated as in Example 6.6 in Chapter 6. Thus, linearizing $g(\mathbf{x}(\mu)) = 0$, we have

$$\sum_i \left(\frac{dg}{dx_i} \frac{dx_i}{d\mu} \right) [\mu^{(v+1)} - \mu^{(v)}] = g - g_{(v)}$$

which reduces to

$$\mu^{(v+1)} = \mu^{(v)} + \frac{g_{(v)}}{Q}, \quad \text{where} \quad Q = \sum_{i \in I} \frac{\varepsilon_i}{\mu^2} \tag{12.22}$$

where I is an index set corresponding to $x_i > x^{\mathrm{L}}$. Once μ is updated, a finite element analysis of the structure is carried out to obtain ε_i. To start the algorithm, a least-squares estimate of μ is obtainable from Eq. (12.19) as

$$\mu = \frac{V^{\mathrm{U}} \sum L_i x_i \varepsilon_i}{\sum (L_i x_i)^2} \tag{12.23}$$

The procedure based on the preceding equations has been implemented in Program TRUSS_OC. The reader can follow the details. The program given in the following can be readily generalized to problems where

$$\text{weight} \propto x_i \text{ and stiffness} \propto x_i^r, \quad \text{where} \quad r \geq 1 \tag{12.24}$$

Of course, for trusses, we have $r = 1$. The case $r = 3$ is used in the following in the context of topology optimization. In the code, move limits on μ are introduced in Eq. (12.22).

Example 12.3
Consider the truss in Fig. E12.3.

Data
Horizontal or vertical distance between any two nodes = 360 in, $E = 10^7$, $V^U = 82{,}947$, $x^L = 1e-6$, $x^U = 100$, $x_i^0 = 10$, number of iterations = 25, $P = 10^5$N.

Solution
Program TRUSS_OC gives compliance = 1e5, $\mu^* = 1e5$, and $\mathbf{x}^* = [57.6, 0.36, 28.8, 28.8, 10^{-6}, 0.36, 0.05, 40.7, 40.7, 0.05]$ in.^{2} From the compliance, we have $\delta = 1$ in.

The optimum areas are roughly indicated in the figure and, in fact, reveal an optimum topology.

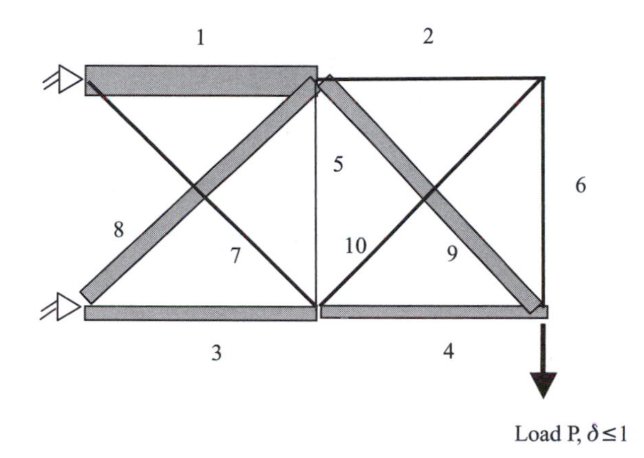

Figure E12.3

It should be noted that optimality criteria methods strive for a near-optimum solution in only a few iterations, even for a large number of variables, and work best for a class of problems: minimization of weight subject to a compliance constraint. Inclusion of stress constraints, etc. slows down the method. However, these methods continue to enjoy popularity and have been used to size aircraft wings over two decades. Optimality criteria methods work well even for large n.

Nonlinear Programming Methods for General Constraints

Consider the general problem in (12.1) with constraints on displacement, stress, resonance frequencies, etc. This can be solved by direct interfacing of a nonlinear programming (NLP) algorithm and a finite element (FE) code. Various NLP codes were discussed in Chapters 5–7 including ZOUTEN, GRG, SQP, PENALTY, Matlab "fmincon" and Excel SOLVER. This interfacing of two different codes is not difficult, as long as duplicate names for variables are avoided. For instance, the NLP code may use **x** for variables and the FE code may use **x** for nodal coordinates. However, care must be taken on computational effort since each function call is an FE analysis. For large size problems, approximate line search techniques within the NLP methods are necessary. For computationally expensive simulations (e.g., crash analysis) with a few variables (say, <25), response surface-based optimization methods have proved successful. We resolve the truss example given earlier, using Matlab fmincon.

Example 12.4

Consider the problem in Example 12.3 but with the objective and constraint interchange. The main routine and subroutines for using Matlab *fmincon* are listed in the following (FE code is suppressed). Analytical gradients are provided to reduce computing time (see also notes on using Matlab in Chapters 1 and 3).

```
function [] = trussopt()
close all; clear all;
global AA BB CC  %FEA variables
global FEAcount

FEAcount=0; XLB(1:NE) = 1e-6*ones(10,1)'; XUB(1:NE)
                        = 100*ones(10,1)';
XDV(1:NE) = 10*ones(10,1)'; A=[];B=[];Aeq=[];Beq=[];
options=optimset('GradObj','on','GradConstr','on');
[XDV,FOPT,EXITFLAG,OUTPUT,LAMBDA] =
fmincon('GETFUN',XDV,A,B,Aeq,Beq,XLB,XUB,...
                                        'NONLCON','OPTIONS')
LAMBDA.ineqnonlin
FEAcount
```

```
function [f,Df] = GETFUN(XDV)
global AA  BB  CC
AREA=XDV;
f=0;
for i=1:NE
  f = f + L(i)*AREA(i);
  Df(i) = L(i);
end
```

```
function [c, ceq, Gc, Gceq] = NONLCON(XDV)
global AA  BB  CC
global FEAcount
FEAcount = FEAcount+1; AREA=XDV; ceq=[]; Gceq=[];
%FEA
(calls to FE subroutines)
%%%Constraint Evaluation
c(1) = F1'*F/COMPLU - 1;
%Gradient Eval --
%column j stores the gradient of constr#j, dimension (nx1)
for i=1:NE
    Gc(i,1)= -EE(i)/AREA(i)/COMPLU;
end
```

Results

Volume = 8.2947e + 004, μ = 8.2947e + 004,

$$\mathbf{x} = [57.6039,\ 10^{-6},\ 28.8,\ 28.8,\ 10^{-6},\ 10^{-6},\ 10^{-6},\ 40.7,\ 40.7,\ 10^{-6}]\,\text{in}^2.$$

`#function calls = #finite element analyses = 385.`

The optimality criteria solution earlier arrived at a near-optimum solution with 25 FE calls compared to 385. As noted earlier, NLP codes can handle more general constraints but take more computing time, even with analytical gradients. Approximate line search strategies need to be integrated into the NLP codes.

12.4 Topology Optimization of Continuum Structures

Topology optimization has more to do with the distribution of material – creation of holes, ribs or stiffeners, creation/deletion of elements, etc. in the structure. By contrast, in shape optimization of continuua, the *genus* of the body is unchanged[1]. Ideally, shape, sizing, and topology optimization should be integrated. However, such a capability for general problems is an area of current research. In this section, we focus on a specific approach and on how to develop a computer code for problem solution.

Topology via Sizing Optimization of Trusses

We saw in Example 12.3 that optimizing the truss cross-sectional areas for minimizing compliance indicated an optimum topology. It has been shown to be an effective

[1] The genus of a body is the number of cuts necessary to separate the body. Thus, a sphere has genus = 1 while a donut has genus = 2.

approach for the synthesis of compliant mechanisms. This is because trusses very often capture the behavior of continuua. Truss elements are introduced so that all possible element-joint connectivities are exhausted. Member areas are treated as design variables. Areas that are at their lower bounds are deleted, thus yielding an optimum topology.

Continuum Structures: Density Approach with Optimality Criteria

We address the general problem in two- or three- dimensional elasticity: given loads and boundary conditions, determine the material distribution which minimizes compliance for a given weight. Compliance is the product of load and displacement, $\mathbf{F}^{\mathrm{T}}\mathbf{U}$. Minimizing compliance is akin to maximizing stiffness. The landmark paper by Bendsoe and Kikuchi [1988] showed how this problem could be solved. Ultimately, the problem is reduced to a parameter optimization problem that can be solved by gradient methods or by optimality criteria methods discussed previously.

The solution approach begins by defining a "material model." Assume that the density of an element in the finite element mesh is proportional to x_i. Evidently, as x_i takes values close to zero, the element has "no material" and as x_i approaches unity, the element is "solid." However, as the material is being made less dense, there must be a reduction in its stiffness, specifically in its Young's modulus. Mlejnek and Schirrmacher [1993] present a simple material model that assumes the density and Young's modulus are functions of the design variable as

$$\begin{array}{lcl} \text{element weight} & \propto & x_i \\ \text{element Young's modulus} & \propto & x_i^{\,r} \end{array} \tag{12.25}$$

where $r = 3$, typically. Use of $r > 1$ results in a sharper division between low and high density values at the optimum. Each finite element (or a subgroup) is parameterized with just one variable, x_i.

More general material models assume a porous microstructure governed by not just a single variable x_i but by two or more variables. These variables define dimensions of a microstructure and homogenization theory is used to extract the functional relations for the orthotropic material properties, namely, E_1, E_2, G_{12}, etc. For certain microstructures, homogenization theory leads to analytical expressions for the material constants as opposed to having to obtain these numerically [Bendsoe and Sigmund 2002]. The optimality criteria-based method and computer implementation presented here can readily be generalized to accommodate these more refined material models.

In the formulation given in the following, if mass fraction M_f (= 0.25 or 25%, say), then the problem posed in words is: determine the stiffest structure with 25%

of the weight of the original ("solid") structure. Maximizing stiffness is achieved by minimizing compliance. Bounds are $0 \le x_i \le 1$. NE equals the number of finite elements in the model. The problem is

$$\begin{aligned} &\text{minimize} \quad \mathbf{F}^{\mathrm{T}}\mathbf{U} \\ &\text{subject to} \quad \frac{\sum_{i=1}^{NE} x_i}{NE} \le M_f \end{aligned} \tag{12.26}$$

The optimality criteria method in Section 12.3 reduces to:

$$\varphi_i \equiv \left(\frac{r\varepsilon_i NE}{\mu x_i}\right) = 1$$

where a finite element code with quadrilateral elements is used to compute ε_i. The resizing scheme is

$$x_i^{(v+1)} = \left(\frac{r\varepsilon_i NE x_i^{(p-1)}}{\mu}\right)^{1/p} \tag{12.27}$$

and the Lagrange multiplier update is

$$\mu^{(v+1)} = \mu^{(v)} + \frac{g_{(v)}}{Q}, \quad \text{where} \quad Q = \sum_{i \in I} \frac{r\varepsilon_i}{\mu^2} \tag{12.28}$$

At the start, a least squares estimate is used as

$$\mu_{L-S} = \frac{r\, NE \sum x_i \varepsilon_i}{\sum x_i^2} \tag{12.29}$$

Program TOPOL has been given in the CD-ROM.

Example 12.5

A plane stress rectangular region is fixed at two points on the left edge and subject to a point load at the center on the right edge as shown. If each element has maximum density and strength, the weight of the structure is 8 units. With $M_f = 0.25$, the stiffest structure with weight equal to 2 units is desired. Program TOPOL with $p = 4$ yields a topology as shown in Fig. E12.5a, with dark elements having density equal to 0.5, and light elements having zero density.

Next, Program TOPOL is used on a (2L × L) rectangular cantilever plate with a downward tip load. With inputs $p = 4$, $M_f = 0.25$, iterations = 25, colors for contour plot ("NCL") = 4, and a (20 × 10) FE mesh, the optimum topology obtained is shown in Fig. E12.5b. Final value of the constraint equals −0.004. Interestingly, this topology is also visible in the truss sizing solution in Fig. E12.3. The topology can also be used to strengthen sheet metal structures.

Finally, on a (40 × 20) mesh, the position of the load P has been shifted vertically upward to yield another topology as shown in Fig. E12.5c.

As an exercise, the reader may add a "filter" subroutine to the code to produce a crisper image, such as based on averaging sensitivity coefficients surrounding each element within a radius R. This also alleviates the problem of 'checker-boarding'.

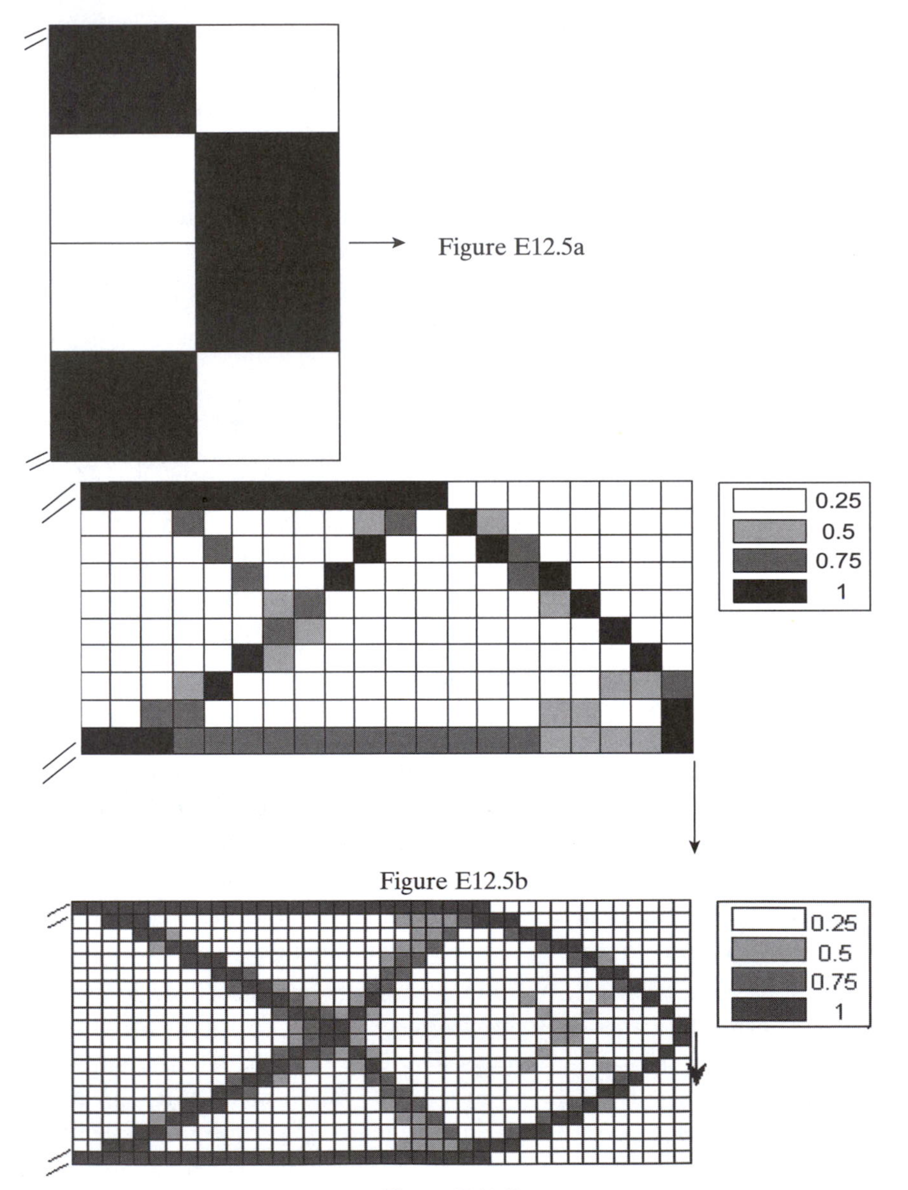

Figure E12.5a

Figure E12.5b

Figure E12.5c

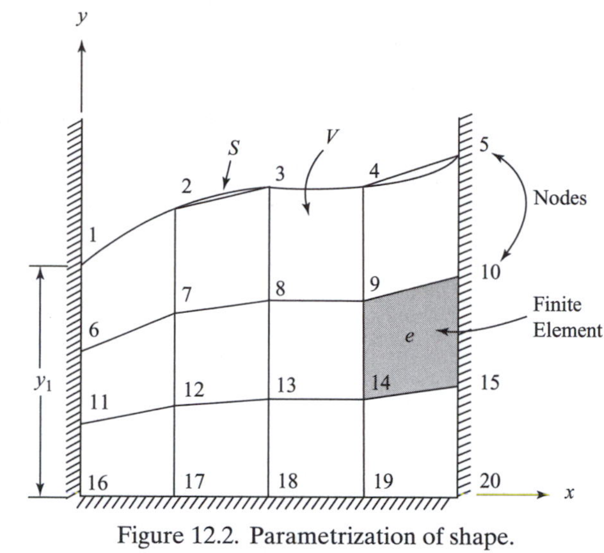

Figure 12.2. Parametrization of shape.

12.5 Shape Optimization

Shape optimization relates to determining the outline of a body, shape and/or size of a hole, fillet dimensions, etc. Here, the X, Y, Z-coordinates of the nodes or grid points in the finite element model are changed iteratively. The main concept involved in shape optimization is *mesh parametrization*. That is, how do we relate the X, Y, Z-coordinates of the grid points to a finite number of parameters x_i, $i = 1, \ldots, n$. This is done by defining *velocity fields* as explained below.

Concept of Velocity Fields

To illustrate the concept, consider the body in Fig. 12.2.

The body, denoted by V, is bounded by three straight sides and one curved top surface. The top surface is denoted by S. In the figure, we also show a finite element mesh consisting of 4-noded quadrilateral elements. It is desired to parameterize S. That is, we must describe changes in the shape of S with a finite number of design variables. Assume that the top right corner of S is fixed by the design engineer ("an user-defined" restriction). Let (X_i, Y_i) represent the coordinates of node i, and $(\Delta X_i, \Delta Y_i)$ be changes in the location of i. Note that $i = 1, \ldots, 20$ as there are 20 nodes or "grid points" in the model. We will now present a simple way to parameterize S. Subsequently, a more general technique will be presented. Here, we will treat *changes* in the y-coordinates of nodes 1, 2, 3 and 4 as our four design variables. Thus,

$$[x_1, x_2, x_3, x_4] \equiv [\Delta Y_1, \Delta Y_2, \Delta Y_3, \Delta Y_4]$$

We have to relate changes in Y_1 to changes in the Y-coordinates of nodes 6 and 11 as well. Otherwise, if S were to move down, the elements on the top will get squeezed and the mesh will very quickly get distorted. Adopting the simple scheme of maintaining equal-spacing of nodes along each vertical line, we have

$$\Delta Y_6 = \tfrac{2}{3}\Delta Y_1 = \tfrac{2}{3}x_1, \quad \Delta Y_{11} = \tfrac{1}{3}\Delta Y_1 = \tfrac{1}{3}x_1, \Delta Y_{16} = 0$$

Let $\mathbf{G}$ represent a (40×1) vector consisting of the X, Y-coordinates as

$$\mathbf{G} = [X_1,\ Y_1,\ X_2,\ Y_2, \ldots,\ X_{20},\ Y_{20}]$$

and $\mathbf{q}^1$ represent the changes in the coordinates due to the change x_1 in Y_1. We see that $\mathbf{q}^1$ is given by

$$\mathbf{q}^1 = [0, 1, 0, 0, 0, 0, 0, 0, 0, 0, 0, 2/3, 0, 0, 0, 0, 0, 0, 0, 0, 0, 1/3, 0, \ldots, 0]^{\mathrm{T}}$$

Thus, change in $\mathbf{G}$ due to x_1 is given by

$$\Delta\mathbf{G} = x_1\mathbf{q}^1$$

or

$$\mathbf{G}(x_1) = \mathbf{G}^{\mathrm{old}} + x_1\mathbf{q}^1$$

where $\mathbf{G}^{\mathrm{old}}$ is the current grid point vector (with no change in x_1). Similarly, we can define $\mathbf{q}^2$ as a (40×1) vector with all zeroes except for $q_4^2 = 1, q_{14}^2 = 2/3$, and $q_{24}^2 = 1/3$. Again, the nonzero components of $\mathbf{q}^3$ are $q_6^3 = 1, q_{16}^3 = 2/3, q_{26}^3 = 1/3$ and of $\mathbf{q}^4$ are $q_8^4 = 1, q_{18}^4 = 2/3, q_{28}^2 = 1/3$. We have

$$\mathbf{G}(\mathbf{x}) = \mathbf{G}^{\mathrm{old}} + \sum_{i=1}^{n} x_i\ \mathbf{q}^i \tag{12.30}$$

where $n = 4$. The stiffness matrix is a function of design variables $\mathbf{x}$ as $\mathbf{K}(\mathbf{G}(\mathbf{x}))$. We refer to $\mathbf{q}^i$ as *velocity field* vectors. The name comes from continuum mechanics where we may think of (12.30) as describing motion of the continuum with x_i as a time parameter with

$$\left.\frac{\partial\mathbf{G}}{\partial x_i}\right|_{x_i=0} = \mathbf{q}^i \tag{12.31}$$

Program Implementation

The optimization problem is exactly in the form (12.1)–(12.2) except that the design variables $\mathbf{x}$ represent *changes* in the coordinates are, therefore, reset to zero at the start of each iteration in the optimization loop.

Subroutine SHAPEOPT has implemented the procedure with 4-noded quadrilateral plane stress/plane strain elements (two-dimensional problems). To understand the code, assume that we are using the subroutine SHAPEOPT with the feasible directions code ZOUTEN that was discussed in Chapter 5. At a typical iteration k, gradients ∇f and ∇g_i, $i \in I_\varepsilon$ are computed. From these gradients, a search direction $\mathbf{d}^k$ is determined and a line search is performed, which involves evaluation of f and all g_i for various values of the step size parameter α. At each α, we compute $\mathbf{x}(\alpha) = \mathbf{x}^k + \alpha\, \mathbf{d}^k = \alpha\, \mathbf{d}^k$ (since $\mathbf{x}^k = \mathbf{0}$), evaluate $\mathbf{G}(\mathbf{x})$ and, hence, the cost and constraint functions. At the end of the line search, the design is updated as $\mathbf{x}^{k+1} = \alpha_k\, \mathbf{d}^k$, the grid is updated as $\mathbf{G}^{\text{new}} = \mathbf{G}^{\text{old}} + \sum_{i=1}^{N} x_i^k \mathbf{q}^i$, $\mathbf{x}^{k+1}$ is reset to zero, $\mathbf{G}^{\text{old}}$ is set equal to $\mathbf{G}^{\text{new}}$, and a new iteration is begun. The calculations performed in subroutine SHAPEOPT as a result of calls from the parent program ZOUTEN, for function and gradient evaluations are given as follows.

Initial Call

(i) set $\mathbf{x}^0 = \mathbf{0}$
(ii) define lower and upper bounds x_i^{L}, x_i^{U} as, say, [−100, 100], respectively (this aspect will be discussed subsequently).
(iii) read file containing velocity field vectors, $\mathbf{q}^i$, $i = 1, \ldots, n$.
(iv) normalize the velocity field vectors so that the largest component in $\mathbf{Q} = [\mathbf{q}^1|\mathbf{q}^2|\cdots|\mathbf{q}^n]$ is, say, 2% of the structural dimension.
(v) Read in finite element data including the current grid, $\mathbf{G}^{\text{old}}$.

Function Evaluation (Given $\mathbf{x}^k$)

(i) Update the grid by: $\mathbf{G}(\mathbf{x}^k) = \mathbf{G}^{\text{old}} + \sum_{i=1}^{N} x_i^k \mathbf{q}^i$
(ii) Form the matrices **K** and **F** using the new grid, solve (12.2), and evaluate f, $\{g_i\}$, $\{h_j\}$. There are no restrictions on the types of cost and constraint functions.

Gradient Evaluation (Given $\mathbf{x}^k$)

(i) Update the grid by: $\mathbf{G}(\mathbf{x}^k) = \mathbf{G}^{\text{old}} + \sum_{i=1}^{N} x_i^k \mathbf{q}^i$
(ii) Reset $\mathbf{G}^{\text{old}} = \mathbf{G}(\mathbf{x}^k)$
(iii) Reset $\mathbf{x}^k = \mathbf{0}$
(iv) Let ψ represent either f or an active constraint g_i. Then, compute $\nabla\psi$ either analytically (discussed later in this section) or by using divided differences as

$$\frac{\partial \psi}{\partial x_i} \approx \frac{(\psi(\mathbf{G}_\varepsilon) - \psi(\mathbf{G}^{\text{old}}))}{\varepsilon} \tag{12.32}$$

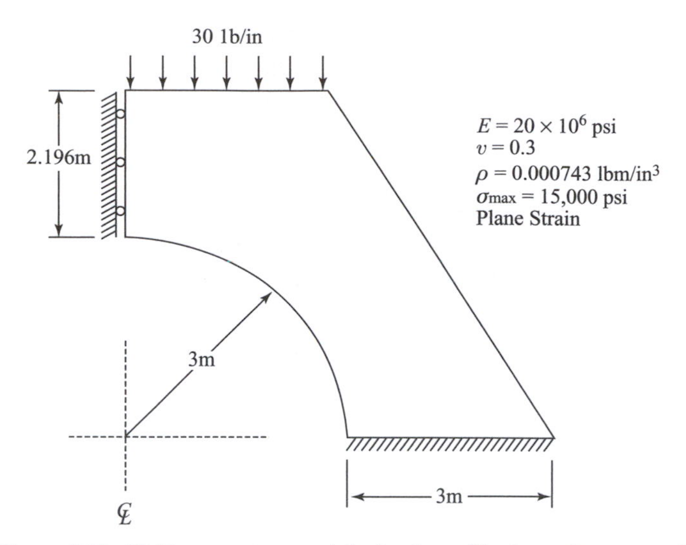

Figure 12.3. Half-symmetry model of culvert ("primary" structure).

where

$$G_\varepsilon = G^{\text{old}} + \varepsilon q^i \tag{12.33}$$

The divided difference parameter ε is chosen so that the largest element of $\varepsilon \mathbf{q}^i =$ 1% of the structural dimension. Note that for an implicit function, computing $\psi(\mathbf{G}_\varepsilon)$ involves solution of $\mathbf{K}(\mathbf{G}_\varepsilon)\ \mathbf{U} = \mathbf{F}(\mathbf{G}_\varepsilon)$ for $\mathbf{U}$.

Use of Deformations to Generate Velocity Fields

The generation of velocity fields $\mathbf{q}^i$ must be possible for both two- and three-dimensional problems without having to look at the detailed appearance of the finite element mesh or on the problem geometry. Consider the ½-symmetry model of a culvert in Fig. 12.3. We refer to the original structure as the "primary" structure. That is, it is this structure with associated loads and boundary conditions that is analyzed for f and g_i. In this problem, the design engineer wishes to optimize the shape of the hole *without changing the shape of the outer boundary*. We now define a fictitious or "auxiliary" structure *merely for the purposes of defining* $\mathbf{q}^i$. Specifically, we introduce the auxiliary structure as shown in Fig. 12.4.

In the auxiliary structure, a set of fictitious loads in the form of specified displacements are applied to nodes on the hole boundary. The loads are applied perpendicular to the boundary to enforce a change in shape. The deformation produced by each set of loads is obtained by a finite element analysis of the auxiliary structure.

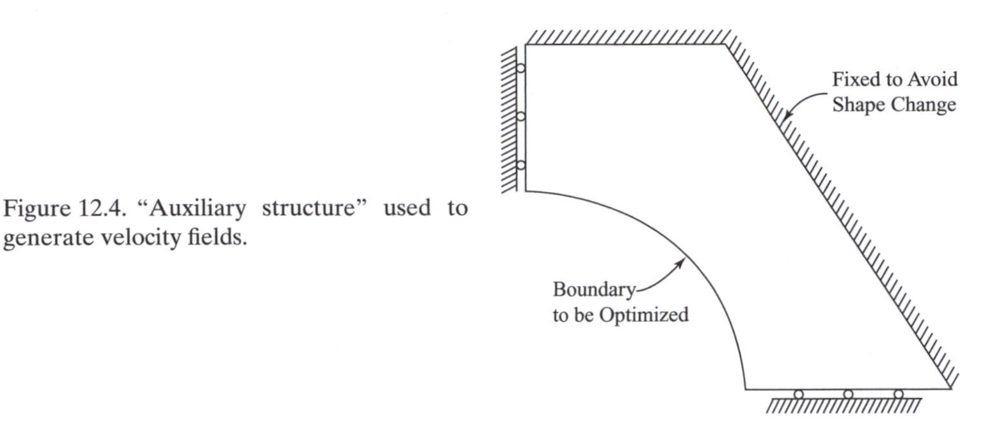

Figure 12.4. "Auxiliary structure" used to generate velocity fields.

The resulting deformation field constitutes $\mathbf{q}^i$. Note that the outer boundary in the auxiliary structure is fixed so that the resulting deformation field does not change the shape of the outer boundary as per the engineer's requirements. To visualize each $\mathbf{q}^i$, we may simply plot $\mathbf{G}^{\text{old}} + s\,\mathbf{q}^i$, where $s =$ a suitable scale factor (a common practice in finite element analysis to visualize deformation fields). The plots of the "basis shapes" are shown in Fig. 12.5 for the culvert problem.

The use of beams along the boundary has shown to produce smooth basis shapes in. The user may fine tune the moment of inertia of the beams to obtain desired smoothness and yet allow shape changes to take place. Further, circular holes may be kept circular during shape optimization by introducing very stiff beams along the hole boundary – nodes on the hole boundary will then move as an unit. Thus, there is considerable leeway in modeling the auxiliary structure to produce satisfactory $\mathbf{q}^i$. The user must be satisfied with the basis shapes *before* executing the optimization program because the final optimum shape is a combination of the basis shapes. Several two- and three-dimensional problems have been solved by this approach and presented by various researchers.

In Fig. 12.6, we show how to generate velocity fields in problems where both inner and outer boundaries are variables.

Distortion Control

It is important that the element distortion is monitored during shape optimization. This topic has been discussed in detail in Zhang and Belegundu [1993]. To illustrate its importance and implementation, consider a finite element model with four noded quadrilateral elements in plane stress. The distortion parameter DP is defined as

$$DP = \frac{4 \det \mathbf{J}_{\min}}{A}$$

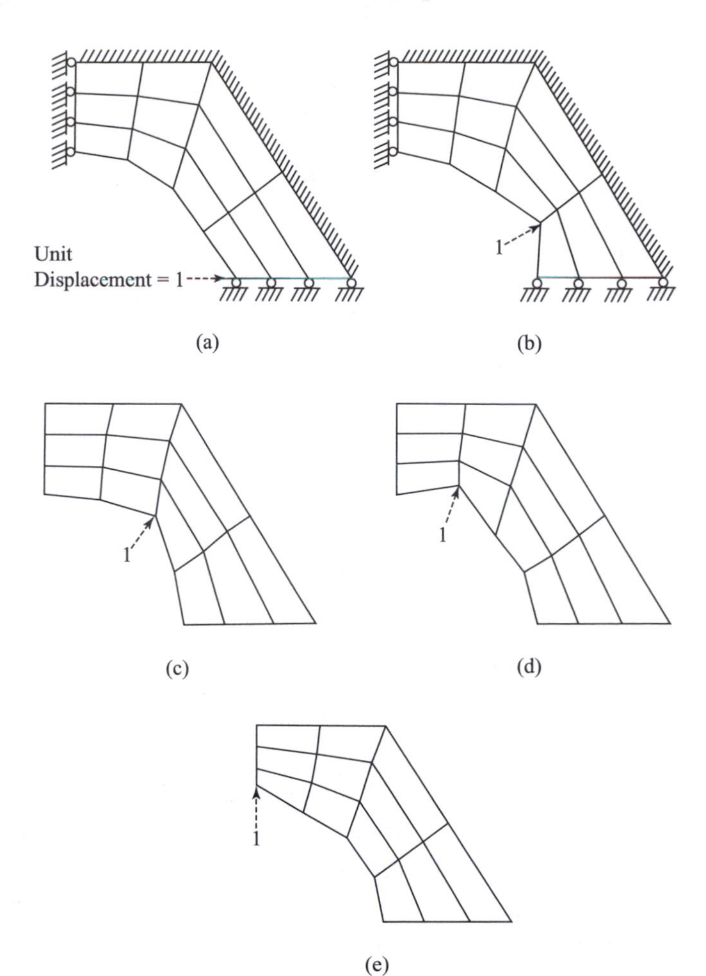

Figure 12.5. (a)–(e): Basis shapes used in culvert example, plotted using $\mathbf{G}^{\text{old}} + \alpha\, \mathbf{q}^i$, $i = 1, \ldots, 5$.

where det $\mathbf{J}_{\min}$ refers to the minimum value of the Jacobian determinant evaluated within the element, and A denotes the element area. For the quad element, we can write

$$DP = 4 \min(DJ_i)/A \quad i = 1, \ldots, 4$$

where $DJ_1 = \frac{1}{4}(x_{21}\, y_{41} - x_{41}\, y_{21})$ with DJ_2 obtained by incrementing the indices as $1 \to 2$, $2 \to 3$, $3 \to 4$ and $4 \to 1$, and so on to obtain DJ_3 and DJ_4. When DP equals zero, then the quadrilateral element degenerates into a triangle (Fig. 12.7). $DP < 0$ implies that the element is no longer convex. It is recommended that a lower limit

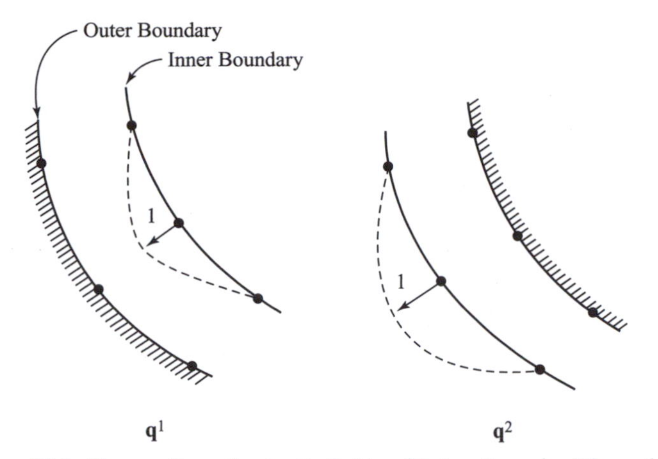

Figure 12.6. Generation of velocity fields with two "moving" boundries.

$DP^{\mathrm{L}} > 0$ be enforced during shape changes. The limit can be directly enforced during line search by restricting the step size. Since $\mathbf{x} = \mathbf{x}^{\mathrm{old}} + \alpha\mathbf{d}$ during line search and $\mathbf{x}^{\mathrm{old}} = \mathbf{0}$, we have

$$\mathbf{G}(\alpha) = \mathbf{G}^{\mathrm{old}} + \alpha \sum_{i=1}^{n} d_i \mathbf{q}^i$$

where $d_i = i$th component of the vector $\mathbf{d}$. The restriction $DP\,(\mathbf{G}(\alpha)) \geq DP^{\mathrm{L}}$ results in the limiting value of α_e being determined by a quadratic equation for each element. We choose $\alpha_{\max} = \min\limits_{e} \alpha_e$, and pass this on to the optimizer as a maximum limit on the step size. The distortion control discussed previously can be applied with other types of finite elements in a similar manner.

Note: In Subroutine SHAPEOPT, we impose distortion control by a simpler scheme: In Program ZOUTEN, the line search begins by first computing an upper limit on step, α_{B}, based on the bounds $\mathbf{x}^{\mathrm{L}} \leq \mathbf{x}^{\mathrm{old}} + \alpha\,\mathbf{d} \leq \mathbf{x}^{\mathrm{U}}$. See Chapter 5 for details. Here, this value of α_{B} is successively reduced, if necessary, until $DP > DP^{\mathrm{L}}$.

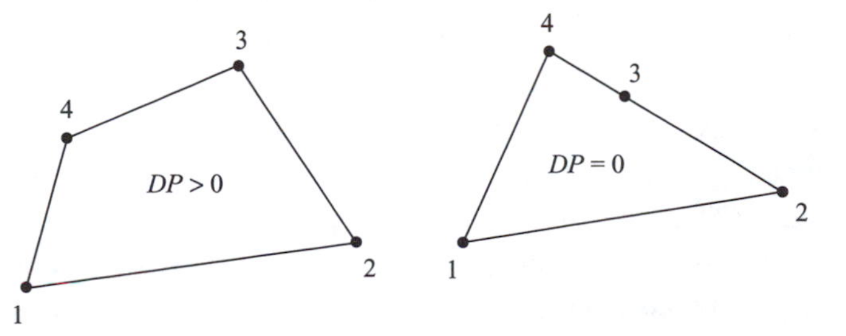

Figure 12.7. Distortion parameter in a four-noded quadrilateral element (plane stress or strain).

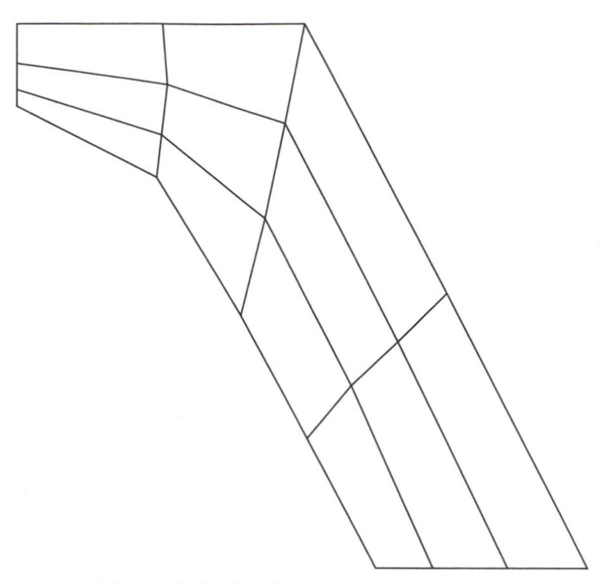

Figure 12.8. Optimum shape of culvert.

"Envelope" constraints of the simple variety $\mathbf{G}^{\mathrm{L}} \leq \mathbf{G} \leq \mathbf{G}^{\mathrm{U}}$ can be handled by limiting the step size based on $\mathbf{G}^{\mathrm{L}} \leq \mathbf{G}(\alpha) \leq \mathbf{G}^{\mathrm{U}}$. Lastly, lower and upper limits on $\mathbf{x}$ are arbitrary: $[-10, +10]$ may be used. The choice of the bounds is not very critical if DP is monitored.

If program termination is due to DP, then a "re-meshing" must be done and the shape further optimized.

Culvert Shape Optimization

The culvert in Fig. 12.3 is optimized for minimum volume (or weight) subject to a limit on the von Mises stress. The von Mises stress in each element is defined to be the *maximum* stress at the four Gauss points in each quadrilateral element. For the simple mesh shown, there are 12 elements, and hence 12 stress constraints.

$DP^{\mathrm{L}} = 0.2$. The magnitude of the distributed load is adjusted so as to make the stress constraint active at the initial design. Thus, the optimum shape is then independent of the actual load and will represent a better distribution of material from the initial. Figure 12.5 shows the basis shapes (plot of the velocity fields). There are 5 design variables. The basis shapes have been obtained by specifying unit displacements to nodes on the hole boundary while restraining the outer boundary as indicated in Fig. 12.4. The optimum obtained using Subroutine SHAPEOPT is shown in Fig. 12.8.

Analytical Shape Sensitivity Analysis

The divided-difference approach for shape sensitivity is implemented based on (12.32)–(12.33). To obtain analytical derivatives, we use (12.31) to obtain

$$\frac{\partial \psi}{\partial x_i} = \frac{\partial \psi}{\partial \mathbf{G}} \mathbf{q}^i \tag{12.34}$$

The dependence of ψ on coordinates $\mathbf{G}$ is known, and hence the derivative can be readily evaluated. For instance, let ψ be the weight of the structure modeled using four-noded elements using 2×2 Gauss quadrature. Then,

$$\psi = \sum_{e=1}^{NE} \sum_{IP=1}^{4} \det J\,(IP\,;\, e)$$

where NE = number of finite elements, and IP denotes the integration point. A closer look at $\det\, J(IP)$ reveals that it is in the form

$$\det J\,(IP) = J_{11}(IP).\, J_{22}(IP) - J_{21}(IP).J_{12}(IP)$$

where

$$J_{11}\,(IP) = \tfrac{1}{4}[\,-(1-\eta)X_1^e + (1-\eta)X_2^e + (1+\eta)X_3^e - (1+\eta)X_4^e]$$

with η is a known constant whose value depends on IP. X_1^e is the X-coordinate of the locally numbered node 1 in element e. Knowing the element connectivity and the preceding expressions, it is a simple matter to obtain $\frac{\partial \psi}{\partial \mathbf{G}}$. The same procedure applies to obtaining the stiffness derivative needed to evaluate derivatives of stress constraints by direct or adjoint methods. In fact, symbolic differentiation is especially attractive to obtain computer code in this case. However, generally speaking, shape optimization is characterized by only a few variables, and hence analytical sensitivity as explained previously is not important. That is, the divided difference approach in Eqs. (12.32)–(12.33) suffice.

Shape optimization is very effective in conjunction with p-Adaptive finite elements [Salagame and Belegundu, 1995]. This is because p-elements are much larger in size than the usual h-elements and consequently allow for large changes in shape to occur before element distortion becomes a problem.

12.6 Optimization with Dynamic Response

Structures that are vibrating as a result of harmonic force excitation are commonly found. Gear boxes, washing machines, automobile engine excitation on the vehicle body, etc. are examples where vibration considerations enter into the design. Dashboard panels resonate at certain frequencies and create an irritating rattle; turbine blades with casings must not resonate axially. Appliance covers, transformer covers, valve covers, aircraft fuselage trim panels, etc. involve shells that radiate noise. Controlling the vibration and/or noise requires avoiding resonant conditions and also reducing functions that depend on the vibration amplitude. Basic equations governing dynamic response are first given in the following. The equations of motion for the vibrating structure are usually expressed as

$$\mathbf{M}\ddot{\mathbf{x}} + (\mathbf{K} + i\mathbf{H})\,\mathbf{x} = \mathbf{f} \tag{12.35}$$

where $\mathbf{M}$ is the $(ndof \times ndof)$ mass matrix of the structure, $\mathbf{K}$ is the stiffness matrix, $\mathbf{x}(t)$ is the displacement vector, $\mathbf{H}$ is the damping matrix and $\mathbf{f}(t)$ is the forcing function. Proportional damping is usually assumed as

$$\mathbf{H} = \alpha\,\mathbf{K} + \beta\,\mathbf{M} \tag{12.36}$$

Corresponding to a harmonic excitation $\mathbf{f} = \mathbf{F}\,\mathrm{e}^{i\omega t}$ where ω is the excitation frequency, the solution is given by $\mathbf{x}(t) = \mathbf{X}\,\mathrm{e}^{\,i\omega\,t}$ where $\mathbf{X}$ is the amplitude of the displacement. Using modal superposition, the amplitude of the nodal velocitiy vector is found to be

$$\dot{\mathbf{X}} = i\omega \sum_{j=1}^{m} y_j \mathbf{Q}^j = i\omega\boldsymbol{\Phi}\mathbf{y} \tag{12.37}$$

where the columns of $\boldsymbol{\Phi}$ consist of the eigenvectors $\mathbf{Q}^{\,j}$. The number of such eigenvectors depends on the frequency range of interest. In the preceding y_j is the modal participation factor determined from

$$y_j = \frac{1}{\left[\left(\omega_j^2 - \omega^2\right) + i\left(\alpha\,\omega_j^2 + \beta\right)\right]}\,\boldsymbol{\Phi}^{j\mathrm{T}}\,\mathbf{F} \tag{12.38}$$

We may also express $\dot{\mathbf{X}}$ in (12.37) in matrix form as

$$\dot{\mathbf{X}} = i\,\omega\;\boldsymbol{\Phi}\,\boldsymbol{\Lambda}^{-1}\boldsymbol{\Phi}^{\mathrm{T}}\,\mathbf{F} \tag{12.39}$$

where $\boldsymbol{\Lambda}$ is a diagonal matrix whose elements are $(\omega_j^2 - \omega^2) + i\,(\alpha\omega_j^2 + \beta)$. The resonance frequencies ω_j and mode shapes $\mathbf{Q}^{\,j}$ are determined by solving the eigenvalue problem

$$\mathbf{K}\,\mathbf{Q}^j = \omega_j^2\,\mathbf{M}\,\mathbf{Q}^j \tag{12.40}$$

Note that the notation ω_j used for resonance frequency must not be confused with ω used to denote excitation frequency. Furthermore, $\mathbf{F}$ and $\dot{\mathbf{X}}$ are complex quantities.

We are now in a position to describe optimization problems. Minimization of maximum vibration or maximum excursion involves a norm of $\mathbf{X}$ or $\dot{\mathbf{X}}$; minimization of acoustic (sound) power involves an expression of the form $\frac{1}{2}\,\dot{\mathbf{X}}^{\mathrm{T}}\,\mathbf{B}\,\dot{\mathbf{X}}^{*}$ where

B is an impedance matrix and $\dot{\mathbf{X}}^*$ is the complex conjugate of $\dot{\mathbf{X}}$; minimization of kinetic energy is a special case of minimizing sound power. Sometimes, the objective function f depends on an excitation frequency that varies within a band. For lightly damped structures, the broadband objective may be replaced by the sum of its peak values that will occur at resonances. Thus, $\underset{\omega^{\mathrm{L}} \le \omega \le \omega^{\mathrm{U}}}{\text{minimize}}\ f \approx \text{minimize} \sum_j f(\omega_j)$. Constraints are usually placed on structural weight and on values of the resonance frequencies ω_j. Design variables are structural parameters that affect stiffness **K** and mass **M**. Vibrating structures are usually modeled using shell elements. Thus, design parameters of special interest are the topology of stiffeners, size and placement of lumped (concentrated) masses, and size and placement of tuned absorbers (point k, m, c). The simulated annealing algorithm is an effective method for optimization for placement. Gradient methods can be used for the parameter optimization. If using gradient methods, we require derivatives of eigenvalues and eigenvectors with respect to design parameters. With these derivatives, we can also obtain derivatives of quantities involving **X** or $\dot{\mathbf{X}}$ (see footnote[2]). Techniques for obtaining these are presented as follows.

Eigenvalue Derivatives

Consider the eigenvalue problem in (12.40) written in the form

$$\mathbf{K}(\mathbf{x})\mathbf{Q} = \lambda\, \mathbf{M}(\mathbf{x})\mathbf{Q} \tag{12.41a}$$

which can be written as

$$[\mathbf{K}(\mathbf{x}) - \lambda\, \mathbf{M}(\mathbf{x})]\, \mathbf{Q} = \mathbf{0} \tag{12.41b}$$

where $\mathbf{Q} \equiv \mathbf{Q}^j$ is the eigenvector under consideration and $\lambda \equiv \lambda_j = \omega_j^2$ is the corresponding eigenvalue. That is, we omit the superscript j for convenience. Matrices **K** and **M** are symmetric in this derivation. The preceding equation is now differentiated with respect to the ith design variable, x_i at the design point $\mathbf{x}^0$. At $\mathbf{x}^0$, let the response be λ_0 and $\mathbf{Q}^0$. For convenience, we introduce the notation that $\delta(.) \equiv \frac{\partial(.)}{\partial x_i} dx_i$ = differential of a quantity due to the differential dx_i. We obtain

$$\mathbf{K}(\mathbf{x})^0\, \delta\mathbf{Q} + \delta\mathbf{K}\, \mathbf{Q}^0 = \delta\lambda\ \mathbf{M}(\mathbf{x}^0)\, \mathbf{Q}^0 + \lambda_0\, \delta\mathbf{M}\, \mathbf{Q}^0 + \lambda_0\, \mathbf{M}(\mathbf{x}^0)\, \delta\mathbf{Q} \tag{12.42}$$

Premultiplying (12.42) by $\mathbf{Q}^{0\,\mathrm{T}}$ and using the fact that $\mathbf{K}(\mathbf{x}^0)\,\mathbf{Q}^0 = \lambda_0\,\mathbf{M}(\mathbf{x}^0)\,\mathbf{Q}^0$ together with the usual normalization relation

$$\mathbf{Q}^{0\,\mathrm{T}} = \mathbf{M}(\mathbf{x}^0)\,\mathbf{Q}^0 = 1 \tag{12.43}$$

[2] While it is possible to obtain the derivative of **X** with respect to x_i by direct differentiation of $-\omega^2$ $\mathbf{MX} + (\mathbf{K} + i\mathbf{H})\,\mathbf{X} = \mathbf{F}$, this method becomes increasingly inaccurate as resonance frequency is approached.

we obtain the derivative of the eigenvalue as $\delta\lambda = \mathbf{Q}^{0\,\mathrm{T}}\,\delta\mathbf{K}\,\mathbf{Q}^0 - \lambda_0\;\mathbf{Q}^{0\,\mathrm{T}}\,\delta\mathbf{M}\;\mathbf{Q}^0$ or

$$\frac{d\lambda}{dx_i} = \mathbf{Q}^{0\mathrm{T}}\frac{\partial\mathbf{K}}{\partial x_i}\mathbf{Q}^0 - \lambda_0\,\mathbf{Q}^{0\mathrm{T}}\frac{\partial\mathbf{M}}{\partial x_i}\mathbf{Q}^0, \quad i = 1, \ldots, \quad n \tag{12.44}$$

Importantly, the derivative of λ_j involves only the pair $(\lambda_j, \mathbf{Q}^j)$. Finally, $\frac{d\omega_j}{dx_i} = \frac{1}{2\omega_j}\frac{d\lambda_j;}{dx_i}$.

Eigenvector Derivatives

Here, the aim is to obtain an expression for the derivative of the jth eigenvector with respect to x_i, or $\mathbf{d}\mathbf{Q}/\mathrm{d}x_i$. See [Sutter et al.] for a study on various methods. We present Nelson's method [Nelson 1976], which is one of the best methods for large systems. The method requires knowledge of only the eigenvalue-eigenvector pair for which the derivative is required. The method is presented here for the case of nonrepeated eigenvalues and for symmetric matrices. From (12.42) we have

$$[\mathbf{K}(\mathbf{x}^0) - \lambda_0\mathbf{M}(\mathbf{x}^0)]\delta\mathbf{Q} = \{-\delta\mathbf{K} + \delta\lambda\mathbf{M}(\mathbf{x}^0) + \lambda_0\delta\mathbf{M}\}\mathbf{Q}^0 \tag{12.45}$$

Using the expression for $\delta\lambda$ in (12.44), the right-hand side of (12.45) is known. However, the coefficient matrix $\mathbf{A} \equiv \left[\mathbf{K}(\mathbf{x}^0)\; - \lambda_0\,\mathbf{M}(\mathbf{x}^0)\right]$ is singular by very construction of an eigensolution. If the eigenvalues are nonrepeated, then the rank $[\mathbf{A}] = ndof$ -1, where $\mathbf{K}$ and $\mathbf{M}$ are $ndof \times ndof$. From matrix theory, we state that the solution of $\mathbf{A}\;\mathbf{z} = \mathbf{b}$ where $\mathbf{A}$ is of dimension $ndof \times ndof$ and possesses rank r, can be expressed as the sum of a particular solution plus $(ndof - r)$ linearly independent solution vectors of the homogeneous solution $\mathbf{A}\;\mathbf{z} = \mathbf{0}$. Here, we have $r = ndof - 1$. Since the jth eigenvector is a solution of the homogeneous solution (see (12.41b)), the solution of (12.45) can be written as

$$\mathbf{dQ}/dx_i = \mathbf{V} + c\mathbf{Q}^0 \tag{12.46}$$

where $\mathbf{V}$ is a particular solution of

$$\mathbf{AV} = \mathbf{R} \tag{12.47}$$

where $\mathbf{R}$ is the right-hand side of (12.45). To obtain $\mathbf{V}$, Nelson suggested to identify an index p such that the pth element of $\mathbf{Q}^0$ has the maximum absolute value. Then, the component V_p is taken equal to zero and the remaining equations are solved for $\mathbf{V}$. The resulting system will appear as

$$\begin{bmatrix} A_{11} & A_{12} & \ldots & 0 & \ldots & A_{1n} \\ A_{21} & A_{22} & \ldots & 0 & \ldots & A_{2n} \\ \ldots & \ldots & \ldots & 0 & \ldots & \ldots \\ 0 & 0 & \ldots & 1 & \ldots & 0 \\ \ldots & \ldots & \ldots & 0 & \ldots & \ldots \\ A_{n1} & A_{n2} & \ldots & 0 & \ldots & A_{nn} \end{bmatrix} \begin{pmatrix} V_1 \\ V_2 \\ \ldots \\ V_p \\ \ldots \\ V_n \end{pmatrix} = \begin{pmatrix} R_1 \\ R_2 \\ \ldots \\ 0 \\ \ldots \\ R_n \end{pmatrix} \tag{12.48}$$

The matrix **A** has the same banded structure as the original eigensystem. Moreover, **A** is the same for differentiation with respect to different x_i. Thus, **A** needs to be factorized only once to obtain all design derivatives. The constant c in (12.46) is obtained as follows. Differentiating (12.43) with respect to x_i, we get

$$2\mathbf{Q}^{\mathrm{T}}\mathbf{M}\,\delta\mathbf{Q} + \mathbf{Q}^{\mathrm{T}}\,\delta\mathbf{M}\,\mathbf{Q} = 0$$

Substituting (12.46) into the preceding equation leads to

$$c = -\tfrac{1}{2}\mathbf{Q}^{\mathrm{T}}\,\partial\mathbf{M}/\partial x_i\,\mathbf{Q} - \mathbf{Q}^{\mathrm{T}}\,\mathbf{M}\,\mathbf{V} \tag{12.49}$$

Thus, (12.46) now yields the derivative of the jth eigenvector with respect to x_i. For each i, a new **R** is assembled and (12.48) is solved for the corresponding **V**. However, **A** is factorized only once.

COMPUTER PROGRAMS

TRUSS_OC, TOPOL, SHAPEOPT

Codes for Example 12.4

Notes on the programs: TOPOL and SHAPE use a quadrilateral plane stress finite element routine. Details regarding the FE routines are given in the authors' FE text cited at the end of this chapter.

PROBLEMS

(Use in-house codes in this and other chapters in the text, or Matlab optimization toolbox, or Excel SOLVER.)

P12.1. Consider the function $f = 10^6\ Q_3$ where **Q** is determined from $\mathbf{K}(\mathbf{x})\,\mathbf{Q} = \mathbf{F}$ as:

$$10^6\begin{bmatrix} 0.75\,x_1 & 0 & 0 \\ 0 & 0.75\,x_4 + 0.384\,x_3 & 0.288\,x_3 \\ 0 & 0.288\,x_3 & x_2 + 0.216\,x_3 \end{bmatrix}\begin{Bmatrix} Q_1 \\ Q_2 \\ Q_3 \end{Bmatrix} = \begin{Bmatrix} 20000 \\ 0 \\ -25000 \end{Bmatrix}$$

Given $\mathbf{x}^0 = [1.0, 1.0, 1.0, 1.0]^{\mathrm{T}}$, determine the gradient $df/d\mathbf{x}$ using: (i) the direct method, and (ii) the adjoint method.

P12.2. Using the data from Example 12.3, use Program TRUSS_OC to optimize the *12-member* truss in Fig. P12.2. That is, optimize for minimum compliance subject to a volume limit.

Data: Horizontal or vertical distance between any two nodes = 360 in, $E = 10^7$, $V^{\mathrm{U}} = 82{,}947$, $x^{\mathrm{L}} = 10^{-6}$, $x^{\mathrm{U}} = 100$. Compare solution with that in Example 12.3.

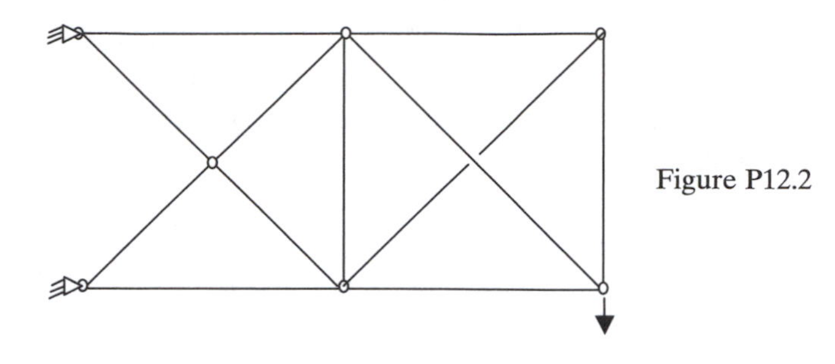

Figure P12.2

P12.3. As in Example 12.4, use Matlab *fmincon* to verify the solution in P12.2.

P12.4. Integrate Truss FEA in Program ZOUTEN and compare with Matlab *fmincon* on P12.2.

P12.5. Compare results in Example 12.3 using various values for p. Study the effect of other algorithmic parameters as well.

P12.6. Compare topology results in Example 12.5(b) with $r = 3$ and 4 (i.e., other than $r = 2$).

P12.7. Use program TOPOL and solve the problem in Fig. P12.7. Rectangle has aspect ratio $Lx/Ly = 4$, and load is at centre on bottom edge. Try 20×10 and 40×20 meshes. (default value of $r = 3$).

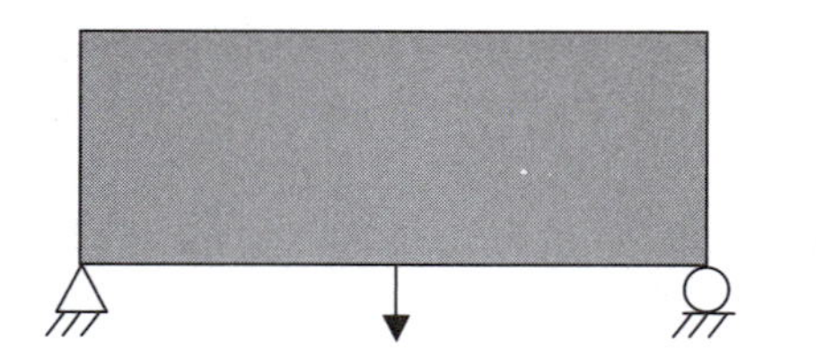

Figure P12.7

P12.8. Do P12.7 with a uniformly distributed load acting downward on the bottom edge.
Note: A distributed load can be replaced by point loads as shown in Fig. P12.8.

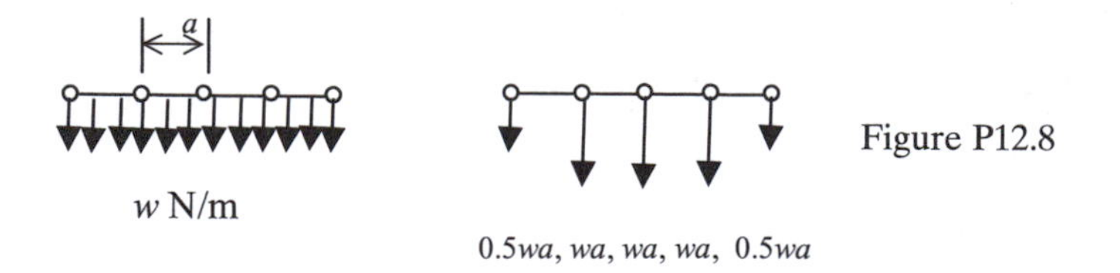

Figure P12.8

P12.9. Use program TOPOL and solve the problem in Fig. P12.9. Dimensions of edges are indicated in the figure. Show that the result is mesh-independent.

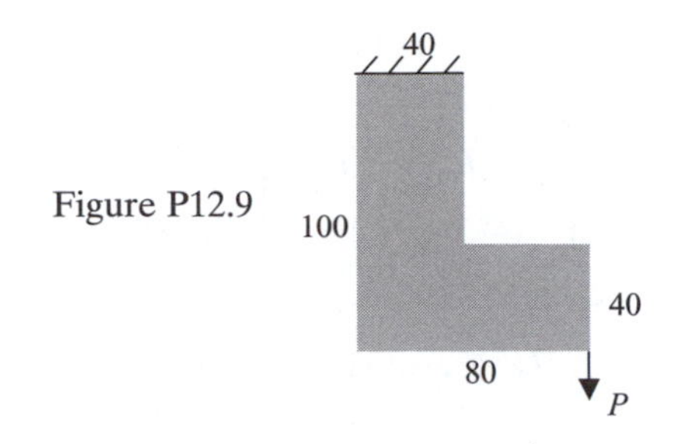

Figure P12.9

P12.10. Reoptimize the culvert problem (see Fig. 12.3) but with a 6th velocity field corresponding to changing the outer boundary as indicated in Fig. P12.10. The boundary should remain a straight line.

Hint: After defining the velocity field in the input file, change the number of variables to 6 in Program SHAPEOPT.

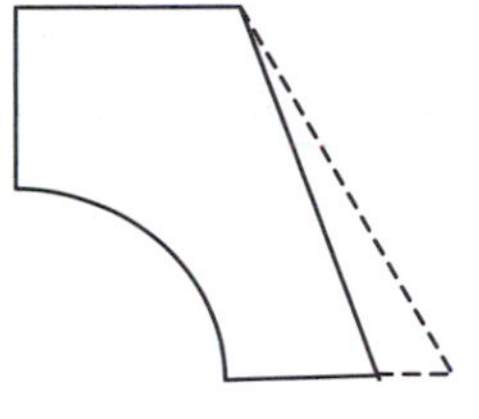

Figure P12.10

P12.11. Optimize the shape of the chain link shown in Fig. P12.11. Use the following guidelines:

(i) use a ¼-symmetry FE model. That is, model the portion in the first quadrant with symmetry boundary conditions in the FE model.

(ii) first, assume the hole dimensions are fixed and optimize the shape of the curve between points A–B. For generating velocity fields, define an auxiliary structure and apply point loads.

(iii) use Program SHAPEOPT

(iv) then, include the diameter of the holes into the optimization problem; that is, generate velocity fields for these too.

(v) part dimensions are your choice (you may look at a bicycle chain, for example); select the thickness so as to have the initial stress equal to the yield limit as was done in the culvert problem in the text.

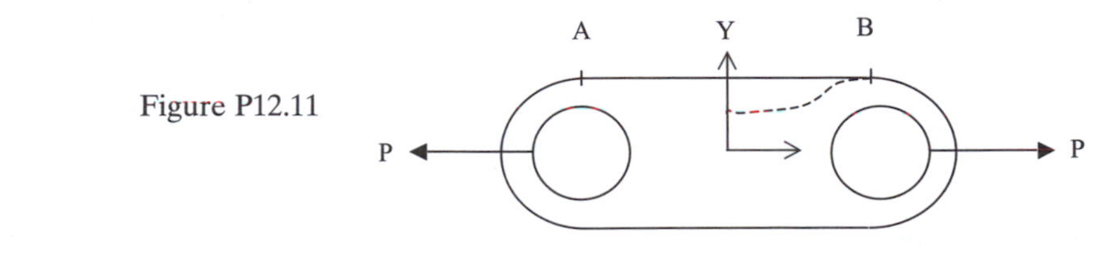

Figure P12.11

P12.12. Consider the eigenvalue problem

$$\begin{bmatrix} \left(\frac{EA_1}{L_1} + \frac{EA_2}{L_2}\right) & -\frac{EA_2}{L_2} \\ -\frac{EA_2}{L_2} & \frac{EA_2}{L_2} \end{bmatrix} \begin{bmatrix} U_1 \\ U_2 \end{bmatrix}$$

$$= \lambda \begin{bmatrix} \rho(A_1 L_1 + A_2 L_2)/3 & \rho(A_2 L_2)/6 \\ \rho(A_2 L_2)/6 & \rho(A_2 L_2)/3 \end{bmatrix} \begin{bmatrix} U_1 \\ U_2 \end{bmatrix}$$

Design variables are areas A_1 and A_2. At the point $A_1 = 1$, $A_2 = 0.5$, and for the data $E = 30 \times 10^6$, $L_1 = 10$, $L_2 = 5$, $\rho = 7.324 \times 10^{-4}$, determine:

(i) the lowest eigenvalue derivative using Eq. (12.44)
(ii) the associated eigenvector derivative using Nelson's method in Eq. (12.46) Verify the solutions using divided differences.

P12.13. The equation of motion for a single degree of freedom mass-spring-damper system is of the form $m\ddot{x} + c\dot{x} + kx = F \sin \omega_d t$, which can be expressed in nondimensional variables $z(t) = x(t)/x_{st}$ where $x_{st} = F/k$ = static displacement as:

$$\ddot{z} + \frac{c}{m}\dot{z} + \frac{k}{m}z = \frac{k}{m}\sin \omega_d t$$

Given the initial conditions $z(0) = \dot{z}(0) = 0$, $m = 10$, $k = 10\,\pi^2$, $\omega_d = 2\,\pi$, $c = \frac{20\,\pi}{\sqrt{2}}$, and the functions

$$\psi_1 = z(t), \quad \psi_2(t) = \int_0^5 [z(t)]^2 \, dt$$

determine, with the help of Matlab functions, *analytical* derivatives for

$$\left[\frac{d\psi_1}{dm}, \frac{d\psi_1}{dk}\right], \left[\frac{d\psi_2}{dm}, \frac{d\psi_2}{dk}\right]$$

Also provide: plots of $\frac{d\psi_1}{dm}$ vs t, $\frac{d\psi_1}{dk}$ vs t and verify solution using divided-differences

P12.14. Consider the plate in Fig. P12.7. Use the following data: material, steel: $E =$ 200e9 Pa, σ_{max} = max. allowable (von Mises) stress = 100 MPa, Lx = 4 m, Ly = 1 m, Load $P = -2.3 \times 10^5$ N (downward acting), choose consistent units throughout this assignment. For eg., load in N, dimensions in m, Yield stress in N/m^2 or Pa. Based on FEA of a solid plate, we find that for a 1 cm thick plate, $\Delta = 0.118\text{e} - 2\,\text{m} =$ 0.118 cm = y-displacement at load, point, von Mises stress equals max. permitted. This solid plate has a volume of 0.04 m^3 and a mass of 0.04*7850 = 314 kg.

Thus, the topology + sizing optimization problem is to design a structure with less mass with max stress and $\Delta \leq$ that in the solid plate.

Step 1: **Topology Optimization**: Determine material distribution for minimum compliance using Program TOPOL (stress does not enter into the problem).

Step 2: **Sizing Optimization**

Identify a **truss** structure from the result in Step 1. Ensure that the truss is stable – that there are no quadrilateral shaped cavities, nor any missing connections, nor should it be, say, a 4–bar linkage. Then,

- minimize: Volume of truss with cross–sectional areas as variables
- constraints: $\frac{|\delta|}{0.118e-2} - 1 \leq 0, \quad \frac{|\sigma_i^{vM}|}{100e6} - 1 \leq 0, i = 1, \ldots, NE$

Provide relevant plots, iteration histories and *physically interpret* and *validate* your design solution.

P12.15. Consider the problem of minimum weight design of a truss structure subject to a minimum limit on the lowest resonant frequency:

$$\begin{aligned} \text{minimize} \quad & f = \sum L_i x_i \\ \text{subject to} \quad & \lambda \geq \lambda^{\mathrm{L}} \\ & x_i \geq x^{\mathrm{L}} \end{aligned}$$

where x_i are the design variables, and λ is the eigenvalue. Develop a resizing and a Lagrange multiplier update formula based on optimality criteria. Develop a computer code and verify a sample problem solution with a nonlinear programming code.

Hint: First define the Lagrangian as $L = \sum_i L_i x_i + \mu\,(1 - \frac{\lambda}{\lambda^{\mathrm{L}}})$. Then, for the derivative, use Eq. (12.44). Since both stiffness and mass of a truss element depend on x_i, we have $\boldsymbol{Q}^{\mathrm{T}} \frac{dK}{dx_i}\, Q = -\frac{\varepsilon_i}{x_i},\ Q^{\mathrm{T}} \frac{dM}{dx_i}\, \boldsymbol{Q} = -\frac{KE_i}{x_i}$. Define the optimality in the form of Eq. (12.19) and proceed. Refer to [Introduction to Finite Elements in Engineering] for expressions for the mass matrix, and use the *eig* command in Matlab. Modify Program TRUSS_OC.

P12.16. Using the hints in P12.15, develop a topology algorithm for maximizing lowest resonance frequency. Modify Program TOPOL.

P12.17. Develop a topology algorithm for shell structures by interfacing a (quadrilateral) shell element FE code instead of a plane stress FE code with Program TOPOL.

REFERENCES

Atkinson, K.E., *An Introduction to Numerical Analysis*, Wiley, New York, 1978.

Banichuk, N.V., *Introduction to Optimization of Structures*, Springer, New York, 1990.

Belegundu, A.D., Interpreting adjoint equations in structural optimization, *Journal of Structural Engineering, ASCE*, **112**(8), 1971–1976, 1986.

Belegundu, A.D. and Arora, J.S., Potential of transformation methods in optimal design, *AIAA Journal*, **19**(10), 1372–1374, 1981.

Belegundu, A.D. and Rajan, S.D., A shape optimization approach based on natural design variables and shape functions, *Journal of Computer Methods in Applied Mechanics and Engineering*, **66**, 87–106, 1988.

Belegundu, A.D. and Salagame, R.R., A general optimization strategy for sound power minimization, *Structural Optimization*, **8**, 113–119, 1994.

Bendsoe, M.P. and Kikuchi, N., Generating optimal topologies in structural design using a homogenization method, *Computer Methods in Applied Mechanics and Engineering*, **71**, 197–224, 1988.

Bendsoe, M. and Sigmund, O., *Topology Optimization*, Springer, New York, 2002.

Chandrupatla, T.R. and Belegundu, A.D., *Introduction to Finite Elements in Engineering*, 2nd edition, Prentice-Hall, Englewood Cliffs, NJ, 1997.

Dorn, W.S., Gomory, R.E., and Greenberg, H.J., Automatic Design of Optimal Structures, *Journal de Mechanique*, **3**(1), 25–52, 1964.

Frecker, M.I., Topological synthesis of compliant mechanisms using multi-criteria optimization, *Journal of Mechanical Design*, 1997.

Gea, H.C., Topology optimization: a new microstructure based design domain method, *Computers and Structures*, **61**(5), 781–788, 1996.

Haftka, R.T. and Gurdal, Z., *Elements of Structural Optimization*, 3rd Edition, Kluwer Academic Publishers, Dordrecht, 1992.

Haug, E.J. and Arora, J.S., *Applied Optimal Design*, Wiley, New York, 1979.

Haug, E.J., Choi, K.K., and Komkov, V., *Design Sensitivity Analysis of Structural Systems*, Academic Press, New York, 1985.

Kirsch, U., Optimal topologies of structures, *Applied Mechanics Reviews*, **42**(8), 223–239, 1989.

Michell, A.G.M., The limits of economy of material in frame structures, *Philosophical Magazine*, **8**(6), 589–597, 1904.

Mlejnek, H. and Schirrmacher, R., An engineering approach to optimal material distribution and shape finding, *Computer Methods in Applied Mechanics and Engineering*, **106**, 1–26, 1993.

Nack, W.V., Belegundu, A.D., shape design sensitivities. Prepared for Structures Group (FESCOE), General Motors Technical Center, Warren, MI, Feb. 1985.

Nelson, R.B., Simplified calculation of eigenvalue derivatives, *AIAA Journal* **14**(9), 1201–1205, 1976.

Olhoff, N., Topology optimization of three-dimensional structures using optimum microstructures, *Proceedings of the Engineering Foundation Conference on Optimization in Industry*, Palm Coast, FL, March 1997.

Phadke, M.S., *Quality Engineering using Robust Design*, Prentice-Hall, Englewood Cliffs, NJ, 1989.

Rajan, S.D. and Belegundu, A.D., Shape optimal design using fictitious loads, *AIAA Journal*, **27**(1), 102–107, 1988.

Rozvany, G.I.N., "Layout theory for grid-type structures," In M.P. Bendsoe, and C.A. Mota Soares (Eds.), *Topology Optimization of Structures*, Kluwer Academic Publishers, Dordrecht, 251–272, 1993.

Salagame, R.R. and Belegundu, A.D., Shape optimization with p-adaptivity, *AIAA Journal*, **33**(12), 2399–2405, 1995.

Schmit, L.A., Structural design by systematic synthesis, *Proceedings of the 2nd Conference on Electronic Computation, ASCE*, New York, 105–122, 1960.

Soto, D.A. and Diaz, A.R., On the modelling of ribbed plates for shape optimization, *Structural Optimization*, **6**, 175–188, 1993.

Sutter, T.R., Camarda, C.J., Walsh, J.L. and Adelman, H.M., "Comparison of several methods for calculating mode shape derivatives", *AIAA Journal*, **26**(12), 1506–1511, 1988.
Vanderplaats, G.N., *Numerical Optimization Techniques for Engineering Design*, McGraw-Hill, 1984.
Venkayya, V.B., Khot, N.S., and Berke, L., Application of optimality criteria approaches to automated design of large practical structures, *2nd Symposium on Structural Optimization*, AGARD-CP-123, 3–1 to 3–19, 1973.
Zhang, S. and Belegundu, A.D., Mesh distortion control in shape optimization, *AIAA Journal*, **31**(7), 1360–1362, 1993.

Index